# MODIFYING THE ROOT ENVIRONMENT TO REDUCE CROP STRESS

Edited by

**G. F. ARKIN**

Professor, Blackland Research Center, Texas Agricultural Experiment Station, Temple, TX

**H. M. TAYLOR**

Professor, Department of Agronomy, Iowa State University, Ames, IA
(Retired Research Soil Scientist, USDA-SEA-AR)

**An ASAE Monograph
Number 4 in a series published by**

American Society of Agricultural Engineers
2950 Niles Road, P.O. Box 410
St. Joseph, Michigan 49085
(phone 616-429-0300)

Manager of Publications: James A. Basselman
1981

# FOREWORD

This monograph is a companion to the ASAE monograph "Modification of the Aerial Environment of Crops." It is intended to complement that monograph by providing a comprehensive reference on modifying the subterranean environment of plants.

It is the editors' intent to provide a monograph that will be used by engineers, agronomists and farmers for selecting soil treatments that likely will reduce soil stresses on plant roots and thereby increase crop yields. Procedures for diagnosing adverse soil environments and methods for correcting them are set forth in almost handbook fashion, when possible. When no particular corrective treatment can be specified, the authors discuss the available information in sufficient detail for the reader to infer an appropriate treatment for the local condition.

A quantitative understanding of the plant's response to its subterranean environment is required if the maximum effectiveness is to be derived from each soil treatment. The root environment is complex and rapidly changing, causing difficulty in making diagnoses and in predicting consequences of soil treatments. The editors attempted to assemble a monograph based on a systems approach and tried to weave this approach throughout the monograph. The consequences of any single corrective action is often mediated by other factors, such as the weather, insect infestations, or other plant or soil stresses. Mathematical simulation models, soundly based on physical, chemical and physiological principles, can account for the effects of these interactions with those of the actions taken to correct the soil stresses. Knowledge about the soil-root system is much more fragmentary and more expensive to obtain than that about the plant aerial environment system, but knowledge is accumulating rapidly.

Finally, the authors were asked to pinpoint, for their particular subject matter, the most pressing research needs. The editors hope that these discussions about gaps in information will inspire and guide future research efforts for modifying the subterranean environment of plants.

Gerald F. Arkin

Howard M. Taylor

# CONTRIBUTORS

F. Adams — Professor of Soils, Department of Agronomy and Soils, Auburn University, Auburn, AL 36849

R. R. Allmaras — Research Leader, Columbia Plateau Conservation Research Center, USDA-SEA-AR, Pendleton, OR 97801

F. C. Boswell — Professor, Department of Agronomy, University of Georgia, Experiment, GA 30212

H. D. Bowen — Professor, Biological & Agricultural Engineering Department, North Carolina State University, Raleigh, NC 27607

R. Q. Cannell — Agricultural Research Council, Letcombe Laboratory, Wantage, OX12 9JT England

H. V. Eck — Soil Scientist, Conservation and Production Laboratory, USDA-SEA-AR, Bushland, TX 79012

R. A. Feddes — Institute for Land and Water Management Research, P.O. Box 35-6700 AA, Wageningen, The Netherlands

G. J. Hoffman — Research Agricultural Engineer, U.S. Salinity Laboratory, USDA-SEA-AR, Riverside, CA 92501

M. B. Jackson — Agricultural Research Council, Letcombe Laboratory, Wantage OX12 9JT England

C. E. Johnson — Professor, Agricultural Engineering Department, Auburn University, Auburn, AL 36849

S. D. Lyda — Professor, Department of Plant Sciences, Texas A&M University, College Station, TX 77843

J. T. Musick — Agricultural Engineer, Conservation and Production Laboratory, USDA-SEA-AR, Bushland, TX 79012

M. E. Sumner — Professor, Department of Agronomy, University of Georgia, Athens, GA 30602

P. W. Unger — Soil Scientist, Conservation and Production Laboratory, USDA-SEA-AR, Bushland, TX 79012

W. B. Voorhees — Soil Scientist, North Central Soil Conservation Research Laboratory, USDA-ARS, Morris, MN 56267

# CONTENTS

## CHAPTER 3— ALLEVIATING PLANT WATER STRESS
### Paul W. Unger, Harold V. Eck and Jack T. Musick

## CHAPTER 4— ALLEVIATING NUTRIENT STRESS
### M. E. Sumner and F. C. Boswell

# CHAPTER 5— ALLEVIATING AERATION STRESSES
R. Q. Cannell and Michael B. Jackson

## CHAPTER 6— ALLEVIATING PATHOGEN STRESS
### Stuart D. Lyda

## CHAPTER 7— ALLEVIATING TEMPERATURE STRESS
### Ward B. Voorhees, R. R. Allmaras and Clarence E. Johnson

## CHAPTER 8— ALLEVIATING CHEMICAL TOXICITIES: LIMING ACID SOILS
### Fred Adams

## CHAPTER 9— ALLEVIATING SALINITY STRESS
### Glenn J. Hoffman

## CHAPTER 10—WATER USE MODELS FOR ASSESSING ROOT ZONE MODIFICATIONS
### R. A. Feddes

# CHAPTER 11—SYSTEMS CONSIDERATIONS AND CONSTRAINTS
Gerald F. Arkin and Howard M. Taylor

# chapter 1

## ROOT ZONE MODIFICATION: FUNDAMENTALS AND ALTERNATIVES

# 1

# ROOT ZONE MODIFICATION: FUNDAMENTALS AND ALTERNATIVES

by  Howard M. Taylor, USDA, SEA-AR, Agronomy
    Department, Iowa State University, Ames, IA
    50011 and Gerald F. Arkin, Blackland Research
    Center, Texas Agricultural Experiment Station,
    Temple, TX 76501

## 1.1 INTRODUCTION

Farmers annually spend vast amounts of money to till, fertilize, lime, irrigate, drain, and apply pesticides. All of these practices modify the plant root environment. Some treatments are applied because experience has taught the farmer that his crop quality or yield is increased. Unfortunately, some practices are recommended to the farmer because they "might do some good." This chapter introduces a monograph designed to improve the probability of increasing yield when soil treatments are applied to decrease crop stress.

### 1.1.1 History of Soil Modification

Men and women first began to modify the plant root environment when they changed from hunters and gatherers to the more sedentary life of planters and harvesters. This change in life-style required that they find the desirable plants in more concentrated and predictable locations. To concentrate these desirable plants, the early cultivators harvested the seeds then planted them into soil disturbed by sticks or stones. This shallow tillage caused changes of the plant root environment, especially for roots of seedlings. Today, machines exist that easily and rapidly modify the plant root environment to a 1.5-m depth. Some of these expensive machines are discussed in Chapter 3.

### 1.1.2 Research Approaches to Assess Need for Modification

A wide spectrum of approaches is used in defining the effects of soil modification on crop yield. On one end of that range, the researcher decides that some specific soil condition probably limits yield then alters that condition and measures only the harvest. On the other end, the researcher decides that the soil probably is limiting yield, determines the range of values for as many soil conditions as possible, modifies those soil conditions that are in possibly limiting ranges and then measures both root function and crop yield. Distinct advantages and disadvantages are present in each approach. The first avenue is simpler, less expensive and requires less technical skill but the results are difficult, if not impossible, to extrapolate to other soils, to other environments, or to other years. This approach usually is called an "extensive" one. The second avenue is more expensive, uses more labor and requires more technical skills but the results are more readily extrapolated. This approach usually is called an "intensive" one.

### 1.1.3 Goals of the Monograph

One goal of this monograph is to encourage researchers who are conducting the simple crop yield experiments (extensive manner) to spend some of their research efforts in the more complex team effort experiments (intensive manner). This monograph is designed to provide the ranges in values of soil conditions that reduce yields of various crops under various management systems and in different environments. Research information continuously accumulates and becomes more precise so this monograph presents the "state of the art, 1981 version."

The intensive method requires information about relationships among root growth, root function, top growth, climate and crop yield. These interactions usually are investigated by crop physiologists. The intensive attack also requires knowledge about the effects of various soil conditions on root growth and function. These soil-root phenomena usually are investigated by scientists with specialized training in soil biology, soil chemistry or soil physics. The intensive method also requires specialized knowledge about methods and machines to alter the soil conditions that limit root growth, top growth or crop yield. Agricultural engineers usually provide these inputs.

A second goal of this monograph is to provide sufficient background information so that a crop physiologist, a soil scientist, or an agricultural engineer assigned to a project using the intensive approach will feel qualified to work with or to lead a group of scientists investigating root zone modification.

A third goal is to set forth diagnostic procedures for diagnosing soil related problems and prescribing corrective treatments that will reduce crop stress. When no clear-cut treatment is apparent, inference information is presented to enable a reasonable course of action to be selected. Agronomists and engineers should be able to use this monograph as a reference source and guide for solving on-farm root zone problems.

### 1.2 Shoot-Root Relationships

The first plant root studies are lost in antiquity, but S. Hales (1727) dug out root systems of cultivated crops and determined their morphology, weight, and length (Böhm, 1979). Root environment research languished for another 150 yr but intensified with the advent of mineral fertilization in the 1870's. Some root studies of the 1870's were conducted in pots, but some were conducted by washing out root systems at a profile wall. Early in the 1900's, pits were dug for continuous *in situ* root examination (Rotmistrov, 1909).

Today, plant root studies are conducted at many locations in the world and with an array of techniques that range from the simple spade to huge excavating machines and radioactive tracers. Some studies are conducted in the field, some in greenhouses, some in growth chambers and some in large sophisticated root observations laboratories. A root observation chamber field installation was recently built at Temple, TX (Arkin et al., 1978) that enables rapid serial observations of root growth and evapotranspiration measurements (Fig. 1.2-1). Plants grown in this installation are in an aerial

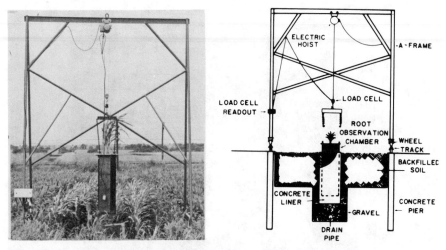

**FIG. 1.2-1 Root observation chamber field installation. The glass sidewall of the chamber and the load cell enables root growth observations and evapotranspiration measurements, respectively.**

environment closely resembling that which they would experience in the field. In contrast to rhizotrons where evapotranspiration is determined from differences in soil water content (Taylor, 1971), evapotranspiration can be obtained as often as desired by directly weighing a compartment of the Temple, TX installation.

*In situ* root growth observations for field grown plants were simplified by

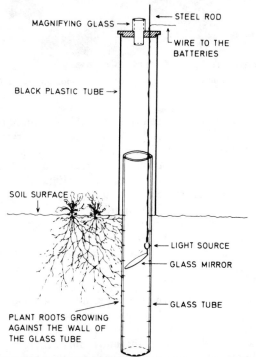

**FIG. 1.2-2 Minirhizotron (glass tube) for *in situ* root observations (Böhm, 1974).**

**FIG. 1.2-3 Installation of a minirhizotron (glass tube 180 cm long and 10 cm inside dia) for** *in situ* **root growth studies.**

the introduction of the minirhizotron (Fig. 1.2-2) by Waddington (1971) and improved by Bőhm (1974) and Arkin (Blackland Conservation Research Center Annual Progress Report, Temple, TX, 1976, unpublished). Arkin used large glass tubes (minirhizotrons) to characterize sorghum and cotton root densities as influenced by tillage treatments (Fig. 1.2-3).

Photographs or videotapes can be made of root growth in the minirhizotrons using fiber optics (Sanders and Brown, 1978) in a duodenoscope (Fig. 1.2-4) or prisms and mirrors in either a borescope (Fig. 1.2-5) or periscope (Gregory, 1979). Photographs and videotapes made in the field this way can be stored and later evaluated quantitatively in the laboratory. Techniques and facilities for various other types of plant root studies are discussed by Bőhm (1979).

Most studies of plant roots and their interaction with soil environments also should include measurements of top growth and yield. These measurements are necessary because roots receive photosynthates and growth hormones from shoots and in return furnish water and essential minerals to the shoots. Root and shoot growth are thus interdependent, and

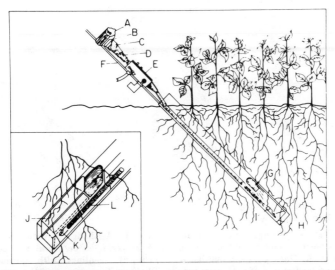

FIG. 1.2-4 Representation of the duodenoscope photographing roots inside the observation tube positioned in the plant row at a 45 deg angle. (A) Honeywell Pentax k-2 camera, (B) 3x Tele-extender, (C) camera adaptor, (D) focus adjust, (E) ACMI F5-A Duodenoscope, (F) plexiglass slide (flat black), (G) Minolta flash (Model CP), (H) observation tube, (I) ring holder, (J) nonreflective rectangular terminal portion of the slide, (K) objective lens, (L) flexible fiber-optic tube, (Sanders and Brown, 1978).

FIG. 1.2-5 Borescope with mirror or prism assembly in use.

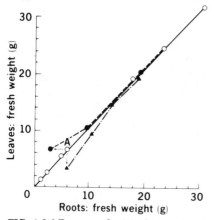

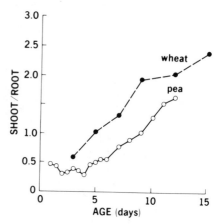

**FIG. 1.2-6 Recovery of original ratio of roots to shoots in beans (*Phaseolus vulgaris*) after mutilation at (A) Open circles: unmutilated plants. Closed circles: half of roots removed. Crosses: half of leaves removed. (Redrawn from Brouwer, 1963)**

**FIG. 1.2-7 Change in shoot: root ratio of wheat (*Triticum aestivum*) and pea (*Pisum sativum*) seedlings with advancing age (Redrawn from Aung, 1974).**

one can proceed in the absence of the other for only a very short time. When part of the shoot is removed, shoot growth accelerates, with respect to root growth, until a balance is again achieved among the plant's supply of photosynthates, growth hormones, water, and minerals. When part of the root system is damaged, root growth accelerates until that balance is achieved (Fig. 1.2-6).

### 1.2.1 Shoot-Root Ratios

During the first few days after wheat (*Triticum aestivum L.*) seeds germinate, the radicle grows faster than the plumule; this relation causes the shoot-root (S:R) ratio by weight to be less than 1.0. Subsequently the plumule grows faster than the radicle and this increases the shoot-root ratio (Aung, 1974). Peas (*Pisum sativum L.*) follow the same general trend (Fig. 1.2-7). By maturity, weight of the shoot may be 15 times that of the roots of corn (*Zea mays L.*) (Foth, 1962). During the grain-filling stage of corn, photosynthates are monopolized by the developing ear and root growth (in weight terms) slows substantially (Cooper, 1955). However, root growth continues in some species because soybean (*Glycine max (L)*. Merr.) roots grow deeper during seed development (Willatt and Taylor, 1978).

### 1.2.2 Environmental Effects on Shoot-Root Ratios

Biological, chemical, and physical factors all alter shoot-root ratios. Chandler (1919) found that pruning shoots increases the S:R ratio of the plant tissue formed immediately afterward. Shank (1945) showed that increases in soil moisture levels from 7.5 to 21 percent increased S:R from 2.5 to 3.4. Crist and Stout (1929) found that long photo-periods decreased the S:R of radishes because taproot development was promoted over shoot development. Shank (1945) showed that increasing amounts of nitrogen (N) and phosphorus (P) resulted in an increased S:R for corn. Eavis and Taylor (1979) indicated that S:R ratios of soybeans were increased both by maintaining an adequate watering regime and by adequate phosphorus and

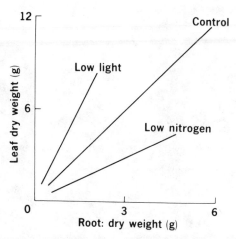

**FIG. 1.2-8 General effect of reduction of light intensity or nitrogen supply below the optimal on root and shoot weight of bean** (*Phaseolus vulgaris*). **(Redrawn from Brouwer and de Wit, 1969).**

potassium levels. A reduction in a plant root stress usually results in an increased S:R ratio, and a reduction in a shoot stress usually results in a decreased S:R ratio (Fig. 1.2-8). For a more complete discussion of S:R ratios, the reader is referred to Russell (1977).

## 1.3 PRELIMINARY DIAGNOSIS OF A SOIL-RELATED PROBLEM

Plant shoots and roots constantly are exposed to changing conditions. Sometimes, the shoot environment limits overall growth while at other times, the root-related stresses reduce growth on the same plant. Low light intensity, excessively low or high aerial temperatures, excessive drying conditions, leaf eating insects, aerial pollutants, and wind damage are examples of shoot stresses. Some of these limitations are discussed in a companion monograph (Barfield and Gerber, 1979). A few key soil measurements, in conjunction with observations of root systems dug out with a spade, allow agricultural technicians to decide on a preliminary basis whether a particular soil-related factor may be reducing top growth and yield. The following discussion of soil constraints hopefully will provide a guide to the more complete discussions of constraints in succeeding chapters.

### 1.3.1 Mechanical Impedance

There are five different ways that mechanical impedance affects crop growth and yield. First, soil crusts may have such high strength that seedling emergence is reduced or prohibited. This condition occurs most often when dicotyledonous crops are planted in structurally unstable soils that are subjected to rapid drying conditions. The usual diagnostic keys are hypocotyls with diameters up to twice normal size but without necrotic spots, many emerged plants with cotyledons broken off or, for monocotyledonous plants, a bent and distorted coleoptile that often does not penetrate the crust. Second, a soil layer may exist that is too hard for any root penetration. In that case, the plant has a root system in which all roots abruptly turn at a particular depth (Fig. 1.3-1). In some cases, part of the roots penetrate, but the remainder abruptly turn at the mechanically impeding layer (Fig. 1.3-2).

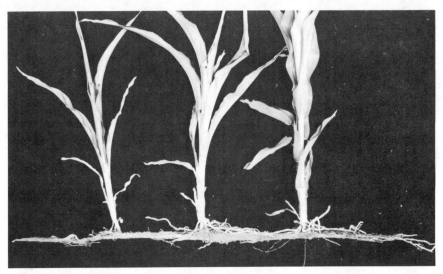

FIG. 1.3-1 Three sorghum plants showing roots confined to a slit caused by the furrow opener of a planter. The condition occurred during a drought year on Pratt fine sandy loam at Woodward, OK (photo courtesy of L. F. Locke).

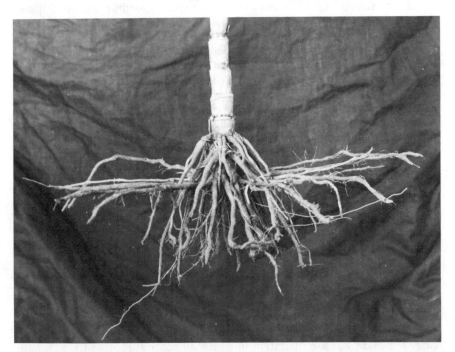

FIG. 1.3-2 Sorghum root system where some roots penetrated the soil pan but others were diverted by it. Note the blunted tips of many roots (Locke et al., 1960).

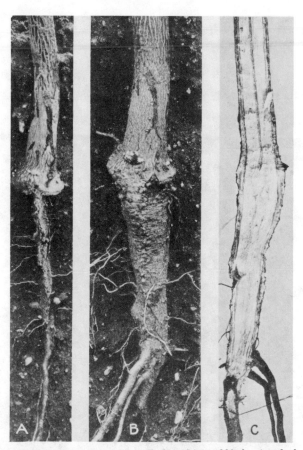

FIG. 1.3-3 A series of photographs taken at Shafter, CA in 1930 showing the base of the stalk and the upper taproot of a cotton plant. (a) View August 15, before irrigation, showing severe taproot constriction with no lateral roots near the surface; (B) View August 21, six days after irrigation, showing taproot greatly increased in diameter and development of many small white lateral roots; (C) View August 26, showing taproot practically normal in size and tough, woody lateral roots with numerous secondary branches (Hubbard, 1932).

Third, a soil layer may exist that will allow root penetration and radial expansion when wet but not when dry. This condition causes rather abrupt decreases in root diameter (Fig. 1.3-3), but when the soil is again wetted and impedance decreases, lateral roots rapidly form and the older entrapped roots expand their radii. Fourth, a soil volume may exist where some plant roots may grow along but not penetrate ped surfaces, but these ped surfaces are too far apart for the plant to fully extract the water and minerals stored there (Fig. 1.3-4). In these cases, both mechanical impedance and aeration usually are limiting the root growth. Finally, some crops flower above ground but develop their seeds below ground. Peanuts (*Arachis hypogaea L.*) and subterranean clover (*Trifolium subterraneum L.*) are examples of such plants. Yield of these plants is considerably improved if fruiting bodies of these plants can be buried, but high-strength crusts reduce burial likelihood (Barley and England, 1970). Peanut plants produce many more gynophores on high-strength, crusted soils than on noncrusted soils (Underwood et al., 1971).

**FIG. 1.3-4 Plant roots located in a vertical shrinkage crack of Houston Black clay. Note that the roots apparently were unable to readily penetrate the vertical face of the crack (Photograph courtesy of E. Burnett).**

One should suspect that mechanical impedance is a soil constraint to crop production whenever a root system, dug out with a spade, exhibits the visual symptoms shown in Figs. 1.3-1, 1.3-2, 1.3-3, or 1.3-4. Similarly, one should suspect mechanical impedance when digging in the surface 30 cm of wet soil requires considerable effort. Chapter 2 provides a further discussion of mechanical impedance.

### 1.3.2 Moisture Stress

The most readily discernible symptoms in plant shoots of lack of moisture are wilting and rolling of leaves and sometimes leaf browning or leaf drop. These symptoms appear when the plant is unable to obtain sufficient water from the soil to transpire at potential rate. Yield (total biomass accumulation) will be reduced when transpiration is reduced by water stress. Using appropriate water balance models, the occurrence and degree of water shortages can be calculated. By assessing the calculated transpiration rates resulting from calculated water shortages, the effect of modifying the root zone can be evaluated. This approach to determining the lack of water and in evaluating the effects on water availability of modifying the root environment is discussed in Chapter 10.

The plant shoots react to the overall lack of water supplied by the roots. Roots, in contrast, react to the water supply in their immediate vicinity. Soybean roots located in soil layers drier than about $-0.1$ to $0.2$ MPa ($-1$ to $-2$ bars) water potential tend to become dark tan to brown. Many of the smaller diameter feeder roots of cotton (*Gossypium hirsutum L.*) also tend to die off and disappear at about $-0.1$ to $-0.2$ MPa potential (Klepper et al., 1973). When the particular soil layer is rewet to higher potentials, white turgid roots often appear within 48 h on the old brown roots of cultivated plants.

Some soil test for water status is the most obvious means to determine if roots in a particular layer are undergoing water stress. A more complete discussion of alleviating water stress in the root zone by soil modification is presented in Chapter 3.

### 1.3.3 Nutrient Stress

Many plants show leaf symptoms characteristic of specific mineral deficiencies. Some of these symptoms are discussed in "Hunger Signs in Crops," a book edited by Sprague (1964). By the time that severe deficiency symptoms appear, however, yield may be reduced substantially. Diagnostic soil and tissue tests have been developed to more completely describe nutrient stresses. These tests are discussed in Chapter 4 along with methods for reducing the effects of the stress.

### 1.3.4 Oxygen Stress

Oxygen stress sometimes occurs in conjunction with mechanical impedance in compacted soils but almost always is caused by excessive soil water contents when occurring in non-compacted soils. The symptoms of oxygen stress are relatively easy to diagnose when caused by a static water table. In that case, white, turgid roots will concentrate in a zone a few centimeters above the water table but will not grow into the water table very far. Symptoms caused by intermittent water tables are much more difficult to diagnose. Chapter 5 provides up-to-date information on diagnosing oxygen stress and methods for alleviating it.

### 1.3.5 Pathogen Effects

Some pathogen effects are easy to diagnose and some are extremely difficult. It is relatively easy to detect the damage caused by root-feeding insects

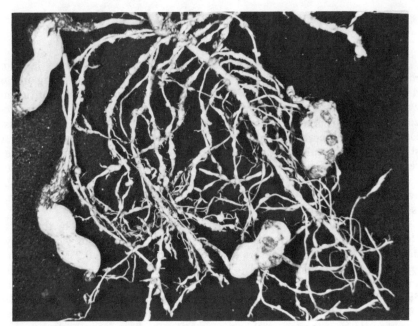

FIG. 1.3-5 **Peanut roots infected with** *Meloidogyne arenaria* **nematodes (Sasser, 1980).**

or the galls caused by certain kinds of nematodes (Fig. 1.3-5). It is much more difficult to pin-point effects on root function when root-invading pathogens are involved. The land-grant university in most states has a service laboratory that will diagnose presence of yield-reducing levels of pathogens and prescribe methods for alleviating the effects of their stresses. If one sees no obvious other cause of reduced shoot growth when a root system is dug, an examination for pathogens is advisable. Further information on these stresses is given in Chapter 6.

### 1.3.6 Temperature Stress

Low soil temperatures generally cause slow growth of crop plants. In addition, low soil temperatures often reduce uptake of minerals and water thus further reducing plant growth. A soil temperature record therefore is the best available technique for detecting low temperature stress.

High temperatures at the soil surface can cause death of young seedlings especially those with very small diameters. Temperatures above about 40 °C often will sear enough material at the ground level to reduce water and photosynthate movement through that portion of the stem. Rapid dehydration of the shoot results from that blockage. Alleviation techniques are to plant when soil surface temperatures likely will not reach 40 °C and to provide surface residue for protecting both the seedling plant and the soil surface around it from direct sunlight.

Temperature stresses and methods of controlling them are discussed more fully in Chapter 7.

### 1.3.7 Aluminum Toxicity

Aluminum toxicity causes similar effects in most plants. Root growth is reduced by levels of aluminum that cause little or no decrease in shoot

**FIG. 1.32-6 Progressively more severe damage to soybean roots (left to right) as aluminum toxicity increases (Devine, 1976).**

growth. When aluminum activity increases in the soil solution the roots tend to become shortened and swollen (Fig. 1.3-6). Lateral root initiation is affected to a lesser extent than lateral root elongation. Necrotic spots appear on some roots while other parts of the same root seem to be healthy.

Soil pH is a quick test to evaluate the possibility that aluminum toxicity is affecting crop yield. If soil pH is below about 5.5, a more careful analysis should be made by a service laboratory. Differences exist among soils in the pH levels that allow soluble aluminum to exist at phytotoxic levels. Differences also exist among plant species in their tolerance to specific levels of soluble aluminum. A more complete discussion of aluminum toxicity is presented in Chapter 8.

### 1.3.8 Salinity Stress

Soil salinity reduces availability of soil water and alters ionic balance within the plant. Slow plant growth caused by salinity thus is difficult to separate from that caused by other stresses. Soil pH above 7.8, a whitish appearance on the surface of an otherwise darker soil, and a high electrical conductivity of the soil solution are soil tests that indicate the possibility of salinity stress. Further details on salinity stress are provided in Chapter 9.

### 1.4 ALTERNATIVES TO SOIL MODIFICATION FOR OVERCOMING PLANT STRESSES

The primary emphasis of this monograph is on reducing plant stresses through soil modification. There are two other possibilities, however, for relieving effects of soil-related constraints on crop production. First, the farmer can change from a sensitive crop to one resistant to a particular stress. Examples are changing from cotton (a crop sensitive to aluminum toxicity) to peanuts (resistant) (Adams and Pearson, 1970), or from an annual to a perennial crop on high-strength soil pans (Fryrear and McCully, 1972), or from a crop sensitive to soil salinity to a resistant one (field beans, *Phaseolus spp.* to barley, *Hordeum vulgare L.*) (Nieman and Shannon, 1976). Second, plant breeders can select genotypes for their resistance to particular soil-related stresses. This selection and breeding approach to overcoming effects of a soil-related constraint has been shown to be feasible to (a) reduce salinity stress for barley (Epstein and Norlyn, 1977), reduce aluminum toxicity for wheat (Moore et al., 1976), reduce manganese toxicity for tropical legumes (Andrew, 1976), and reduce water stress for sorghum (Jordan and Miller, 1980). Wide diversity exists in genetic materials of most crop species. The selection and breeding approach is a long-term one, but should be applicable for most crop species. In the long term, it often will be more economical to adopt the plant breeding approach than the soil modification one for solving the plant stress problem. Soil modification will be the most feasible alternative where soil-related constraints are too severe to solve through selecting alternate crops or through selecting resistant genotypes. Soil modification also is more desirable than plant breeding on a short-term basis because at least a decade is necessary to develop and release a new cultivar somewhat tolerant to a particular soil constraint.

### 1.5 Systems Considerations and Constraints

Assessing the impact of root zone modification is difficult because (a) subterranean crop growth observations are costly, time consuming, and destructive and (b) other environmental and management factors frequently mask the crop response. As a result, alternative production practices that attempt to capitalize on an improved root environment often cannot yet be adequately evaluated for general utility.

Modeling, a systematic approach for quantifying crop responses to changes in the environment, could prove useful for this evaluation. Computer simulations derived from models can be used to separate crop response to root zone modifications from other confounding responses. Impact and cost-benefit could then be evaluated quantitatively from year to year and location to location. Crop-environment interactions, modeling approaches and alternative production practices are discussed in Chapter 11.

### 1.6 SUMMARY

This chapter discussed (a) history of soil modification, (b) effects of shoot and root environments on the shoot-root ratios, (c) techniques for the preliminary diagnosis of a soil-related constraint, (d) alternatives to soil modification for relieving effects of a soil and (c) systems considerations and constraints in root zone modification. Relatively simple soil tests and root system appearances are used to direct the reader's attention to full chapters, which provide a more complete discussion of each soil constraint.

### References

1  Adams, F. and R. W. Pearson. 1970. Differential response of cotton and peanuts to subsoil acidity. Agron. J. 62:9-12.

2  Andrew, C. S. 1976. Screening tropical legumes for manganese tolerance. In: M. J. Wright (ed). Plant adaptation to mineral stress in problem soils. A special publication of Cornell Agricultural Experiment Station, Ithaca, NY.

3  Arkin, G. F., A. Blum and E. Burnett. 1978. A root observation chamber field installation. Texas Agric. Exp. Sta. Misc. Publ. 1386, 10 p.

4  Aung, L. H. 1974. Root-shoot relationships. In: E. W. Carson (ed.) The plant root and its environment, Univ. of Va. Press, Charlottesville. Pages 29-61.

5  Barfield, B. J. and J. F. Gerber. 1979. Modification of the aerial environment of plants. 538 pp. ASAE, St. Joseph, MI 49085.

6  Barley, K. P. and P. J. England. 1970. Mechanics of burr burial by subterranean

clover. Proceedings XI International Grassland Congress: 103-107 Brisbane, Australia.

7   Böhm, W. 1974. Mini-rhizotrons for root observations under field conditions. Z. Acker- und Pflanzenbau 140:282-287.

8   Böhm, W. 1979. Methods of studying root systems. Ecological studies 33. Springer-Verlag Berlin. 188 p.

9   Brouwer, R. 1963. Some aspects of the equilibrium between overground and underground plant parts. Jaarb. I.B.S. (Wageningen), 31-39.

10   Brouwer, R. and C. T. de Wit. 1969. A simulation model of plant growth with special attention to root growth and its consequences. p. 224-244. In: Root growth. W. J. Whittington (ed.) Butterworths, London.

11   Cooper, A. J. 1955. Further observations on the growth of the root and the shoot of the tomato plant. Proc. 14th Int. Hort. Congress 589-595.

12   Chandler, W. H. 1919. Some results as to the response of fruit trees to pruning. Proc. Am. Soc. Hort Sci. 16:88-101.

13   Crist, J. W. and G. J. Stout. 1929. Relation between top size and root size in herbaceous plants. Plant Physiol. 4:63-85.

14   Devine, T. E. 1976. Genetic potentials for solving problems of soil mineral stress: aluminum and manganese toxicities in legumes. p. 65-72, In: M. J. Wright (ed) Plant adaptation to mineral stress in problem soils. A special publication of Cornell Agricultural Expt. Stn., Ithaca, NY.

15   Eavis, B. W. and H. M. Taylor. 1979. Transpiration of soybeans as related to leaf area, root length and soil water content. Agron. J. 71:441-445.

16   Epstein, E. and J. D. Norlyn. 1977. Seawater-based crop production: a feasibility study. Sci. 197:249-251.

17   Foth, H. D. 1962. Root and top growth of corn. Agron. J. 54:49-52.

18   Fryrear, D. W. and W. G. McCully. 1972. Development of grass root systems as influenced by soil compaction. J. Range Mgt. 25:254-257.

19   Gregory, P. J. 1979. A periscope method for observing root growth and distribution in field soil. J. Exp. Bot. 30:205-214.

20   Hales, S. 1727. Vegetable staticks, London. 1927. Reprint, McDonald, London, 1961.

21   Hubbard, J. W. 1932. Root constriction of cotton plants in the San Joaquin Valley of California. J. Agr. Res. 44:39-47.

22   Jordan, W. R. and F. R. Miller. 1980. Genetic variability in sorghum root systems: implications for drought tolerance. p. 383-399. In: N. C. Turner and P. J. Kramer, Adaptation of plants to water and high temperature stress. Wiley, Interscience, N.Y.

23   Klepper, B., H. M. Taylor, M. G. Huck and E. L. Fiscus. 1973. Water relations and growth of cotton in drying soil. Agron. J. 65:307-310.

24   Locke, L. F., H. V. Eck and H. J. Haas. 1960. Plowpan investigations at the Great Plains Field Stations, Woodward, Oklahoma and Mandan, N.D. USDA Production Research Report #40. 33 p.

25   Moore, D. P., W. E. Kronstad and R. J. Metzger. 1976. Screening wheat for aluminum tolerance. p. 287-295. In: M. J. Wright (ed.) Plant adaptation to mineral stress in problem soils. A special publication of Cornell Agric. Expt. Stn., Ithaca, NY.

26   Nieman, R. H. and M. C. Shannon. 1976. Screening plants for salinity tolerance. p. 359-367. In: M. J. Wright (ed.), Plant adaptation to mineral stress in problem soils. A special publication of Cornell Agr. Exp. Stn., Ithaca, NY.

27   Rotmistrov, V. 1909. Root system of cultivated plants of one year's growth. pp. 57. South Russian Printing Co., Odessa, Russia.

28   Russell, R. Scott. 1977. Plant root systems: Their function and interaction with the soil. McGraw-Hill Book Co., London. 298 pp.

29   Sanders, J. C. and D. A. Brown. 1978. A new fiber optic technique for measuring root growth of soybeans under field conditions. Agron. J. 70:1073-1076.

30   Sasser, J. N. 1980. Root-knot nematodes: a global menance to crop production. Plant Disease 64:36-41.

31   Shank, D. B. 1945. Effects of phosphorus, nitrogen and soil moisture on top-root ratios of inbred and hybrid maize. J. Agr. Res. 70:365-377.

32   Sprague, H. B. 1964. Hunger signs in crops, third edition. Published by David McKay Co., NY, NY. 461 p.

33   Taylor, H. M. 1971. The rhizotron at Auburn, AL: design and three year's use. Proc. 3rd Intl. Seminar for Hydrology Professors. Purdue Univ., Lafayette, IN, June 1971.

34   Underwood, C. V., H. M. Taylor and C. S. Hoveland. 1971. Soil physical factors affecting peanut pod development. Agron. J. 63:953-954.

35   Waddington, J. 1971. Observation of plant roots in situ. Can. J. Bot. 49:1850-1852.

36   Willatt, S. T. and H. M. Taylor. 1978. Water uptake by soya-bean roots as affected by their depth and by soil water content. J. Agric. Sci. (Camb) 90:205-213.

# chapter 2

**ALLEVIATING MECHANICAL IMPEDANCE**

**2**

**2**

# ALLEVIATING MECHANICAL IMPEDANCE

by    H. D. Bowen, Professor of Cotton Mechanization, Department of Biological and Agricultural Engineering, North Carolina State University, Raleigh, NC 27607

## 2.1 INTRODUCTION

The term mechanical impedance usually refers to the growth response of some plant part as it is mechanically constrained by a volume of soil. There are vast areas in the world where mechanical properties in the top 0.3 m of soil reduce potential crop growth and yield. Rocks, shales, fragipans and other permanently hard materials cause the problem in part of this area. Those impedances will not be discussed because most of these areas, for the foreseeable future, will remain in permanent vegetation due to the cost of modification. Impedance caused by excessive soil compaction and by cohesive but massive soil layers will be discussed. These impedances, if severe, usually justify the cost of modification. This chapter covers four such growth responses—the impedance of a root to further exploration of the soil volume, the radial constraint of a root to prevent its continued expansion in diameter, the impedance of a shoot to emergence and the impedance of a fruiting body when its entry into the soil is required for fruit development.

Agricultural tractors and equipment are increasing in mass, power and speed. These factors combine to allow the farmer to perform more frequent and more intense operations on the field, often when the soil is too wet. It seems likely that the economic pressures for labor efficiency will increase unless there is overwhelming evidence that soil structure is being permanently damaged. In the absence of this evidence, the trend to larger equipment will continue (Wismer et al., 1968). This fact means that soil compaction, destruction of soil structure and increases in mechanical impedance will continue and even accelerate unless preventative measures are employed.

Although root impeding layers can reduce anchorage and allow crop plants to topple, their usual effect is to reduce the quantities of water and essential nutrient elements available to the plant. Therefore interactions among mechanical impedance, soil water contents and nutrient concentrations in the soil solution must be discussed. When possible, reference to other chapters of this monograph or to the ASAE monograph Compaction of Agricultural Soils (Barnes et al., 1971), will be made to avoid redundancy.

### 2.1.1 Definitions

The total volume of soil is made up of the volume of soil solid materials (including vegetative matter) and the volume of pores between and among solid materials. The ratio of the volume of voids to volume of solids is the void ratio while the ratio of volume of voids to total volume of soil is the porosity. The mass of oven-dry solids contained in a unit volume of soil is the bulk density. These three parameters all are measures of soil compactness, a state variable. The act of increasing (but not of decreasing) the state of compactness is called soil compaction, which is a dynamic behavior of soil.

No discussion about alleviating mechanical impedance and other effects of soil compaction is meaningful without a clear definition of soil texture and soil structure as the soils engineer and the agronomist see them.

Soil texture refers to the relative proportions of primary particles in a soil mass (Donahue et al., 1977). The term structure, however, is somewhat ambiguous. Structure to a soil mineralogist deals with the arrangement of the various building materials contained in a clay mineral. Structure to the soils engineer and the agronomist is the geometric arrangement of particles in the soil matrix. Structure is an independent entity—the fortuitous arrangement of aggregates and primary particles as influenced by total past history (Gill and Vanden Berg, 1967). When used in this sense, structure does not include the dynamic behavior of soil under load. Soil structure, as defined and used here, is often called soil fabric.

### 2.1.2 History of Mechanical Impedance Studies

There is a long history of mechanical impedance studies. Cato the Elder (234-149 B.C.) wrote that the first principle of good crop husbandry was to plow well and the second principle was to plow again (Weir, 1936). Presumably, these principles provided a mellow seedbed. King (1895) was more explicit when he wrote, "a mellow seedbed with its many well-aerated pores allows roots to grow unhindered in any and every direction and to place their absorbing surfaces in vital touch with the soil grains and soil moisture. In this way, nourishment in the seed provides the maximum root surface in the shortest time."

During the intervening years from 1895 to 1945, many tillage experiments were conducted to alleviate mechanical impedance and thus to increase yield. These experiments often gave inconsistent and inconclusive results, primarily because the soil environment created by tillage was not measured.

Three events occurred in the late 1940's and early 1950's to intensify mechanical impedance studies. Firstly, Veihmeyer and Hendrickson (1948) conclusively showed that increases in bulk density reduced root growth even in soils where aeration should not have been a problem. Secondly, Lutz (1952) stressed that major gaps existed in our knowledge about mechanical impedance and plant growth and provided an outline for bridging those gaps. Thirdly, Gill and Bolt (1955) reviewed the root growth pressure studies of Pfeffer (1893) which showed that plants can exert pressures up to 2500 kPa during growth. These articles renewed interest in mechanical impedance and served to focus attention on unanswered questions. During the 1960's and 1970's, considerable progress was made in mechanical impedance research. This chapter attempts to summarize that progress.

### 2.1.3 Basic Principles of Mechanical Impedance

Plant organs grow in soil when new cells are formed and the turgor pressure inside these cells is sufficient to overcome constraint of the cell walls and any external constraint caused by the surrounding soil matrix. The difference between turgor pressure and cell wall constraint (for roots) is defined as root growth pressure by Gill and Bolt (1955). The growth pressure must be greater than the impedance acting on the cross-section of the plant organ if it is to grow. Maximum root growth pressures were first measured by Pfeffer (1893), who found values of 700 to 2500 kPa for horsebean (*Vicia faba* L.) corn (*Zea mays* L.), common vetch (*Vicia sativa* L.) and horse chestnut (*Aesculus hippocastanum* L.). Since that time, Stolzy and Barley (1968), Eavis et al. (1969) and Taylor and Ratliff (1969a) have confirmed that maximum axial (longitudinal) root growth pressures of several crops range from 900 to 1500 kPa. Kibreab and Danielson (1977) have shown that radish (*Raphanus sativas* L.) roots also cease enlarging their diameter when subjected to radial constraints of about 850 kPa.

Scott Russell (1977) has pointed out that agriculturalists should be more concerned with the effects of partial constraints than with those that prohibit growth. Russell and Goss (1974) showed that a 20 kPa pressure applied to a glass bead system reduced elongation rate of barley (*Hordeum vulgare* L.) roots by 50 percent and a 50 kPa pressure reduced the rate by 80 percent (Fig. 2.1-1). As the pressure increased, the roots became shorter and larger in diameter. The relationship between root elongation rate and applied pressure was not affected by pore diameters between 16 and 157 microns in the Russell and Goss (1974) experiment.

Pore diameters sometimes do affect root elongation rates, however. If the pore is larger than the root tip, the root can enter even a rigid matrix (Wiersum, 1957; Aubertin and Kardos, 1965a, b). If a material is easily

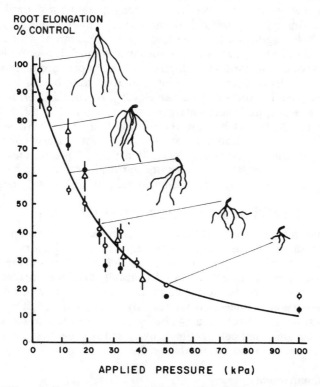

FIG. 2.1-1 Effect of applied pressure on the rate of elongation of seminal axes of barley plants grown for 6 days in beds of ballotini. Open circles: pore diameter 157 μm; closed circles: pore diameter 69 μm; triangles: pore diameter 16 μm. Redrawn from Russell and Goss (1974).

deformable, roots enter and grow until some factor other than mechanical impedance controls elongation rate even if the material is nonporous (Taylor and Gardner, 1960). In most situations, roots grow partly through existing pore spaces and partly by moving aside soil particles.

There is no reason to believe that growth pressure of shoot cells differs substantially from that of root cells. Shoots push upward through the soil displacing soil particles and aggregates from their paths. Many soils form mechanically strong surface crusts or seals that affect shoot growth. These crusts form both under natural rainfall or irrigation conditions.

Water breaks the cohesive bonds between soil particles and causes the average size of the structural units in the surface layer to be reduced. Splash, flow and sedimentation processes then result in resorting and repacking the units until the surface is levelled and covered with fine, closely packed particles. During drying, this surface layer hardens and may crack depending on the type and amount of clay. If the crust forms and dries after planting but before emergence, stand establishment is adversely affected.

Seedlings emerge through these crusts either by increasing their diameter thus increasing the force brought to bear on the crust or by the shoot bending then growing parallel to the crust surface until it encounters a crack through which to emerge.

Radial constraint of roots or stems is caused by root or shoot growth through a layer at low cohesive strength. Afterwards, the layer dries and forms a cohesive, high strength layer. Usually, a rain or irrigation, which rewets the cohesive layer, will reduce the strength temporarily until the soil dries again. This type of mechanical impedance is more severe in semiarid areas where soil organic matter is low and soil drying is rapid. Although very damaging in local areas, this type of impedance is much less prevalent than root impedance due to compacted layers or shoot impedance due to crusts.

Soil crusts also reduce the likelihood that fruiting bodies of peanuts (*Arachis hypogaea* L.) and subterranean clover (*Trifolium subterraneum* L.) will penetrate to depths where enlargement can occur. The difficulty with penetration usually occurs because the fruiting body is not sufficiently anchored to exert its maximum force onto the crust surface (Barley and England, 1970; Underwood et al., 1971). Sometimes, however, anchorage is sufficient but a combination of tractor tire traffic and soil drying has caused the crust to be too strong even for well-anchored fruiting bodies to penetrate.

## 2.2 MECHANICAL PROPERTIES OF SOILS

The four general classes of impedance discussed in this chapter—root elongation, radial root constraint, shoot emergence and fruit entry into soil—all are associated with the development of significant levels of soil mechanical strength. This strength, in cultivated soils, usually develops because (a) the soil has been compacted due to forces applied during traction or tillage, (b) particle aggregation has been partially lost by excessive tillage or by organic matter losses, or (c) soil cohesion has been increased by loss of soil water in the impending zone. Soil compaction is discussed first because methods for its alleviation are the most straightforward.

### 2.2.1 Soil Compaction

The resistance of a soil to compaction (i.e. compression or shear) is determined by its mechanical strength. Soil mechanical strength consists of two components—cohesive strength and frictional strength. The values of these two components vary considerably depending on water content, particle size distribution, particle shape including roughness, aggregate size, ionic composition and concentration in the soil solution, organic matter, type of clay mineral, and previous history of the sample. These factors, and others, exist in so many combinations that at the present time the actual strength of a given soil in a given condition must be measured if precise values are required. Later in this section current attempts to simulate soil compaction by tractor tire traffic will be discussed.

### 2.2.2 Causes of Soil Compaction

Chancellor (1976) lists four causes of soil compaction as:
(a) Natural consolidation during soil forming processes.
(b) Trampling by animals, including humans.
(c) Natural shrinkage of soils upon drying.
(d) Soil response to pressures and deformations imposed by wheels, tracks and soil engaging tools.
Another cause is:
(e) Actions of overburden and of water droplets on water weakened aggregates during rainfall, sprinkler irrigation or flood irrigation (Keller, 1967, 1970a, 1970b; Ghavami et al., 1974).
Humans can readily influence factors (b), (d), and (e) through avoidance of the force application or by management decisions about operations.

### 2.2.3 Consequences of Soil Compaction

When soil is subjected to an applied load that is sufficient to cause soil compaction, there are four possible factors which may have changed. These are a reduction in the liquid and gas contents in the pores, a rearrangement of soil particles, a change in pressure within the liquid and gases in the pore space and a reduction in the volume of the solid particles. The pressure increases usually are rapidly dissipated by flows of materials from the compacted zone and solid particles usually compress to a much lesser extent than the soil pores. When force is applied to a soil, therefore, the major actions within the soil are a rearrangement of particles and a reduction in pore space, especially of large pores (Harris, 1971). This particle rearrangement and pore space reduction increases both soil compactness and soil coherence. In addition, particle rearrangement and pore space reduction cause changes in aeration, nutrient, heat and water relations of the soil mass but these factors are discussed in other chapters of this monograph and also in Barnes et al. (1971).

### 2.2.4 Soil Compaction Simulation

A sense of urgency exists worldwide to find a simple and reliable way of predicting the amount of compaction that may be expected from operating heavy wheeled and tracked vehicles on agricultural soils. Tractor wheel traffic is the worst offender in causing excessive compaction in modern power farming and results when pressures on the soil exceed bearing capacity and shear strength of the soils. The risk of excessive compaction increases as the wheel loading and draft increases, the tire pressure increases, the number of passes increases, and the soil moisture increases. The heavier the tractor the deeper the compaction (Taylor et al., 1978).

Although these generalizations are helpful guidelines they are not precise enough for the farm producer-manager who must maximize labor efficiency, while reducing the rate of accumulation of excessive compaction. The farm manager can make a wise decision on whether to purchase the larger of two tractors only if he knows the compaction that will result from operation of each of the two tractors. In the daily decision of crop production, he may have to decide whether to go to the field to harvest when the field is a little too wet or risk losing a crop from a predicted weather front that is moving into the area. He needs to know the consequences of his decision. The availability of an easy-to-use, inexpensive computer simulation for calculating the compaction by different size tractors operating at various moisture contents on his soil would allow him to enhance his ability to make decisions involving compaction.

### 2.2.4.1 History of Simulation Attempts

In the 1950's W. Soehne made the first comprehensive study of soil compaction under tractor tires. Soehne used the approaches of soil mechanics developed by civil engineers for use in predicting the soil pressures under circular and strip footings for buildings and dams. Soehne (1958) used Boussinesq's equation

$$\sigma_r = \frac{3Q}{2\Pi r^2} \, Cos\theta \quad \dots\dots\dots\dots\dots\dots\dots\dots\dots\dots\dots\dots\dots\dots \quad [2.2\text{-}1]$$

to calculate the polar principal stress, $\sigma_r$, and Boussinesq's equation

$$\sigma_z = \frac{3Q}{2\Pi r^2} \, Cos^3\theta \quad \dots\dots\dots\dots\dots\dots\dots\dots\dots\dots\dots\dots\dots \quad [2.2\text{-}2]$$

to calculate the vertical compressive stress, $\sigma_z$, where Q is the load, r is the radius from the point of load application and $\theta$ is the vertical angle from the point of load application to the volume element.

Froehlich (1934) modified Boussinesq's equation to introduce a load concentration factor. After selecting a "suitable" concentration factor, $\mu$, Soehne (1958) used the formula

$$\sigma_z = (\frac{\mu Q}{2\Pi z^2}) \, Cos^{\mu+2}\theta = k(\frac{Q}{z^2}) \quad \dots\dots\dots\dots\dots\dots\dots\dots \quad [2.2\text{-}3]$$

where $\sigma_z$ is the vertical stress and z is the depth.

Soehne (1958), testing soils under a number of condition, concluded that equation [2.2-3] gave a reasonable estimate of the major principal stress, $\sigma_1$, that occurs under wheel traffic in many situations. That assumption seemingly was justified by the soil mechanic and foundation engineering literature of the time. The relation of the major principal stress, $\sigma_1$, to soil compaction was determined by loading the soil in a confined compression test, which consisted of obtaining the porosity of the soil as a function of the average pressure developed by a close-fitting piston compressing soil in a cylindrical vessel.

When the piston has a small clearance with the cylinder, the soil is confined on all sides and shear stresses and lateral strains are assumed to be prevented. In this apparatus all resultant axial strain is presumed to be due to the axial stress, $\sigma_1$, imposed upon the soil by the moving piston. The intermediate and minor principal stresses, $\sigma_2$, and $\sigma_3$, are equal and are due to passive reaction of the vessel walls to the soil lateral stress generated by the axial stress, $\sigma_1$. From laboratory tests on each soil type, it is possible to construct semi-log plots of the porosity versus $\log_{10}$ pressure for several levels of soil water and obtain a family of curves similar to Fig. 2.2-1.

The curves of porosity versus $\log_{10}$ pressure are roughly straight lines over a wide range of pressure for most soils at pressures greater than about 40 kPa (Fig. 2.2-1) so that porosity at each water content can be written as:

$$\eta = - A \log_{10} P + C, \quad \dots\dots\dots\dots\dots\dots\dots\dots\dots\dots\dots\dots\dots [2.2\text{-}4]$$

where $\eta$ is the porosity of the soil, P is the pressure from the confined compression test, A is the slope of the semi-log plot of porosity vs P, and C is the intercept of the curve at P = 1.0. Thus $\sigma_1$ can be calculated from the wheel loading using equation [2.2-3] and related to soil compaction through the confined compression tests at any place where $\sigma_1$ uniquely determines compaction.

Throughout the late 1960's and the early 1970's engineers and soil scientists continued to investigate various aspects of the mechanics of soil compaction. Vanden Berg (1966) proposed a theory and performed tests that related soil compaction to the mean normal stress, $\sigma_m$ and the maximum natural shear strain, $\gamma_{max}$ rather than to the principal normal streaa, $\sigma_1$, as assumed by Soehne. Vanden Berg's equation for bulk density, BD, was:

$$BD = A \log (\sigma_m + \sigma_m \, \bar{\gamma}_{max}) + B \quad \dots\dots\dots\dots\dots\dots\dots\dots\dots\dots [2.2\text{-}5]$$

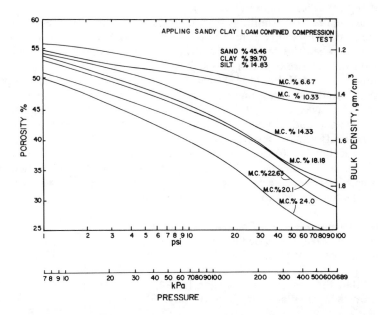

FIG. 2.2-1 The effect of applied axial stress, $\sigma_1$, on porosity and bulk density of Appling sandy clay loam compressed at various water contents. After Jaafari (1978).

where A is the slope of the curve and B is the intercept of the curve on the BD-axis. The mean normal stress, $\sigma_m$, was given by $\sigma_m = 1/3(\sigma_1 + \sigma_2 + \sigma_3)$, and $\gamma_{max}$ was determined from tests using triaxial apparatus. Dunlap and Weber (1971) showed that under certain loading paths $\sigma_2$ rather than $\sigma_1$ or $\sigma_m$ controlled soil compaction. Bailey (1971) showed that on one artificial soil, samples compacted to the same bulk density by different loading paths had different yield strengths. Taylor et al. (1978) showed that Froelich's modification to the Boussinesq equation approximately described the soil pressure distribution in both a sandy soil and a clay soil with uniform density profiles. However when a compacted layer was inserted into the density profile, the Froelich equation was no longer suitable.

Chancellor et al. (1962) and Jaafari (1978) reported that the maximum bulk density under a tire does not occur at the tire-soil interface where the maximum principal stress, $\sigma_1$, occurs, but rather at a depth approximately half the tire width below the tire-soil interface on the centerline below the tire track. At greater depths along the centerline, equation [2.2-1] gave a satisfactory value for the $\sigma_1$ to calculate the bulk density using equation [2.2-4] for the porosity and equation [2.2-6] for converting porosity to bulk density. Schmertmann (1970) using finite element techniques showed that the maximum vertical strain under the center of a loaded area is not at the surface where the pressure is a maximum, but has a maximum at a depth of between 0.72 and 1.2 of the radius of the foundation for surface loadings of 24 kPa and 192 kPa, respectively, the depth to the position of the maximum vertical strain increasing with increased surface loading. Jaafari (1978) showed that for no-draft operation of three tractors on a Wagram loamy sand and an Appling sandy clay loam, equation [2.2-1] could be used with a multiplier factor for each depth to predict soil bulk density from the tire-soil interface to the subsoil.

Chancellor and Schmidt (1962) found that when the depth-to-width ratio for the tire rut is small, as from operation of a wide low pressure tire on medium strength soil, the maximum vertical strain occurs at a shallower depth than when the depth-to-width ratio of the rut is large, as from a high pressure narrow tire operating on soft soil.

Bodman and Rubin (1948) have shown that shear strain superimposed during confined compression tests results in more compaction than compression alone. Davies et al. (1973) and Raghavan and McKyes (1977) showed that draft load and slippage in the field results in greater compaction than when no draft and no slippage occurs.

Chancellor (1966) reported that soils vary in sensitivity to slippage and shear strains. He indicated that the sensitive soils could be identified from the breadth of the soil grain size distribution. The wider the distribution the more sensitive the soil to shearing deformation.

### 2.2.4.2 An Example of a Compaction Simulation

Bowen (1975) developed a computer program to provide a visual display of bulk density that results from the passage of a tire over the soil surface. He used equation [2.2-3] to calculate the pressure, $\sigma_1$, under the tire and substituted this result into equation [2.2-4] to obtain the porosity, $\eta$. The program then calculated dry bulk density, BD, from:

$$BD = 2.65(1 - \eta/100) \quad \dots\dots\dots\dots\dots\dots\dots\dots\dots\dots \quad [2.2\text{-}6]$$

SOIL PROFILE - BULK DENSITY DISTRIBUTION

```
FFFFFFFFFFFFFFFFFFFFFFFFFFFFFFFFFFFFFFFFFFFFFFFFFFFFFFFFFFFFFFFF
FFFFFFFFFFFFFFFFFFFFFFFFFFFFFFFFFFFFFFFFFFFFFFFFFFFFFFFFFFFFFFFF
FFFFFFFFFFFFFFFFFFFFFFFFFFFFFFFFFFFFFFFFFFFFFFFFFFFFFFFFFFFFFFFF
FFFFFFFFFFFFFFFFFFFFFFFFFFFFFFFFFFFFFFFFFFFFFFFFFFFFFFFFFFFFFFFF
LLLLLLLLLLLLLLLMMMMMMMMMLLLLLLLLLLLLLLLLLLLLLLLLLLLLLLLLLLLLLLLL
FFFFFFFFFFGHIKLMNNNNNMLKIHGFFFFFFFFFFFFFFFFFFFFFFFFFFFFFFFFFFFFF
FFFFFFFFFGHIJKLMNOOONMLKJIHGFFFFFFFFFFFFFFFFFFFFFFFFFFFFFFFFFFFF
FFFFFFFFFGHIJKLMMNONMMLKJIHGFFFFFFFFFFFFFFFFFFFFFFFFFFFFFFFFFFFF
FFFFFFFFFGHHIJKLIMMMMMLIKJIHHGFFFFFFFFFFFFFFFFFFFFFFFFFFFFFFFFFF
FFFFFFFFFGHIIJKKLIMMMLIKKJIIHGFFFFFFFFFFFFFFFFFFFFFFFFFFFFFFFFFF
FFFFFFFFGGHIIJJKKKKKLMLKKKJJIIHGGFFFFFFFFFFFFFFFFFFFFFFFFFFFFFFF
FFFFFFGHHIIJJJKKKKKKKKKKJJJIIHHGGFFFFFFFFFFFFFFFFFFFFFFFFFFFFFFF
FFFFFFGGHHIIIJJJKKKKKJJJIIIHHGGFFFFFFFFFFFFFFFFFFFFFFFFFFFFFFFFF
FFFFFFGGHHIIIJJJJJJJJJIIIHHGGFFFFFFFFFFFFFFFFFFFFFFFFFFFFFFFFFFF
FFFFFFGGHHHIIIJJJJJJJIIIIHHHGGFFFFFFFFFFFFFFFFFFFFFFFFFFFFFFFFFF
FFFFFFGGHHHIIIIJJJJJJIIIIHHHGGFFFFFFFFFFFFFFFFFFFFFFFFFFFFFFFFFF
FFFFFFGGHHHHIIIIIIIIIIIIIHHHHGGFFFFFFFFFFFFFFFFFFFFFFFFFFFFFFFFF
FFFFFFGGHHHIIIIIIIIIIIIIIHHHGGFFFFFFFFFFFFFFFFFFFFFFFFFFFFFFFFFF
FFFFFFGGGHHHHIIIIIIIIIIHHHHGGGFFFFFFFFFFFFFFFFFFFFFFFFFFFFFFFFFF
FFFFFFGGGHHHHHIIIIIIIIHHHHHGGGFFFFFFFFFFFFFFFFFFFFFFFFFFFFFFFFFF
FFFFFFGGGGHHHHHHHIIHHHHHHHGGGGFFFFFFFFFFFFFFFFFFFFFFFFFFFFFFFFFF
FFFFFFGGGGHHHHHHHHHHHHHHHGGGGGFFFFFFFFFFFFFFFFFFFFFFFFFFFFFFFFFF
FFFFFFGGGGGHHHHHHHHHHHHHHGGGGGFFFFFFFFFFFFFFFFFFFFFFFFFFFFFFFFFF
```

| CODE FOR BULK DENSITY DISTRIBUTION | | INPUTS | |
|---|---|---|---|
| 1.00 = A | 1.05 | DISK | |
| 1.05 = B | 1.10 | DRAFT | 568 kg (1250 pds) |
| 1.10 = C | 1.15 | WIDTH | 3.0 m (10 ft) |
| 1.15 = D | 1.20 | WEIGHT | 687 kg (1500 pds) |
| 1.20 = E | 1.25 | DEPTH | 10 cm (4 inch) |
| 1.25 = F | 1.30 | | |
| 1.30 = G | 1.35 | TRACTOR | |
| 1.35 = H | 1.40 | WEIGHT | 1818 kg (4000 pds) |
| 1.40 = I | 1.45 | WHEEL SPACING | 2.0 m (80 inch) |
| 1.45 = J | 1.50 | TIRE SIZE | 31.5 x 71.1 cm |
| 1.50 = K | 1.55 | INFLA. PRES. | 83 kPa (12 psi) |
| 1.55 = L | 1.60 | RUT DEPTH | 7.5 cm (3 inch) |
| 1.60 = M | 1.65 | | |
| 1.65 = N | 1.70 | SOIL | |
| 1.70 = O | 1.75 | TYPE | Wagram loamy sand |
| | | FACTOR μ | 6 for non-cohesive |

**FIG. 2.2-2 Bulk density distribution in Wagram loamy sand as predicted by the computer simulation of Bowen (1975).**

Fig. 2.2-2 shows a computer printout of the visual display and bulk density code for one tractor pass over a Wagram loamy sand soil with the tire marks erased by disking to a 10-cm depth. Inputs for the program (Bowen, 1975) are indicated on Fig. 2.2-2 and except for the soil factors are readily available. The major soil factors are relationships among soil bulk density, applied principal stress, and soil water content (Fig. 2.2-3).

Larson et al. (1980) determined the relationships among bulk density, applied principal stress and water content for 36 agricultural soils collected from Venezuela, Brazil, Nigeria, and eight states of the USA. They found that the bulk density (or porosity) versus logarithm of applied principal stress, $\sigma_1$, relationship was linear, at a specific water content, over a range of stresses form 100 to 1000 kPa. At different water contents, the relationships between BD and log $\sigma_1$ were parallel to each other over a range of soil water potentials of -5 to -100 kPa. The slope of the BD - log $\sigma_1$ curve increased approximately linearly as clay content increased to 33 percent, and then remained about constant as clay content increased further. Larson et al. (1980) suggested a procedure whereby, if the compression curve at one water content is known, the BD - log relationship can be predicted for other water contents. The procedures of Bowen (1975) and Larson et al. (1980) are logical extensions of Soehne's approach to predict the pattern of bulk density with depth and the distance from the tire track under non-uniform water and soil

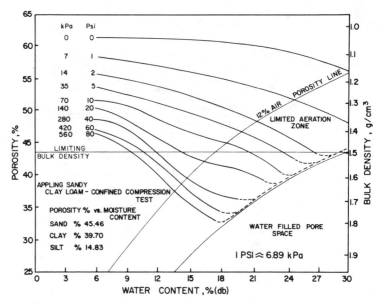

**FIG. 2.2-3 The relations among applied principal stress, $\sigma_1$, and water content at the time of loading on soil bulk density and, soil porosity. Lines showing assumed critical airfilled porosity and soil bulk density are drawn on the figure. After Jaafari (1978).**

conditions.

Raghavan and McKyes (1978) developed mathematical models that account for slip, number of passes and other factors not accounted for in Soehne's model.

### 2.2.4.3 The Simulation Output

The computer simulation provides a visual pattern of the predicted soil bulk densities that occur when any specific loading device, such as a tire, is applied to the soil at a specific water content. If auxiliary experimental data showing that some threshold level of bulk density (say 1.50 g/cm³) or of airfilled porosity (say 12 percent) exists that critically reduces rooting, then the maximum load that can be applied to the soil safely at that water content can also be determined.

### 2.2.4.4 Assessment of Prospects for a General Use Compaction Simulation

A computer program can be developed to provide the soil compaction information shown in this article at a reasonable cost. Use of the program is within the computer facilities available to farmers in most areas of the United States.

The computer simulation of soil compaction is only as good as the soil compression data used as an input. The most exact simulation will occur when confined compression tests are conducted on soils of the specific farm at a range of soil water contents. Obviously, this procedure is unrealistic for general use. It seems likely, however, that compression data can be obtained at one water content on a few benchmark soil types of a state or land resource area and then the soils of a specific farm can be matched with the benchmark soils. Appropriate matching parameters could be (a) a soil mechanical analysis, (b) known sensitivity of the soil to packing, or (c) the presence of an easily compacted layer.

### 2.2.5 Devices to Measure Mechanical Properties of Compacted Soils

In the previous section, I discussed a computer simulation of soil bulk density as influenced by wheel loading and water content at the time of loading. Soil water contents are never static. Soil volumes lose water by evaporation from the land surface, by deep drainage and by plant root extraction. Water is gained from flow into zones of low potential from zones of higher potential. All of these water content changes affect soil mechanical properties simultaneously.

Several devices have been used to evaluate the effects of soil bulk density and water content on soil mechanical properties. These include soil penetrometers (Davidson, 1965), shear vanes (Freitag, 1971), unconfined compressive strength machines (Sallberg, 1965), modulus of rupture (Reeve, 1965) and tensile strength machines (Gill and Vanden Berg, 1967). Each of these techniques provide empirical results when used in mechanical impedance studies. The results should be calibrated with actual plant growth measurements under impeding conditions before being used in predicting severity of impedance.

Soil penetrometers have been used to a greater extent than the other devices for determining the effects of soil bulk density and water content on root penetration. As soil bulk density increases or as soil water contents decrease, penetrometer values increase. However, the exact change in penetrometer value as bulk density or soil water content changes is soil dependent. For example, at - 33 kPa soil water potential and a penetrometer resistance value of 2 MPa, Columbia loam soil was at a bulk density of 1.55 g/cm³ while Miles fine sandy loam was at a bulk density of 1.80 g/cm³ (Fig. 2.2-4).

Penetrometers of various sizes and shapes have been used in mechanical impedance studies. The three main groups of penetrometers are (a) those which measure the pressure required to push the tip a specific distance into the soil volume (usually called a static-tip penetrometer), (b) those which measure the pressure (or force) required to move the tip through the soil at a

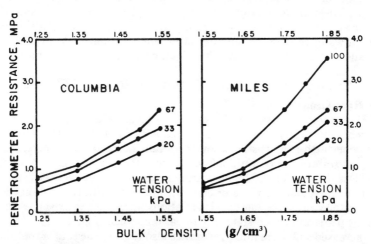

FIG. 2.2-4 Relations among the bulk density, soil water tension and penetrometer resistance of two soils. Redrawn from Taylor and Bruce (1968).

more or less constant rate (called a moving-tip penetrometer) and (c) those which record the number of blows required to drive the penetrometer tip through a specific depth of soil (called an impact penetrometer). ASAE has published the standard for the Cone Index Penetrometer, a moving-tip type. It is useful in field studies of mechanical impedance. Barley et al. (1965), have described a small diameter penetrometer useful in laboratory experiments. The Barley et al. (1965) penetrometer is constructed in such a manner that a point load component and a skin friction component of soil mechanical strength can be separated.

### 2.2.6 Predicting Penetrometer Resistance

Barley et al. (1965) postulated that the base of the penetrometer was represented by the limits of a sphere under internal pressure and that the radial stress was a function of radial distance from the apex of the probe and the internal stress in the sphere. The stress-strain relation was represented by an experimentally determined curve showing the change in void ratio during compressional loading with shear. Finally, the resistance of the soil to penetration was calculated as the sum of the internal pressure required to expand the sphere and the frictional resistance generated at the probe point.

The possibility exists that this relatively simple theory can be used to predict penetrometer resistance on the same soils where soil compaction is predicted. Thus it may be possible to simulate the range in penetrometer resistances that plant roots will encounter while they grow into and through soil compacted by wheel traffic. Presumably, the simulation can be developed along these lines. First, the bulk density (or porosity) versus log $\sigma_1$ curve will be developed at some specific water content for the soil in question. The procedure of Larson et al. (1980) will be used to simulate a family of bulk density versus log $\sigma_1$ curves at other water contents. These curves will be used in the procedure of Bowen (1975) to predict soil bulk density as a function of wheel load and soil water content. Curves also will be developed for the compression under shear. These latter curves will be used in the procedure of Barley et al. (1965) to predict penetrometer resistance. It seems likely that one or a few of the experimentally derived curves will be necessary to predict penetrometer resistances as function of soil bulk density and water content. These curves can be used to predict penetrometer resistance as the compacted soils gain or lose water or as the roots grow into soil volumes at different water contents or bulk densities. Such a simulation is vital in order to predict the effect of a given wheel load on root mechanical impedance.

## 2.3 LIMITING VALUES OF MECHANICAL IMPEDANCE

### 2.3.1 Root Growth Limits

Plant roots are less likely to enter soil layers with massive structure as soil mechanical resistances increase (Taylor et al., 1966). Almost all cotton roots penetrated low resistance soil cores but the proportion decreased with increased soil resistance until no roots penetrated at 3000 kPa soil resistance regardless of whether the high resistance was caused by increased soil bulk density or by reduced soil water content. For soil textures that ranged from silt loam to loamy sand, one general relationship seemed to exist between cotton root penetration and soil mechanical resistance (Fig. 2.3-1).

Increased soil mechanical resistance also reduces the rate of root elongation. A penetrometer resistance of 700 kPa reduced cotton root elongation

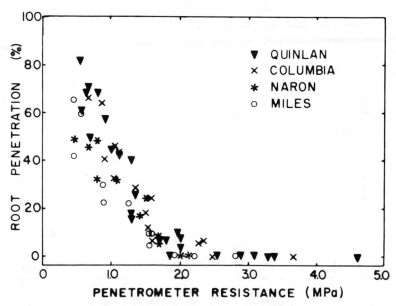

FIG. 2.3-1 The effect of penetrometer resistance on the percentage of cotton (*Gossypium hirsutum L.*) tap roots that penetrated through 0.025 m thick cores of four soils. Redrawn from Taylor et al. (1966).

rate 50 percent but a 2000 kPa penetrometer resistance was required to reduce peanut root elongation by 50 percent (Taylor and Ratliff, 1969b). A resistance of 1100 kPa reduced pea (*Pisum sativum* L.) rates by 50 percent (Cockroft et al., 1969). Peanut root elongation rates thus are reduced less than cotton or pea root penetration by a specific level of penetrometer resistance.

Many studies have shown that high strength soil layers reduce the quantity of roots growing in field soil profiles. Among these studies are those by Fiskell et al. (1968) on maize rooting in Florida, USA; by Taylor and Burnett (1964) and Taylor et al. (1964) on cotton and grain sorghum (*Sorghum bicolor* L. Moench.) grown in the Southern Great Plains, USA; by Reijerink (1973) on asparagus (*Asparagus officinalis* L.) in the Netherlands; by Kumar et al. (1971) on rice (*Oryza sativa* L.) in India; by Cohen and Sharabani (1964) on grapes (*Vitis* spp.) in Israel; by Monteith and Banath (1965) on sugar cane (*Saccharum officinarum* L.) in Queensland, Australia; by Davis et al. (1968) on cantaloupe (*Cucumis melo* L.) in California, USA; by Bennie and Laker (1975) on wheat (*Triticum aestivum* L.) and cotton in Orange Free State, Republic of South Africa; and by Phillips and Kirkham (1962) on maize in Iowa, USA. A great many other studies are reported in the literature; those cited here illustrate the widespread occurrence of mechanically impeding soils and their effects on root development.

### 2.3.2 Seedling Emergence Limits

The emergence of seedlings is affected by mechanical properties of the soil above the seeds. Bowen (1966) showed that soil mechanical impedance (measured as the pressure required to expand toy balloons within the soil) above 75 kPa reduced the percentage of cotton seedlings that emerged (Fig.

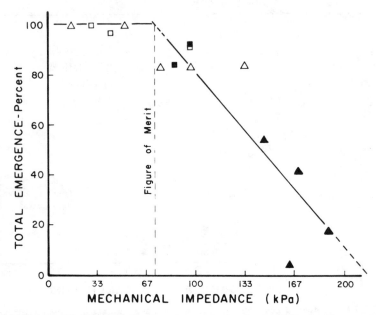

**FIG. 2.3-2 Relationship between soil resistance to expansion (by tiny ballons) and emergence of cotton seedling. Note that emergence is only affected by impedance levels over 70 kPa. Redrawn from Bowen (1966).**

2.3-2). No emergence occurred when mechanical impedance was greater than about 200 kPa.

Indentation type penetrometers were used to measure mechanical impedance of crusts by Parker and Taylor (1965). Emergence of grain sorghum after 10 days was affected only slightly be penetrometer resistance to 1000 kPa (Fig. 2.3-3). Soil temperatures of 21 and 35 °C affected the rate at which emergence occurred but did not affect the final emergence percentage. No emergence occurred at penetrometer resistances greater than about 1600 kPa. Small seeded crops such as onions (*Allium cepa* L.) are more easily mechanically impeded than crops such as cotton and grain sorghum (Taylor et al., 1966).

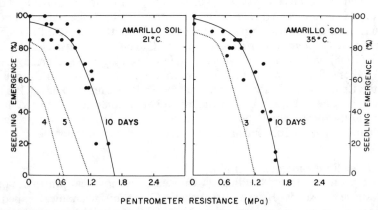

**FIG. 2.3-3 Emergence of grain sorghum (*Sorghum bicolor* Moench) as affected by penetrometer resistance, time after planting, and soil temperature. Redrawn from Parker and Taylor (1965).**

Wanjura (1973) has developed and partly verified a model to predict cotton emergence as a function of soil temperature, soil water content, soil mechanical impedance, planting depth and time. Output of the model is the fraction of live seeds (as determined from a standard germination test) that emerge as a function of time after planting and of treatment.

### 2.3.3 Limiting Values of Mechanical Impedance for Crop Yield
### 2.3.3.1 Root Impedance

It is much easier to show that a direct cause-effect relationship exists between the presence of mechanically impeding layers and root growth or shoot emergence than it is to show that such a relationship exists between mechanical impedance and crop yield. There are several reasons for this. First, mechanical impedance of root growth does not, of itself, reduce yield. Plants require water, essential minerals, and anchorage from the soil. If the impeding layers of soil volume do not increase plant stresses because of a lack of these items at any time between emergence and physiological maturity, impedance will not affect yield. Second, the definitive experiments showing that mechanical impedance reduces root growth were conducted in the laboratory under conditions of closely controlled uniformity. These conditions usually are not achieved in fields where most yield trials are conducted.

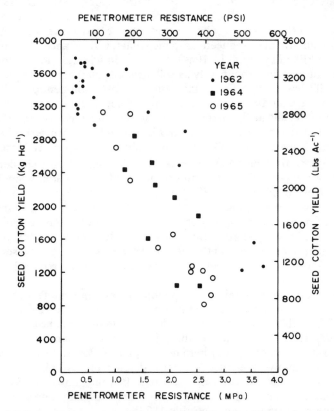

**FIG. 2.3-4 Relationship between penetrometer resistance and yield of seed cotton. Data from Carter et al. (1965) and Carter and Tavernetti (1968).**

Finally, all commonly used strength sensing devices integrate their measurements over soil volumes substantially larger than the size of the plant root. In addition, the devices either follow a rigid path (penetrometers) or cause a preordained failure pattern (shear vanes and compressive strength machines). Small, flexible plant roots can penetrate soil layers through soil cracks, worm holes, root channels, and other voids that do not substantially affect results obtained with strength-sensing devices (Nash and Balingar, 1974; Davis et al., 1968).

Despite these difficulties, many experiments have shown that crop yields decrease as the strength of soil layers or volumes increase. In an irrigation experiment, Carter et al. (1965) found that seed cotton yield decreased linearly from 3600 kg/ha where penetrometer resistance measured at 'field capacity' (cone penetrometer) was 0.3 MPa to 1450 kg/ha where resistance was 4.0 MPa (Fig. 2.3-4). Thus, yield of irrigated cotton grown in the arid environment found in the San Joaquin Valley of California increased as mechanical impedance was reduced. Research by Grimes et al. (1975) showed that water uptake rates ($cm^3H_2O/cm^3$ soil·day) increase with rooting density on these San Joaquin Valley soils (Typic Torriorthents and Typic Xerorthents) so that plant water stress probably decreased, even under irrigation, as mechanical impedance was reduced.

In the semi-arid environment of the Southern Great Plains, yield of nonirrigated cotton decreased as penetrometer resistance (static-tip penetrometer) increased. Yield of lint cotton was 560 kg/ha where no soil pan existed and 280 kg/ha where penetrometer resistance was 2500 kPa. Yield was not further reduced as penetrometer resistance was increased above 2500 kPa (Taylor et al., 1964). Grain yield of sorghum (Taylor et al., 1964) also decreased curvilinearly as soil strength increased, but yields were substantially lower on a Pratt fine sandy loam (Psammentic Haplustalfs) than on an Amarillo fine sandy loam (Aridic Paleustalfs) when soil pans were created in both soils at Bushland, Texas. Reseeded range grasses (Barton et al., 1966) and sugar beets (*Beta vulgaris* L.)(Taylor and Bruce, 1968) followed the same general trend of decreased yield with increased soil strength.

In the subhumid environment of Alabama, seed cotton yield decreased from 180 grams per cylinder (an oil drum buried in field soil) when no pan existed in Norfolk loamy sand (Typic Paleudults) to about 70, 45, and 35 grams when soil pans with 800 kPa penetrometer resistance existed at 30, 20, and 10 cm, respectively (Lowery et al., 1970). Soybeans (*Glycine max* (L.) Merr.) followed the same general pattern of a reduced yield as penetrometer resistance increased (Rogers and Thurlow, 1973). Plant water stresses induced by the soil pans were thought to be the reason for the reduced yields in both cases.

A soil profile that contains an impeding layer of only moderate thickness is less likely to reduce yield in perennial than in annual crops. Barton et al. (1966) found that the first year yield of giant cenchrus (*Cenchrus myosuroides* H.B.K.) grown on Amarillo fine sandy loam at Big Spring, Texas, was about 3400 kg/ha where no soil pan existed and was about 1050 kg/ha where a pan with a penetrometer resistance of 3400 kPa existed. The rooting pattern also was severaly distorted. In time, however, grass roots found planes of weakness in the fine sandy loam soil and permeated most of the soil volume (Fryrear and McCully, 1972).

In many instances, roots encounter compacted layers at field capacity water contents and therefore at their lowest mechanical impedance. When that occurs, mechanical impedance can be estimated from soil bulk density. A general rule-of-thumb (with many exceptions) is that bulk densities of 1.55, 1.65, 1.80 and 1.85 g/cm³ will severely impede root growth and thus will reduce crop yields on clay loams, silt loams, fine sandy loams and loamy fine sands, respectively. Obviously these general guides should be verified for the specific growth conditions.

### 2.3.3.2 Seedling Emergence, Radial Restraint and Fruiting Body Penetration

No definitive experiments (to the author's knowledge) have been conducted to evaluate the impedance levels that reduce yields through root radial restraint or the lack of fruiting body penetration. Many factors affect stand establishment. Mechanical impedance is only one of these factors. Definitive relationship between mechanical impedance of seedlings and crop yield also are extremely difficult to evaluate.

## 2.4 EVADING ROOT IMPEDANCE

Several methods have been used to reduce the effects of mechanical impedance on crop production. Each situation must be examined closely in order to determine the course of action most likely to succeed. Each course must be evaluated for costs, equipment availability, operator expertise, timeliness of operations and social acceptance.

### 2.4.1 Controlled Traffic

Evading mechanical impedance on most of the land surface by controlled traffic has been somewhat successful (Dumas et al., 1973). In this technique, certain narrow strips of the field are set aside for all of the vehicular tire traffic. These areas often become so compact and have such high strength that no plant roots grow in the surface soil horizons. Consequently, the water and fertilizer nutrients in these soil volumes are not available to crops (Trouse et al., 1975). In all soil volumes not compressed by tire traffic, water and nutrients are freely available. In a 3-year study on a Norfolk loamy sand soil in Alabama, Young and Browning (1977) found that seed cotton yields averaged 2335 kg/ha when tractors and sprayers were not controlled to traffic strips and 2540 kg/ha when controlled to strips. Dumas et al. (1973) also found that controlling traffic increased cotton yields in Alabama. Sheesley et al. (1974) found that alfalfa (*Medicago sativa* L.) yields and stand longevity decreased as the proportion of ground surface covered by tire traffic increased. Even though controlled traffic increased infiltration of irrigation water in California (Abernathy et al., 1975), cotton yields did not increase in the San Joaquin Valley (L.M. Carter quoted by Chancellor, 1976).

### 2.4.2 Water Content Control

Crops grown in semiarid and arid parts of the world often are planted when the soil in the plow layer is at a water content of less than field capacity but still high enough for germination. Mechanical impedance is a function of both soil bulk density and soil water content. These low-water-content layers sometimes impede plant roots until a rain occurs (Taylor et al., 1964b) or until irrigation water is added. When the water is added, soil strength decreases and root penetration increases. Additional irrigation or runoff water diverted onto a cropped area are management tools that not only tend to increase the water supply in the profile but also tend to make that water more available by increasing root penetration through high strength soil layers.

### 2.4.3 Soil Pan Shattering by Plants

The research literature contains many references to the pan-shattering abilities of various plants such as alfalfa, sweet clover (*Melilotus alba* Medik) and guar (*Cyamopsis tetragonolobus* L.). The reports usually are clear about the effects of these plants on succeeding crops but are not clear about the exact mechanism that causes the yield increase, i.e., increased root penetration, nitrogen supply, aeration, infiltration, etc. for the succeeding crop. Elkins et al. (1977), however, have produced unambiguous results showing that bahiagrass (*Paspalum notatun* Flugge 'Pensacola') roots will penetrate soil layers that mechanically impede cotton roots. They found that bahiagrass increased the number of pores greater than 1.0 mm in diameter and thus increased cotton rooting densities to a depth of at least 60 cm (Fig. 2.4-1). Cotton grown where bahiagrass sod had been plowed under 3 years before still yielded more than cotton grown where the soil had been chiseled to 35 cm deep. C. M. Peterson (quoted by Elkins et al., 1977) found that bahigrass roots possess a fibrous sheath beneath the epidermis. This sheath, which is absent in most plants, is probably responsible for the additional penetrating ability of bahiagrass roots.

### 2.4.4 Altering Soil to Reduce Mechanical Strength
### 2.4.4.1 Additions of Organic Matter

The incorporation of organic matter has several effects on crop growth including reducing overall soil bulk density, tieing-up or releasing nitrogen, developing aggregating and aggregate stabilizing materials in the soil, increasing $CO_2$ accumulation in the soil and in the air above the soil, increasing earthworm and other fauna populations, and changing soil pH.

Raw organic matter added as green or dried plant top growth always

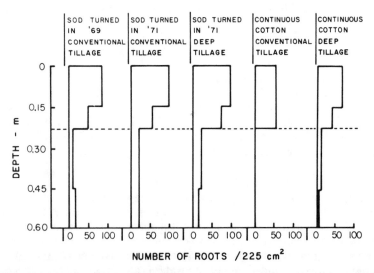

FIG. 2.4-1 Effect of Pensacola bahiagrass and deep tillage on cotton root penetration of a high-strength layer in Cahaba loamy sand soil, Tallassee, AL. The dotted line shows depth to the high-strength layer. From Elkins et al. (1977).

reduces the bulk density of mineral soils because the bulk density of organic matter (O.M.) is less than that of soil displaced by the organic matter. The residual humus materials, if any, have densities of about 1.4 g/cm³ as compared to specific densities of 2.65 to 2.70 g/cm³ for most mineral soil materials.

On 80 forest soils under natural conditions in the prairie-forest transition of southern and western Minnesota and 56 samples from the Itasca State Park region, Drew (1973) found that the relationship of bulk density (BD) to organic matter was

$$1/BD = 0.6268 + 0.0361 \ (\% \ O.M.) \quad \dots\dots\dots\dots\dots\dots\dots \ [2.4\text{-}1]$$

with a correlation of 0.842.

Since medieval times, organic matter levels have been suspected of contributing to fertility and increased yields. According to E. W. Russell (1977) soil organic matter contributes to soil fertility in several quite distinct ways. It helps to stabilize structure. It increases the cation exchange capacity in sandy soils. On soils containing appreciable amounts of active iron and aluminum hydroxides it helps increase available phosphate. A small part of the organic matter is dispersed in the soil solution and contains trace amounts of chelated copper and iron which may serve as important sources of the elements for the crop. It takes approximately 7 years under a pasture or ley to restore a long cultivated soil to a structure near that of soil under virgin sod (E. W. Russell, 1977). Cultivation reduces structure in 1 to 3 years.

### 2.4.4.2 Addition of Polyelectrolytes

Polyelectrolytes such as IBMA (Taylor and Vomocil, 1959) VAMA and HPAN (Allison, 1965) and Krilium (Martin et al., 1952) increase aggregation and aid the soil in resisting soil compaction. They also reduce cohesion of soil subjected to specific loads. The effects of the polyelectrolytes may last several years but the expense of their addition often is too great for field applications. These materials may be economical when applied in low concentrations to crusting soils to enhance emergence of such crops as carrots (*Daucus carotts* L.) and onions (*Allium cepa* L.) (Rawitz and Hazen, 1978).

### 2.4.4.3 Earthworm Management

Earthworms provide channels through soil layers that might otherwise impede root development. According to Minnich (1977) there are some 1800 catalogued species of earthworms on earth. The two most useful species for earthworm tillage in field and garden under natural conditions were imported from Europe. They are the common night crawler, *Lumbricus terrestria* and the common field worm, *Allolobophora caliginosa*. A third species of lumbricid, *Eisenia foetida* or manure worm, also imported from Europe, is found in piles of soil with high organic matter or manure and is useful in converting raw organic matter into a highly fertile mulch.

The above three species are well distributed across much of the world and because of their natural preferences of depth, temperature, moisture, food, and feeding habits, they can be used in various combinations to accomplish many tillage functions currently being carried on by machinery—but they require longer time periods and a higher level of management.

The night crawler *L. terrestris* works from the surface to 1 to 2 meters depth and can be fed from the surface with mulches of non-acid leaves, straw, corn stover, or hay. The worm builds permanent channels of about its diameter at the rate of half to two-thirds its own length per day. Mature night crawlers range from 10 to 30 cm in length (Minnich, 1977). The animals prefer a loamy soil but will tolerate sands and heavy clay if pH is above 5.5 (Minnich, 1977) and a minimum of about 4 tons/ha of organic matter is available on the surface. The night crawler has the greatest potential for alleviating mechanical impedance of roots in soil below 30 cm depth (by tunneling and transfer of organic matter from the surface) in loams and heavier soil, but with good management will operate effectively in all soil types.

The common field worm, *A. caliginosa*, prefers living in the surface 30 cm, but will burrow to deeper depths if temperature goes too high or too low. They can tolerate higher temperatures and lower moistures than night crawlers. They are valuable for developing and maintaining aeration, infiltration, and drainage in grain, pasture and hay crops as they feed primarily on freshly dead roots of growing plants.

The manure worm, *E. foetida,* has a potential for manufacturing organic fertilizer from wastes and can be useful in garden and greenhouse growth mediums. This species has improved gardens, flower beds, and orchards under controlled management. The "domesticated" earthworm probably is an improved strain of *E. foetida* (Minnich, 1977).

Slater and Hopp (1947) showed that earthworm populations were either higher or lower on cultivated land depending on the amount of fall protection provided the worms by a mulch of grass, hay, straw, burlap bags, or paper sacks. Infiltration rates on cropped land conformed to levels of earthworm populations at all but one site where high infiltration rates were due to soil cracks and fissures rathre than to holes made by earthworms.

### 2.4.4.4 Additions of Lime

Lime as $CaCo_3$, $Ca(OH)_2$, $CaMg(CO_3)_2$ or $CaO$ has been used for many years. Agricultural lime additions caused easier draft, better strength at equal water contents, higher plastic indexes and higher plastic limits. Very high rates of lime (3 to 30 percent by weight) have been used in recent years for stabilizing soils in roadbeds, earth dams, and under foundations of buildings by soil engineers.

Quicklime $Ca(OH)_2$ is superior to $CaCO_3$ for engineering purposes (Eriksson et al., 1974). Pettry and Rich (1971) found that $Ca(OH)_2$ reduced swelling pressure of 5 heavy clays and decreased the plasticity index values in all treated samples, especially those clays dominated by montmorillonite.

The unconfined maximum compressive strength of soils dominated by montmorillonite, kaolinite or illite clays in their natural state seems to be about 200 kPa, (Eriksson et al., 1974) unless the soils have been flocculated from a very thin mixture of soil and water with an excess of Ca. In the latter case laboratory tests on drained samples show maximum deviator strengths of 800 to 1000 kPa for unconfined compression (Dickson and Smart, 1978). It is unlikely that complete flocculation is obtained under field conditions and unconfined compression strengths seldom, if ever, exceed 200 kPa even when improved by the addition of lime for agricultural purposes.

Eriksson et al. (1974) reported dramatic effects on structure that resulted from applying 0 to 50 tons/ha of burnt or slaked lime ($CaO$ or $Ca(OH)_2$). Liming reduced soil compaction that resulted from cultivation

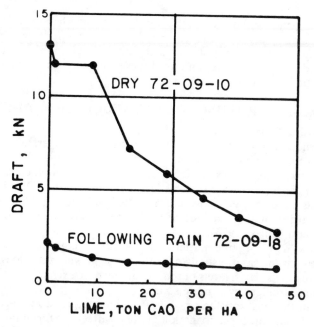

FIG. 2.4-2 Relationship between lime application and draft requirements at time of fall plowing. Measurements were partly done under dry conditions. 1972-09-10; partly under wet conditions after a rain, 1972-09-18. Brunna, Vastmanlands County. Heavy clay. Redrawn from Eriksson (1974).

and harvest traffic. Burnt lime (CaO) applied at rates of 50 T/ha reduced draft on a plow to less than a third of that of unlimed heavy clay (Fig. 2.4-2). Ground limestone, $CaCO_3$, had essentially no effect on the physical characteristics of their soil.

Kamprath (1971) indicates older highly weathered soils of the tropics and the southeastern U.S. containing mostly kaolinite respond somewhat differently to lime applications than less weathered soils. The hydroxy-Al and hydroxy-Fe coatings on clays regulate to a large degree the chemical and physical properties of Ultisols and Oxisols. The hydroxy cations which are OH acceptors are responsible for pH dependent cation exchange capacity and for the absorption of anions and for soil aggregation. Liming of Ultisols and Oxisols to pH 7 and above often decreases plant growth due to (a) a reduction in aggregate size with a consequent reduction in rate of infiltration and an increase in susceptibility to erosion (b) induced deficiencies of phosphorus, boron, manganese, and zinc. Liming rates to increase yield on Ultisols and Oxisols should be based on quantities required to neutralize exchangeable Al and, in some instances, to reduce soluble Mn rather than to bring pH to neutrality. This effect of liming on availability of other ions is discussed in greater detail by Adams (Chapter 8).

Lime rates high enough to significantly affect flocculation and aggregation and to increase strength of weathered clay soils may reduce plant growth. This reduced growth sometimes can be overcome by applying nutrient elements in the years immediately following liming. Structure can be changed more in montmorillonite and illite containing soils with less danger of creating deficiencies than that which results from liming the kaolinite containing Utilsols and Oxisols, but close attention must be paid to form of lime, fineness of grinding, and to organic matter content when experimenting on the use of lime to improve soil structure. No direct comparisons of the unconfined compression strength of soils with and without additions of lime have been found in the literature, and until these comparisons are made no cost-benefit ratios can be developed.

## 2.5 MANAGING CRUSTING SOILS TO REDUCE SHOOT IMPEDANCE

### 2.5.1 Evading Crusts

A structure develops in the uppermost few millimeters of almost every soil that differs substantially from the underlying material. This uppermost layer has fewer large pores, a higher bulk density and often contains a more platy arrangement of particles. This layer often is harder than the rest of the soil because it dries faster and has a higher bulk density. This crust layer often is hard enough to reduce or prohibit seedling emergence through it (Kemper and Miller, 1974).

Soils with low organic matter and high silt (or high sand with sufficient clay to cement the sand) usually have the low aggregate stabilities associated with severe crusts. High levels of sodium on the base exchange complex make the crusting problem even more severe.

One of the most satisfactory ways for managing crusting soils is to evade the crust formation. Since irrigation methods affect the formation of surface crusts (Keller, 1967; 1970a, 1970b), an irrigation method should be selected that will not saturate the soil covering the crop seeds. Furrow or rill irrigation causes less crusting than border irrigation because less soil is saturated. Similarly, low intensity sprinkler irrigations cause less crusting than high intensity sprinklers (Keller, 1967).

Another means to evade the crust problem is by leaving the soil surface protected by residues of the previous crop (Duley, 1939). This residue increases infiltration rate and reduces the chances that soil saturation, with its attendant soil aggregate destruction, will occur. The residue also reduces raindrop impact, which tends to destroy aggregates.

The thickest and strongest crusts generally develop at the bottom of depressions or on flat surfaces where the soil is more likely to become saturated. Seedling emergence can be improved by avoiding these locations and planting the seed on the sloping side if the soil is saline or on top of the ridge in nonsaline soils (Kemper and Miller, 1974).

Various soil amendments have been used to reduce the tendency of soils to crust. Polyelectrolytes, surfactants, fatty alcohols, Portland cement, phosphoric acid and gypsum have been used but usually are prohibitively expensive except for high value crops (Miller and Gifford, 1974).

### 2.5.2 Increasing Emergence Where Crusts Have Formed

If the surface of a crusted soil can be kept moist, the crust strength usually remains low enough for emergence to occur. Sprinkler systems with fine spray nozzles and low application rates can be used for this purpose. Sometimes where seeds have been planted on the sides of beds, furrows can be irrigated in such a manner that crusts directly over the seeds can be wetted by unsaturated flow of water upward from the furrow. Crust strength will then be reduced.

Trickle irrigation also can be used to increase emergence on high value crops through impeding crusts. This system delivers water at a slow rate through plastic tubing and maintains moist soil over part of the soil surface. The moist unsaturated soil does not form crusts as readily as with other irrigation systems and the soil crusts stay more moist thus reducing their strength (Kemper and Miller, 1974).

Crusts often are mechanically destroyed in attempts to improve emergence. However, such treatments can result in injury to seedlings from crushing or from uprooting as soil is moved during the operation. Various types of "crust-breakers", (such as rows of narrow, outside rings supported by a smaller supporting tube, hardware cloth welded to two wheels spaced up to 25 cm apart, or long spikes welded to wheels that are pulled over the planted seeds) are being used. Crust breaking tillage sometimes is used after seedlings have emerged in order to improve aeration. Such tillage usually is undesirable because only wet crusts reduce aeration sufficiently to warrant tillage and crust breakers are not effective on wet crusts.

## 2.6 EQUIPMENT FOR MODIFYING MECHANICAL IMPEDANCE

The quickest and often the most complete means for modifying a soil zone that potentially will impede plant growth is some mechanical tillage method. Some of the equipment to modify impeding zones and the action of that equipment in the soil will now be discussed.

### 2.6.1 Moldboard Plows

The most efficient tool for soil loosening in the 0 to 30 cm depth range is the moldboard plow according to Gill and McCreery (1960) where the ratio of the work on the soil, $W_s$, versus the work on the tool, $W_t$, is the criterion. Gill and McCreery (1960) determined the work on the soil as the energy required to break up the soil by dropping clods from various heights. Two factors account for most of the plow's effectiveness: (a) the furrow is not confined and therefore the cutting draft is less than it would be otherwise and (b) the entire soil slice is subjected to shear and bending strains as the furrow slice is lifted and inverted. The initial pulverization results from shear failure of the soil from the primary and secondary shear planes resulting from the moldboard curvature (Nichols and Kummer, 1932).

Using scratch traces on plows that were coated with nitrocellulose lacquer, O'Callaghan and McCoy (1965) showed that the paths taken by the soil particles were higher on the moldboard at 11.67 km/h than at 3.33 km/h. The component of draft due to the moldboard was much more dependent on the speed of travel than that due to the share while the disk coulter draft remained constant with speed. The major portion of the increase in draft with increase in speed of travel was due to the action of the moldboard on the furrow slice (O'Callaghan and McCoy, 1965).

As much as 30 percent of the draft on a moldboard plow is due to friction on the moldboard (Wismer et al., 1968). Cooper and McCreery (1961) showed that a teflon* covered moldboard reduced this draft up to 23 percent. Fox and Bockhop (1965) showed that teflon or teflon with glass filler reduced draft by 6 to 38 percent depending on clay content and moisture content of the soil. The study also showed that teflon or teflon with glass filler wore eight to ten times as rapidly as steel, and Wismer et al. (1968) indicated that

---

*Brand name mention for convenience of reader and does not constitute endorsement by North Carolina State University or ASAE to the exclusion of any similar product.

20 ha in sand and 120 ha in clay soil was about the wear life of 0.5 cm thick teflon coatings.

Plowing under granular material broadcast uniformly on the surface results in poor distribution because most of the material is placed near the bottom of the tilled section (Hulburt and Menzel, 1953). However, broadcasting half the grains before and half after plowing, or plowing, or applying them through a fertilizer hopper during plowing results in a fairly even vertical distribution, but an unsatisfactory horizontal distribution in comparison with rotary tillage (Hulburt and Menzel, 1953).

If draft can be reduced by the use of economical wear-resistent low friction materials, the moldboard plow will continue to be a primary tool for loosening compacted soil and for alleviating mechanical impedance, because it is the most efficient tool yet devised for loosening and pulverizing the soil and for covering residues.

### 2.6.2 The Standard Disk Plow

Standard disk plows with individually inclined disk blades mounted on a frame supported by wheels have lost popularity in the last 30 years because they are less efficient than moldboard plows. Heavy duty moldboard plows have been developed for use in soils that once could only have been tilled with a disk plow. The disk plow leaves a rougher surface and less uniform pulverization which may or may not be desirable. The unit draft of the disk plow is usually higher than that of the moldboard plow. Disk plows must be forced into the soil by their weight instead of by suction as with a moldboard plow. Disk plows are very heavy and the downward force on the soil adds to soil compaction. They leave a rough plow sole and a rough soil surface and do not invert the soil or cover residues as well as moldboard plows. More secondary tillage often is required after a disk plow than after a moldboard plow. If it is desirable to leave crop residue in and on the soil surface for erosion control and better infiltration, the disk plow may be the preferred tool. Disk plows traditionally have been used where moldboard plows were ineffective such as on land with rocks, stumps, tree roots, extremely hard soil, and extremely sticky soils. The disk plow will continue to be used where other methods fail, but it is less efficient than the moldboard plow in normal soil conditions.

### 2.6.3 Vertical Disk Plow

The vertical disk plow, known as the disk tiller harrow plow, wheatland plow or one-way disk, has uniformly spaced disks on a common axle or gang bolt with spools for spacing the disks. Vertical disk plows are used in the grain growing regions of the U.S. and Canada for shallow plowing (often only 8 to 10 cm) and for mixing stubble with the soil (Kepner et al., 1978). These harrows are effective for loosening the soil for killing weeds and for cutting and partially covering surface mulches to aid in planting and in soil and water conservation. The 50 to 60 cm disks are smaller than those of the standard disk plow and larger than those of the disk harrow and are spaced from 20 to 25 cm on the gang bolt. They are built primarily for relatively shallow plowing so they are lighter than standard disk plows but about twice as heavy as a disk harrow. The entire frame is angled to increase disk penetration and to aid the pulverzing action. They inherently leave soil ridges at the bottom of the tilled zone but the ridges are less severe as the angle increases.

Vertical disk plows are used for fast tillage to increase machine efficiency and labor productivity (compared to moldboard plowing) where deep tillage is not required. The vertical reaction of any disk tool creates a compacted layer immediately below tillage depth. This compacted zone may be beneficial or harmfull depending on the soil type and the soil conditions at the time of tillage. Vertical disk plows can cover large land areas and, though not as popular as in the early 1950's, are useful tools for destroying crusts, killing weeds and incorporating residue into the surface.

### 2.6.4 Subsoilers

Conventional subsoilers are high draft rigid tine implements that loosen a triangular shaped trench in a firm soil (Fig. 2.6-1). The apex of the triangle is in the tine slot that forms immediately below the loosened soil. Some excellent experimental and theoretical work was carried out in the 1950's and 1960's on subsoilers and other tined implements (Payne, 1956; Nichols and Reaves, 1958; Moller, 1959; Payne and Tanner, 1959; Tanner, 1960; O'Callaghan and Farrelly, 1964; Osman, 1964; O'Callaghan and McCullen, 1965; Spoor, 1969a; Spoor, 1969b).

These early studies showed the importance of (a) keeping rake angle of the tine between 20 and 45 deg; (b) keeping the width of the tine, w, relative to the depth, d, greater than $w/d = 0.2$ in order to maximize soil loosening and minimize draft, and (c) taking account of the fact that the more plastic the soil, the deeper the tine slot developed for a given breakup of soil (which normally breaks out at about 45 deg to either side of the tine slot). As a result subsoiling was recommended to be carried out when the soils were dry to obtain the deepest shattering (Trouse and Humbert, 1959). In tests with conventional subsoilers in South Africa, Boshoff (1973) recommended that subsoilers should run at about 1 1/2 times the expected depth of tilled soil in order for the tine to develop enough force to cause shear failure in the triangular trench above the tine slot. The South African tests showed that similar shattering occurred at moisture contents suitable for good plowing and also at much drier soil moisture levels, but if the soil at the point depth was not too plastic, then the draft was greatly reduced at the higher moisture level. A further discussion of tests in South Africa is described in Section 2.6.10.

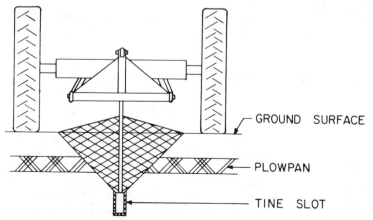

GROUND SURFACE

PLOWPAN

TINE SLOT

FIG. 2.6-1 Illustration of principles regarding the depth of operation for subsoil tines. Redrawn from Boshoff (1973).

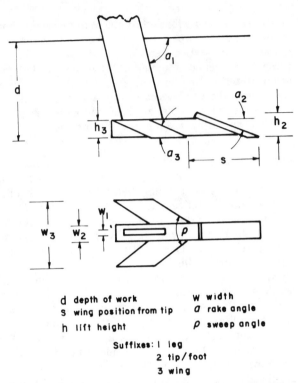

d  depth of work          W  width
S  wing position from tip  α  rake angle
h  lift height             ρ  sweep angle

Suffixes: 1  leg
          2  tip/foot
          3  wing

FIG. 2.6-2 Subsoiler foot and wind variables. Redrawn from Spoor and Godwin, 1978.

Spoor and Godwin (1978) showed that there is a critical working depth for all rigid tines, below which depth the tine foot cannot develop sufficient lifting force to shear the soil on the sides of the triangular trench of shattered soil of Fig. 2.6-1 without the development of a tine slot. There will be essentially no tine slot formed below the shattered soil for operation at or less than the critical depth. As soon as the critical depth is reached, however, further increases in tine depth results in an increased tine slot without an increase in the depth of width of the triangular trench of shattered soil.

Spoor and Godwin (1978) reported that with conventional subsoilers operating at or above their critical depths the soil is displaced forward,

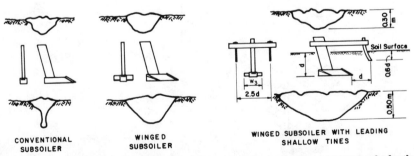

CONVENTIONAL          WINGED          WINGED SUBSOILER WITH LEADING
SUBSOILER           SUBSOILER              SHALLOW TINES

FIG. 2.6-3 Comparative patterns of soil disturbance for 0.3 m and 0.5 m working depths for the conventional and modified subsoilers. Adapted from Spoor and Godwin, 1978.

sideways, and upwards (crescent failure) failing along well defined rupture planes which radiate from just above the tine tip to the surface at angles of approximately 45 degree to the horizontal. Crescent failures continue with increasing working depth until at a certain depth, the critical depth, the soil at the tine base begins to flow forward and sideways only (lateral failure) creating compaction at depth.

The critical working depth is primarily controlled by the width inclination and lift height of the tine foot and by the moisture and density status of the soil (Spoor and Godwin, 1978). The attachment of wings to the tine foot (Fig. 2.6-2) and the use of shallow tines to loosen the surface layers of soil ahead of the deep tine (Fig. 2.6-3) increases soil disturbance particularly at depth, reduces specific resistance, increases the critical depth and allows more effective soil rearrangement. Spoor and Godwin (1978) reported that a complete soil loosening at depth, rather than a deep tine slot, could be achieved by selecting an appropriate tine spacing and working at or above the critical working depth. Both the critical working depth and the tine spacing can be increased through the use of wings and shallow leading tines.

The test work of Boshoff (1973) summarizes conventional operation of the subsoiler in which a tine slot is formed below the triangular trench of shattered soil. Typically the tine is operated at 0.45 m depth and the triangular trench of loosened soil has a depth of about 0.3 m. The tine cuts through the pans below the plow layer and into the subsoil where roots may gain access to the subsoil nutrients and moisture.

The work of Spoor and Godwin (1978) indicates that the soil volume loosened by the subsoiler can be greatly increased by increasing the width of the tine foot, proper tine spacing, the use of wings, and the use of shallow leading tines. They show that proper selection of the above factors results in a complete loosening of the soil to a depth of 0.5 m or more without the formation of a tine slot and with a minimum of compaction beneath where the tine foot was operated. This increased volume of loosened soil is achieved with essentially the same total energy input and draft as is required for the conventional operation of the subsoiler, thus resulting in a greatly reduced energy and draft per unit volume of loosened soil.

The above discussion suggests that the typical operation of conventional subsoilers is at depths substantially in excess of the critical depth resulting in excessive draft and energy per unit of disturbed soil. In addition all tine slots are compacted and slicked on the sides and bottoms because they are the result of operation below the critical depth. This compaction of the slot may prevent penetration by roots and full exploitation of the subsoil for nutrients and water.

The subsoiler with wings and leading shallow tines brings a whole new dimension to tillage. If incorporation of amendments can be successfully accomplished the subsoiler and other tine tools will rival the moldboard plow for efficiency of granulation and may often substitute for the moldboard plow except where the inversion of crop residue is required. Some form of active element may be needed on the subsoiler, if soil amendments are to be introduced into the surface 0.3 m of soil after primary loosening of the subsoiler (Cooper, 1971).

### 2.6.5 Disk Harrows

The tandem disk harrow is the most common tillage tool in the midwest and the humid southeastern United States because it has many uses. The disk harrow, however, is often not the best tool for any one tillage operation.

The disk harrow is used in preparing a reasonably loose seedbed to break up clods, to kill weeds, to mix broadcast lime, fertilizer and herbicides and to chop and incorporate organic matter into soil.

The heavy duty tandem disk harrow is the most popular tillage tool today because of its many uses but tillage depth is limited to about 15 cm and a compacted layer is always left just beneath the tilled depth. When it is used in the low organic matter sandy soils of the southeast, the disk harrow compacts a 3 to 5 cm layer that often severely restricts root growth. Soil compaction is usually much less in clays and soils with high organic matter contents. The disk harrow is not as good as a rotary tiller for mixing fertilizers, lime and herbicides into the surface soil. Poor weed control results from poor mixing of herbicides with disk harrows. One pass with rotary powered equipment usually mixes herbicides into the soil better than two crosswise or parallel passes of a tandem disk (Mayfield et al., 1978).

Disk harrow operating speeds have been increased in order to increase the amount of land tilled. Operating depths for most disk harrows tend to lessen as speeds approach 9 km/h (Krasnoshchekov, 1962). Most disk harrows pulverize the soil excessively and because of the increased energy input onto the soil slice at speeds above 9 km/h, the draft resistance of the tool increases.

Krasnoshchekov (1962) showed that energy requirements can be reduced, soil pulverization can be lessened and draft reduced by braking the disks so they no longer roll free. He indicated that the braked disks are more stable and will stay in the ground at a deeper depth and do a satisfactory job of disking at speeds up to 14 to 15 km/h. The disk gangs are connected to the wheels or to the P.T.O. shaft to rotate at 55 to 60 percent of the speed when rolling free. He also showed that notched disks produced a higher quality of work at speeds up to 11 km/h than solid disks and were especially advantageous for preseeding tillage.

Kulebakin and Arzhanykh (1966) found that inclined (60 deg) flat disks have advantages over spherical disks at higher speeds with less energy required, less throwing of the soil, better penetration, and less pulverization. The inclined flat disks are difficult to mount on the frame but will permit satisfactory operation in the 14 to 15 km/h speed range.

Young (1975) tested disk gangs powered in the forward direction at disk peripheral rotation speed ratios of 0.8 to 1.6 of the forward speed of machine. The powered disks required less total horsepower at speeds up to 9 km/h but more horsepower from 9 to 11 km/h than free wheeling unpowered disk gangs. The powered disks gave more cutting and mixing of the mulch with the soil, gave more pulverization, gave better penetration, had less slippage of tractor wheels, worked better in heavy mulches at high gang angles, and had fewer limitations due to adverse soil moisture conditions.

### 2.6.6 Springtooth Harrows

Springtooth harrows are not popular tools now but they are satisfactory for light tillage with little surface penetration and without much accompanying soil compaction. Springtooth harrows are effective in loosening the surface of coarse-textured soils (sands and loams) and have a light draft requirement. They are effective at speeds up to 10 km/h. The springtooth harrow was the forerunner of the chisel plow and the spring action field cultivator which have grown in popularity.

### 2.6.7 Spiketooth Harrows

Spiketooth harrows have light draft and do not pack the soil much. They are useful to pulverize cloddy soils already in a friable state and for breaking up a soil crust. They sometimes are used in tandem with a plow or disk harrow for a one pass primary and secondary tilage combination.

### 2.6.8 Rollers

Rollers, both smooth and barred have been used to crush clods and to pack the surface soil. They were especially useful for horsedrawn equipment because of their low draft. Culpin (1937) showed that soil bulk density was significantly increased by rolling, but the bulk densities averaged just under 1.0 g/cm³. Moisture contents averaged slightly higher where soil was rolled, but these differences were not significant. Rollers crushed surface clods and tended to smooth the surface by pressing down high spots and filling in low spots. Bulk densities were increased at depths up to 0.18 m after the passage of a roller with bars but the changes were greatest in the top 0.1 m (Culpin, 1937).

### 2.6.9 Rotary Tillers and Power Harrows

Powered rotary tillers and powered harrows have been used extensively in Europe (Ewing and Butterworth, 1979). Rotary tillers agitate and mix the soil from surface to bottom of the tillage zone and are used both for primary and secondary tillage. Power harrows have horizontally rotating, oscillating, or reciprocating tines that pulverize and stir the soil within the working depth but do not mix the surface soil with that at lower strata. Power harrows leave the surface residue intermixed with the tilled soil. Power harrows are less aggressive than rotary tillers and requires less power per unit width (Ewing and Butterworth, 1979).

Rotary tillers are good mixers, but present designs often pulverize the soil more than is desirable. Too many fine particles are formed leaving a soil that is fluffy. The pulverized soil quickly develops a surface seal that prevents infiltration and that crusts severely when the soil surface dries.

Wismer et al. (1968) indicate that the presently available tractors and power takeoffs must be redesigned to successfully operate rotary tillers. This design change is needed if most of the power is used through the PTO instead of for draft because the rotary tiller converts about 25 percent of the energy into forward thrust and that puts a damaging reverse stress on the tractor transmission. Wismer et al. (1968) also indicated that, with proper design of the rotor blades and proper size of bite, up to 90 percent of the horsepower could be absorbed through the tractor PTO rather than by traction through the driving wheels. They estimated that the tractor weight could probably be reduced by 50 percent if this were done. However, a better understanding of functional relations such as depth of cut, draft, speed, bite length and the resulting reactions of negative draft, life, and rotor torque requirements would be required before such a tractor could be designed.

Power harrows, like rotary tillers, require much PTO power but not much draft. Currently available power harrows tend to be limited in speed to about 8 km/h, which is much too slow for many operators. The possibility of reducing tractor weight by reducing draft requirement is a viable solution for avoiding excessively deep compaction on many soils.

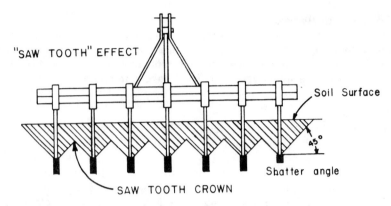

**FIG. 2.6-4 The saw-tooth effect of passing a multi-tine implement through a shattering soil. Redrawn from Carter-Brown (1975).**

### 2.6.10 Chisel Plows, Rippers and Field Cultivators

The following discussion of chisel plows, rippers, and field cultivators is primarily abstracted from an introduction to a series of official reports of tests on tine implements carried out by personnel of the Division of Agricultural Engineering Republic of South Africa (Carter-Brown, 1975).

A wide variety of tine implements is on the market, ranging from multi-tine subsoilers to shallow working chiesel plows. Rippers fall between subsoilers and chisel plows. These rippers do not usually penetrate as deeply as subsoilers and are used to loosen the surface soil. This ripping action is an alternative to conventional plowing with moldboard or disk plows. Rippers are robust implements with a tine spacing of not more than 1 1/2 times the maximum working depth. Multi-tine subsoilers on the other hand are used to shatter soil pans, as they have a working depth of more than 0.45 m. These

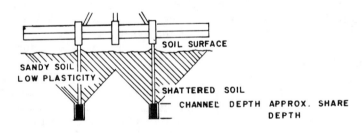

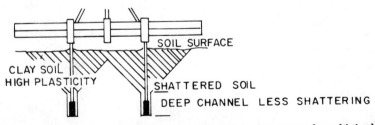

**FIG. 2.6-5 A comparison of the amount of shattering caused by the passage of a multi-tined implement in a low plasticity sandy and a high plasticity clay soil. Redrawn from Carter-Brown (1975).**

subsoilers can be used as rippers provided their tine spacing is not greater than 1 1/2 times the working depth. Chisel plows are more lightly constructed than rippers and have a maximum working depth of 0.3 m. The working depth of a chisel plow is more or less equal to that of a conventional plow, usually in the region of 0.25 m. The tines of chisel plows are usually spring-loaded or made of spring steel. The majority of chisel plows have interchangeable shares which makes it possible to use them as primary or secondary tillage implements. Field cultivators may also be chisel-type implements depending on the type of shovel—sweep or chisel point—that is used for the particular application. Field cultivators are used primarily for secondary tillage and weed control rather than primary tillage, but the principles of tine implements are applicable when chisel points are mounted on the standards. It must be noted though that certain implements, because of their tine length, tine spacing or share design can be used in a dual role, i.e. as a subsoiler as well as a ripper or as a ripper as well as a chisel plow.

The degree to which tine implements break up the soil is determined by tine spacing, working depth, the type and width of the share and the condition of the soil. Other important factors contributing towards ease of operation in lands with plant residue are the frame design and clearance.

Tine implements break the soil through a shattering action. The soil is not inverted or pulverized to the same extent resulting from moldboard and disk-plow operation. The action of the tine passing through the soil instead causes the soil to shatter along the shear planes. These shear planes form an angle of approximately 45 deg from the horizontal for most soils. As a result of this shattering action the sub-surface breakup of the soil can be said to have a "saw-tooth" profile (see Fig. 2.6-4).

It will be found that in most types of soil the tine share tends to cut a channel at the bottom of the "saw-tooth" which is slightly broader than its own width. The depth to which this channel is cut depends upon the type of soil being worked. This channel occurs because of the soil plasticity. With clay soils, the degree of plasticity is higher than in sands and the channel cut by the share can be expected to be deeper; as a result the amount of shattering in the soil will be smaller (Fig. 2.6-5).

The "saw-tooth" crowns formed in sandy loam soil are not as sharply pointed in practice as those illustrated in Fig. 2.6-4, but are flatter and rounder. Implements fitted with broad sweep or half sweep shares shatter more soil and the crowns are flatter still.

The spacing of tines has an important bearing on shatter profile and surface breakup. If the tine spacing is too great, the "saw-tooth" crowns becomes an abstract concept and surface breakup is then incomplete. A too small tine spacing can, on the other hand, result in continual blockages if plant residue is present on the soil surface. If the soil shear plane is taken as 45 deg, then the depth of the crowns below the surface for a given tine spacing and working depth may be estimated by means of the rule-of-thumb: Crown depth = (the working depth) - (1/2 the tine spacing) (assuming the degree of plasticity of the soil is not too large) (see Fig. 2.6-5).

The condition of the soil has a considerable bearing on soil breakup. When the soil is hard and dry the implements tend to leave the surface very rough and strewn with large clods. When the soil is moist less disturbance of the soil surface is noted. Provided the soil is not too moist, the same degree of shattering is achieved at both soil water contents. As the moisture content of the soil increases, so does the plasticity of the soil, with the result that less shattering is achieved. The question now arises: "when is the soil too moist to be worked"? The easiest method of answering this question is to dig with a spade at any spot where the implement has already worked and to study the shatter profile. The farmers' experience will tell him whether or not the "saw-tooth" crowns are deep enough. If the crowns are too high even when a reasonable tine spacing is being used and working depth achieved then the soil clearly is too wet. Experience has shown that if conditions are suitable for moldboard plowing then a ripper also can be used successfully.

It must be kept in mind that while ripping of very hard dry soils appears to result in good breakup, it is more economical to rip when the soil is slightly moist. When hard dry soils are ripped several further tillage operations may be required to break down large clods left on the surface; only then may a reasonable seed bed be achieved. The draft of tine implements increases in proportion to soil hardness and thus more power is required to pull the implement, leading to higher fuel costs and greater tine wear.

### 2.6.11 Tillage With Vibration

Vibratory tillage to alleviate soil compaction has been investigated extensively (Hendrick, 1979). Hendrick and Buchele (1963) reported that a rigid tine had a minimum draft reduction of 35 percent when vibrated at frequencies slightly higher than the natural frequencies of soil shearing action. Total energy is usually higher than for draft drawn tools because more energy is used in accelerating soil. Vibratory tillage usually results in finer soil particles than draft drawn tillage.

Dubrovskii (1968), in a review of Russian studies with vibratory plows, showed a 20 to 35 percent decrease in tractive resistance compared to non-vibratory plows while tilling at a speed of 1 m/s and a frequency of 30 Hz. Tests with an experimental vibration mole plow (Dubrovskii, 1968) show the possibility of achieving substantial power reductions. The maximum power reduction of 50 percent corresponds to a rectilinear velocity of 0.3 m/s of the implement and a frequency of 33 Hz. Further increases in rectilinear velocity cause vibratory effeciency to fall to zero at 1.3 m/s. Vibration reduced tractive resistance on a trencher from a minimum of 15 percent in peat to 55 percent in a dense pebbly loam (Dubrovskii, 1968).

Vibration of cultivators by oscillation in horizontal planes (Dubrovskii, 1968) reduced tractive resistance up to 60 percent and total energy consumption by 40 percent. Vibration also results in freedom from hang-up of weeds. Cultivator shovel wear is reduced 3 to 10-fold by vibration as compared to rigid shovels. The tests were conducted with a vibration amplitude of 7 mm, frequency of 32 Hz and an average tractor speed of 4.5 km/h. The depth of cultivation was 0.15 m. Vibratory potato plows reduced the draft of the potato combine by 26 to 53 percent (Dubrovskii, 1968).

Self-excited vibrations about horizontal and vertical axes owing to non-uniform specific resistance of soil and flat spring suspension of the tooth resulted in improved shedding of the soil and better weed control (Dubrovskii, 1968). Tractive resistance of vibratory teeth was less than resistance for non-vibratory teeth at some moisture contents, but not at others (Dubrovskii, 1968). Moller (1959) reported tests where spring cultivator shanks gave draft reductions of 20 to 30 percent over rigid shanks due to self-excitation vibrations.

Many tests on all kinds of vibratory machines indicate that their tractive effort is reduced by 25 to 50 percent, quality of tillage (as judged by looks) is improved and total energy use probably reflects the greater reduction in granule size of the vibratory tilled soils.

What is left unsaid in vibratory tillage reviews, but is of crucial importance, is that vibratory tillage is limited in speed for plowing, subsoiling, trenching, etc., to about 1 m/s (= 3.6 km/h) if these reductions in draft are to be realized. This speed may be too slow to satisy the American farmers, who are accustomed to operating at 6 to 9 km/h and would prefer to operate at 12 to 15 km/h.

Forward speed can be increased with increased vibratory frequencies, but 3.6 km/h is about the maximum practical speed because about 33 Hz is about the state of the art maximum frequency for heavy loads and large amplitudes without incurring excessive maintenance and prohibitive manufacturing costs. Thus, the main limitations to the adoption of vibratory tillage are the low forward speed, the high first cost, and the high maintenance. The increasing cost of fuel will probably accelerate the research and development of higher frequency, higher forward speed vibratory equipment.

### 2.6-12 Trenching Machines

Large wheel-type trenching machines have been used to modify profiles to greater than normal depths. These machines throughly mix the soil profile to the depth trenched but are prohibitively expensive for use in field crop production. Trencher use may be economically feasible for growing special purpose crops such as grapes on mechanically impeding soils. Since trenchers are used only to modify soil profiles to greater than 0.3 m depths, their use is described by Unger et al. (Chapter 3).

### References

1  Abernathy, G. H., M. D. Cannon, L. M. Carter and W. J. Chancellor. 1975. Tillage systems for cotton—a comparison in the US western region. Bull. 870. Berkeley: Division of Agricultural Sciences, Univ. of CA.

2  Allison, L. E. 1965. Soil and plant response to VAMA and HPAN soil conditioners in the presence of high exchangeable sodium. Soil Sci. Soc. Amer. Proc. 20:147-151.

3  Aubertin, G. M. and L. T. Kardos. 1965a. Root growth through porous media under controlled conditions. I. Effect of pore size and rigidity. Soil Sci. Soc. Amer. Proc. 29:290-293.

4  Aubertin, G. M. and L. T. Kardos. 1965b. Root growth through porous media under controlled conditions. II. Effect of aeration levels and ridigity. Soil Sci. Soc. Amer. Proc. 29:363-365.

5  Bailey, A. C. 1971. Compaction and shear in compacted soils. TRANSACTIONS of th ASAE 14(1):201-205.

6  Barley, K. P. and P. J. England. 1970. Mechanics of burr burial by subterranean clover. Proceedings XI International Grassland Congress 2:103-107.

7  Barley, K. P., D. A. Farrell and E. L. Greacen. 1965. The influence of soil strength on the penetration of a loam by plant roots. Aust. J. Soil Res. 3:69-79.

8  Barnes, K. K., W. M. Carleton, H. M. Taylor, R. I. Throckmorton and G. E. Vanden Berg. 1971. Compaction of Agricultural Soils. American Soc. Agr. Engr., St. Joseph, MI 49085. 471 pp.

9  Barton, H., W. G. McCully, H. M. Taylor and J. E. Box. 1966. Influence of soil compaction on emergence and first-year growth of seeded grasses. J. Range Mgt. 19:118-121.

10  Bennie, A. T. P. and M. C. Laker. 1975. The influence of soil strength on plant growth in red apedal soils. Proc., 6th Congress, Soil Sci. Soc. South Africa. Blydepoort, RSA.

11  Bodman, G. B. and J. Rubin. 1948. Soil puddling. Soil Sci. Soc. Amer. Proc. 13:27-36.

12  Boshoff, B. V. D. 1973. The selection and use of subsoilers. Official test reports 7322 to 7328. Div. Agr. Eng., Silverton, Transvaal, Republic of South Africa. 20 pp.

13  Bowen, H. D. 1966. Measurement of edaphic factors for determining planter specifications. TRANSACTIONS of the ASAE 9(5):725-735.

14  Bowen, H. D. 1975. Simulation of soil compaction under tractor-implement traffic. ASAE Paper No. 75-1569, ASAE, St. Joseph, MI 49085.

15  Carter, L. M., J. R. Stockton, J. R. Tavernetti, and R. F. Colwick. 1965. Precision tillage for Cotton Production. TRANSACTIONS of the ASAE 8(2):177-179.

16  Carter, L. M. and J. R. Tavernetti. 1968. Influence of precision tillage and soil compaction on cotton yields. TRANSACTIONS of the ASAE 11(1):65-67, 73.

17  Carter-Brown, D. H. 1975. Official Test Reports Nos. 7511 to 7521: Tine Implements. Division of Agricultural Engineering, Department of Agricultural Technical Services. Transvaal, Republic of South Africa. 23 pp.

18  Chancellor, W. J. 1966. Combined hypotheses for anticipating soil strains beneath surface impressions. TRANSACTIONS of the ASAE 9(6):887-892, 895.

19  Chancellor, W. J. 1976. Compaction of soil by agricultural equipment. Division of Agr. Sci. Univ. of California. Bulletin 1881, 53 pp.

20  Chancellor, W. J. and R. H. Schmidt. 1962. Soil deformation beneath surface loads. TRANSACTIONS of the ASAE 5(2):240-246, 249.

21  Chancellor, W. J., R. H. Schmidt, and W. H. Soehne. 1962. Laboratory measurement of soil compaction and plastic flow. TRANSACTIONS of the ASAE 5:235-239.

22  Cockroft, B., K. P. Barley and E. L. Greacen. 1969. The penetration of clays by fine probes and root tips. Aust. J. Soil Res. 7:333-348.

23  Cohen, O. P. and N. Sharabani. 1964. Moisture extraction by grape vines from chalk. Israel J. Agric. Res. 14:179-185.

24  Cooper, A. W. 1971. Effects of tillage on soil compaction. In: Compaction of Agricultural Soils pp. 315-364. K. K. Barnes et al. Amer. Soc. of Agr. Engrs., St. Joseph, MI 49085.

25  Cooper, A. W. and W. F. McCreery. 1961. Plastic surfaces for tillage tools. ASAE Paper No. 61-649, ASAE, St. Joseph, MI 49085.

26  Culpin, C. 1937. Studies on the relation between cultivation implements, soil structure and the crop. III. Rollers, an account of methods employed in a study of their actions on the soil. J. Agr. Science 27:432-446.

27  Davidson, D. T. 1965. Penetrometer Measurements pp. 472-484 in C.A. Black et al. Ed. Methods of Soil Anaysis. Agronomy Monograph 9, part 1. Amer. Soc. Agron. Modison, WI.

28  Davies, D. B., J. B. Finney and S. J. Richardson. 1973. Relative effects of tractor weight and wheel-slip in causing soil compaction. J. Soil Sci. 24:399-409.

29  Davis, R. M., P. E. Martin, A. W. Fry, L. M. Carter, M. B. Zahara and R. M. Hagan. 1968. Plant growth as a function of soil texture in the Hanford series. Hilgardia 39:107-120.

30  Dickson, J. W. and P. Smart. 1978. Some interactions between stress and microstructure of kaolin. In: Modification of Soil Structure pp. 53-57. W. W. Emerson et al. Ed. John Wiley and Son, Chichester, England.

31  Donahue, R. L., R. W. Miller and J. C. Shickluna. 1977. Soils: An introduction to soils and plant growth, 4th Edition. Prentice Hall, Inc. Englewood Cliffs, NJ. 626 pp.

32  Drew, L. A. 1973. Bulk density estimation based on organic matter content of some Minnesota soils. Minnesota Forestry Research Notes No. 243. 4 p.

33  Dubrovskii, A. A. 1968. Vibration engineering in agriculture. Published for the USDA ARS and National Science Foundation, Washington, DC by the Indian National Scientific Documentation Centre, New Delhi. TT71-51036.

34  Duley, F. L. 1939. Surface factors affecting the rate of intake of water by soils. Soil Sci. Soc. Amer. Proc. 4:60-64.

35  Dumas, W. T., A. C. Trouse, L. A. Smith, F. A. Kummer and W. R. Gill. 1973. Development and Evaluation of tillage and other cultural practices in a controlled traffic system for cotton in the southern coastal plains. TRANSACTIONS of the ASAE 16(5):872-875. 880.

36  Dunlap, W. H. and J. A. Weber. 1971. Compaction of an unsaturated soil under a general state of stress. TRANSACTIONS of the ASAE 14(4):601-607, 611.

37  Eavis, B. W., L. F. Ratliff and H. M. Taylor. 1969. Use of a dead-load technique to determine axial root growth pressure. Agron. J. 61:640-643.

38  Elkins, C. B., R. L. Haaland and C. S. Hoveland. 1977. Grass roots as a tool for penetrating soil hardpans and increasing crop yields. pp. 21-26. Proceedings, 34th Southern Pasture and Forage crop Improvement Conference. Auburn, AL.

39  Eriksson, J., I. Hakansson and B. Danfors. 1974. The effect of soil compaction on soil structure and crop yields. Swedish Inst. Agr. Engr. Bull 354. (English translation J. K.

Aase, ARS-SEA, USDA, Sidney, Montana). publ. Sven-Uno Skarp, Uppsala, Sweden. 101 pp.

40  Ewing, C. G. and B. Butterworth. 1979. Scanning the power harrows. Implement and Tractor 94:34-37.

41  Fiskell, J. G. A., V. W. Carlisle, R. Kashirad and C. E. Hutton. 1968. Effect of soil strength on root penetration in coarse-textured soils. Transactions, 9th Int. Cong. Soil Science. I:793-802.

42  Fox, W. R. and C. W. Bockhop. 1965. Characteristics of a teflon-covered simple tillage tool. TRANSACTIONS of the ASAE 8(2):227-229.

43  Freitag, D. R. 1971. Methods of measuring soil compaction. pp. 47-103, In: K. K. Barnes et al. Compaction of agricultural soils. ASAE, St. Joseph, MI 49085.

44  Froehlich, O. K. 1934. Druckverteilung in Baugrunde (Formulas of Boussinesq). Vienna, Austria. (Cited in Soehne, 1958).

45  Fryrear, D. W. and W. G. McCully. 1972. Development of grass root systems as influenced by soil compaction. J. Range Mgt. 25:254-257.

46  Ghavami, M., J. Keller and I. S. Dunn. 1974. Predicting soil density following irrigation. TRANSACTIONS of the ASAE 17(1):166-171.

47  Gill, W. R. and G. H. Bolt. 1955. Pfeffer's studies of the root growth pressures exerted by plants. Agron. J. 47:166-168.

48  Gill, W. R. and W. F. McCreery. 1960. Relation of size of cut to tillage tool efficiency. Agricultural Engineering 41:372-374, 381.

49  Gill, W. R. and G. E. Vanden Berg. 1967. Soil dynamics in tillage and traction. Agriculture Handbook 316. US Government Printing Office. Washington, DC. 511 pp.

50  Grimes, D. W., R. J. Miller, and P. L. Wiley. 1975. Cotton and corn root development in two field soils of different strength characteristics. Agron. J. 67:519-523.

51  Harris, W. L. 1971. The soil compaction process pp. 9-44 in K. K. Barnes et al. Compaction of Agricultural Soils. Pub. ASAE. St. Joseph, MI 49085.

52  Hulburt, W. C. and R. G. Menzel. 1953. Soil mixing characteristics of tillage implements. Agricultural Engineering 34(10):702-704, 706, 708.

53  Hendrick, J. G. 1979. An annotated bibliography on vibratory soil dynamics (1969-1979). National Tillage Machinery Laboratory, USDA. Auburn, AL. 57 pp.

54  Hendrick, J. G. and W. F. Buchele. 1963. Tillage energy of a vibrating tillage tool. TRANSACTIONS of the ASAE 6(3):213-216.

55  Jaafari, H. 1978. Simulation of soil compaction under tractor-traffic. PhD. Thesis, Dept. of Biological and Agricultural Engineering. North Carolina State University. Raleigh, NC. 63 pp.

56  Kamprath, E. J. 1971. Potential detrimental effects from liming highly weathered soils to neutrality. Soil and Crop Sci. Soc. of Fla. Proc. 31:200-203.

57  Keller, J. 1967. The effect of water application on soil tilth. Utah Sci. 9:93-96.

58  Keller, J. 1970a. Sprinkler intensity and soil tilth. TRANSACTIONS of the ASAE 13(1):118-125.

59  Keller, J. 1970b. Control of soil moisture during sprinkler irrigation. TRANSACTIONS of the ASAE 13(6):885-890.

60  Kemper, W. D. and D. E. Miller. 1974. Management of crusting soils: Some practical possibilities pp. 1-6. In: J. E. Cary and D. D. Evans, Ed. Soil-Crusts Tech. Bull. 214, Arizona Agr. Ecpt. Sta. Tucson, AZ.

61  Kepner, R. A., R. Bainer and E. L. Barger. 1978. Principles of farm machinery. Third Edition. AVI Publishing Co. Inc., Westport, CT. 527 pp.

62  Kibreab, T. and R. E. Danielson. 1977. Measurement of radish root enlargement under mechanical stress. Agron. J. 69:857-860.

63  King, F. H. 1895. The soil p. 277. The MacMillan Co., New York, NY.

64  Krasnoshchekov, N. V. 1962. Disk implement for operation at high speeds. Mechanization and Electrification of Socialist Agriculture. No. 4, pp. 22-23. (Translated by W. R. Gill, NTML, Auburn, AL).

65  Kulebakin, P. G. and A. I. Arzhanykh. 1966. Increased speed of disk tools. Scientific Reports Siberian Branch of the All Union Research Institute of Mechanization and Electrification of Agriculture. No. 3, pp. 204-208. (Translated by W. R. Gill, NTML, Auburn, AL).

66  Kumar, V., K. T. Mahajan, S. B. Varade and B. P. Ghildyal. 1971. Growth responses of rice (oryza sativa L.) to submergence, soil aeration and soil strength. Indian J. Agric. Sci. 41:527-534.

67  Larson, W. E., S. C. Gupta and R. A. Useche. 1980. Compression of agricultural soils from eight soil orders. Soil Sci. Soc. Amer. J. 44:450-457.

68  Lowry, F. E., H. M. Taylor and M. G. Huck. 1970. Growth rate and yield of cotton as influenced by depth and bulk density of soil pans. Soil Sci. Soc. Amer. Proc. 34:306-309.

69    Lutz, J. F. 1952. Mechanical impedance, pp. 43-71, In: B. T. Shaw, Ed. Soil physical conditions and plant growth. Amer. Soc. Agron. Madison, WI.
70    Martin, W. P., G. S. Taylor, J. C. Engibous and E. Burnett. 1952. Soil and crop responses from field applications of soil conditioners. Soil Sci. 73:455-471.
71    Mayfield, W., J. W. Everest and T. Whitwell. 1978. Incorporating herbicides. Agr. and Natural Resources Information Bulletin M78-3. Alabama Agr. Expt. Sta., Auburn.
72    Miller, D. E. and R. O. Gifford. 1974. Modification of soil crusts for plant growth pp. 7-16. In J. W. Cary and D. D. Evans, Ed. Soil crusts. Tech. Bull. 214, Arizona, Agr. Expt. Sta., Tucson.
73    Minnich, J. 1977. The earthworm book; How to raise and use earthworms for your farm and garden. Rodale Press, Emmaus, PA.
74    Moller, R. 1959. Draft requirements and working efficiency of rigid and spring cultivator tines. Grundlagen der Landtechnik 11:85-94.
75    Monteith, N. H. and C. L. Banath. 1965. The effect of soil strength on sugar-cane root growth. Tropical Agriculture 42:293-296.
76    Nash, V. E. and V. C. Baligar. 1974. The growth of soybean (Glycine max.) roots in relation to soil micromorphology. Plant and Soil 41:81-89.
77    Nichols, M. L. and T. H. Kummer. 1932. The dynamic properties of soil: IV. A method of analysis of plow moldboard design based on dynamic properties of soil. Agricultural Engineering 13(11):279-285.
78    Nichols, M. L. and C. A. Reaves. 1958. Soil reaction to subsoiling equipment. Agricultural Engineering 39(6):340-343.
79    O'Callaghan, J. R. and K. M. Farrelly. 1964. Cleavage of soil by tined implements. J. Agr. Eng. Res. 9:259-270.
80    O'Callaghan, J. R. and J. G. McCoy. 1965. The handling of soil by mouldboard ploughs. J. Agr. Eng. Res. 10:23-35.
81    O'Callaghan, J. R. and P. J. McCullen. 1965. Cleavage of soil by inclined and wedge-shaped tines. J. Agr. Eng. Res. 10:248-254.
82    Osman, M. S. 1964. The mechanics of soil-cutting blades. J. Agr. Eng. Res. 9:313-328.
83    Parker, J. J. Jr. and H. M. Taylor. 1965. Soil strength and seedling emergence relations I. Soil type, moisture tension, temperature, and planting depth effects. Agron. J. 57:289-291.
84    Payne, P. C. J. 1956. The relationship between the mechanical properties of soil and the performance of simple cultivation implements. J. Agr. Eng. Res. 1:23-50.
85    Payne, P. C. J. and D. W. Tanner. 1959. The relationship between rake angle and the performance of simple cultivation implements. J. Agr. Eng. Res. 4:312-325.
86    Pettry, D. E. and C. I. Rich. 1971. Modification of certain soils by calcium hydroxide stabilization. Soil Sci. Soc. Amer. Proc. 35:834-838.
87    Pfeffer, W. 1893. Druck-und arbeitsleistrung durch wachsende pflanzen. Abh. Sachs. Ges. (Akad.) Wiss. 33:235-474.
88    Phillips, R. E. and D. Kirkham. 1962. Soil compaction in the field and corn growth. Agron. J. 54:29-34.
89    Raghavan, G. S. V. and E. McKyes. 1977. Laboratory study to determine the effect of slip-generated shear in soil compaction. Canadian Agr. Eng. 19:40-42.
90    Raghaven, G. S. V. and E. McKyes. 1978. Statistical models for predicting compaction generated by off-road vehicular traffic in different soil types. J. of Terramechanics 15:1-14.
91    Rawitz, E. and A. Hazan. 1978. The effect of stabilized, hydrophobic aggregate layer properties on soil water regime and seedling emergence. Soil Sci. Soc. Amer. Proc. 42:787-793.
92    Reeve, R. C. 1965. Modulus of rupture pp. 466-471, In C. A. Black et al. Methods of Soil Analysis. Agronomy Monograph 9, part 1, Amer. Soc. Agron., Madison, WI.
93    Reijmerink, A. 1973. Microstructure, soil strength and root development of asparagus on loamy sands in the Netherlands. Neth. J. Agric. Sci. 21:24-43.
94    Rogers, H. T. and D. L. Thurlow. 1973. Soybeans restricted by soil compaction. Highlights of Agr. Research 20:10, Auburn University Agricultural Experiment Station, Auburn, AL.
95    Russell, E. W. 1977. The role of organic matter in soil fertility. Phil. Trans. R. Soc. Lond. B 281:209-219.
96    Russell, R. S. 1977. Plant Root Systems: Their function and interaction with the soil McGraw-Hill Book Co. (U.K.) Ltd., London. 298 p.
97    Russell, R. S. and M. J. Goss. 1974. Physical aspects of soil fertility—The response of roots to mechanical impedance. Neth. J. Agric. Sci. 22:305-318.
98    Schmertmann, J. H. 1970. Static cone to compute static settlement over sand. J. of the Soil Mechanics and Foundations Division, Proc. of the Amer. Soc. of Civil Engrs. SM 3:1011-1043.

99   Sallberg, J. R. 1965. Shear strength. pp. 431-447. In, C. A. Black et al. Methods of Soil Analysis. Amer. Soc. Agron. Monograph 9, Part 1. Madison, WI.
100   Sheesley, R., D. W. Grimes, W. D. McClellan, C. G. Summers and V. Marble. 1974. Influence of wheel traffic on yield and stand longevity of alfalfa. California Agr. 28:6-8.
101   Slater, C. S. and H. Hopp. 1947. Relation of fall protection to earthworm populations and soil physical conditions. Soil Sci. Soc. Amer. Proc. 12:508-511.
102   Soehne, W. 1958. Fundamentals of pressure distribution and soil compaction under tractor tires. Agricultural Engineering 39(5):276-281, 290.
103   Spoor, G. 1969a. Design of soil engaging implements. Part I. Farm Machinery Design Engineering. September.
104   Spoor, G. 1969b. Design of soil engaging implements. Part II. Farm Machinery Design Engineering. December.
105   Spoor, G. and R. J. Godwin. 1978. An experimental investigation into the deep loosening of soil by rigid tines. J. Agric. Engineering Res. 23:243-258.
106   Stolzy, L. H. and K. P. Barley. 1968. Mechanical resistance encountered by roots entering compact soils. Soil Sci. 105:297-301.
107   Tanner, D. W. 1960. Further work on the relationship between angle and the performance of simple cultivation implements. J. Agr. Eng. Res. 5:307-315.
108   Taylor, H. M. and R. R. Bruce. 1968. Effects of soil strength on root growth and crop yield in the Southern United States. Trans, 9th Int. Congress Soil Sci. 1:803-811.
109   Taylor, H. M. and E. Burnett. 1964. Influence of soil strength on the root growth habits of plants. Soil Sci. 98:174-180.
110   Taylor, H. M. and H. R. Gardner. 1960. Use of wax substrates in root penetration studies. Soil Sci. Soc. Amer. Proc. 24:79-81.
111   Taylor, H. M., L. F. Locke and J. E. Box. 1964. Pans in southern great plains soils. III. Their effects on yield of cotton and grain sorghum. Agron. J. 56:542-545.
112   Taylor, H. M., A. C. Mathers and F. B. Lotspeich. 1964b. Pans in southern great plains soils. I. Why root-restricting pans occur. Agron. J. 56:328-332.
113   Taylor, H. M., J. J. Parker, Jr. and G. M. Roberson. 1966. Soil strength and seedling emergence relations. II. A generalized relation for Gramineae. Agron. J. 58:393-395.
114   Taylor, H. M. and L. F. Ratliff. 1969a. Root growth pressures of cotton, peas, and peanuts. Agron. J. 61:398-402.
115   Taylor, H. M. and L. F. Ratliff. 1969b. Root elongation rates of cotton and peanuts as a function of soil strength and soil water content. Soil Sci. 108:113-119.
116   Taylor, H. M., G. M. Roberson and J. J. Parker, Jr. 1966. Soil strength—root penetration relations for medium to coarse-textured soil materials. Soil Sci. 102:18-22.
117   Taylor, H. M. and J. A. Vomocil. 1959. Changes in soil compressibility associated with polyelectrolyte treatments. Soil Sci. Soc. Amer. Proc. 23:181-183.
118   Taylor, J. H., E. C. Burt, and A. C. Bailey. 1978. Traction and compaction of big tractors. ASAE Paper No. 78-1029, ASAE, St. Joseph, MI 49085.
119   Trouse, A. C., Jr., W. T. Dumas, L. A. Smith, F. A. Kummer, and W. R. Gill. 1975. Residual effects of barrier removal for root development. ASAE Paper No. 75-2534. ASAE, St. Joseph, MI 49085.
120   Trouse, A. C., Jr., and R. P. Humbert. 1959. Deep tillage in Hawaii. I. Subsoiling. Soil Sci. 88:150-158.
121   Underwood, C. V., H. M. Taylor and C. S. Hoveland. 1971. Soil physical factors affecting peanut pod development. Agron. J. 63:953-954.
122   Vanden Berg, G. E. 1966. Triaxial measurements of shear strain and compaction in unsaturated soil. TRANSACTIONS of the ASAE 9(4):460-463, 467.
123   Veihmeyer, F. J. and A. H. Hendrickson. 1948. Soil density and root penetration. Soil Sci. 65:487-493.
124   Wanjura, D. F. 1973. Effects of physical soil properties on cotton emergence. USDA Technical Bulletin 1481. 20 pp.
125   Weir, W. W. 1936. Soil Science. Its principles and practice. Lippincott, Chicago.
126   Wiersum, L. K. 1957. The relationship of the size and structural rigidity of pores to their penetration by roots. Plant and Soil 9:75-85.
127   Wismer, R. D., E. L. Wegscheid, H. J. Luth, and B. E. Romig. 1968. Energy application in tillage and earth moving. Soc. Auto. Engr. Trans. 77:2486-2494.
128   Young, P. E. 1975. A machine to increase productivity of a tillage operation. ASAE Paper No. 75-7503, 19 pp.,ASAE, St. Joseph, MI 49085.
129   Young, R. E. and V. D. Browning. 1977. Soil moisture response to deep tillage controlled traffic cultural practices. Highlights of Agr. Res. 24(3):12. Auburn Univ. Agr. Expt. Sta., Auburn, AL.

# chapter 3 ▮▮▮▮▮▮▮▮▮▮▮▮

## ALLEVIATING PLANT WATER STRESS

**3**

# 3

# ALLEVIATING PLANT WATER STRESS

by    Paul W. Unger, Soil Scientist, Harold V. Eck, Soil
      Scientist and Jack T. Musick, Agricultural Engi-
      neer,  USDA  Conservation  and  Production
      Research Laboratory, Bushland, TX 79012

## 3.1 INTRODUCTION

Root zone modifications for alleviating plant water stress include treatments that change the physical and chemical environment of soils so that more water is made available for plant growth. This chapter emphasizes root zone modification for alleviating plant water stress, but treatments that affect root-zone soil water also affect other soil conditions and related plant responses, which are discussed in other chapters. Conversely, modification for other reasons will also affect soil water because some effects are inseparable. We will strive to limit our discussion to modification effects on soil water and plant responses to that water. We may, however, occasionally mention related effects of modification that have a direct bearing on the availability of soil water. Emphasis will be on treatments at soil depths greater than 0.3 m.

## 3.2 SOIL AND PLANT RELATIONSHIPS

The distribution of vegetation on the earth's surface is controlled more by the availability of water than by any other single factor (Kramer, 1969). Where precipitation is abundant and evenly distributed during the growing season, plant growth is lush, as in the rain forests of the tropics, the Olympic Peninsula of the northwestern United States, and the cove forests of the Southern Appalachians. As precipitation becomes less abundant and less frequent, forests are replaced by grasslands, grasslands by scattered shrubs in semiarid regions, and finally semiarid regions by deserts (Kramer, 1969).

As precipitation amount and frequency decrease, the need for storing adequate water in the soil for plant use between precipitation events becomes increasingly important. Water stress is normally most severe in semiarid and arid regions because of long rainless periods and limited soil water storage. Plant water stress, however, may also occur in subhumid and humid regions due to short term drought, low soil water storage capacity, or the inability of plant roots to grow into or proliferate within soil zones containing water.

Contribution from Agricultural Research, Science and Education Administration, USDA, in cooperation with The Texas Agricultural Experiment Station, Texas A&M University.

The adverse effects of plant water stress resulting from lack of precipitation can be alleviated by irrigation. Irrigation is widely used for crop production in the western United States and has become an increasingly important production practice in the more humid eastern United States. Although yields are increased by irrigation, competition for water by agricultural, industrial, urban, and recreational users demands that water for crop production be used efficiently. For efficient water use to occur in crop production, water, whether from irrigation or precipitation, must be effectively stored in soil, and plants must be capable of extracting it easily for growth and reproductive purposes.

Factors affecting soil water storage include soil texture (sand, silt, and clay content), organic matter content, depth to impervious or slowly permeable layers, density, and structure. These factors influence the amount of water that a soil can hold, the rate at which the water reservoir is replenished by precipitation or irrigation, and the rate and extent to which plant root systems can use the stored soil water. The goal of root zone modification to alleviate water stress is to provide conditions so that the soil retains more water, the soil is more readily refilled with water, and plant roots more readily grow in the soil to extract the water.

### 3.2.1 Soil Water Availability to Plants

Water from precipitation or irrigation is stored in pore spaces between soil particles and aggregates. The pore system forms a reservoir for the water that enables plants to survive and grow, even though the water additions are intermittent (Gardner, 1968).

Not all water that enters a soil becomes available for plant use. Some that may be temporarily held in large pores percolates from the soil if drainage is not restricted. Other water may be so tightly held in fine pores, within clay particles, or on soil particle surfaces that it cannot be extracted by plants, or it may be lost from the soil surface by evaporation.

#### 3.2.1.1 Range in Water Availability

"Water availability" to plants has been discussed in numerous reports since the concepts of field capacity and wilting coefficient or permanent wilting point were introduced early in this century. These terms are widely used to indicate the upper and lower limits of water availability for plants. While appropriate for establishing general limits, they are not precise and must be defined when used in a quantitative sense (Gardner, 1968).

The upper limit of water availability to plants (field capacity) generally is based on water contents after saturated soil has freely drained for 2 or 3 days or by subjecting wetted soil to pressures in the range from 5 to 30 kPa (kilopascal) (0.05 to 0.3 bar) in pressure membrane or pressure plate equipment. The lower value is generally applicable to sandy soils and the higher value to clay soils. While the soil is draining to field capacity, growing plants may use some of the water above field capacity. The lower limit (permanent wilting point) is estimated by determining the water content at which indicator plants growing in the soil wilt and then fail to recover turgor when subjected overnight to a humid atmosphere, or by determining the equilibrium water content of wetted soil subjected to pressures of 1500 kPa (15 bars) in appropriate equipment (Kramer, 1969; Peters, 1965). The range of available water is the difference in water content between field capacity and the permanent wilting point. This concept, however, does not provide for consideration of "how available" the water might be for plant use (Gardner, 1968).

The concept of potential energy of soil water has formed the basis for the modern understanding of water availability to plants, and is discussed in detail in reviews by Gardner (1968), Kramer (1969), and Newman (1974). Several phenomena contribute to the total potential energy of soil water. The three most common terms included in total potential energy are osmotic, gravitational, and matric potentials. Osmotic potential is a measure of the effects of dissolved compounds on soil water potential, gravitational potential arises from the differences in elevation between the water in question and some reference elevation, and matric potential arises from forces associated with water adsorption on particle surfaces and within the soil matrix.

The relation between soil water content and the matric potential, termed the soil water characteristic, is important. At saturation, all soils have a matric potential of zero. As soil water content decreases, the potential decreases, that is, it becomes more negative, but the rate of change between water content and matric potential is greatly different for different soils. For example, the water content of sandy soils, which retain less water than clay soils at −30 kPa (−0.3 bar) matric potential, decreases much more rapidly with decreases in potential than that of clay soils. As the potential decreases, plants must expend more energy to extract water from soil. Water held at potentials less than −1500 kPa (−15 bars) is generally considered of little value to plants, but plants can extract some water at potentials below −1500 kPa (Haise et al., 1955; Slatyer, 1957).

The soil water content-matric potential relationship is strongly influenced by soil texture (sand, silt, and clay content) and bulk density through their influences on the number and size distribution of soil pores and the surface area of soil particles. The matric potential limits of water availability to plants, especially the lower limit, apparently can be altered only slightly by root zone modification. The soil water characteristic, however, is subject to alteration by modifications that change the number and size distribution of soil pores. Possible treatments are mixing high-sand surface horizons with clay from deeper in the profile to increase the plant available water holding capacity of the sandy soil near the surface; disrupting dense horizons that retain much of their water at potentials above −1500 kPa, but are too dense to permit root proliferation; and adding organic materials that retain water and decrease the density of problem soil layers.

### 3.2.1.2 Amount of Water Available to Plants

The matric potential indicates the energy by which water is retained in soil, but gives no indication of the amount of water retained at that energy. Although much water drains from sandy soils at high potentials (small negative values), the following example illustrates the differences in water retention at the same potentials by soils of different texture. A sandy soil retained 5.7 and 3.4 percent water by volume at −33 and −1500 kPa matric potentials, respectively. At the same potentials, a clay soil retained 47.0 and 35.1 percent water (Unger, 1975). The differences showed a five-fold advantage for the clay soil. Because clay soils retain more water, plants growing in such soils experience stress later than those growing in sandy soils if the profiles are of similar depths and the matric potentials are similar at the beginning of the growth period. Under actual field conditions, however, the difference would not be as great as the example showed because the field capacity of sandy soils is at a potential of around −5 to −10 kPa rather than at −33 kPa as used by Unger (1975). Although clay content significantly influences soil water retention, the amount retained is also influenced by soil organic matter content and bulk density (Unger, 1975).

Beneficial root zone modifications for alleviating water stress are those that (a) increase the water retained in each increment of the soil profile, (b) increase the effective depth of root extension and proliferation, and (c) result in more complete filling of the soil with water. Possible treatments include (a) disrupting dense- or high-clay content layers, (b) mixing high sand surface layers with clay from deeper in the profile, and (c) installing barriers in sandy soils. Disrupting dense soil layers may enhance water infiltration, penetration, storage, leaching, and drainage, and allow increased root penetration and proliferation. Mixing sandy surface layers with clay may increase water retention in the surface layer by changing the texture. Barriers restrict drainage from the root zone.

### 3.2.2 Soil Water Uptake by Plants
### 3.2.2.1 Mechanism and Rate of Uptake

Water moves in the liquid phase of the soil-plant system in response to differences in the potential energy of water in the system. For water to move from soil into the plant, the potential in the plant must always be lower than that in the soil (Gardner, 1968). The potential gradient in the soil-plant system results from water losses by transpiration at plant leaf surfaces. As transpiration progresses, plant water must be replenished by water uptake from soil through roots. Because uptake normally lags behind transpirational losses, most terrestrial plants experience dehydration and a water potential decrease as the radiant energy load increases during early daylight hours. Decreases in the energy load in the afternoon result in decreased transpiration and, if water uptake by roots is adequate, plant tissues rehydrate until they regain turgor during the night (Taylor and Klepper, 1978).

Water enters plants through epidermal cells of roots that are in contact with moist soil. The water then passes in turn through cortical cells, the endodermis, pericyclic cells, and into the xylem elements, which transport it to aerial portions of the plants (Trouse, 1971). Water transport through the soil-plant system has been discussed in numerous books and reports, including those by Black (1960), Gardner (1968), Kramer (1969), Newman (1974), and Trouse (1971). We, therefore, will not present a detailed discussion of water transport in a soil-plant system, but will briefly discuss what occurs in soil in the vicinity of plant roots.

The soil-root system is dynamic. The water content at the soil-root interface decreases when plants extract water or increases due to precipitation, irrigation, or water flow from surrounding soil. Likewise, the roots either increase in length and density as they grow into moist soil or decrease due to drought, insect or disease damage, toxic substances, or natural dieback. The permeability of roots presumably decreases with age (Klepper and Taylor, 1979); therefore, it is important that new roots be constantly produced to assure adequate absorbing surfaces for sufficient water uptake.

New absorbing surfaces are produced as roots elongate and as lateral roots are formed. The latter may develop from the elongating roots or from older roots that have a reduced ability to absorb water. Root hairs from epidermal cells further increase the root's absorbing surface.

Water losses from leaf surfaces vary widely depending upon plant, soil, and environmental factors, but rates of 8 mm/day for irrigated crops, such as grain sorghum (*Sorghum bicolor* L. (Moench)) and corn (*Zea mays* L.), correspond closely with potential evaporation rates in the Southern Great Plains. Assuming that the absorbing zone of roots moves downward in soil as a uniform plane, which is by no means realistic, downward root growth rates per unit area would need to be 348 and 67 mm/day in the sandy and clay soils mentioned previously to meet the plant's water needs, provided all available water was extracted by the roots and water held at potentials above −33 kPa was not considered (see comments concerning upper limits for sandy soils in the previous section).

The above water requirements were for irrigated crops during high-demand periods. Such requirements prevail only under limited circumstances and not throughout the growing season. However, even 50 percent lower requirements would be difficult to meet, especially for the sandy soil, because root elongation rates frequently range from only 30 to 50 mm/-day (Pearson, 1974; Sivakumar et al., 1977; Trouse, 1971). For adequate water supply to plants, it is important that (a) the soil root zone be large enough to supply the plants' needs between additions by precipitation or irrigation, (b) roots proliferate freely in the soil, (c) the soil water supply be replenished frequently enough to avoid plant stress, (d) the added water freely infiltrate the soil, but not leach nutrients to depths below the root zone, and (e) an effective water-absorbing root system be maintained throughout the plant's life cycle. The latter is influenced not only by soil water supply, but by soil nutrients, aeration, pathogens, temperatures, and toxic chemicals, which are subjects for other chapters of this monograph.

The root zone modifications discussed in the previous section should be effective for improving the uptake of soil water. In addition, thorough mixing of swelling clays with soil from other horizons may be beneficial by reducing soil cracking, which breaks roots. In addition, roots shrink about 50 percent as the water potential decreases from −200 to −1000 kPa. Because root shrinking may not coincide with soil shrinking (Klepper and Taylor, 1979), the thorough mixing of swelling clays with nonswelling material may decrease the formation of gaps between roots and soil. These gaps could reduce water uptake by roots. Soil cracking, however, may be desirable in some cases because cracked soils readily absorb water.

### 3.2.2.2 Plant Rooting Patterns

Rooting patterns are different for different plant genotypes and cultivars, but the soil environment usually limits rooting to depths much less than the normal growth habit of the plant allows (Pearson, 1974).

Plants most readily absorb water from soil zones where the water potential is highest. After the soil profile is recharged, water absorption normally is from near the surface. However, plants can absorb water at similar rates from near the surface and from deep in the profile if densities and soil water contents are similar and if root xylem resistances to water flow are low (Pearson, 1974).

Where the water content is not uniform, roots develop in most parts of the soil that they can reach if the matric potential is above −1500 kPa (Taylor and Klepper, 1974). Rooting in soil with matric potentials of less than −1500 kPa apparently is negligible (Kramer, 1969). If aeration is adequate, root development occurs in very moist soil (Trouse, 1971) or even in nutrient solutions (Kramer, 1969).

If the plant's physiological needs can be supplied, its roots can grow in infertile soil and through weak zones in relatively dense materials. However, severely-compacted zones, poorly-aerated zones, water-saturated soil, droughty soil lenses, cold horizons, highly-saline horizons, and zones with toxic substances decrease root elongation rates and often prevent root growth completely (Trouse, 1971). Because these factors affect root growth, they all affect plant root patterns.

The "optimum" root system has not yet been defined. For crop production that depends on irregularly occurring precipitation, a deep, readily penetrable soil zone with high water storage capacity is desirable. The required recharge frequency is directly related to root zone depth and water storage capacity. Using again the sand and clay soils mentioned previously as examples, the sandy soil would need to be recharged with water at about 3- to 6-day intervals to 1- and 2-m depths, respectively, to provide the high water requirements for grain sorghum and corn. On the clay soil, the respective recharge intervals would be about 15 and 30 days. These values assume high water requirements, complete withdrawal of available water, and complete recharge of the profile. At many locations, complete recharge of the profile from precipitation is highly unlikely. In practice, more frequent recharge would be necessary for optimum crop production because growth and yields decrease well before water contents throughout the root zone approach the lower limit of availability. This illustration, however, emphasizes the value of a deep root zone where water is added at irregular intervals.

With frequent irrigation and intensive fertilization, the value of a deep root system is questionable, especially for vegetative crops, because root growth requires the diversion of energy from potential aboveground plant growth (Pearson, 1974). Possible benefits from a restricted root system are also suggested by studies involving nutrient solutions, but these have not been demonstrated under field conditions because of interacting water, nutrient, and oxygen supply problems (Pearson, 1974). Irrigated crops with restricted root systems are more susceptible to plant water and nutrient stresses than those with deep root systems.

### 3.2.3 Plant Water Stress
### 3.2.3.1 Water Loss From Plants

Plants lose water mainly by transpiration through stomates. When water loss exceeds uptake during daylight hours, water deficits occur that lead to plant stress. When uptake again exceeds loss, as at night, stress decreases.

Transpiration is an energy-driven process with incoming radiation furnishing the required energy. When transpiration occurs, water moves from the bulk soil into plant roots through the plant's vascular system to the surrounding bulk air along gradients of decreasing (greater negative) water potential. Potential in cropped soil is less than $-100$ kPa much of the time and is seldom lower than $-1500$ kPa (permanent wilting point), whereas the vapor pressure deficit of bulk air provides diffusion driving forces of several thousand kPa (a wet bulb depression of 1 °C approximates a water potential of $-8000$ kPa). When the soil water potential is above a threshold, evaporative demand controls tranpiration, whereas below the threshold, transpiration is controlled by water uptake by roots (Ritchie and Jordan, 1972). The soil water potential at which plants wilt is related to evaporative demand (Denmead and Shaw, 1962). Increasing demand causes actual transpiration rate to drop below the potential rate and plant stress to develop at higher soil water potentials.

### 3.2.3.2 Plant Water Stress Development

Essentially all plant processes occur in a water environment, but all are not equally sensitive to stress. For example, Hsiao (1973) ranked cell growth as very sensitive and sugar accumulation as relatively insensitive to stress. He ranked other processes between these extremes.

As water stress develops, transpiration is limited primarily by stomatal closure, but water loss in some plants is reduced by mechanisms that reduce interception of radiant energy. These include (a) curling and rolling of leaves to reduce the area exposed to sunlight, (b) movement of leaves to parallel the sun rays, and (c) shedding of older leaves.

### 3.2.3.3 Growth and Yield Responses

The stage of development when stress occurs can have a major effect on plant growth and yields (Salter and Goode, 1967). Because cell growth is very sensitive to turgor pressure, stress during vegetative growth reduces shoot extension and leaf expansion, which reduces dry matter accumulation. If stress is not severe, rewatered plants recover normal growth and produce near-normal seed yield. However, stress during the vegetative stage hastens development rates and maturity in many species. For example, dryland wheat normally flowers and matures about 5 to 10 days earlier than irrigated wheat in the Southern High Plains. Root growth is rapid during the vegetative growth stage, and stress during this stage slows top growth more than root growth. However, as stress severity increases, the root extension rate progressively decreases and the area of roots actively taking up water decreases (Slatyer, 1973).

In many species, stress during the flowering-pollination period has the greatest effect on yields. In some drought-tolerant species, however, the pollination stage is not particularly stress sensitive and stress during the time that fruiting structures develop may be the most critical because stress decreases the number and size of "yield containers". Examples are tiller abortion on small grains, which reduces heads per unit area; abortion of fruiting squares on cotton; and dwarfing of seed heads on grain sorghum.

During fruiting or seed filling, proteins, sugars, etc. accumulate in the fruit or seed; translocation to roots is limited; and root extension with depth may slow greatly or even stop (Mengel and Barber, 1974; McClure and Harvey, 1962). However, root growth may continue until physiological maturity of soybeans (Kaspar et al., 1978). In areas subject to stress, it is generally desirable to have plants with deep, well-developed root systems by the fruiting stage. If major root development occurred at high soil water contents, root systems may be short and highly branched (Burstrom, 1965). Normal profile drying between rewetting cycles enhances root extension with depth and increases the plant's ability to extract water from greater depths and withstand stress later in the season (Ritchie, 1974).

### 3.2.3.4 Alleviation Methods

Plant water stress is alleviated by decreasing the evaporative demand, increasing the wetting of soil by rainfall or irrigation, and increasing the storage and availability of soil water to plants. Thus, root zone modification is beneficial if it increases water storage and root development in the profile or alleviates such problem conditions as poor drainage or high salinity. Profile modification changes soil water storage primarily through changing water infiltration and hydraulic conductivity. Changes in water availability result from increased water storage and from proliferation of plant roots in greater soil volumes.

## 3.3 EFFECT OF MODIFICATION

For soil profile modification to alleviate plant water stress, it must improve a condition or conditions that limit water movement, root growth, or water-holding capacity. The benefits to be expected from profile modification depend on many factors. Some of the more important factors are the severity of the problem, climatic conditions including precipitation and temperature, management including methods and amounts of irrigation, and the crop grown. Certainly, unless a recognizable problem exists that profile modification can correct, there is little chance that it will be worthwhile.

Some problems that can possibly be altered by root zone modification are low infiltration and high runoff rates from relatively small amounts of precipitation or irrigation, restricted plant and root growth, plant water stress relatively soon after water additions, compact or dense layers in the profile, or deep sandy soils. Generally, these conditions are obvious from observations of water behavior after precipitation or irrigation, from plant response to applied water, from root examinations and growth patterns, and from physical examinations of the soil profile. Because the soil-water-plant system is highly dynamic, it is difficult to establish clear-cut limits beyond which profile modification will be beneficial. One limit that apparently has considerable merit is that of soil resistance to root penetration. Taylor and Gardner (1963) showed that cotton (*Gossypium hirsutum* L.) root penetration was stopped when soil strength, as measured with a penetrometer, reached about 3000 kPa. Thus, if root zone volume is important, as in soils infrequently recharged with water, profile modification should be beneficial if soil resistance is near 3000 kPa for any moist zone near the soil surface. However, even the 3000 kPa limit is arbitrary because root penetration progressively decreased as soil resistance increased from about 500 to 3000 kPa (Taylor and Gardner, 1963).

### 3.3.1 Removing or Disrupting Restricting Zones

Profile modification studies involving disruption of growth-limiting layers, such as plowpans, fragipans, claypans, high clay horizons, or salty or alkaline horizons, have been conducted at numerous locations. Usually, profile modification without incorporation of materials with different water-holding characteristics had little effect on the waterholding characteristics of the soil material, but it often increased the amount of water available to plants by increasing infiltration rates, reducing runoff, or changing the distribution of water in the profile. Also, profile modification disrupted or removed root-impeding layers, which allowed more complete exploration of the soil by roots and, therefore, greater extraction of water from the soil profile.

After reviewing the literature concerning deep tillage, Burnett and Hauser (1967) concluded that plant growth can be increased by deep plowing only where root development and perhaps water movement are restricted by dense, compact, or fine-textured layers in the profile. Where water is not limiting, plant growth is usually not increased by profile modification, even though root development is restricted by dense soil layers.

Profile modification has been studied as a means of increasing water and root penetration on most soils where root penetration is restricted. These include (a) soils with a slowly permeable, fine-textured B horizon below normal tillage depth, (b) claypan soils, (c) clay soils in which all horizons are slowly permeable, (d) fragipan soils, (e) medium- to coarse-textured soils with a plowpan, (f) poorly-drained stratified and clay soils, and (g) salt-affected soils.

### 3.3.1.1 Soils with a Slowly Permeable, Fine-textured B Horizon

Profile modification has been studied rather extensively on Pullman clay loam (fine, mixed, thermic Torrertic Paleustoll). Pullman and associated soils make up about 5 million ha of arable soils in the Southern Great Plains. The moderately permeable surface soil is underlain by a dense, very slowly permeable montmorillonitic clay horizon (B22t) that extends from the 0.2-m depth to the 0.5- to 0.6-m depth. Below this depth, the soil is again more permeable than the B22t horizon. Depth to the highly calcareous "caliche" layer varies from 1.2 to 1.5 m. A detailed description of the soil is available (Taylor et al., 1963). Major problems of the soil are low water infiltration rates and low water extraction by some crops from deep in the profile. During the first 4 h of irrigation, 50 to 100 mm of water usually enters the soil, but the infiltration rate at the end of 4 h is usually less than 2.5 mm/h and declines to about 1.2 mm/h after 10 h. Water use from deep in the profile varies with crops grown. On unmodified soil, sunflower (*Helianthus annuus* L.) extracted water to a depth of 3.0 m (Unger, 1978), whereas grain sorghum extracted water to only about 1.2 m on Pullman clay loam. In contrast, sorghum extracted water to a depth of about 2 m on Richfield silty clay loam (fine, montmorillonitic, mesic Aridic Agriustoll) in Kansas (Musick and Sletten, 1966).

Hauser and Taylor (1964) studied the effects of various profile modification treatments on the water infiltration rate of Pullman clay loam. Compared with the disking 0.075 m deep (check), disk plowing 0.6 m deep and vertical mulching 0.6 m deep on 2-m centers increased the rate of soil wetting. In 20 min, water penetrated to a depth of 1.8 m on plowed plots but to only 0.3 m on the check plots. In the 3-yr study, profile modification increased grain sorghum yields in only 1 yr. In that year, grain sorghum plants wilted before the first irrigation on all except the deep-plowed plots.

Eck and Taylor (1969) mixed the Pullman profile to depths of 0.9 or 1.5 m in November 1964 with a wheel-type ditching machine. The soil water data for 1967 (Fig. 3.3-1), typical of those for the study period, are from plots that received 250 mm of irrigation water before planting and no further irrigation. Growing-period rainfall totaled 116 mm. The data showed that profile modification increased water storage in the lower portion of the profile. The slowly permeable horizon of the unmodified profile prevented deep wetting, even though a rather large quantity of water was applied. Grain sorghum extracted water from a depth of about 2.4 m on modified soil and from a depth of about 1.3 m on unmodified soil. Because the 1.3- to 1.5-m zone in unmodified soil contained more water at harvest than similar zones in modified soil, the authors concluded that deep root activity was limited on unmodified soil as compared with the modified soil. Detailed measurements of water in-filtration were not made; however, sharp differences in infiltration were observed. At the preplant irrigation in 1967, 152 mm of applied water entered the modified plots in 1 h, but about 6 h were required for the same amount of water to enter the unmodified plots. With preplant irrigation only, profile modification to depths of 0.9 and 1.5 m increased average annual grain yields 66.2 and 80.1 percent, respectively. With adequate irrigation, profile modification did not affect grain yields.

In 1967, Unger (1970) found trends toward higher water contents at saturation in the modified soil, and higher water contents at −33 and −1500 kPa matric potential in the unmodified soil. However, plant-available water was nearly identical for all treatments. Profile modification significantly decreased soil bulk density and strength and significantly increased soil porosity. At an irrigation in 1969, 203 mm of water entered the soil in 6.5 and 5.3 h on 0.9 and 1.5-m deep modified plots, respectively, but 33 h were re-quired for the same quantity of water to enter unmodified soil.

An alfalfa (*Medicago sativa*) study was conducted on the plots from 1970 through 1975 (Eck et al., 1977). With single irrigations between harvests, profile modification increased dry-matter yields 40 percent (0.9-modified) and 60 percent (1.5-m modified) over those for the un-modified check. With two irrigations between harvests, profile modification increased yields 30 percent (14.1 to 18.4 t/ha). In 1975, sustained water in-filtration rates were measured on each of the three profiles. From 4 to 14 h after inundation, average water infiltration rates were 1.6, 3.3, and 4.0 mm/h for the unmodified, 0.9-m modified, and 1.5-m modified treatments, respectively. The differences in infiltration and those measured in bulk den-sity and surface elevation (Eck et al., 1977) showed that the effects of modification were still evident 12 yr after treatment.

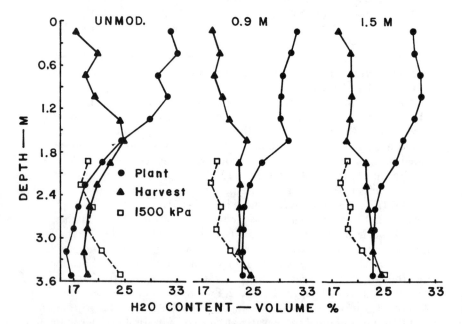

FIG. 3.3-1 Soil water content as affected by profile modification of Pullman clay loam. Sampled at grain sorghum planting and harvest (redrawn from Eck and Taylor, 1969).

Musick and Dusek (1975) measured water infiltration, retention, and depletion on furrow-irrigated Pullman clay loam that had been moldboard plowed 0.2, 0.4, 0.6, or 0.8 m deep in 1966. Fig. 3.3-2 shows total soil water with depth before and after irrigation on 8 July 1970. Deep tillage increased infiltration from 102 mm on 0.2-m plowed soil to 172 mm on 0.8-m plowed soil. Soil samples taken 2 wk after irrigation showed that the soil was wetted into the 0.3- to 0.6-m depth on 0.2-m plowed soil; into the 0.6- to 0.9-m depth on 0.4-m plowed soil; and into the 0.9- to 1.2-m depth on 0.6- and 0.8-m plowed soil. Storage efficiency was increased by deep tillage if the soil was dry when irrigated. Storage in a 1.8-m profile ranged from 68 percent on the 0.2-m plowed plots to 87 percent on the 0.8-m plowed plots. However, if the profile contained appreciable water when irrigated, some losses to deep percolation occurred from plots plowed through the B22t-layer (0.6 and 0.8 m deep). Depth of tillage had little effect on soil water extraction by grain sorghum and winter wheat during major drying cycles. Deep-plowed profiles were not dried to deeper depths or lower water contents than those which were not deep-plowed. They supplied more water to plants only when they contained more water at beginning of drying cycles.

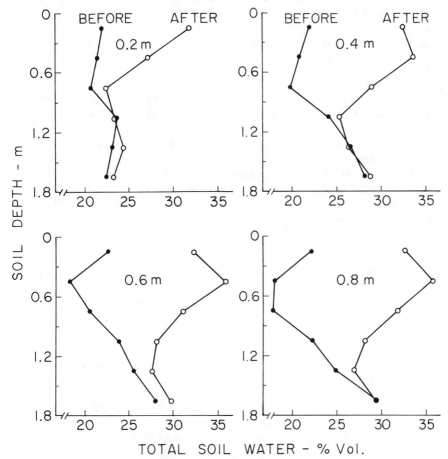

FIG. 3.3-2 Soil water content as affected by depth of plowing of Pullman clay loam. Sampled before and 2 wk after grain sorghum emergence irrigation (redrawn from Musick and Dusek, 1975).

Deep tillage increased 4-yr average grain sorghum yields from 3390 to 5650 kg/ha when one preplant and one seasonal irrigation were applied and from 4660 to 4940 kg/ha when one preplant and two seasonal irrigations were applied. With more irrigation, the entire yield increase resulted from the first increase in tillage depth (0.4-m vs. 0.2-m). With less irrigation, most of the yield increase resulted from the first increase in tillage depth, but there were additional yield increases with tillage depth increases to 0.8 m (Musick and Dusek, 1975). The authors concluded that deep tillage of slowly permeable surface irrigated Pullman clay loam can increase irrigation water infiltration, wetting depth, and crop yields under limited irrigation. Because a normal-sized irrigation of conventionally-tilled Pullman soil wets the profile only about 0.6 m deep, deep tillage permits greater utilization of the soil profile for available water storage and use by wetting the soil deeper. They cautioned, however, that deep tillage should not completely penetrate the slowly permeable B22t horizon. Complete disruption of that layer allowed excessive infiltration and some water loss to percolation below the root zone. They recommended deep tillage for use under limited irrigation where increased water storage is important, but not where frequent irrigation is practiced for high yields, except for crops that respond to reduced bulk density and improved aeration associated with deep tillage.

Freeman silt loam (fine-silty, mixed, mesic Mollic Haploxeralf) in eastern Washington and northern Idaho responded to profile modification by backhoe mixing and moldboard plowing 0.9 m deep (Mech et al., 1967; Cary et al., 1967). The soil has a silt loam A horizon about 0.3-m thick and a well-developed A2 horizon from the 0.30- to 0.46-m depth overlying a dense silty clay loam B horizon. Cary et al. (1967) reported the chemical and physical characteristics of the soil. Mech et al. (1967) reported that the B horizon is almost impervious to roots. When it was loosened and mixed, root proliferation and withdrawal of water were greater in the deeper depths. In the backhoe mixing study, wheat used 114, 145, 290, and 250 mm of water from the top 1.5 m during the growing season under conventional tillage; soil mixed to 0.46 m; topsoil and subsoil removed to a depth of 1.2 m, mixed, and replaced in original position; and topsoil and subsoil mixed to a depth of 1.2 m, respectively. Wheat yields were almost doubled by profile modification to 1.2 m (2000 to 3880 kg/ha) as were the 3-yr average alfalfa hay yields (3230 to 5940 kg/ha).

Deep moldboard plowing of Freeman silt loam slowed water infiltration rate because the clayey subsoil was deposited on the soil surface. Deep-plowed soil, however, accumulated 53 mm more water in the surface 0.9 m from precipitation than the conventionally-tilled soil (Mech et al., 1967). The authors hypothesized that water that entered the surface of the conventionally-tilled soil may have been lost by seepage along the A2 horizon. Infiltration into the undisturbed B2 horizon, with the A and A2 horizons removed, dropped to less than 0.25 mm/h after it was wetted to about the 0.05-m depth.

Mech et al. (1967) also deep moldboard plowed a Palouse silt loam (fine-silty, mixed, mesic Pachic Ultic Haploxeroll), which did not contain a slowly permeable horizon, but yields were not increased on this soil.

### 3.3.1.2 Claypan Soils

Efforts to improve the claypan soils of the Central States through deep tillage and subsoil fertilization have met with only limited success. Woodruff and Smith (1946) and Smith (1951) conducted studies on Mexico silt loam (fine, montmorillonitic, mesic Udollic Ochraqualf), a typical claypan soil near McCredie, MO. In most areas underlain by a claypan or a zone of deep plastic clay, subsoiling did not improve the soil because the tillage machines could not penetrate through the plastic clay horizon. The subsoil was shattered satisfactorily only when it was dry (water content 16 percent or less). The shattering changed the root development of sweet clover (*Melilotus* sp.) on Mexico silt loam from that typical on claypan soils to that typical of deep pervious soils. Deep liming further enhanced deep rooting of the clover. Generally, shattering was of little value unless the soil was treated with lime at a rate of 7 to 11 t/ha. Jamison and Thornton (1960), in summarizing the research at McCredie, indicated that yield increases from subsoil shattering and fertilization were small and concluded that when good fertilization and management practices are used on the surface plow layer, it is doubtful whether expensive subsurface treatments give economic yield increases for corn and alfalfa.

Fehrenbacher et al. (1958) mixed Weir silt loam (fine, montmorillonitic, mesic Typic Ochraqualf), a claypan soil at Carbondale, IL, to depths of 0.23, 0.46, 0.7, or 0.9 m. Mixing the silty A horizon with the clayey B horizon did not alter the available water storage capacity, and depth of mixing alone did not affect corn yields. However, fertilizer in addition to deep tillage increased root penetration, especially when the fertilizer was mixed throughout the tilled layer.

### 3.3.1.3 Slowly Permeable Clay Soils

Profile modification increased water infiltration, soil water depletion, and crop yields on Blackland soils of Central Texas and clay soils in the lower Rio Grande Valley of Texas. Burnett and Tackett (1968) mixed Houston Black clay (fine, montmorillonitic, thermic Udic Pellustert) to the 0.6-m depth by rototilling and to 0.6- and 1.2-m depths with a ditching machine. On modified profiles, water was normally used from greater depths earlier in the season. Four years after rototilling and 3 yr after mixing with a ditching machine, bulk density of the loosened soil was still lower than that at similar depths in conventionally-tilled plots. Three-year average yields of lint cotton and grain sorghum were increased by rototilling or soil mixing to 0.6 m. Soil mixing to 1.2 m further increased yields of lint cotton in only 1 yr, but did not affect grain sorghum yields. The treatments affected both the depth and texture of grain sorghum root systems. Photograph A (Fig. 3.3-3) depicts a grain sorghum root system from conventionally-tilled soil. The root system in Photograph B developed in finely-divided profile-modified soil (rototilled 0.6 m deep) and that in Photograph C developed in coarsely-divided profile-modified soil (ditching machine 1.2 m deep).

In another study on Houston Black clay, Burnett et al. (1974) found that 0.6-m deep moldboard plowing resulted in cotton plants that were taller, had greater leaf area, and accumulated more dry matter than plants on conventionally plowed plots (plowed 0.2 m deep). The better plant growth was attributed to greater root proliferation within a larger soil volume (Fig. 3.3-4).

Heilman and Gonzalez (1973) studied the effects of trenches in Harl-

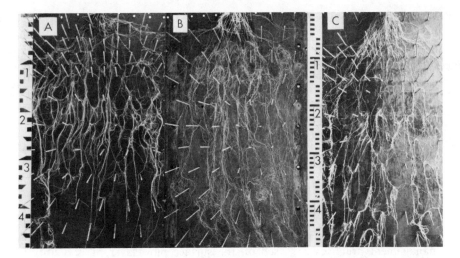

FIG. 3.3-3 Grain sorghum root systems grown in Houston Black clay. (A) From conventional tillage plot. (B) From plot rototilled 0.6 m deep. (C) From plot profile modified to 1.2 m with ditching machine (from Burnett and Tackett, 1968).

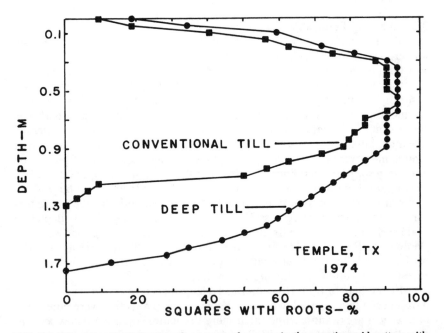

FIG. 3.3-4 Root density (defined as the percent of squares in the counting grid pattern with roots) distribution measured on 23-24 July 1974 in the profiles of conventional- and deep-tilled Houston Black clay (redrawn from Burnett et al., 1974).

ingen clay (very-fine, montmorillonitic, hyperthermic Entic Chromustert) on cotton lint yield, and on water infiltration rate and depletion. Cotton was planted over backfilled trenches 0.1 m wide and 0.6 or 1.0 m deep. The 0.6-m deep trenches were backfilled either with vermiculite from 0.15 to 0.6 m and soil from 0 to 0.15 m (T-2) or with soil (T-3). The deep trenches were backfilled with soil (T-4). There was also a check treatment (T-1). Cotton lint yields were highest on the T-2 and next highest on the T-4 treatment plots. Initial infiltration rates were 127, 104, 91, and 180 mm/h on T-1, T-2, T-3, and T-4 treatment plots, respectively. After 20 min, infiltration rates were consistently greater for trench treatments than for the check treatment. At 5 h, infiltration rates were 0.4, 1.2, 1.6, and 1.1 mm/h for the respective treatments. Cotton extracted much more water from the 0.6- to 0.9-m depth of the T-4 treatment than from the same depth increment of the T-1 treatment. More than 83 percent of cotton roots were recovered from the surface 0.3 m of nontrenched soil, whereas roots penetrated to greater depths of the trenched soil.

### 3.3.1.4 Fragipan Soils

Profile modification improved fragipan soils when it extended through the fragipan layer. In Missouri, Bradford and Blanchar (1977) modified the profile of Hobson silt loam (fine-loamy, siliceous, mesic Typic Fragiudalf) with a trenching machine. This and related soils with fragipan horizons cover about 2.8 million ha in Missouri, Arkansas, and Oklahoma. Typically, these soils have a brown, acid, silt loam or loam surface layer and a brown, very strongly acid, clay loam B2t horizon underlain by a dense sandy fragipan at about the 0.5- to 0.7-m depth. The fragipan layer ranges in thickness from about 0.3 to 0.8 m and is underlain by weathered sandstone. The soil was thoroughly mixed in trenches 0.5 m wide and 1.5 m deep. Trenches were 1.5 m apart. Soil water depletion by grain sorghum was greatest from trenches, less between trenches on trenched plots, and least from nontrenched plots. The greater depletion from the undisturbed zones between trenches than from nontrenched areas was attributed to lateral movement of water to roots growing in the trenches. Trenching alone increased grain sorghum yields by 34 percent (from 3230 to 4320 kg/ha) and trenching + lime + fertilizer (mixed in the trench) increased yields by 59 percent (to 5150 kg/ha). An additional treatment, for which sawdust, lime, and fertilizer were mixed in the trenches, increased yields by 85 percent (to 5990 kg/ha). Profile modification increased the amount of water available to plant roots by increasing total pore space, increasing the saturated hydraulic conductivity, and decreasing the mechanical resistance to root growth. Grain sorghum roots did not penetrate the fragipan horizon, but grew into the trenched soil. Water depletion was greater on trenched soil to which limestone was added than on unamended soil, indicating that liming improved the conditions for root growth in the acid soil material. The pH of the unamended soil was 4.2 (saturated paste).

Van Doren and Haynes (1961) found that deep tillage of fragipan soils in Ohio did not increase crop yields. However, their maximum depth of plowing was 0.3 m and their depth of chiseling was 0.45 m. The fragipans in their soils extended from 0.35 to 0.9 m. Thus, their tillage treatments did not penetrate through the fragipan layers.

### 3.3.1.5 Plowpan Soils

Compaction by tillage machinery and traffic by heavy equipment have formed "plowpans" in some soils. These layers, as well as naturally-formed dense or slowly permeable layers, impede water infiltration and root penetration. Experiments involving destruction of these layers have attained varying degrees of success. Taylor and Gardner (1963) dispelled much of the confusion surrounding the effects of plowpans when they showed that soil strength, not soil bulk density or any other physical feature of soil, controls the penetration of roots through sandy soil pans. Both soil water and bulk density affect soil strength, but strength is the determining factor. Plowpans are most likely to occur on fine sandy loams or other soils that do not swell and shrink while wetting and drying.

In general, destruction of plowpans by tillage has been more beneficial in the South and Southwest than in the North. This probably results from the destruction of plowpans in the North by freezing and thawing and from rapid and more complete soil drying in the Southwest, where the plowpans become stronger more quickly after rains. Locke et al. (1960) described plowpans at Woodward, OK, and at Mandan, ND. The soils at both locations were fine sandy loams. The plowpan at Woodward hindered root penetration, reduced water infiltration, increased soil erosion, and reduced crop yields, but the one at Mandan did not influence yields. During a field experiment at Woodward from 1958 to 1963 when precipitation was near average, yields were not affected by the plowpan. However, at Big Spring, TX, on a sandy clay loam soil, cotton yield decreased as soil strength increased. Few roots penetrated soil with strength above 2500 kPa at field capacity (Taylor et al., 1964).

In Louisiana, 0.3-m deep moldboard plowing or 0.35-m deep chiseling increased yields on three of four silt loam soils. The soil that did not respond had only a slight plowpan. The other three soils had high bulk density layers just below plow depth (Patrick et al., 1959). Saveson and Lund (1958), also in Louisina; Campbell et al. (1974) and Doty et al. (1975) in South Carolina; and Bruce (1960), Bruce and Jones (1963), Grissom et al. (1955, 1956), and Raney et al. (1954) in Mississippi obtained positive responses from deep chiseling or plowing of soils containing either traffic-induced or naturally high-strength layers. Responses generally were greatest in dry years and when the chiseling was done when the soil was relatively dry. When soil horizons below the compacted zones were infertile or acidic, deep application of fertilizer or line was beneficial (Patrick et al., 1959).

Robertson et al. (1976), who summarized soil profile modification studies in Florida, stated that profile modification did not consistently give desired benefits and that, in general, widespread use of intensive modification on deep horizons in Florida soils did not seem necessary. They also stated that on soils where tillage pan development was identified as a serious problem, favorable returns could be expected by breaking the pan with a chisel subsoiler for deeply placing lime and fertilizer.

Larson et al. (1960) reported on a series of subsoiling experiments in Iowa and Illinois. In 12 experiments conducted over 3 yr on seven important soil types in Iowa, subsoiling 0.4 or 0.6 m deep did not increase corn yields significantly. In Illinois, subsoiling 0.45 m deep increased corn yields in only one of eight experiments on five soil types. None of the soils had pronounced plowpan layers, even though subsoiling was a common crop production practice on these soils.

Hobbs et al. (1961) conducted deep tillage studies in Kansas on eight soil types that had plowpans or naturally dense layers. Deep tillage lowered the bulk density and increased the permeability in dense layers, but it did not increase water storage in the soil and seldom increased crop yields.

In Michigan, deep tillage improved crop yields only on problem soils and had little effect under average conditions. Deep tillage improved crop yields only when the compact zones were immediately below the plow layer. Yield responses to deep tillage were greater on soils with artificially produced compact zones than on naturally compact soils, possible because the natural compact zones were too thick for tillage tools to penetrate them completely (Robertson et al., 1977). In Connecticut, plowpans had drastic effects on root development of tobacco (*Nicotiana* sp.) plants but shattering of the plowpan produced only an 8 percent increase in yields. Deep application of fertilizer with deep shattering resulted in a 24 percent yield increase above the control treatment (De Roo, 1961).

### 3.3.1.6 Poorly Drained Soils

In California, profile modification with chisels, moldboard plows, and slip plows improved drainage and leaching, reduced soil strength, and increased plant growth (Kaddah, 1976). The slip plow is a modified soil chisel with a flat plate 0.2 to 0.3 m wide and 3 to 3.5 m long attached to its back from the chisel point to the aboveground portion. The plate extends upward at a 60- to 70-degree angle from the chisel point. As the shank moves through the soil, the upper end of the inclined plane rides even with the surface. The slip plow mixes the soil material somewhat in addition to lifting and shattering the soil layers.

Willardson and Kaddah (1969) compared conventional tillage, deep moldboard plowing, slip plowing, and chiseling to determine the effects of profile modification on leaching in a Holtville silty clay loam (clayey over loamy, montmorillonitic (calcareous), hyperthermic Typic Torrifluvent). The silty loam surface soil is underlain by coarser-textured materials. This soil had serious salination and crop growth problems. Deep moldboard plowing, slip plowing, and chiseling were to a depth of 1.1 m. Chiseling was done on 1-m centers and slip plowing was on 2.4-m centers. During the settling irrigation, deep-plowed plots absorbed 240 mm of water and slip-plowed and chiseled plots absorbed 173 mm, whereas the conventionally-tilled plots absorbed only 100 mm of water. During a leaching period, conventionally-plowed plots absorbed 230 mm of water, chiseled plots 410 mm, moldboard-plowed plots 250 mm, and slip-plowed plots 300 mm. The authors stated that these responses were different from those during the initial irrigation and indicated the importance of continued evaluation of the treatments.

Robinson and Luthin (1968) studied the effects of slip plowing on the drainage characteristics of a nonstratified clay soil. Slip plowing to the 1.2-m depth opened the soil profile enough to allow leaching from all points at the same depths to proceed at nearly equal rates. Chiseling 0.56 m deep did not accomplish this. Also, slip plowing increased water infiltration rates more than chiseling.

Kaddah (1976) subsoil chiseled and slip plowed Rositas loamy fine sand (mixed, hyperthermic Typic Torripsamment) that naturally overlies Imperial silty clay (fine, montmorillonitic (calcareous), hyperthermic Vertic Torrifluvent) at about a 1.0-m depth. Along the slits formed by subsoiling or slip plowing 0.9 m deep, the stratified sandy soil layers were more disturbed, root and shoot growth and water infiltration rate were greater, soil strength was lower, and bulk density was lower in some layers but not in others, when compared with conditions in areas disked 0.2 m deep. The differences in root penetration and distribution caused differences in soil water tension during irrigation cycles. Before irrigation, water tension usually was 50 to 80 kPa higher in slits formed by the deep tillage implements than in areas between the slits or in the control treatment. Wheat yields were increased from 4.5 t/ha on disked plots to 6.3 t/ha on plots slip-plowed in two directions.

### 3.3.1.7 Salt-affected Soils

Effects of modification for alleviating salinity will be discussed more thoroughly in Chapter 8. However, it should be mentioned here that modification has improved water storage and depletion and increased yields on salt-affected soils. Modification has been effective on Rhoades (fine, montmorillonitic Leptic Natriboroll) and Belfield (fine, montmorillonitic Glossic Natriboroll) soils (silt loam-silty clay complex) in North Dakota, Chilcott (fine, montmorillonitic, mesic Abruptic Xerollic Duragid) and Sebree (fine-silty, mixed, mesic Xerollic Nadurargid) soils (slick spot complex) in Idaho and Oregon, and on similar soils in Canada. These soils have a dense, high-sodium subsoil (B horizon) underlain by a calcareous soil zone (formerly called Solonetz soils). The dense, high-sodium subsoil restricted water infiltration and root development, which resulted in poor crop yields. In North Dakota, Sandoval et al. (1972) found increased water storage during fallow and increased extraction during cropping on a soil plowed to the 0.6-m depth as compared with plowing to the 0.15- or 0.3-m depth. The 3-yr average soil water extractions were 70, 194, and 104 mm on soils plowed to 0.15-, 0.3,-, and 0.6-m depths, respectively. Although not measured, greater root penetration undoubtedly was a factor in the increased soil water extraction (Sandoval, 1978). In the fifth year after deep tillage, spring wheat yields were 1310, 1760, and 2110 kg/ha with the respective plowing depths. Rasmussen et al. (1964) reported five-fold increases in alfalfa hay yields from mixing Sebree slick spot soil to a depth of 0.76 m. Treatments in which the soil was mixed to 0.56 m with gypsum applied at 44.8 t/ha increased the terminal infiltration rate (infiltration rate near the end of time required for a 100-mm irrigation to be absorbed) from 1.8 to 33 mm/h. Water penetrated to the full depth of mixing or plowing on all plots within irrigation periods of 10 to 12 h. In further research, these authors showed that application of gypsum was not necessary for reclamation of this soil because the profile contained adequate natural gypsum and concluded that this soil could be most economically reclaimed by deep plowing alone.

In Canada, Cairns and Bowser (1977) obtained a 56 percent increase in hay crop yields from plowing a Solonetz soil (Hemaruka loam) deep enough to penetrate the C horizon to a depth that provided a 1:1:1 mixture of the A, B, and C soil horizons. In general, this is about 0.4 to 0.6 m deep on most Canadian Solonetzic soils.

### 3.3.2 Increasing Root Zone Depth

Usually, where profile modification increases crop yields, the expanded root zone is important for increasing the yields. Taylor and Bruce (1968) investigated the effect of soil strength on cotton taproot growth (Fig. 2.3-1). Root elongation through soil decreased as penetrometer resistance (soil strength) increased. At about 2000 kPa penetrometer resistance, no cotton roots entered compacted soil layers. Taylor and Burnett (1964) found that if soil strength is greater than 3000 kPa, no roots penetrate compacted soil layers. If the strength is less than 3000 kPa, the percentage of the roots that penetrate the soil is inversely proportional to soil strength. Taylor (1976) found that, under favorable conditions, cotton roots grew at a rate of more than 40 mm/day. Thus, plants have the genetic potential for rapid root elongation if conditions for growth are ideal. The magnitude of the yield increase that tillage causes varies with the intensity and depth of the root-restricting layer under any particular climatic environment. If the tillage operation does not make more water available to the plant, it generally will not increase yield (Taylor, 1976). Also, if conditions other than soil strength prevent root exploration (toxic substances, poor aeration, poor drainage not alleviated by tillage, etc.), tillage alone cannot increase yields.

### 3.3.2.1 Soils with a Slowly Permeable, Fine-Textured B Horizon

On Pullman clay loam, the principal effect of profile modification is to allow more water to enter the soil. When adequate water is available, deep tillage does not increase crop yields unless the crop responds to the reduced soil bulk density and improved aeration associated with deep tillage. Eck and Taylor (1969) and Musick and Dusek (1975) found no yield responses to deep tillage on fully irrigated grain sorghum; however, Mathers et al. (1971) obtained yield responses to deep tillage on fully irrigated sugarbeets and Eck and Davis (1971) reported yield responses to deep tillage on fully irrigated sugarbeets, sudangrass (*Sorghum sudanense* Stapf.), and soybeans (*Glycine max* L.). Mathers et al. (1971) attributed yield increases to increased water infiltration, improved aeration, reduced bulk density, less root rot, and increased response to nitrogen. Eck and Davis (1971) measured root yields of sugarbeets, sudangrass, soybeans, and cabbage (*Brassica oleracea*) to a 0.9-m depth on 0.2-m deep rototilled and 0.9-m deep moldboard plowed Pullman clay loam. All four crops produced higher top yields in proportion to root yields on moldboard plowed plots than on rototilled plots. Thus, top/root ratios were increased and, apparently, root effectiveness was increased by deep plowing. Root yields of all four crops decreased with soil depth on the rototilled treatment and those of sudangrass, sugarbeets, and soybeans decreased with soil depth on the deep-plowed treatment.

Eck et al. (1977) measured a 30 percent increase in alfalfa yields on profile-modified Pullman clay loam. Roots were larger and more plentiful in the modified soil, but differences in rooting depth were not as great as were anticipated from differences in water infiltration rates and soil water accretion and depletion patterns (Fig. 3.3-5). In addition to increasing alfalfa yields, profile modification simplified management. Because of the low infiltration rate on unmodified plots, ponding was sometimes a problem and the plots had to be drained to preserve the stand. As a result of differences in water infiltration and surface drying, the surface of unmodified plots remained wet longer than that of modified soil. The longer wet period weakened the alfalfa stand and allowed barnyardgrass (*Echinochloa crusgalli* (L) Beauv.) to become established on unmodified plots.

Rakov and Eck (1975) showed that when the B22t layer of Pullman clay was not disturbed, grain sorghum root growth and root exploration was much less in the B22t layer and in the undisturbed subsoil below than when this layer was disturbed (Fig. 3.3-6). They attributed reduced water use on undisturbed soil to incomplete root exploration.

FIG. 3.3-5 Effect of profile modification of Pullman clay loam on alfalfa rooting. Profiles from left to right are: unmodified, modified to the 0.9 m depth, and modified to the 1.5 m depth. Arrows in photographs indicate depths of profile modification (from Eck et al., 1977).

### 3.3.3 Increasing Water Holding Capacity of Sandy Soils

Sandy soils have a low plant-available water holding capacity. The high-sand-content material, usually with more than 70 percent sand, may be in the surface horizons, where it is underlain by finer-textured materials, or it may extend to great depths. For alleviating plant water stress on sandy soils, any modification treatment that increases the retention of plant-available water within the root zone should be beneficial. Water retention in sandy soils can be increased by adding soil or non-soil materials that retain more water than sand alone.

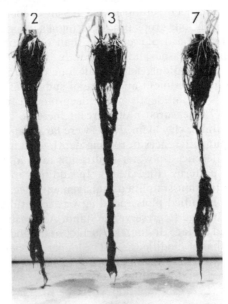

FIG. 3.3-6 Grain sorghum roots washed from cores of Pullman clay loam in which the B22t horizon was coarsely divided (2), finely divided (3), and left undisturbed (7) (from Rakov and Eck, 1975).

**FIG. 3.4-1 Disk plowing Pullman clay loam to the 0.6 m depth at Bushland, TX.**

### 3.3.3.1 Adding Fine-textured Materials

Fine-textured materials from outside sources or from deeper in the profile can be used to increase the water holding capacity of sandy surface horizons. Adding materials from outside sources may be practical on limited areas, but not on large areas because of the large amount of material needed and the expense of transporting the materials. Within the realm of practicality, however, is the mixing of finer materials from deeper in the profile with the coarser materials at the surface.

The Hezel (sandy over loamy, mixed, nonacid, mesic Xeric Torriorthent) and similar soils in central Washington have coarse wind-deposited materials overlying high-silt, water-deposited materials. The Hezel soil is sandy loam to the 0.46-m depth and silt loam at the 0.46- to 0.90-m depth. In 1970, Miller and Aarstad (1972) moldboard plowed the soil 1 m deep in an attempt to increase the water holding capacity by bringing finer materials to the surface and by turning under the coarse surface material, which would restrict downward water movement from the finer soil brought to the surface. After plowing, the texture was silt loam at the 0- to 0.15-m depth and loam at the 0.15- to 0.30-m depth. The textures after plowing resulted from thoroughly mixing the surface 0.30 m of soil across the furrow slice. The available water holding capacity of the surface 0.30 m of soil was 36 mm before plowing and 61 mm after plowing, a 69 percent increase, which was largely the result of the change in texture. The sand content of the surface layer was reduced from about 70 percent to 40 or 50 percent, with corresponding increases in clay and silt. Below the 0.30-m depth, the difference in water holding capacity between unplowed and deep-plowed soil was slight.

Increasing the water holding capacity and the plant nutrient supply of the surface soil layer, which would aid seedling establishment, and decreasing the wind erosion potential were goals of deep plowing studies on sandy soils at five locations in Oklahoma (Harper and Brensing, 1950). Before deep plowing, the surface soil contained from 86.8 to 94.8 percent sand and from 2.7 to 3.7 percent clay. Initial clay content of the subsurface soil and exact plowing depths were not specified, but the plowing depth ranged from about 0.25 to 0.40 m. After plowing, the surface soil contained from 64.5 to 82.5 percent sand and from 10.3 to 17.7 percent clay. Deep plowing generally increased crop yields, but the reason was not apparent because soil water and plant nutrient contents were not measured. Undoubtedly, the water content of the surface soil was a factor. Unger (1975) showed that plant available water (based on $-33$ and $-1500$ kPa matric potentials) in soil cores increased about 0.09 percent by weight for each percentage unit increase in soil clay content. For sieved soil, the increase was about 0.28 percent per percentage unit, suggesting that modification treatments that thoroughly disrupt the structure would increase the water holding capacity of soils over that for unmodified soils. Such thorough disruption, however, is impractical under large-scale field conditions.

The clay content of modified soil depends on initial clay contents of the soil layers and the modification depth. Although increasing the clay content of sandy soils is often desirable for increasing water retention within the surface layers, optimum clay contents have not been established. However, if the clay content in the surface layer of sandy soils can be increased to over 8 percent, subsequent cultivation causes clods that resist wind erosion (Harper and Brensing, 1950).

The distribution of sand, silt, and clay in a soil profile after deep plowing depends largely upon the initial distribution of these soil materials and the nature of action of the implement used. Miller and Aarstad (1972) found high variability in water retention below the surface layer because this layer was not cross-tilled after deep plowing. Harper and Brensing (1950) advised against deep plowing sandy soils that had subsurface layers with very high clay percentages because it places too much clay in the surface layer, which could decrease water infiltration. They suggested that this condition can be avoided by shallower plowing or by using a disk plow, which causes less soil inversion than a moldboard plow. Lyles et al. (1963), however, showed little difference in soil mixing by disk and moldboard plows used for deep plowing, but recognized that other makes of plows may invert the soil more or less than the ones used. For even greater homogenization of a soil profile, Eck and Taylor (1969) modified Pullman clay loam to 1.2- or 1.5-m depths with a wheel-type ditching machine. The resultant profiles had more uniform sand, silt, and clay contents than the unmodified soil (Unger, 1970).

If the modification treatment results in thorough soil mixing, if the initial clay contents of the surface and subsurface layers are known, if the surface layer depth is known, and if the desired clay content after modification is specified, then the required penetration of the subsoil to obtain the approximate specified clay content in the modified soil can be obtained from

$$(A_d \times A_{\%c}) + (B_d \times B_{\%c}) = (C_d \times C_{\%c}), \dots\dots\dots\dots\dots\dots\dots [3.3\text{-}1]$$

where

$A_d$ = depth of surface layer (known),

$B_d$ = depth of subsurface layer to be penetrated by modification operation (to be solved for),

$C_d$ = depth of modified soil ($A_d + B_d$, with volume increase ignored),

$A_{\%c}$ = % clay in surface layer (known),

$B_{\%c}$ = % clay in subsurface layer (known), and

$C_{\%c}$ = % clay in modified soil (specified).

Substituting $(A_d + B_d)$ for $C_d$ and rearranging equation [3.3-1] results in

$$B_d = \frac{A_d\,(C_{\%c} - A_{\%c})}{(B_{\%c} - C_{\%c})} \cdot \quad\dots\dots\dots\dots\dots\dots\dots\dots\dots\dots\dots\dots\dots\dots\quad [3.3\text{-}2]$$

After solving for $B_d$, the modification depth is obtained from

$$C_d = A_d + B_d \cdot \quad\dots\dots\dots\dots\dots\dots\dots\dots\dots\dots\dots\dots\dots\dots\dots\dots\quad [3.3\text{-}3]$$

The above equations imply that water retention is a function only of soil clay content. This, however, is not the case because soil silt content also influences water water retention (Jamison, 1953; Unger, 1975). While equations [3.3-1], [3.3-2], and [3.3-3] illustrate the technique for determining modification depths, similar equations can be developed for determining modification depths where silt or silt and clay contents are high in the subsurface layer.

### 3.3.3.2 Adding Organic Materials

Increasing a soil's organic matter content has been considered to be an effective method for increasing its available water holding capacity (Jamison, 1953). This belief is based on the high water retention capacity of organic matter and the generally high correlation between organic matter content and water retention in various soils (Jamison, 1953; Lyon et al., 1952).

Unger (1975), using multiple regression analyses, showed that plant available soil water (volume basis) increased about 1.8 percentage units for each percentage unit increase in organic matter content for soils ranging in texture from sand to clay. Jamison (1953) showed that organic matter increased water retention only in coarse-textured soils that had very low capacities to store available water. He showed a high positive correlation between soil water retention and organic matter content for sandy soils and soils with less than 15 percent clay. The benefits in sandy soils resulted from diluting the droughty sandy material with material having a greater water retention capacity than the undiluted material.

In fine-textured soils, an increase in organic matter content does not increase the available water capacity as much as in sandy soils (Jamison, 1953). In fine-textured soils, the benefits from increased organic matter contents result from improved soil structure, which reduces soil crusting and surface sealing, permits greater water infiltration, and increases root proliferation throughout the soil profile (Jamison, 1953). Although available water storage capacity *per se* is not greatly increased in these soils by organic matter, the result of adding the organic matter is that plants use the soil water more effectively.

Because water retention in sandy soils increases with increases in organic matter (Jamison, 1953; Peele et al., 1948), adding organic materials has potential for alleviating plant water stress in these soils. The gain in available water retention by adding organic matter to sandy soils may be small (Allison, 1973), but even small gains can greatly increase the amount of water retained in these soils. If, for example, a one percentage unit increase in organic matter content increases available water retention by 1.8 percentage units, as shown by Unger (1975), the available water retained in the sandy soil previously discussed would increase from 2.3 to 4.1 percentage units, a 78 percent increase in water retention. Such an increase would significantly affect crop production on sandy soils. It is, however, difficult to increase the organic matter content enough to be of much value. To raise the soil organic matter content by one percentage unit to a 0.3-m depth would require about 45 t/ha of material. Such amount is greatly above annual production by most crops. Thus, concentration from outside sources is needed, which is highly impractical for large areas. Smaller additions resulting in smaller increases in water retention may be practical, especially if the materials are grown in place. Returning most or all residues from well-managed, high-residue crops to the soil should maintain or gradually increase the soil organic matter content (Allison, 1973). Additional benefits result from decomposing roots and from adding materials such as manure, woody materials (bark, chips, sawdust), peat, and composts (Allison, 1973). Because not all added materials become stable organic compounds in a soil, repeated applications of organic materials are needed to maintain or increase soil organic matter.

### 3.3.4 Installing Barriers to Deep Drainage in Sandy Soils

In the previous section, we discussed increasing water retention in sandy soils by increasing the content of fine-textured material or of organic matter in the sandy zone of the soil. For deep sandy soils, mixing fine-textured materials with the sandy horizon is not possible because the fine material may be too far below the surface or there may be no fine material in the profile. Such soils can be modified by installing physical barriers that restrict or prevent the downward movement of water. The theory concerning the influence of subsurface barriers on water movement in sandy soils was discussed by Erickson et al. (1968). Water near the edges of the barrier drains rapidly below the barrier, thus creating capillary discontinuities, which further slow the drainage of low tension water from above the barrier.

In 1966, Erickson et al. (1968) installed asphalt barriers at 0.55- to 0.60-m depths in 61- by 61-m plots on sandy soils in Montcalm and Allegan Counties, MI. Vegetable crops were then grown with sprinkler irrigation or without irrigation on treated and on control areas. Soil water contents were determined during the growing season. At the 0.3- to 0.6-m depth, the water content was greater above the barriers, except for the nonirrigated plots in Montcalm County where not enough rain fell to fill the water storage reservoir above the barrier. Barriers at the 0.6-m depth had little effect on soil water content at the 0- to 0.15-m depth because water readily drained below this depth in the sandy soils. In Allegan County, 85 mm of rain fell on 15 August. After 3 days of drainage, soil at the 0.3- to 0.6-m depth contained 23.7 percent water with the barrier and 15.5 percent without the barrier, a difference of 25 mm in water retention.

The difference in rainfall distribution at the two locations resulted in different responses in vegetable crop yields (Table 3.3-1). Favorable rainfall in Allegan County increased cucumber (*Cucumis sativus*) and cabbage yields with asphalt barriers, but irrigation decreased cabbage yields, possibly due

TABLE 3.3-1. EFFECT OF ASPHALT BARRIERS AND
IRRIGATION ON CROP YIELDS ON SANDY SOIL
WITH FAVORABLE AND UNFAVORABLE RAINFALL
(from Erickson et al., 1968)

| Rainfall, location, and crop | Treatments | | | |
|---|---|---|---|---|
| | Control | Control with irriga- tion | Asphalt layer | Asphalt layer with irrigation |
| | | | Yield | |
| Favorable rainfall— Allegan County, MI | | | metric tons/ha | |
| Cucumbers | 18.7b* | 19.5b | 24.9a | 25.5a |
| Cabbage | 23.0b | 18.5c | 32.7a | 23.8b |
| Unfavorable rainfall— Montcalm County, MI | | | | |
| Potatoes | 24.3b | 37.3a | 25.6b | 38.6a |
| Beans | 13.8b | 25.1a | 14.4b | 25.3a |

*Row values followed by the same letter are not significantly differ-
ent at the 5 percent level (Duncan's Multiple Range Test).

to nutrient leaching. With less favorable rainfall in Montcalm County,
asphalt barriers did not increase potato (*Solanum tuberosum*) and bean
(*Phaseolus* sp.) yields without irrigation (Erickson et al., 1968). These results
showed that the extra water retention capacity increased yields when the
rainfall distribution or irrigation frequency allowed the increased storage
capacity to decrease drought injury to crops.

In Taiwan, a sandy soil without asphalt barriers was considered un-
suitable for paddy rice (*Oryza sativa*). Installing asphalt barriers significantly
increased yields (Table 3.3-2). Paddies without asphalt required seven times
more water to keep them flooded than those with the barriers. Depth to barrier had little effect on rice yield (Erickson et al., 1968). In India, subsurface
barriers on bitumen (asphalt) or concrete significantly increased rice yields,
decreased the water requirement (Table 3.3-3), and increased water-use efficiency. As in Taiwan, depth to barrier had little effect on rice yields (Rao et
al., 1972).

TABLE 3.3.2. RICE YIELDS AS AFFECTED
BY AN ASPHALT BARRIER IN SOIL,
TAIWAN, 1967 (from Erickson et al., 1968)

| Treatment | Yield |
|---|---|
| | metric tons/ha |
| No asphalt | 0.40 |
| Asphalt at 0.2-m depth | 4.32 |
| Asphalt at 0.3-m depth | 4.79 |
| Asphalt at 0.4-m depth | 4.85 |
| Asphalt at 0.6-m depth | 5.38 |
| L.S.D. (0.01) | 0.56 |

TABLE 3.3-3. EFFECT OF SUBSURFACE BARRIERS ON
YIELDS AND WATER REQUIREMENTS FOR RICE IN INDIA
(from Rao et al., 1972)

| Barrier depth | Yield | | Water used | |
|---|---|---|---|---|
| | Early crop | Late crop | Early crop | Late crop |
| m | kg/ha | | mm | |
| None (check) | 4,980c* | 3,920a | 3,170a | 706a |
| 0.2 | 7,200a | 4,260a | 965b | 485b |
| 0.3 | 6,310b | 4,320a | 854b | 472b |
| 0.4 | 7,600a | 4,420a | 869b | 472b |

*Column values followed by the same letter are not significantly dif-
ferent at the 5% level (Duncan's Multiple Range Test).

Studies with an asphalt barrier at a 0.6-m depth in Lakeland fine sand
(thermic, coated Typic Quartzipsamment) were initiated in Florida in 1967.
Saxena et al. (1973) studied tomato (*Lycopersicon esculentum* Mill.) and
corn rooting in treated and untreated plots. The asphalt decreased rooting
depth, but soil above the asphalt contained about 110 mm of water, about
the same as in 0.9 m of soil without asphalt. The water above the asphalt was
maintained at higher potentials, which, along with the increased root concentration, increased the availability of water to plants. The increased
availability of water, both with and without irrigation, generally increased
crop yields (Saxena et al., 1969, 1971).

In 1971, corn on the asphalt barrier plots yielded more fodder than on
check and irrigation treatment plots during the first 11 wk when water stress
was moderate. An 11-day drought after that period caused severe plant water
stress on check and asphalt treatments (without irrigation) which significantly decreased final fodder and grain yields. The asphalt barrier decreased the
need for supplemental water and increased water-use efficiency in the well-
drained sandy soil (Robertson et al., 1973). Although some roots penetrated
the asphalt barrier, 4 yr of continuous cropping with a variety of crops did
not decrease the effectiveness of the asphalt barrier for increasing water
storage in the Florida soil (Saxena et al., 1973).

### 3.3.5 Summary of Soil Modification Effects

The responses obtained from the various soil modification treatments
discussed in the foregoing sections are summarized in Table 3.3-4. Modification treatments generally were beneficial (positive yield response) when the
problem horizons, such as slowly permeable layers, dense materials,
claypans, plowpans, fragipans, poorly drained layers, salt-affected layers,
and deep sands were adequately disrupted or altered. Benefits increased as
intensity and depth of modification increased, and were generally most
beneficial when limited soil water availability to plants was associated with
the soil problem. The benefits resulted from providing more water for plant
growth, either by conserving more of the applied water or by permitting root
proliferation within a larger soil volume so that the water in soil was used
more effectively. Modification treatments were of no or limited usefulness
when the water supply was plentiful or when the modification treatments did
not effectively alter the soil. No negative responses to modification were
reported.

TABLE 3.3-4. SUMMARY OF REPORTED RESPONSES TO MODIFICATION TREATMENTS
ON SOME PROBLEM SOILS AT VARIOUS LOCATIONS

| Soil | Location | Problem | Treatment | Crop | Response* |
|---|---|---|---|---|---|
| Pullman clay loam | Bushland, TX | Slowly permeable B horizon | Chiseling—0.6 m deep, 2-m centers | Grain sorghum | 0 |
| Pullman clay loam | Bushland, TX | Slowly permeable B horizon | Vertical mulching— 0.6 m deep, 2-m centers | Grain sorghum | 0 |
| Pullman clay loam | Bushland, TX | Slowly permeable B horizon | Disk plowing—0.6 m deep | Grain sorghum | + |
| Pullman clay loam | Bushland, TX | Slowly permeable B horizon | Profile mixing with ditching machine— 0.9 or 1.5 m deep | Grain sorghum, alfalfa | + |
| Pullman clay loam | Bushland, TX | Slowly permeable B horizon | Moldboard plowing— 0.4, 0.6, or 0.8 m deep | Grain sorghum, sugarbeets | + |
| Pullman clay loam | Bushland, TX | Slowly permeable B horizon | Moldboard plowing— 0.9 m deep | Sudangrass, soybeans, sugarbeets, cabbage | + |
| Freeman silt loam | Washington & Idaho | Dense B horizon | Mixing or moldboard plowing—0.9 m deep | Wheat, alfalfa | + |
| Mexico silt loam | McCredie, MO | Claypan | Subsoiling | Corn, alfalfa | 0 |
| Weir silt loam | Carbondale, IL | Claypan | Mixing—0.2, 0.5, 0.7, or 0.9 m deep | Corn | 0 |
| Houston Black clay | Temple, TX | Dense clay | Mixing, plowing—0.6 or 1.2 m deep | Grain sorghum, cotton | + |
| Harlingen clay | Rio Grande Valley, TX | Dense clay | Trenching—0.6 or 1.0 m deep | Cotton | + |
| Hobson silt loam | Missouri | Fragipan | Mixing—1.5 m deep | Grain sorghum | + |
| — | Ohio | Fragipan | Plowing—0.3 m; chiseling—0.45 m deep | — | 0 |
| (Fine sandy loams) | Oklahoma | Plowpan | Loosening | — | + |
| (Fine sandy loams) | North Dakota | Plowpan (not severe) | Loosening | — | 0 |
| (Sandy clay loam) | Big Spring, TX | Plowpan | Loosening | Cotton | + |
| (Silt loams) | Louisiana & Mississippi | Plowpan | Chiseling—0.35 m; plowing—0.3 m deep | Cotton | 0 to + |
| Holtville silty clay loam | California | Poor drainage and leaching | Moldboard or slip plowing, chiseling— 1.1 m deep | — | + |
| A nonstratified clay | California | Poor leaching | Slip plowing—1.2 m deep | — | + |
| Rositas loamy fine sand | California | Poor infiltration, dense soil | Subsoiling or slip plowing—0.9 m deep | Wheat | + |
| Rhoades, Belfield, Sebree, Chilcott, & Hemaruka | North Dakota, Idaho, Oregon, & Canada | Salt affected | Plowing or mixing— 0.3 to 0.76 m deep | Wheat, alfalfa | + |
| Hezel (sandy soil) | Washington | Low water retention (sand over clay) | Moldboard plowing— 1 m deep | — | + |
| (Sandy soils) | Oklahoma | Low water retention and high wind erosion (sand over clay) | Moldboard or disk plowing 0.25 to 0.4 m deep | — | + |
| (Deep sand) | Michigan | Low water retention | Installing subsurface barrier | Cucumber, potato, beans, cabbage | 0 to + |
| (Deep sands) | Taiwan & India | Low water retention, high water use | Installing subsurface barrier | Rice | + |
| Lakeland sand | Florida | Low water retention | Installing subsurface barrier | Tomato, corn | + |

*Responses: + = positive, 0 = no effect.

# 3.4 METHODS OF MODIFICATION

To effectively correct a soil problem that limits water utilization by plants, the equipment used must reach the problem zone in the soil. For maximum effectiveness, the equipment selected should have the capacity to alter the soil condition, and the operation should be performed when the soil water content is at or near the optimum for the equipment used.

## 3.4.1 Equipment for Modification

In this chapter, we have been concerned with root zone modification at depths of 0.3 m or more. Modifying soils at such depths requires the use of

TABLE 3.4.1. IMPLEMENTS FOR ROOT ZONE MODIFICATION

| Implement | Operating depth, m* | Action | Optimum soil water content |
|---|---|---|---|
| Moldboard plow | 1.1 | Inversion and mixing | Mid to upper plastic range |
| Disk plow | 0.7-0.8 | Inversion and mixing | Mid to upper plastic range |
| Slip plow | 2.0 | Shattering, considerable mixing | Around lower plastic limit |
| Chisel, subsoiler, or ripper | 0.9-2.0 | Shattering, little mixing | Around lower plastic limit |
| Trenching machine | 1.5 | Complete mixing | Mid plastic range |
| Barrier installer | 0.6 | Undercutting, lifting, applying barrier | Around upper plastic limit |

*Based on depths reported in the literature.

larger equipment than that used for tillage at more normal depths ($< 0.3$ m). As power for tillage has increased, the size and operating depth of equipment has also increased. Equipment and power sources are now available to modify soils to almost any reasonable depth concerned with crop production.

Table 3.4-1 lists some implements, along with depths to which they have been operated, their nature of action in soils, and soil water conditions for optimum operation.

Moldboard and disk plows, which mix the plowed layer, effectively disrupt dense horizons and bring deep soil materials, such as clay in sandy soils, to the surface. These plows should be operated when the soil water content is at the mid- to upper-plastic limit. Plowing when the soil water content is at or below the lower-plastic limit, especially in high clay content soils, results in many large clods that require much secondary tillage for seedbed preparation. Plowing under such conditions also causes soil penetration problems and requires more energy than when the soil contains more water. Plowing when the soil is too wet may cause problems with traction, soil adhering to the plows, and soil puddling. Disk plows perform more effectively than moldboard plows when soil adheres to the plows (Harper and Brensing, 1950).

Slip plows lift and shatter impermeable layers and interrupt any pronounced stratification. Since soil shattering and mixing are objectives of slip plowing, the plowing should be done when the soil water content is at or below the lower-plastic limit. Little shattering and mixing would occur in a wet soil.

Chisels, subsoilers, and rippers shatter dense and impermeable soil layers, but they mix the soil only slightly. For maximum effectiveness, chiseling and subsoiling should be done when the soil water content is low (Bruce, 1960; Bruce and Jones, 1963; Grissom et al., 1955). These implements cause little or no soil shattering in wet soil. Soil water content has little effect on ripping of hardpans (rock).

Trenching machines have been used on limited areas, mostly for research, to completely homogenize the soil profile to the depth of operation. After removal and mixing by the machines, the soil is returned to the trench in a secondary operation. Trenching machines can operate within a wide range of soil water contents. Operation at the mid-plastic range, however, seems desirable because it avoids the possibility of puddling, which may occur in wet soil, and the need for excessive power, which would be needed in dry soil.

For installing subsurface barriers that restrict deep percolation of water, as in sandy soils, machines have been developed which undercut the surface, lift the soil, and spray asphalt or similar substances on the soil at the depth of operation (Erickson et al., 1968; Saxena et al., 1973). Operation at the upper-plastic limit decreases power requirements and the possibility of soil shattering, which would cause a roughened, uneven zone where the asphalt is applied. Such conditions would cause an irregular barrier, which is less effective than a smooth, even barrier.

**3.4.2 Equipment Needs**

Equipment is available for performing almost any reasonable root zone modification operation, as discussed in the foregoing section. Hence, there appears to be little need for developing additional equipment, but some equipment modifications are needed. We also need a better definition of the extent of root zone modification necessary for optimum soil water use by plants. Modifying a portion of the total area might be as effective as complete modification. This is now done to a limited extent by chiseling, subsoiling, and ripping under the planted row and by slip plowing at wide spacings. Reducing the areal extent would reduce the high energy requirement of most deep root zone modification operations.

Deep plowing with disk and moldboard plows results in rough, ridged surfaces (Fig. 3.4-1 and 3.4-2) that require considerable amounts of secondary tillage to obtain suitable seedbeds. Development of accessory equipment on the plows to do some of the smoothing would reduce the need for secondary tillage. While this would increase the power needed for deep plowing, total power input would be less than where deep plowing is followed by several separate operations. Possibilities for obtaining improved surface conditions after deep plowing include use of a blade at the surface to laterally

FIG. 3.4-2 Moldboard plowing Pullman clay loam to the 0.9 m depth at Bushland, TX.

TABLE 3.4-2. ENERGY USED FOR DEEP PLOWING IN
ISRAEL (from Hadas et al., 1978)

| Site | Soil | Operation and soil condition | Depth, cm | Energy input/unit area, $kg/m/m^2$ |
|------|------|------------------------------|-----------|------------------------------------|
| Gevim | Sandy clay loam | Deep plow (dry) | 39-42 | 4,660 |
| | | Deep plow (moist) | 39-42 | 4,330 |
| | | Subsoil (dry) | 45-50 | 2,100 |
| | | Vibrating subsoiler (dry) | 27-31 | 1,920 |
| Revadim | Clay | Deep plow (dry) | 35-40 | 4,190 |
| | | Deep plow (moist) | 35-40 | 1,870 |
| | | Subsoil (dry) | 38-43 | 2,210 |
| Gan Shemuel | Clay loam | Deep plow (dry) | 35-37 | 4,420 |
| | | Deep plow (moist) | 35-37 | 1,900 |
| | | Subsoil (dry) | 39-42 | 2,060 |

shift the soil to reduce ridge height, use of twisted shanks in the plow layer to more thoroughly mix and spread the loosened soil, and redesign of plows to improve mixing and spreading of the loosened soil.

### 3.4.3 Economics of Different Systems

Root zone modification operations are energy intensive and costly. In determining their economic feasibility, their costs must be weighed against the expected benefits of the operations. When benefits from the operations occur for several years, the costs can be prorated over the period for which benefits are expected.

Costs of root zone modification operations vary widely and are strongly influenced by type and depth of operation, soil type, and soil water content. However, little comparative information is available regarding costs under varying conditions. A comparison involving energy input per unit area would be most desirable, but such information is too limited to warrant this type of analysis. Therefore, we will present a few examples to illustrate probable costs or energy inputs for performing various types of operations.

Based on 1977 values in California, subsoiling cost $50/ha (Kaddah, 1977). Relative costs were 1.0, 1.2, and 3.4, respectively, for subsoiling (at 1-m centers), slip plowing (at 2-m centers), and moldboard plowing, all at the 0.9-m depth. The most costly operations resulted in higher yields, but inadequate information was given for making a valid cost-return analysis.

Hadas et al. (1978) measured energy inputs for deep plowing at three locations in Israel (Table 3.4-2). Soil water content and the implement used had a major effect on energy input. However, differences among sites for the same implements were slight, except for deep plowing in moist soil.

Several investigators gave information concerning the power sources used for deep plowing. Hadas et al. (1978) used tractors having power ratings from 37 x 10³ to 86 x 10³ W (50 to 115 drawbar hp). Pullman clay loam at Bushland, TX, was moldboard plowed 0.8 m deep with two tractors having a combined power rating of 205 x 10³ W (275 hp) (Fig. 3.4-2) (personal communication, R. R. Allen, Bushland, TX). Similar power sources were recommended for disk and moldboard plowing 0.6 m deep in North Dakota (Sandoval and Jacober, 1977). For sandy soils in Oklahoma, 97 x 10³ W (130

FIG. 3.4-3 Tractor-mounted ripper used to loosen hardpans in California soils. Depending on conditions, one, two, or three shanks are used.

FIG. 3.4-4 Four tractors operating as a unit to deeply rip California soils using a ripper similar to the one shown in Fig. 3.4-3.

hp) tractors were recommended for moldboard and disk plowing 0.5 to 1.1 m deep (Harper and Brensing, 1950). In each situation, the energy input per unit would be strongly influenced by conditions (depth, soil water, etc.) prevailing at the time of operation.

In 1973, the cost of installing an asphalt barrier at the 0.6-m depth in sandy soil in Florida was estimated to be about $618/ha (Saxena et al., 1973). No power requirements were given, but a tractor with a power rating similar to those recommended for deep plowing sandy soils in Oklahoma undoubtedly would be needed.

A very intensive type of root zone modification is practiced on some soils in California where a hardpan greatly restricts water penetration and root development. To fracture the indurated hardpan, the soil is ripped to depths of 1.5 to 2.1 m at 0.9- to 1.2-m spacings. The ripper (Fig. 3.4-3) is pulled and pushed by up to four tractors operated as one unit (Fig. 3.4-4) having a combined power rating of over $1,100 \times 10^3$ W (1,500 hp) (Anon, 1975). Land modification rates ranged from 8 ha/day (600 acres/30 days) (Anon., 1975) to 12 to 16 ha/day (30 to 40 acres/day) (comments by ripping contractor). All rates are based on 24 h/day of operation. Cost of ripping and secondary operations for preparing the land for crops was about $1,235/ha ($500/acre) in November 1978.

## 3.5 GENERAL SUMMARY

Crop production in subhumid and semiarid regions often is low because of low and erratic precipitation and because the precipitation is not effectively stored in soil or subsequently removed from the soil by crops. Even in higher precipitation areas, plants experience stress during short-term droughts because of low soil water content or the inability of plant roots to grow into soil zones containing water. Although irrigation reduces plant dependence on precipitation, water available for irrigation is often limited, thus the available water should be used as efficiently as possible.

Soil conditions that reduce water infiltration, distribution and retention in the profile, and availability to plants, and that restrict root growth and proliferation within the profile affect the efficiency with which water is used by plants. Some conditions that adversely affect soil water relations and root growth are slowly permeable horizons, claypans, dense clays, fragipans, and plowpans. Water availability and root exploration may also be restricted on poorly-drained and salt-affected soils. Low water content often is a problem on deep sandy soils.

Numerous deep tillage and profile modification studies have been conducted to improve crop performance on soils with these problems. In general, root zone modification improved soil water relations and crop yields when the operations increased the soil water supply for plant growth. The added water was made available through increased storage of rainfall and irrigation water or through increased root penetration and proliferation. Degree of improvement generally increased as the intensity and depth of disrupting the problem soil layers increased. For example, deep plowing or mixing a soil generally was more beneficial than chiseling or subsoiling. On soils with slight or no definite problems, root zone modification usually will not improve crop production.

Equipment is available for performing almost any reasonable type of root zone modification. Included are power sources for pulling various types of chisels, subsoilers, disk and moldboard plows, slip plows, and rippers. Common operating depths are 0.6 to 1.2 m, but slip plows and rippers have been used to depths of 2.1 m. Only general guides are available regarding the economics of root zone modification. If a known soil problem severely restricts crop production because of poor soil water relations, then root zone modification most likely will be beneficial and economical, provided the problem zone is effectively altered by the treatment. Cost of treatment increases with intensity of treatment, but more intensive treatments generally have longer-lasting effects. For example, plowing or mixing Pullman clay loam to 0.4- to 1.5-m depths is still effective more than 12 yr after the operations were performed. The costs of root zone modification can be prorated over the period that the treatments are beneficial. Soil modification is generally more economically feasible for high value than for low value crops. However, the nature of the problem, cost of operation, expected response, and crop value must be considered when contemplating a root zone modification operation.

### References

1  Allison, F. E. 1973. Soil organic matter and its role in crop production. Elsevier Scientific Publishing Co., Amsterdam, London, New York. p. 346-359, 417-444.

2  Anon. 1975. Ripping the unrippable. Reprinted from California Builder & Engineer. March 14, 1975. 4 pp.

3  Black, C. A. 1960. Soil-plant relationships. John Wiley and Sons, Inc., New York. p. 39-87.

4  Bradford, J. M., and R. W. Blanchar. 1977. Profile modification of a Fragiudalf to increase crop production. Soil Sci. Soc. Am. J. 41:127-131.

5  Bruce, R. R. 1960. Deep tillage of dry soil gives better returns. Mississippi Farm Research 23(10):1, 6.

6  Bruce, R. R., and T. N. Jones. 1963. Effects of soil chiseling on crop yields—Tests on moisture limits and depths of breaking hardpans. Mississippi Farm Research 26(2):2.

7  Burnett, E., G. F. Arkin, and D. L. Reddell. 1974. Deep tillage effects on crop physiological response to water deficits. ASAE Paper No. 74-2533, ASAE, St. Joseph, MI 49085.

8  Burnett, E., and V. L. Hauser. 1967. Deep tillage and soil-plant-water relationships. p. 47-52. In: Tillage for Greater Crop Production Conf. Proc., Dec. 1967. ASAE, St. Joseph, MI 49085.

9  Burnett, E., and J. L. Tackett. 1968. Effect of soil profile modification on plant root development. Int. Congr. Soil Sci. Trans. 9th (Adelaide, Australia) III:329-337.

10  Burstrom, H. 1965. The physiology of plant roots. In: K. F. Baker et al. (eds.) Ecology of soil-bourne plant pathogens. p. 154-169. Univ. California Press, Berkeley.

11  Cairns, R. R., and W. E. Bowser. 1977. Solonetzic soils and their management. Canada Dept. Agric. Pub. 1391. 37 pp.

12  Campbell, R. B., D. C. Reicosky, and C. W. Doty. 1974. Physical properties and tillage of Paleudults in the southeastern Coastal Plains. J. Soil Water Conserv. 29:220-224.

13  Cary, E. E., G. M. Horner, and S. J. Mech. 1967. Relationship of tillage and fertilization to the yield of alfalfa on Freeman silt loam. Agron. J. 59:165-168.

14  Denmead, O. T., and R. H. Shaw. 1962. Availability of soil water to plants as affected by soil moisture content and meteorological conditions. Agron. J. 54:385-389.

15  De Roo, H. C. 1961. Deep tillage and root growth. Connecticut Agric. Exp. Stn. Bull. 644. 48 pp.

16  Doty, C. W., R. B. Campbell, and D. C. Reicosky. 1975. Crop responses to chiseling and irrigation in soils with a compact A2 horizon. TRANSACTIONS of the ASAE 18(4):668-672.

17  Eck, H. V., and R. G. Davis. 1971. Profile modification and root yield, distribution, and activity. Agron. J. 63:934-937.

18  Eck, H. V., T. Martinez, and G. C. Wilson. 1977. Alfalfa production on a profile modified slowly permeable soil. Soil Sci. Soc. Am. J. 41:1181-1186.

19  Eck, H. V., and H. M. Taylor. 1969. Profile modification of a slowly permeable soil. Soil Sci. Soc. Am. Proc. 33:779-783.

20  Erickson, A. E., C. M. Hansen, and A. J. M. Smucker. 1968. The influence of subsurface asphalt barriers on the water properties and the productivity of sand soils. Int. Congr. Soil. Sci. Trans. 9th (Adelaide, Australia) I:331-337.

21  Fehrenbacher, J. B., J. P. Vavra, and A. L. Long. 1958. Deep tillage and deep fertilization experiments on a claypan soil. Soil Sci. Soc. Am. Proc. 22:553-557.

22  Gardner, W. R. 1968. Availability and measurement of soil water. p 107-135. In: T. T. Kozlowski (ed.) Water deficits and plant growth, Vol. I, Development, control, and measurements. Academic Press, New York and London.

23  Grissom, P., E. B. Williamson, O. B. Wooten, F. E. Fulgham, and W. A. Raney. 1955. The influence of deep tillage on cotton production in the Yazoo-Mississippi Delta. Mississippi State College Agric. Exp. Sta. Info. Sheet 507.

24  Grissom, P., E. B. Williamson, O. B. Wooten, F. Fulgham, and W. A. Raney. 1956. 1955 results of subsoiling tests of Delta Station. Mississippi Farm Research 19(4):1, 8.

25  Hadas, A., D. Wolf, and I. Meirson. 1978. Tillage implements—soil structure relationships and their effects on crop stands. Soil Sci. Soc. Am. J. 42:632-637.

26  Haise, H. R., H. J. Haas, and L. R. Jensen. 1955. Moisture studies of some Great Plains soils: II. Field capacity as related to 1/3-atmosphere percentage, and "minimum point" as related to 15- and 26-atmosphere percentages. Soil Sci. Soc. Am. Proc. 15:20-25.

27  Harper, J., and O. H. Brensing. 1950. Deep plowing to improve sandy land. Oklahoma Agric. Exp. Stn. Bull. No. B-362. 28 pp.

28  Hauser, V. L., and H. M. Taylor. 1964. Evaluation of deep-tillage treatments on a slowly permeable soil. TRANSACTIONS of the ASAE 7(2):135-136, 141.

29  Heilman, M. D., and C. L. Gonzalez. 1973. Effect of narrow trenching in Harlingen clay soil on plant growth, rooting depth, and salinity. Agron. J. 65:816-819.

30  Hobbs, J. A., R. B. Herring, D. E. Peaslee, W. W. Harris, and G. E. Fairbanks. 1961. Deep-tillage effects on soils and crops. Agron. J. 53:313-136.

31  Hsiao, T. C. 1973. Plant responses to water stress. Ann. Rev. Plant Physiol. 24:519-570.

32  Jamison, V. C. 1953. Changes in air-water relationships due to structural improvement of soils. Soil Sci. 76:143-151.

33  Jamison, V. C., and J. F. Thornton. 1960. Results of deep fertilization and subsoiling on a claypan soil. Agron. J. 52:193-195.

34  Kaddah, M. T. 1976. Subsoil chiseling and slip plowing effects on soil properties and wheat grown on a stratified fine sandy soil. Agron. J. 68:36-39.

35  Kaddah, M. T. 1977. Conservation tillage in the southwest. p. 57-62. In: Conservation tillage: Problems and potentials. Spec. Publ. No. 20. Soil Conserv. Soc. Am., Ankeny, IA.

36  Kaspar, T. C., C. D. Stanley, and H. M. Taylor. 1978. Soybean root growth during the reproductive stages of development. Agron. J. 70:1105-1107.

37  Klepper, B., and H. M. Taylor. 1979. Limitations to current models describing water uptake by plant root systems. p. 51-65. In: J. L. Harley and R. Scott Russell (eds.) The soil-root interface. Academic Press, Inc., London.

38  Kramer, P. J. 1969. Plant and soil water relationships—A modern synthesis. McGraw-Hill Book Co., New York. 482 pp.

39  Larson, W. E., W. G. Lovely, J. T. Pesek, and R. E. Burwell. 1960. Effects of subsoiling and deep fertilizer placement on yields of corn in Iowa and Illinois. Agron. J. 52:185-189.

40  Locke, L. F., H. V. Eck, B. A. Stewart, and H. J. Haas. 1960. Plowpan investigations at the Great Plains Field Stations, Woodward, Okla., and Mandan, N. Dak. USDA-ARS Prod. Res. Rep. No. 40. 33 pp.

41  Lyles, L., M. D. Heilman, and J. R. Thomas. 1963. Soil-mixing characteristics of three deep-tillage plows. J. Soil Water Conserv. 18:150-152.

42  Lyon, T. L., H. O. Buckman, and N. C. Brady. 1952. The nature and properties of soils (5th ed.). The MacMillan Co., New York. p. 182-217.

43  Mathers, A. C., G. C. Wilson, A. D. Schneider, and P. Scott. 1971. Sugarbeet response to deep tillage, nitrogen, and phosphorus on Pullman clay loam. Agron. J. 63:474-477.

44  McClure, J. W., and C. Harvey. 1962. Use of radiophosphorus in measuring root growth of sorghums. Agron. J. 54:457-459.

45  Mech, S. J., G. M. Horner, L. M. Cox, and E. E. Cary. 1967. Soil profile modification by backhoe mixing and deep plowing. TRANSACTIONS of the ASAE 10(6):775-779.

46  Mengel, D. B., and S. A. Barber. 1974. Development and distribution of the corn root system under field conditions. Agron. J. 66:341-344.

47  Miller, D. E., and J. S. Aarstad. 1972. Effect of deep plowing on the physical characteristics of Hezel soil. Washington Agric. Exp. Stn. Circ. 556. 9 pp.

48  Musick, J. T., and D. A. Dusek. 1975. Deep tillage of graded-furrow-irrigated

Pullman clay loam. TRANSACTIONS of the ASAE 18(2):263-269.
    49   Musick, J. T., and W. H. Sletten. 1966. Grain sorghum irrigation-water management on Richfield and Pullman soils. TRANSACTIONS of the ASAE 9(3):369-371, 373.
    50   Newman, E. I. 1974. Root and soil water relations. p. 363-440. In: E. W. Carson (ed.) The plant root and its environment. University Press of Virginia, Charlottsville.
    51   Patrick, W. H., Jr., L. W. Sloane, and S. A. Phillips. 1959. Response of cotton and corn to deep placement of fertilizer and deep tillage. Soil Sci. Soc. Am. Proc. 23:307-310.
    52   Peele, T. C., O. W. Beale, and F. F. Lesesne. 1948. Irrigation requirements of South Carolina soils. AGRICULTURAL ENGINEERING 29(4):157-158, 161.
    53   Pearson, R. W. 1974. Significance of rooting pattern to crop production and some problems of root research. p. 247-267. In: E. W. Carson (ed.) The plant root and its environment. University Press of Virginia, Charlottsville.
    54   Peters, D. B. 1965. Water availability. p. 279-285. In: C. A. Black (ed.-in-chief) Methods of soil analysis, Part I, Physical and mineralogical properties, including statistics of measurement and sampling. Am. Soc. Agron., Madison, WI.
    55   Rakov, K., and H. V. Eck. 1975. Effect of degree of soil profile disruption on plant growth and soil water extraction. Soil Sci. Soc. Am. Proc. 39:744-746.
    56   Raney, W. A., P. H. Grissom, E. B. Williamson, O. B. Wooten, and T. N. Jones. 1954. The effects of deep breaking on Mississippi soils. Mississippi State College Agric. Exp. Sta. Info. Sheet 492.
    57   Rao, K. V. P., S. B. Varade, and H. K. Pande. 1972. Influence of subsurface barrier on growth, yield, nutrient uptake, and water requirement of rice (Oryza sativa). Agron. J. 64:578-580.
    58   Rasmussen, W. W., G. C. Lewis, and M. A. Fosberg. 1964. Improvement of the Chilcott-Sebree slick spot soils in Southwestern Idaho. USDA-ARS 41-91. 39 pp.
    59   Ritchie, J. T. 1974. Atmospheric and soil water influence on the plant water balance. Agric. Meteorol. 14:183-198.
    60   Ritchie, J. T., and W. R. Jordan. 1972. Dryland evaporative flux in a subhumid climate. IV. Relation to plant water status. Agron. J. 64:173-176.
    61   Robertson, L. S., A. E. Erickson, and C. M. Hansen. 1977. Tillage systems for Michigan soils and crops. Part I. Deep, primary, supplemental and no-till. Michigan Agric. Ext. Stn. Bull. E. 1041. 8 pp.
    62   Robertson, W. K., L. C. Hammond, G. K. Saxena, and H. W. Lundy. 1973. Influence of water management through irrigation and a subsurface asphalt layer on seasonal growth and nutrient uptake of corn. Agron. J. 65:866-870.
    63   Robertson, W. K., G. M. Volk, L. C. Hammond, and J. G. A. Fiskell. 1976. Soil profile modification studies in Florida. Soil and Crop Sci. Soc. Florida Proc. 35:144-150.
    64   Robinson, F. E., and J. N. Luthin. 1968. Slip plowing in non-stratified clay. California Agric. 22(11):8-9.
    65   Salter, P. J., and J. E. Goode. 1967. Crop responses to water at different stages of growth. Commonwealth Agric. Bur. Farnham Royal, Bucks, England. 246 pp.
    66   Sandoval, F. M. 1978. Deep plowing improves sodic claypan soils. North Dakota Agric. Stn. Rep. No. 919 (from March-April 1978 Farm Research 35(4):15-18).
    67   Sandoval, F. M., J. J. Bond, and G. A. Reichman. 1972. Deep plowing and chemical amendment effect on a sodic claypan soil. TRANSACTIONS of the ASAE 15(4):681-684, 687.
    68   Sandoval, F. M., and F. C. Jacober. 1977. Deep plowing—cure for sodic claypan. Crops Soils 29(7):9-10.
    69   Saveson, I. L., and Z. F. Lund. 1958. Deep tillage improves delta "Hot Spot" areas. Crops Soils 10(5):15.
    70   Saxena, G. K., L. C. Hammond, and H. W. Lundy. 1969. Yields and water-use efficiency of vegetables as influenced by a soil moisture barrier. Florida State Hort. Soc. Proc. 82:168-172.
    71   Saxena, G. K., L. C. Hammond, and H. W. Lundy. 1971. Effect of an asphalt barrier on soil water and on yields and water use by tomato and cabbage. J. Am. Soc. Hort. Sci. 92:218-222.
    72   Saxena, G. K., L. C. Hammond, and W. K. Robertson. 1973. Effects of subsurface asphalt layers on corn and tomato root systems. Agron. J. 65:191-194.
    73   Sivakumar, M. V. K., H. M. Taylor, and R. H. Shaw. 1977. Top and root relations of field-grown soybeans. Agron. J. 69:470-473.
    74   Slatyer, R. O. 1957. The influence of progressive increases in total soil moisture stress on transpiration, growth, and internal water relationships of plants. Australian J. Biol. Sci. 10:320-336.
    75   Slatyer, R. O. 1973. The effect of internal water status on plant growth, development,

and yield. p. 177-191. In: Plant response to climatic factors. UNESCO Proc. Uppsala Symp., 1970.

76   Smith, D. D. 1951. Subsoil conditioning on clay pans for water conservation. Agric. Eng. 31:503-505, 532.

77   Taylor, H. M. 1976. The effect of soil compaction on rooting patterns and water uptake of cotton. Proc. 7th Conf. Int. Soil Tillage Research Organization, Uppsala, Sweden. Paper No. 43. p. 1-4.

78   Taylor, H. M., and R. R. Bruce. 1968. Effects of soil strength on the root growth and crop yield in the southern United States. Int. Congr. Soil Sci. Trans. 9th. (Adelaide, Australia) I:803-811.

79   Taylor, H. M., and E. Burnett. 1964. Influence of soil strength on the root-growth habits of plants. Soil Sci. 98:174-180.

80   Taylor, H. M., and H. R. Gardner. 1963. Penetration of cotton seedling taproots as influenced by bulk density, moisture content and strength,of soil. Soil Sci. 96:153-156.

81   Taylor, H. M., and B. Klepper. 1974. Water relations of cotton. I. Root growth and water use as related to top growth and soil water content. Agron. J. 66:584-588.

82   Taylor, H. M., and B. Klepper. 1978. The role of rooting characteristics in the supply of water to plants. Adv. Agron. 30:99-128.

83   Taylor, H. M., L. F. Locke, and J. E. Box. 1964. Pans in southern Great Plains soils: III. Their effects on yield of cotton and grain sorghum. Agron. J. 56:542-545.

84   Taylor, H. M., C. E. Van Doren, C. L. Godfrey, and J. R. Coover. 1963. Soils of the Southwestern Great Plains Field Station. Texas Agric. Exp. Stn. Misc. Pub. No. 669. 14 pp.

85   Trouse, A. C. 1971. Soil conditions as they affect plant establishment, root development, and yield. p. 221-312. In: K. K. Barnes, W. M. Carleton, H. M. Taylor, R. I. Throckmorton, G. E. Vanden Berg (eds.) Compaction of agricultural soils. ASAE, St. Joseph, MI 49085.

86   Unger, P. W. 1970. Water relations of a profile-modified slowly permeable soil. Soil Sci. Soc. Am. Proc. 34:492-495.

87   Unger, P. W. 1975. Relationships between water retention, texture, density and organic matter content of west and south central Texas soils. Texas Agric. Exp. Stn. Misc. Pub. MP-1192C. 20 pp.

88   Unger, P. W. 1978. Effect of irrigation frequency and timing on sunflower growth and yield. Proc. 8th Int. Sunflower Conf., Minneapolis, MN, July 1978. p. 117-129.

89   Van Doren, D. M., Jr., and J. L. Haynes. 1961. Subsoil tillage on fragipan soils in Ohio. Ohio Agric. Exp. Stn. Res. Circ. 95. 4 pp.

90   Willardson, L. S., and M. T. Kaddah. 1969. Soil profile modification to improve leaching. ASAE Paper No. PCR-69-126, ASAE, St. Joseph, MI 49085.

91   Woodruff, C. M., and D. D. Smith. 1946. Subsoil shattering and subsoil liming for crop production on clay pan soils. Soil Sci. Soc. Am. Proc. 11:539-542.

# chapter 4

## CHAPTER 4— ALLEVIATING NUTRIENT STRESS

**4**

# 4

## ALLEVIATING NUTRIENT STRESS

by    M. E. Sumner, Professor and F. C. Boswell, Professor, Department of Agronomy, University of Georgia, Athens, GA 30602 and Experiment, GA 30212, U.S.A., respectively

## 4.1 INTRODUCTION

Soil has been defined as the unconsolidated mineral material on the immediate surface of the earth that serves as a natural medium for the growth of land plants (Anon., 1975). However, when soil management practices are imposed and thus influenced by genetic and environmental factors, many modifications occur. Modification effects on soil moisture, temperature, oxygen, chemical aspects and nutrient stress or toxicities may be immense with interactions occurring among these parameters. Other than interacting effects, the following discussion will be limited to root zone modification as related to alleviating nutrient stress with only a limited discussion of toxicities which are treated elsewhere (Chapter 8).

Tillage systems often modify root zones appreciably. Such operations are performed for a number of reasons such as to loosen soil for root penetration, bury the residue from previous crops, provide a suitable environment for seed, improve water infiltration, provide aeration, and in certain instances, control weeds. Tradition, esthetics and certain documented advantages have motivated growers to practice certain tillage and cultural practices that effect root zone modifications. Such approaches were more feasible when energy sources, from fossil fuel especially, were considered more abundant and economical. The energy use concept has changed drastically in recent years, especially for agricultural production. On-farm usage accounts for about 3 percent of the nation's energy with over half of this used for tillage (Anon., 1976; Reid, 1978). Because of energy shortages and thus increased costs and concern for soil conservation practices, minimum tillage systems or reduced tillage operations have received considerable attention in recent years (Adams et al., 1973; Bennett et al., 1973; Lutz and Lillard, 1973; Mock and Erbach, 1977; Shear and Moschler, 1969). These systems influence root zone modifications and may have a significant effect on nutrient stress.

Valid data about the influence of root zone modification on nutrient stress are difficult and costly to obtain. Between and within site heterogeniety and complex interactions between factors lead to difficulties in interpretation of data particularly if replications in time are not studied. Nevertheless it is possible to alter and, in certain instances, to overcome nutrient stress by root zone modification.

In order to introduce the techniques available to improve the fertility regime of a soil for alleviating nutrient stress, it is essential to first delve into the nature of the problem. To achieve this, one must have a working knowledge of the various equilibria in the soil which regulate the supply of nutrients to the plant root. Once this information has been assimilated, it becomes necessary to evaluate the precision and prognostic value of the methods which are available to assay the fertility status of a soil. This allows one to establish the reliability with which nutrient stress can be diagnosed in any particular case. Thereafter various approaches to alleviating a diagnosed nutrient stress can be explored to maximize crop response to the ameliorative treatments.

Much of the above procedure is easier stated than done. There are many problems and pitfalls in diagnosing the nature and severity of the problem let alone the difficulties of curing or circumventing the particular stress. Many treatises have been written on the subject of soil fertility and the diagnosis of nutrient stresses covering many tomes. It would be totally beyond the scope of the present chapter to delve at great depth into the processes governing the equilibria and the approaches for making diagnoses. Therefore, sufficient background material will be briefly outlined to enable a non-fertility specialist to obtain a reasonable grasp of the philosophy and principles involved. In addition, frequent citations of source material will be made so that the interested reader can explore the subject further.

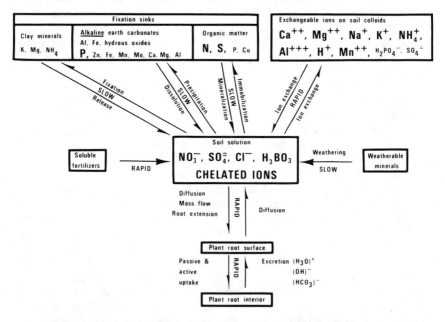

FIG. 4.2-1 Equilibria involved in the supply of nutrients to plant roots.

TABLE 4.2-1. COMPARISON OF THE AMOUNTS OF CATIONS PRESENT IN
EXCHANGEABLE AND SOLUTION FORM FOR A NUMBER OF SOIL ORDERS

| Soil order | Exchangeable cations | | | | Solution cations | | | | $\dfrac{\text{Solution}}{\text{Exchangeable}} \times 100$ | | | |
|---|---|---|---|---|---|---|---|---|---|---|---|---|
| | Ca | Mg | K | Na | Ca | Mg | K | Na | Ca | Mg | K | Na |
| | ---------------meq/100g----------- | | | | | | | | ---------------%---------- | | | |
| Oxisol | 1.3 | 1.7 | 0.5 | 0.1 | 0.009 | 0.016 | 0.010 | 0.007 | 0.7 | 0.9 | 2.0 | 7.0 |
| Ultisol | 3.8 | 3.9 | 0.3 | 0.2 | 0.011 | 0.028 | 0.005 | 0.015 | 0.3 | 0.7 | 1.7 | 7.5 |
| Alifisol | 8.7 | 5.9 | 1.0 | 0.1 | 0.016 | 0.024 | 0.016 | 0.014 | 0.2 | 0.4 | 1.6 | 14.0 |
| Vertisol | 13.5 | 10.4 | 0.4 | 0.2 | 0.036 | 0.057 | 0.003 | 0.026 | 0.3 | 0.6 | 0.8 | 13.0 |

Based on: le Roux, J. 1966. Studies on ionic equilibria in Natal soils. Ph.D. Thesis, University of Natal, Pietermaritzburg, South Africa.

In no way can this material be considered as an exhaustive in-depth treatment of the subject but rather a handbook for the scientist who is not a specialist in soil fertility and plant nutrition.

## 4.2 SOIL PLANT RELATIONSHIPS

In addition to being the rooting anchor for plants, soil supplies all the essential elements required by the crop with the exception of $CO_2$ and $O_2$ which come from the atmosphere. Because water, oxygen and lime are dealt with elsewhere, this discussion will be limited to the so-called essential mineral elements.

As a backdrop, the interaction between the solid and liquid phases in supplying essential nutrients from the soil to plant roots, is illustrated in Fig. 4.2-1. An attempt has been made to pictorially illustrate the various chemical equilibria and processes involved in the transfer of plant nutrients. In each category various essential elements are listed either in bold or regular print, the former indicating the dominant source of supply.

Because it is generally agreed that plants absorb most of their nutrients directly from the soil solution, this compartment should be the focus of the discussion. The concentration of the soil solution is always dilute, seldom exceeding 10 mM except under saline conditions. The soil solution is in dynamic equilibrium with the solid phase of the soil which represents the major storehouse for nutrients. This is illustrated in Table 4.2-1 which shows that only a very small percentage of the available pool of cations exists in the solution phase.

### 4.2.1 Nutrient Supply and Availability

To better comprehend the equilibria illustrated in Fig. 4.2-1, it is necessary to explore the concepts of availability and supply of nutrients to plants. The term availability is rather loose and ill-defined, but has been taken to mean the ease with which the plant is able to obtain a nutrient. For example, ions in the soil solution are readily available but the total quantity is low. Therefore, the continued uptake of a nutrient from the soil solution depends on the rate at which its concentration can be replenished from the storehouse located on the solid phase. It is generally true therefore, that the first increment of a nutrient taken up is more readily available than subsequent increments because the energy with which the solid phase holds nutrients usually increases as the quantity present decreases. This is a situation analogous to the relationship between the potential or free energy of water in a soil (availability) and the amount present (supply). It is well known that as the quantity of water in a soil decreases its availability also decreases. The same is essentially true for plant nutrients. Therefore, in order to adequately describe the nutrient status of a soil one needs ideally, to characterize the relationship between the chemical potential or free energy level of the nutrient in the soil solution (intensity factor), and the amount present on the solid phase (quantity factor). The ability of the system to replenish the soil solution is measured by the capacity factor which is the ratio of the change in quantity factor to unit change in intensity factor. Such a characterization is possible but often laborious and, at best, requires at least two analyses per sample; separate measures of the solution concentration of a particular nutrient as well as the amount which is labile are required.

Having a particular nutrient in the liquid phase is only part of the solution to the problem. It then has to be supplied to the root surface at a rate sufficient to satisy the needs of the plant. Nutrient stress would arise when this requirement is not fulfilled. Current knowledge (Barber, 1962, 1966, 1968; Barber et al., 1963) indicates that nutrients arrive at the root surface via one of the following mechanisms:

### 4.2.1 Mass Flow

This is the process whereby ions are swept along to the root surface in the water moving towards and into the root as a result of transpiration. The quantity of nutrients so transported can be calculated as the product of the volume of water transpired and the concentration of ions in the soil solution. There are a number of assumptions in such a calculation which may introduce errors; for example, the composition of the soil solution may and often does change with time. Therefore, a single measure of the soil solution concentration is unlikely to reflect the conditions over the entire growing season. In addition, transpiration is an intermittent process whereas nutrient uptake continues when transpiration is slow or absent. The reverse is often true for subsoils from which much water and little nutrient is absorbed. Not-

TABLE 4.2.2. SOIL SOLUTION COMPOSITION AND QUANTITIES OF NUTRIENTS
SUPPLIED TO CORN ROOT SURFACE BY MASS FLOW

| Soil order | Ca | Mg | K | Na | P | S | Ca kg | Ca % | Mg kg | Mg % | K kg | K % | P kg | P % | S kg | S % |
|---|---|---|---|---|---|---|---|---|---|---|---|---|---|---|---|---|
| | Soil solution composition | | | | | | Amounts and percentage of crop requirement reaching root in transpiration stream | | | | | | | | | |
| | ----------ppm---------- | | | | | | | | | | | | | | | |
| Oxisol | 4.4 | 4.6 | 9.8 | 3.7 | 0.007 | — | 18 | 68 | 18 | 59 | 39 | 22 | 0.03 | 0.08 | — | — |
| Ultisol | 7.2 | 10.8 | 5.9 | 13.0 | 0.02 | — | 29 | 111 | 43 | 139 | 23 | 13 | 0.08 | 0.22 | — | — |
| Alfisol | 223 | 43 | 33 | | 0.49 | 59 | 892 | 3430 | 172 | 555 | 132 | 75 | 1.96 | 5.44 | 236 | 874 |
| Vertisol | 80 | 35 | 15 | 26 | 0.12 | | 320 | 1231 | 140 | 452 | 60 | 34 | 0.48 | 1.33 | — | — |
| Entisol | 110 | 67 | 16 | | 0.49 | | 440 | 1692 | 268 | 865 | 64 | 36 | 1.96 | 5.44 | — | — |
| Mollisol | 196 | 65 | 27 | 23 | 0.29 | | 784 | 3015 | 260 | 839 | 108 | 61 | 1.16 | 3.22 | — | — |
| Aridisol | 208 | 55 | 82 | — | 0.59 | 49 | 832 | 3200 | 220 | 710 | 328 | 186 | 2.36 | 6.56 | 196 | 726 |
| Amount of nutrients taken up by 8000 kg/ha corn crop | | | | | | | 26 | — | 31 | | 176 | | 36 | — | 27 | — |

Based on: Burgess (1922), le Roux and Sumner (1967).

withstanding these difficulties, it is possible to obtain an approximation of the quantities of nutrients likely to arrive at the root surface by mass flow as the product of the volume of moisture transpired and the average composition of the soil solution. The results of such a calculation are presented in Table 4.2-2 for a number of soils from different soil orders. In order to carry out these calculations, it was necessary to assume that an 8000 kg/ha corn (*Zea mays* L.) crop transpired $4 \times 10^6$ L of water/ha and assimilated into its aerial parts 26 kg Ca, 31 kg Mg, 176 kg K, 36 kg P and 27 kg S during the growing season.

Mass flow as a means of ion transport to root surfaces is completely inadequate to account for the quantities of P and K required by a good corn crop. However, in most cases it can adequately supply the Ca, Mg and S needs of above ground portions of the crop with the exception of the highly leached and unlimed Oxisol. The data in Table 4.2-2 indicate that in the base rich soils, Ca and Mg are likely to accumulate around roots (Barber, 1974). It seems to be generally agreed that mass flow is an important mechanism in conducting $Ca^{++}$, $Mg^{++}$, $SO_4^{--}$, $H_3BO_3$, $Cu^{++}$, $Zn^{++}$, $Fe^{++}$, $Cl^-$ and $NO_3^-$ to root surfaces. However, for many of the other essential elements the process of diffusion is likely to be the dominant mechanism.

### 4.2.1.2 Diffusion

When the uptake of a nutrient by a root is faster than the supply to the root surface by mass flow, a concentration gradient is set up which causes diffusion of the nutrient from regions of high (bulk of soil) to low (root surfact) concentration. Diffusion is described by Fick's Law which, in the form suitable for application to soils, reads:

$$\frac{dQ}{dt} = - \epsilon aD \frac{\delta C}{\delta x} \quad \dots\dots\dots\dots\dots\dots\dots\dots\dots\dots\dots\dots\dots\dots\dots \text{[4.2-1]}$$

where

$dQ/dt$ = quantity of nutrient diffusing to the root in unit time
$D$ = diffusion coefficient
$\delta C/\delta x$ = concentration gradient of nutrient
$\epsilon$ = fractional porosity of the soil filled with water through which the nutrient diffuses; it also includes a tortuosity factor which takes into account the fact that soil pores are not uniform straight parallel tubes but are interconnected tortuous passages.
$a$ = cross sectional area which is assumed to represent the total root surface available for absorption.

The degree to which a concentration gradient forms depends on such factors as the rate of nutrient uptake, the diffusion coefficient of the particular ion, the relative contribution of mass flow in moving the nutrient to the root surface and the capacity of the soil to replenish the pool from which the nutrient is diffusing. The flux of nutrients is highly dependent on the moisture content of the soil as reflected by the fractional porosity and tortuosity factor in the above equation. All other things being equal, the flux decreases with decreasing moisture content. At the same soil water potential, diffusion would be faster in fine than coarse textured soils because of their greater water holding capacities. The total absorption surface of the root system is important in diffusion controlled processes. The greater the total absorbing surface, the slower the rate per unit surface can be in order for the same quantity of nutrient to be absorbed.

For a given set of environmental conditions, a diffusion volume around the root exists from which ions are just satisfying the plant's demand by diffusion. This diffusion volume is a function of the diffusion coefficient of the ion in the soil system, the concentration gradient and the capacity of the soil to maintain the concentration gradient. The diffusion volume is inversely proportional to rate of nutrient requirement of the plant since, if the rate is low, there is more time for diffusion and ions can diffuse greater distances to meet the needs of the plant.

### 4.2.1.3 Root Effects

When mass flow is unable to meet demand, the extension of a root into new, undepleted soil areas will increase the rate of nutrient supply. This situation represents an increase in rooting volume from which nutrients are able to diffuse to the root surface.

The size and morphology of a root system can have a profound effect on the extraction of nutrients from soil. For a given mass of root tissue, long, thin roots have a larger surface area than short thick roots and would be expected to explore the same soil volume more effectively by reducing the mean diffusion pathway through the soil. For single roots considered in isolation the resistance of the soil to ion transfer is relatively much more important in nutrient uptake than for whole root systems in which rooting density has an overriding effect on the rate at which the soil is depleted. For whole root systems with high root density per unit volume of soil, the resistance to transfer is small except for the most immobile elements.

Root exudates of one kind or another are often excreted by roots into the soil. Suggestions have been made that these exudates can influence the form of combination and therefore the availability of certain nutrients in soils. Little positive evidence is available to document these claims. However, root exudates often form the energy source for microorganisms in the rhizosphere which may stimulate root extension as a result of growth substances secreted by the microbes. In addition, they may play some role in rendering relatively insoluble forms of plant nutrients more available to the root. Mycorrhizal associations between a fungus and a plant root are widespread in nature and appear to be important to the mineral nutrition of the plant. The improved nutritional status of the higher plant appears to be the result of increased exposed surface on the roots in the form of the fungal hyphae.

### 4.2.2 Definition of Problem

Having briefly discussed the concept of availability and the various mechanisms by which nutrients arrive at the root surface, we are now in a position to return to Fig. 4.2-1 in order to discuss, in a little more detail, the equilibria involved in supplying nutrients to plants. This will be done as a prelude to defining the problem which one faces in diagnosing whether or not nutrient stress exists in a particular case and which elements are responsible. Such a diagnosis is a necessary precursor to any ameliorative treatments required to remedy the stress.

As mentioned previously, all ions reaching the root surface pass through the soil solution which becomes depleted as uptake proceeds. Exchange sites which represent the major reservoir of ions such as Ca, Mg, K, and $NH_4$ are in very rapid equilibrium with the soil solution. Provided that the quantity of these ions is sufficiently high, transport problems seldom arise in supplying the needs of roots under normal conditions. Therefore, quantity parameters are often selected as preferred measures for these ions. Organic matter is often the major storehouse for N, S and some P which are incorporated into the structure in insoluble forms. The release of these elements from organic matter depends on its rate of decomposition which in some cases is too slow for an adequate supply rate. It is extremely difficult to assay accurately the quantities of these elements which are likely to be released to the soil solution during a season because mineralization is a function of many factors which vary widely over the growth of the crop. Most soils contain surfaces composed of either iron and aluminum hydrous oxides or alkaline earth carbonates or both. Such surfaces represent the major immobilization sinks for nutrients such as P, Zn, Fe, Mn, Mo, Ca, Mg, and Al. Consequently, problems of dissolution can sometimes form the rate limiting step in regulating the supply to the roots. Clay minerals such as the micas are a further storehouse for K and Mg. However, these ions are held in such an immobile form that it is doubtful whether quantities are substantial enough to sustain adequate growth. Soluble fertilizers represent the biggest single input to the system, usually entering the equilibria via the soil solution.

Both the introduction (fertilizers and crop residues) and removal (crop uptake and leaching) result in a disturbance of the equilibria which are constantly changing. Therefore, the picture presented in Fig. 4.2-1 should be considered as a dynamic and not static equilibrium.

Because the system must remain electrically neutral the plant must either take up anions and cations in stoichiometrically equivalent quantities or excrete ions of the same charge as the ion being absorbed. These excretions usually take the form of $H_3O^+$, $HCO_3^-$, and $OH^-$ but because plants usually take up greater quantities of cations than anions, soils under cropping usually become more acid with time. A discussion of the mechanisms of nutrient uptake is beyond the scope of this chapter. The interested reader is referred to some of the many excellent reviews on this topic (Barley, 1970; Russell, 1978).

The above discussion has been rather too simplistic; the real situation in the field is exceedingly complex and difficult to characterize. The problem facing the specialist in soil fertility and plant nutrition is to characterize the complex state of affairs illustrated in Fig. 4.2-1 by means of inexpensive, rapid, yet reliable tests for the nutrient status of the soil before the crop is planted. Chemical tests which are capable of reflecting the supply pattern of a soil integrated over the entire growing season are required. This requirement is difficult to completely fulfill by analyzing a single soil sample taken before the crop is planted. However it is amazing how well soil tests do reflect the position in spite of the complexity and problems involved. The extent to which success in this direction can be expected will be evaluated in the following section.

## 4.3 DIAGNOSIS OF NUTRIENT STRESS

Having discussed the reactions and equilibria which govern the supply of nutrients to the root, the question now arises: "how can the fertility status of the soil be assessed in order to estimate the quantity of fertilizer or other treatment required to remedy any nutrient stress?"

The yield and quality of a crop, and by implication the growth and performance of its root system, are the resultants of the efficiency with which vital biochemical processes take place within the plant cells. These processes which result in dry matter accumulation and thus yield, are dependent on various environmental (radiation, temperature, rainfall, day length, soil physical and chemical properties, parasites and symbiotic organisms), cultural (tillage, herbicides, irrigation, fertilizer and lime, population) and genetic (cultivar) factors over which man may or may not have some degree of control. The interrelationships between these factors and their effect on crop growth need to be calibrated in order to make meaningful and reliable diagnoses of the fertilizer and other treatments required to increase the chances of obtaining a high yield or improved quality.

Cultural practices are in a large measure controllable by man; nevertheless they can and do influence the nutrition of the crop and consequently the recommendations which are made. For example, broadcasting or banding phosphate at planting influences the crop differently and should be taken into account in making recommendations. Similarly, whether the crop is to be irrigated or not influences the potential yield level which has a bearing on the quantity of nutrients to be applied. Selection of a suitable adapted cultivar is essential to good results. Man has some control in this regard in being able to breed adapted, disease and insect resistant varieties. The special nutrient needs of a particular cultivar must be taken into account in making recommendations. Because crops differ widely in nutrient requirements and ability to forage for nutrients the type of crop to be grown is an important consideration. As far as the soil is conserned, man can change some of its chemical and biological properties rather easily by addition of fertilizers, fungicides, nematicides and insecticides; but the physical properties are much more difficult to alter. Thus, in making recommendations, one must bear in mind the limitations to growth imposed by soil physical defects which are not readily remedied. This, in turn, dictates the level of fertilization to which economic yield response will be obtained. Environmental factors such as radiation, temperature, rainfall, etc. are not readily controllable in a field situation. Nevertheless, they must be taken into consideration because they are, or ultimately become, the most limiting factors in relation to crop productivity in mnay situations and therefore can influence the reaction of the crop to applied fertilizer.

Thus, assessing the nutrient requirements of a crop is therefore only one facet of improving productivity under a given set of often variable circumstances. Although not always possible, diagnosis of nutrient stress and its correction should be considered in the context of the whole dynamic soil-plant-atmosphere continuum. Because of the complexity of this system the greater the number of meaningful and calibrated parameters available from which to make a diagnosis, the higher will be the chances of making an appropriate recommendation likely to result in a yield increase.

The nutrient status of a particular production field is usually assayed by means of a soil test and/or tissue analysis. The following discussion will be limited to making diagnoses of nutrient stress in field situations where root zone modification can be effected.

### 4.3.1 Soil Testing

In addition to the actual chemical analysis of soil, the term soil testing in

its broadest context and commonest usage refers to interpretation, evaluation and fertilizer recommendation based on the analysis. It is unfortunate
that this broadening of context has occurred because it is important to dif-
ferentiate between diagnosis and recommendation as they are not uniquely
linked. On the basis of the analytical data a diagnosis of the quantity of
nutrients required is made but other factors such as soil physical properties,
the presence or absence of irrigation, economic considerations, etc. determine the form of fertilizer to be used as well as the method of application and
incorporation.

Approximately 14 elements essential for plant growth are supplied from
the soil. However, until very recently soils were usually only tested for P, K,
Ca, Mg and acidity because these are often the most frequent limiting factors
in many soils. Soils are seldom tested for N because of problems in developing a rapid, convenient and meaningful test to measure the quantity of N
likely to be mineralized from the organic fraction of the soil over the growing
season. Recently, certain laboratories have instituted tests for minor and
secondary elements but their reliability and relationship to proven yield
responses is usually inferior to that of the major elements on which most attention has been focused.

A soil testing program is usually composed of four phases, namely,
sampling, analysis, interpretation of analytical results and recommendations
of ameliorative fertilizers.

### 4.3.1.1 Sampling

In general, it would be true to say that at least one composite soil sample
(25 to 30 cores) should be taken to represent an area which can or is likely to
be treated as a separate unit. This may be a part or an entire field. Fields in
which fertilizers have been band placed for a number of years present particular problems in sampling due to the heterogeneity so induced. The only
effective way of overcoming this difficulty is to increase the number of cores
taken but this is seldom practised because of the substantially greater effort
required. Therefore, the value of analytical data from such inadequately
sampled fields is questionable. In sampling, one should bear in mind that the
final few grams of sample which are chemically assayed for available
nutrients must be representative of the field from which the sample
originated. Therefore, extreme care should be exercised in taking a representative sample, in reducing sample volume before analysis and preventing
contamination from other sources such as fertilizer, etc. A soil test is no better than the final sample used for analysis!

The depth to which soil should be sampled is an important consideration. Most often only topsoils are sampled; however, in an increasing number
of cases, the need for deeper samples has become apparent. Subsoil sampling
is much more laborious and unless clear benefits are likely to emanate the
practice is usually avoided. Nevertheless, in the context of this treatise subsoil sampling can often shed light on rooting problems arising from nutrient
stress or chemical toxicities in the lower horizons of the soil. In cases of
duplex soils where the subsoil contains greater quantities of clay and a
greater accumulation of nutrients than the sandy topsoil, subsoil sampling
can be useful in explaining the lack of response to fertilizer additions based
on topsoil samples alone. Therefore, soil sampling should be undertaken in
such a way as to ensure that the rooting zone is adequately characterized in

terms of the quantities of nutrients likely to become available to the plant over the growing season.

### 4.3.1.2 Soil Analysis

Analyses of samples in a soil testing program attempt to measure the quantities of nutrients likely to enter the labile pool available to the crop during the season. The term "available" is not readily covered by any single definition but because the reference organism in this case is the plant, only it alone can really decide what nutrient is available. In assessing availability, soil test procedures achieve various degrees of success. Over the past century, soil scientists have been seeking the "universal" extractant which will simulate the growing plant but with little success. All extractants yield empirical results and are of little value in themselves. To become useful they must be calibrated in terms of the yield response to fertilizer levels obtained in well defined experiments. Correlation is the key to a successful soil testing program. Any soil test method for a particular nutrient must at least preferentially assay that fraction which is relatively labile and likely to enter the pool from which the plant feeds during the growing season.

The extractants used in most soil testing programs are either dilute acids or strong salt solutions which tend to extract that fraction of a particular nutrient which is highly correlated with the quantity factor mentioned above in section 4.2.1. In general, they do not give a measure of the intensity factor (level of availability) which can sometimes result in poor prognoses. For example, if the same norms are used for soils of variable amount and type of clay, problems can arise. Because at a given soil test value for K, a loam has a higher intensity level than a clay soil with similar mineralogy, the availability of K to the plant would be lower with the clay and stress could result whereas on the loam, plant growth might be adequate. This problem is often overcome by establishing separate calibrations for soils of different textures. Where the soil texture is not taken into account, as frequently occurs, some error can be encountered in the amount of nutrient recommended to alleviate an insufficiency.

### 4.3.1.3 Interpretation and Recommendation

The interpretation of the analytical results of a soil test depends very heavily on the background information used to calibrate the soil test in terms of crop yield. In the case of many soil testing laboratories, calibration of the various empirical procedures has been conducted and often updated. Nevertheless, this is not a universal truth. Therefore, before relying heavily on the recommendations of a soil test program, particularly for high yield crops, one should ascertain whether or not adequate calibration has been conducted in establishing the norms. In the past few years relatively little correlation work has been conducted and it would appear that present calibrations in many instances are not sufficiently precise for use with confidence at high yield levels. Work on soil test calibration needs to be continuously conducted in order to maintain and enhance the confidence in these programs.

Before briefly outlining how an interpretation or diagnosis is made on a given sample from a farmer's field, it should be clearly understood that the final recommendation is based on the interpretation of data from a single sample. If the sample does not represent the field, positive yield responses to an applied fertilizer can hardly be expected because there is no assurance

TABLE 4.3-1. HYPOTHETICAL SOIL TEST
RECOMMENDATIONS FOR CORN AND
ALFALFA AT VARIOUS SOIL TEST
RATINGS FOR P AND K

| Soil test rating | Nutrient required | | | |
|---|---|---|---|---|
| | P | | K | |
| | corn | alfalfa | corn | alfalfa |
| | - - - - - - - - - - - -kg/ha - - - - - - - - - - - | | | |
| low | 40 | 45 | 80 | 225 |
| medium | 20 | 34 | 40 | 180 |
| high | 10 | 24 | 20 | 135 |
| very high | 0 | 0 | 0 | 0 |

that a correct diagnosis was made in the first place. Because analyses are usually not conducted in duplicate, analytical errors may contribute to an incorrect diagnosis. Furthermore, other factors such as season, weeds, diseases, water stress, applied pesticides, etc. may influence the response of the crop to the applied fertilizer in such a way that a yield increase is not observed. Therefore, the diagnosis and subsequent recommendation does little more than increase the chances of obtaining an improved yield if all other factors are not limiting.

Usually the soil test norms are categorized as low, medium, high and very high with the chances of obtaining a positive response to the particular nutrient decreasing in this order. By comparing the analysis of the farmer's sample with the norm, one establishes the soil test rating which usually cor-

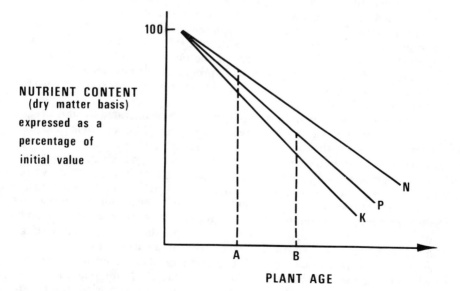

FIG. 4.3-1 Diagrammatic representation of the variation in nutrient content of tissue with plant age.

responds with the recommendation of a specific amount of nutrient to be added to a particular crop. This is illustrated by the hypothetical example in Table 4.3-1 which gives the quantities of P and K required to be added in order to grow corn for grain and alfalfa (*Medicago sativa* L.) for hay starting with different initial soil test ratings. There are substantial differences in the fertilizer programs recommended for the two crops at the same soil test rating which largely reflect the differences in their nutritional requirements.

Because each element in the above example is considered in isolation, synergistic and antagonistic effects between nutrients are neglected and the importance of nutritional balance is not taken into consideration. Systems have been proposed and are in use which strive to obtain a balance between nutrients in the soil. These systems usually attempt to bring the nutrients in the soil within ranges such that a favorable balance is maintained. For the cations such as Ca, Mg, K and Na, a range of percentage saturations of the cation exchange capacity is often used. This approach may be somewhat better than the former but usually has the disadvantage that additional determinations are required in order to calculate the percentage saturations.

It is impossible in a short discussion such as this to present soil test norms for various regions, soils and crops. Therefore, interested parties are referred to two excellent texts which give a comprehensive discussion of the topic as well as presenting useful interpretive norms (FAO, 1973; Walsh and Beaton, 1973). In addition, state and private soil testing laboratories have bulletins and other material available which can be invaluable.

### 4.3.1.4 Utility of Information Obtained from Soil Tests

In addition to the problems mentioned above, soil testing programs are constrained by the fact that soils vary widely from region to region. Therefore, soil test calibrations made in one area can seldom be applied to another where different soils occur. Thus, soil test programs with a single set of calibrated norms are only valid over relatively limited areas.

Furthermore, because in many programs only a few of the essential elements are assayed, the value of soil testing is limited in that meaningful recommendations can only be made for the nutrients considered. Recommendations for other nutrients in such programs are either based on experience or guesswork and invariably few data to substantiate the recommendations are available.

In many soil testing programs, particularly those operated by state agencies, computers are used to interpret the analytical data and make recommendations. While this is often necessary by virtue of the sheer volume of samples processed in a year, it is far from desirable as it is virtually impossible to program a computer with the experience of a trained soil fertility extension specialist who is able to bring many factors to bear on the recommendation. In addition, a visit to the field in question is of immense importance in tailoring a recommendation to the requirements of a particular farm and farmer. It is in this respect that many reputable private soil testing laboratories are able to offer a better rounded service to the farmer. In general, private testing laboratories maintain a file on each farmer and in so doing are able to chart the changes in fertility status over the years. In addition to being a built-in check on analytical and sampling precision, this trend with time is invaluable in fine tuning recommendations particularly if yield data for the field under consideration are available. Because of the way state testing laboratories are structured, these benefits can seldom be attained.

As mentioned earlier, soil test variables are only a few of the parameters which determine crop yield. Therefore, if additional information is available and can be used in making the diagnosis and subsequent recommendation, better interpretation may result. Such complementary information can be supplied by plant and tissue analysis which will be discussed below. In general, soil test levels indicate the amounts of nutrients available to the plant, whereas plant analysis reflects the extent to which the plant has been able to utilize the proffered supply.

### 4.3.2 Plant Analysis

The analysis of various tissues from crop plants has been used for many years as an aid in determining the nutrient status of the plant. As yield levels and associated technology continue to increase, diagnostic procedures based on soil testing will be progressively complemented by plant analysis. With the advent of emission and x-ray spectrographs capable of analyzing large numbers of tissue samples for many essential elements simultaneously, the cost and labor involved in plant analysis has sharply decreased making plant analysis a much more attractive criterion for diagnostic purposes.

There are two categories of plant analysis, namely, total which measures the total element content and partial which measures the unassimilated, soluble elements in the cell sap. Both approaches are extensively used in diagnostic work the former being exclusively a laboratory assay while the latter is usually used in the field as a rapid test on fresh tissue.

Plant analyses are conducted for a number of reasons such as investigating nutritional disorders, interpreting the results of field experiments and diagnosing insufficiencies, excesses and imbalances for fertilizer advisory purposes. The following discussion will be confined the the latter purpose.

Soil testing and plant analysis are complementary tools which, when possible, should be used together to obtain the maximum amount of information in order to formulate a diagnosis. Soil is usually tested before the crop is planted while plant analysis can be used to monitor the crop during the growing season to uncover incipient insufficiencies or imbalances which might be corrected. In addition, plant analysis conducted late in the season can also be used as a guide to the next season's fertilization provided it is accompanied by a soil test. It should be clearly understood that while plant analysis is capable of making a valid diagnosis of the nutrients limiting growth, the results of a soil test are usually required to determine whether elements diagnosed as being limiting in the tissue are, in fact, in short supply in the soil or whether the insufficiency in the tissue might have been induced by some other factor. Furthermore, the soil test is required to determine the quantity of nutrient which must be added. In cases where soil tests are not available for all essential elements, diagnoses based on plant analysis alone can be used to formulate a fertilizer program on the premise that what is lacking in the plant is also lacking in the soil. This is not always the case and care should be exercised in this regard. Many cases have been documented where leaf analysis indicates that a nutrient is deficient but the cause has not been a lack of the nutrient in the soil but rather some other factor such as drought, nematodes, soil acidity, etc.

Tissue tests in the field assay the levels of soluble nutrients in the cell sap. Because the soluble nutrients are usually vitally involved in cell chemistry, they are probably the form of nutrient in the plant which is most likely to be related to net photosynthetic activity and thus yield. However, tissue testing can only be carried out on fresh material which places a serious constraint on the technique in that the test system must be such that instant results can be obtained in the field. Many such systems have been devised and the interested reader is referred to a review by Syltie et al. (1972). In general, the tests are based on colorimetric methods for freshly expressed cell sap, the results being qualitative or semi-quantitative in nature. Although the actual testing in the field is relatively simple only requiring test papers, powders and standard color charts, the interpretation of the semi-quantitative results is much more difficult relying heavily on the experience of the person conducting the test. The reason for this difficulty lies in the lack of a calibrated set of norms for all species which can easily be consulted in the field. Nevertheless, rapid kit tests have considerable merit in trouble shooting in the field particularly when growth in part of a field is inferior. In such a case, comparisons between good and poor areas can often identify the particular nutrient causing the stress. They have the distinct advantage over total tissue analysis in producing instant results which can be used for formulating corrective strategy immediately. In addition, they assay the most labile fraction of nutrient in the plant whereas total analysis often tends to dilute the active fraction with large quantities of relatively inert nutrient already immobilized in some sink from which return is slow or nonexistent. An example would be Ca and Mg where only a relatively small proportion of the total amount present is in soluble form at any one time.

In spite of this very obvious advantage tissue testing has not found wide favor with diagnosticians who have generally preferred total plant analysis because the data are more reproducible and easily manipulated. Therefore, the remainder of this discussion will be confined to total plant analysis.

Plant analysis involves various steps such as sampling, sample preparation and analysis and finally interpretation of the results.

### 4.3.2.1 Sampling

Sampling procedures for different crops are quite varied depending largely on the nature of the species involved. As in soil sampling the importance of obtaining a representative sample cannot be over-emphasized. In general, at least 25 to 30 plants should be sampled in order to represent a particular field. For diagnostic purposes leaves and/or petioles have been favored as the tissue to be sampled although in certain circumstances even whole young plants can be used. Care should be exercised in sampling tissue which is as free from disease or insect damage as the field to be represented. Because the concentration of nutrients in tissue changes with age and position on the plant, samples of a specific diagnostic tissue should be taken at the same physiological stage or age as were the samples used to develop the comparable diagnostic norms. In general, the plant part sampled for diagnostic purposes has been selected on the basis of its ease of identification rather than on the basis of the relevance of its composition to yield. Although it is unfortunate that the most informative tissues have not always been selected, one cannot now readily change because of the large amount of useful information already collected.

### 4.3.2.2 Sample Preparation and Analysis

Because this facet of the diagnostic process is usually handled by a laboratory well versed in plant analysis, only a brief summary of this part of the procedure will be presented.

The tissue sampled should be dried at a temperature of 60 to 70 °C as soon as practicable. Thereafter, it should be ground exercising precautions to minimize contamination from grinding mills, soil, etc. The tissue is then analyzed by standard wet chemical or spectrographic procedures. Should sample storage be necessary for subsequent analysis, the sample should be placed in a contaminant free container with air-tight fitted lid. Because the plant material is analyzed for total content the same results should be obtained irrespective of procedure adopted. This is a major advantage of total plant over soil analysis.

### 4.3.2.3 Interpretation of Results

Plant analysis can be interpreted in a number of ways using various criteria developed by a number of schools of thought on the subject. The most commonly used approach involves the use of critical, normal or threshold values and sufficiency ranges, while more recently developed systems, in which nutritional balance is central to the diagnosis, have gained some prominence. Before proceeding to outline how these approaches are used in a diagnosis, a short comment on the form of expression of the analytical results is pertinent.

Historically, the nutrient content of plant tissue has been mainly expressed as a percentage of the dry matter content. This form of expression continues to find favor despite the fact that the denominator usually increases markedly with age. As a result percent plant nutrient contents vary with the age of the plant (Fig. 4.3-1) which leads to complications in making diagnoses as will be discussed later. For most plant nutrients, contents expressed on a dry matter basis decrease with age. However, in some cases such as Ca and Mn which accumulate in tissues, contents can increase with age.

Other forms of expression have been used such as percentage expression on a fresh matter basis or ratios of nutrients expressed either in percentages or milliequivalents per 100 g. These ratios have the advantage that the influence of the changing dry matter content with age is largely eliminated. For example, N (%) and P (%) both have dry matter in the denominator whereas in the ratio $N/P = N/DM \times DM/P$, the dry matter component cancels out. The fresh matter form of expression reduces the variation in nutrient content with time because the moisture content of the tissue varies relatively less than dry matter during the season. Although the fresh matter form of expression is not convenient for every day use, ratios can readily be calculated from any data for plant analysis and their use should be encouraged.

In order to give the non-fertility specialist an understanding of how a crop responds to the level of nutrient in its tissue, a brief discussion of the general type of relationship between nutrient content in a tissue and yield is pertinent. A diagrammatic representation of this relationship is presented in Fig. 4.3-2. Zone A represents the yield increase obtained when the nutrient content increases from insufficient to adequate levels. With further increase a zone is reached where no further yield response is obtained (Zone B). Thereafter, at high nutrient contents, yield depressions are often observed (Zone C). The relative magnitudes and slopes of Zones A, B and C can vary with nutrient and species considered. In addition, synergistic or antagonistic effects of other nutrients can also affect the slope of the relationship. In Fig. 4.3-2, curve ABC might correspond to a situation where a small response is obtained to K when P is also limiting whereas curve AB'C might be obtained if the insufficiency of P was corrected before K was varied. For all the essential elements Zone A is usually found when the particular element is limiting while Zone B is usually present but its length is highly variable. In some cases Zone C is not observed. This diagram points to the sensitivity of crop yield to nutritional balance which is well documented and therefore should be considered in making diagnoses. In terms of nutrient balance an excess of one nutrient (Zone C) usually corresponds to an insufficiency of another (Zone A). It is also generally true that at low yield levels the nutrient content of a particular element in a tissue can vary much more widely than at higher yields (Sumner, 1977c).

### 4.3.2.3.1 Critical Value Approach

The critical value is defined as that level of nutrient in a plant tissue below which growth rate, yield or quality decline "significantly" which is taken to mean by 5 to 10 percent. This critical level is usually obtained by studying the yield response to varying an element which is limiting when all the others are present in adequate amounts. It assumes no interaction between nutrients whereas one knows that interactions, whether synergistic or antagonistic, do occur. From Fig. 4.3-2, it is clear that in the case of curve ABC when a nutrient other than that being considered becomes limiting a lower critical value would be obtained than when the level of the other limiting factor had been raised (curve AB'C).

Nevertheless, critical values have been and still are useful in making diagnoses provided that they are used within the limits set in obtaining the norms. This is illustrated by the data presented in Table 4.3-2 for N, P and K in corn sampled at different ages. Although only N, P and K are discussed here, it should be pointed out that the principles apply equally to all essential elements and that for many crops a full set of critical values are available (Walsh and Beaton, 1973). The critical values for corn ear leaf at silking published by Melsted et al. (1969) namely 3.0 percent for N, 0.25 percent for P and 1.9 percent for K will be used to make diagnoses. When the values in Table 4.3-2 are compared with the norms, a diagnosis before day 85 is not possible because the N, P and K contents of the leaves are all above the critical value. However, after day 85 one would diagnose N and K as deficient and after day 114, N, P and K. As pointed out above the critical values used in making the diagnoses were developed for corn at the silking stage which for the data in Table 4.2-2 occurred between day 79 and 88. From these data the importance of taking samples of age corresponding to the norms is clear because had the samples been taken earlier than day 79 no diagnosis would have been possible whereas if one had waited too long all elements might have been diagnosed as being deficient.

One way of overcoming this problem has been the establishment of sets of critical values for different stages of growth. However, this requires that the age of the crop be recorded when the sample is taken which is not always possible or convenient.

An additional disadvantage of the critical value approach is that it does not take nutritional balance into consideration which in many cases is just as

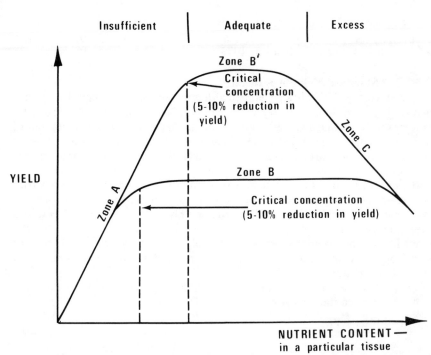

**FIG. 4.3-2 Diagrammatic representation of the relationship between yield and nutrient content of plant tissue.**

important as nutrient content. Ways of overcoming this difficulty are suggested later.

Sufficiency ranges corresponding to Zone B in Fig. 4.3-2 are often used for diagnostic purposes but have little or no advantage over critical values. If the nutrient content of a sample lies within the sufficiency range, it is deemed to be adequately nourished. However, because the ranges for each nutrient have been established independently, it is quite possible that the nutrition of a plant falling within the sufficiency ranges for all nutrients might be completely out of balance and likely to respond to improved nutrition. In other words, the diagnostic precision of the sufficiency range concept is poor (Sumner, 1979).

TABLE 4.3-2. EFFECT OF AGE OF PLANT SAMPLED ON NUTRIENT CONTENT OF FULLY DEVELOPED CORN LEAVES AND ON DIAGNOSIS OF ELEMENTS LIKELY TO BE LIMITING GROWTH

| Days after plant-ing | Leaf composition N,% | P,% | K,% | Elements diagnosed as deficient by critical value | Kenworthy N | P | K | Order | DRIS N | P | K | Order | DRIS using critical values N | P | K | Order |
|---|---|---|---|---|---|---|---|---|---|---|---|---|---|---|---|---|
| 48 | 3.40 | 0.36 | 2.2 | none | 111 | 132 | 104 | K > N > P* | -1 | 4 | - 3 | K > N > P | -10 | 14 | - 4 | N > K > P |
| 64 | 3.10 | 0.32 | 2.0 | none | 103 | 123 | 104 | N > K > P* | 0 | 3 | - 3 | K > N > P | - 9 | 13 | - 4 | N > K > P |
| 79 | 3.00 | 0.31 | 1.9 | none | 100 | 119 | 100 | K > N > P* | 0 | 3 | - 3 | K > N > P | - 8 | 13 | - 5 | N > K > P |
| 85 | 2.75 | 0.28 | 1.6 | N, K | 93 | 110 | 88 | K > N > P | 3 | 4 | - 7 | K > N > P | - 6 | 14 | - 8 | K > N > P |
| 99 | 2.70 | 0.29 | 1.5 | N, K | 92 | 113 | 82 | K > N > P | 2 | 9 | -10 | K > N > P | - 7 | 19 | -12 | K > N > P |
| 114 | 2.30 | 0.24 | 1.5 | N, P, K | 81 | 97 | 82 | N > K > P | -1 | 3 | - 2 | K > N > P | - 9 | 13 | - 4 | N > K > P |
| 133 | 2.00 | 0.20 | 1.3 | N, P, K | 72 | 84 | 75 | N > K > P | 1 | 1 | - 2 | K > N > P | - 7 | 10 | - 3 | N > K > P |

*All diagnosed as apparently more than adequate.

Critical values and sufficiency ranges useful in diagnostic work on many crops are presented in various publications (Chapman et al., 1966; Melsted et al., 1969; Walsh and Beaton, 1973).

#### 4.3.2.3.2 Standard Value Approach

Kenworthy (1973) has developed a diagnostic technique based on standard values for various fruit tree crops which takes nutrient balance into consideration. A standard value is defined as the mean of the nutrient content of a particular tissue, sampled at a particular stage from a population of crops having desirable horticultural performance. Presumably desirable in this context is taken to mean producing a high yield of good quality. In addition, sufficient background data must be available to determine the normal variation (coefficient of variation) for each element, specie, environment, etc.

In order to make a diagnosis by this method the leaf composition values are expressed as percentages of the standard values and adjusted for the variation in the composition of normal plants by using the coefficient of variation. This computation yields the balance index which is obtained from the following equations:

(i) If x the sample value is smaller than S the standard value, the balance index B is given by

$$B = \frac{100x + V(S-x)}{S} \quad \dots\dots\dots\dots\dots\dots\dots\dots\dots\dots\dots\dots\dots\dots\dots \quad [4.3\text{-}1]$$

(ii) If x is greater than S, then

$$B = \frac{100x - V(x-S)}{S} \quad \dots\dots\dots\dots\dots\dots\dots\dots\dots\dots\dots\dots\dots\dots \quad [4.3\text{-}2]$$

where V is the coefficient of variation.

The results for such calculations are presented in Table 4.3-2 under the heading of Kenworthy indices. For purely illustrative purposes the critical values presented above have been used as the standard values together with values of 16, 19 and 23 percent for the coefficients of variation for N, P and K respectively. Using these balance indices one is able to rank the nutrients in accordance with their deviation from the optimum value of 100. These indices also indicate the relative imbalance between the nutrients which would give the order of their requirement by the plant. Although it is possible to rank them in this order at any sampling date only diagnoses for the growth stage corresponding to that at which the standard values were derived can be considered to be appropriate and valid. While this approach takes balance into consideration, it is unable to diagnose insufficiencies over a range of tissue age. As a result diagnoses can only be made at growth stages for which standard values are available. This inflexibility is largely overcome by the approach to be discussed next. Although standard values for this approach have been published for a limited number of fruit crops, there seems to be no reason why well established critical values for other crops cannot be substituted for standard values in order to incorporate an element of balance into the critical value approach.

#### 4.3.2.3.3 The DRIS Approach

This approach originally developed by Beaufils (1957, 1959) as Physiological Diagnosis has been renamed the DRIS (Diagnosis and Recommendation Integrated System) approach (Beaufils, 1971, 1973) so that all factors which govern yield can ultimately be calibrated and incorporated into the system. However, this discussion will be limited exclusively to the aspect dealing with plant analysis. This approach also utilizes indices in order to make a diagnosis but differs from the Kenworthy approach in that valid diagnosis can be made over a wide range of tissue ages and the norms are derived in a different manner.

Because complete details of how the norms are derived have been published elsewhere (Beaufils, 1973; Sumner, 1977a, b, c) only a brief outline will be presented here. Basically, the norms are derived as the mean values for various forms of expression (e.g., N%, N/P, P/N, N/K, K/N, etc.) for a population of high yielding plants with which no nutritional or other visible fault can be found. Those forms of expression, which discriminate significantly between the low and high yielding sub-populations of plants, are used in a set of equations together with their respective coefficients of variation in order to calculate the indices. These indices which measure the relative deviations from the mean value and thus from the desired composition for a particular form of expression have either positive or negative signs corresponding to relative excesses or insufficiencies. Because they measure relative deviations from the same point of origin, their sum is zero. The indices are calculated from the following equations for N, P and K in corn leaves:

$$N \text{ index} = [\frac{f(N/P) + f(N/K)}{2}] \dotfill [4.3\text{-}3]$$

$$P \text{ index} = -[\frac{f(N/P) + f(K/P)}{2}] \dotfill [4.3\text{-}4]$$

$$K \text{ index} = [\frac{f(K/P) - f(N/K)}{2}] \dotfill [4.3\text{-}5]$$

$$\text{where } f(N/P) = 100 \, (\frac{N/P}{n/p} - 1) \frac{10}{cv} \text{ when } N/P > n/p \dotfill [4.3\text{-}6]$$

$$f(N/K) = (\frac{N/K}{n/k} - 1) \frac{10}{cv} \text{ when } N/K > n/k \dotfill [4.3\text{-}7]$$

$$f(K/P) = (\frac{K/P}{k/p} - 1) \frac{10}{cv} \text{ when } K/P > k/p \dotfill [4.3\text{-}8]$$

$$\text{or} \quad f(N/P) = 100 \, (1 - \frac{n/p}{N/P}) \frac{10}{cv} \text{ when } N/P < n/p \dotfill [4.3\text{-}9]$$

$$f(N/K) = 100 \, (1 - \frac{n/k}{N/K}) \frac{10}{cv} \text{ when } N/K < n/k \dotfill [4.3\text{-}10]$$

$$f(K/P) = 100 \, (1 - \frac{k/p}{K/P}) \frac{10}{cv} \text{ when } K/P < k/p \dotfill [4.3\text{-}11]$$

in which N/P, N/K, K/P are the values of the ratios in the leaf under consideration; n/p, n/k, k/p are the mean values of the ratios for the population of high yielding plants and cv is the coefficient of variation for each particular ratio for the high yielding population.

In order to calculate DRIS indices for the data in Table 4.3-2 the following mean values and coefficients of variation were used: n/p = 10.04, cv = 15 percent; n/k = 1.49, cv = 22 percent; k/p = 6.74, cv = 25 percent.

The results in Table 4.3-2 illustrate the facility of the DRIS approach in making the same diagnoses over a wide range of plant age. Irrespective of age the order or requirement is diagnosed as K>N>P which is very much more consistent than the other two approaches. The reason for the ability of the DRIS approach to make diagnoses over a wide range of ages lies in the fact that the effect of dry matter cancels out when ratios are used as diagnostic parameters. This is further illustrated in Fig. 4.3-1. Ratios of all combinations of N, P and K computed at ages A and B would lie within narrow limits because curves for N, P and K are nearly parallel.

DRIS norms have been developed for a number of crops and the interested reader is referred to the following references: corn (Sumner, 1977d), rubber *(Hevea brasiliensis)* (Beaufils, 1957, 1959), wheat *(Triticum aestivum L.)* (Sumner, 1977b), sugarcane *(Saccharum officinarum L.)* (Beaufils and Sumner, 1976), soybeans *(Glycine max L.)* (Sumner, 1977a), potatoes *(Solanum tuberosum L.)* (Meldal-Johnsen and Sumner, 1980), sorghum *(Sorghum bicolor L.)* (J. O. Arogun, 1978). Application of the DRIS system to sorghum and millet (M. S. Thesis, University of Wisconsin, Madison, WI 1978).

In cases where no DRIS norms are available an approximation can be made by computing the various ratios from well established critical or standard values. This has been done in Table 4.3-2 using the critical values of Melsted et al. (1969). As a result substantial improvement in diagnostic flexibility over the critical value approach has been achieved. It is now possible to diagnose that P is present in relatively excessive amounts at all stages while at the appropriate stage (85 days) the DRIS and DRIS/critical value approaches make the same diagnosis. However, computing indices from the critical values does not give the same diagnosis as the DRIS approach at all stages of growth because the critical values and DRIS norms are not coincident. However, a word of warning on using this approach should be sounded in that this might prove to be a dangerous practice particularly if the critical values used were not suitably derived. Furthermore, without the test of significance for discrimination between populations for the various forms of expression one may choose ratios which are not suitable diagnostic parameters.

### 4.3.2.3.4 Utility of Information Obtained

In order to effectively utilize plant analysis in diagnostic work, it is essential that a multi-element approach be adopted. All too often, researchers, particularly, only assay tissue for the nutrient(s) being varied in their experiments, thus losing much useful information on interaction and balance between nutrients. At all costs this approach should be avoided when plant analysis forms part of a diagnostic program.

Because the plant can be considered as the integrator of its environment, plant analysis reflects the resultant effects of all the growth factors external to the plant and how they interact with internal processes. Thus plant analysis is a measure of the nutrient *status quo* within the plant. It gives no direct evidence of the cause of any insufficiency, excess or imbalance. Thus a given element in the tissue might be diagnosed as limiting growth but its insufficiency is not necessarily caused by a low level in the soil. It might very well result from a relative excess of another element or may be caused by some impairment of root activity by some toxic factor or parasite. Thus plant analysis alone without any information about the soil should seldom be used to formulate a fertilizer program as there is always a danger that the cause of the problem has not been correctly identified. Because plant analysis and soil testing complement one another their combined use should be encouraged in making diagnoses and remedial fertilizer recommendations which have the highest chances of increasing yield and quality of the crop.

Having discussed the pros and cons of various diagnostic systems one is now in a position to evaluate the effects of various techniques of modifying the root zone in overcoming any diagnosed nutrient stress.

## 4.4 METHODS OF MODIFICATION

Modification of the root zone to improve the supply and availability of nutrients to plants can be brought about in a number of ways. Having diagnosed the cause of the problem reflected as nutrient stress in the plant, the experienced soil fertility advisor is in a position to make a recommendation as to how the required fertilizer should be applied and what other treatments are necessary to effectively utilize the improved nutrient status of the soil.

In most cases nutrient stress in the soil is alleviated by simply applying the correct fertilizer and incorporating it in one of a number of ways. In addition, physical modification of the root zone may sometimes be necessary to reap the full benefit of the treatment. The possibilities of using this and other types of technology in alleviating nutrient stress will now be discussed.

### 4.4.1 Fertilizer Application

One cannot discuss root zone modification as related to alleviating nutrient stress without emphasizing that proper fertilizer application rates, sources, time of application and soil fertility interactions with many other soil factors (moisture, structure, microbial populations, oxygen levels, etc.) are prerequisites to proper plant growth. Although soil is the primary source of plant nutrients, the supply is often inadequate for the efficient production of crops requiring amendments to supplement the natural supply. Essential nutrients that perform functions in plant metabolism, structure and reproduction, are often required in amounts and ratios that can be attained only by supplemental amendments using various means of application. These fertilizer applications may often have a greater impact on plant growth and production than modification of the root zone by tillage equipment, or the selection of improved genetic material.

Fertilizer is usually applied either over the entire soil surface by the method commonly referred to as "broadcasting" or in a localized area below the surface close to the seed or plant, usually termed "banding" or row placement. However in the last few years an additional method has been introduced for certain nutrients, commonly referred to as "fertigation". Fertigation is a term used to describe the introduction and application of soluble fertilizer material to the land through irrigation water. A fourth method of fertilizer application that has received attention in recent years is "foliar feeding" which involves the application of essential nutrients in a limited quantity of water directly to the leaves of the crop for immediate assimilation.

When much of the available fertilizer was of low grade, growers tended to use the banded (and starter) application method most. With the advent of higher analysis materials and with higher rates of fertilizer other application methods have become common. However, current high costs due to energy stresses are encouraging reconsideration of band application.

Specific crops vary in their need for certain nutrients which may influence their response to methods of application. A plant that has a high phosphate requirement may respond to banding of fertilizer while other species may fail to respond. Certain species may be very sensitive to fertilizer salts when they some into contact with roots. Excessive amounts of fertilizer should not be placed directly under the seeds of plants that normally develop tap roots.

As indicated earlier, tillage practices should be such that roots may develop freely in regions where the fertilizer is placed. Lack of aeration or moisture in the root and fertilizer zones may restrict the availability of essential plant nutrients.

### 4.4.1.1 Broadcast

Of the various methods of fertilizer application, broadcasting is probably the most widely used because of its convenience. This method also has the advantages that it minimizes seed injury, distributes the nutrients over a wider area for plant root contact, and may result in greater uniformity of growth.

Early studies (Coe, 1926; Truog et al., 1925) to evaluate various methods of applying fertilizer indicated that the initial application of essential elements should be maintained in the soil by broadcast application in order to prevent or minimize injury to germinating seedlings. However, it was pointed out that hill or row drilled fertilizer was often desirable for promoting early growth, hastening maturity and increasing yield. Nelson et al. (1949), using radioactive P, reported that less fertilizer P was absorbed by corn plants when broadcast than when mixed in the row or placed with the seed. However, yield and total P content of the grain were equal irrespective of placement method. The most profitable method of application may depend upon certain characteristics of the soil. Dumenil et al. (1965) concluded that broadcast and plowed-under P appears to be the best method of application for well-drained soils in western Iowa. Welch et al. (1966) found the broadcast method to be better suited for Illionis soils that had a higher initial P content and that the relative efficiency of broadcast P varied with rate. Olsen et al. (1961) reported that P uptake by corn seedlings was inversely related to soil moisture tension. Such a relationship is to be expected in view of the fact that P supply to the root is a diffusion controlled process which would depend on moisture status of the soil. For the same reason broadcast and plowed-down P may be more readily absorbed during dry periods than that placed in a band which would tend to dry out more rapidly (greater concentration of roots) than the bulk soil. Broadcast applications of K are more desirable for most plants because on the one hand banding may result in luxury consumption while fertigation or foliar sprays may injure leaves on the other. Because secondary and microelements are often applied in mixed fertilizers broadcasting is more convenient. Exceptions to this are foliar sprays and certain materials applied through irrigation systems, which will be discussed later.

In general, the broadcast method of application would be selected when either large quantities of fertilizer must be applied, or the initial level of the nutrient in the soil is relatively high or the plant is likely to experience a drought stress during the season.

### 4.4.1.2 Banding and Starter

Fertilizer application by banding and/or starter methods is generally employed when either rapid initial growth is desired, or the fertility level is very low and fertilizer supply is limited because of availability/cost, or plants are growing in soils that have high fixation characteristics. Although occasionally K fixation may be a problem in certain soils containing vermiculite or weathered mica, the major concern is with P which is fixed to a greater or lesser degree in most soils. Phosphorus is sorbed by many mineral surfaces in soils such as the hydrous oxides of Fe, Al and Mn, the silicate clay minerals and $CaCO_3$. In addition, P can be precipitated by soluble Al, Mn and Ca in soils (Fig. 4.2-1). The availability of P immobilized in any of these ways generally decreases fairly rapidly with time as more stable combinations are formed. Whenever soils severely fix nutrients fertilizer banding may be desirable. Under cool, wet soil sonditions, when root proliferation may be limited, starter fertilizer applications could be advantageous.

Fertilizer may be placed in single or double bands on either side of the seed with little consistent difference in yield in favor of either method. Fertilizer bands which vary from 2 to 5 cm in width are usually placed a similar horizontal distance from the seed to decrease the hazards of seed germination and decrease root injury. In the case of starter fertilizer much smaller application rates placed with or near the seed are used for accelerating early growth of the crop. The only real difference between banding and starter fertilizer lies in the application rate.

Because fertilizer placed in bands influences a smaller volume of soil than that broadcast, roots near bands will experience higher salt concentrations and higher localized levels of nutrients than those associated with broadcast applications. Depending on conditions this may or may not be beneficial in terms of growth. Duncan and Ohlrogge (1958) found greater length and number of small corn roots associated with band applications of N and P while Miller and Ohlrogge (1958) showed an interaction for these two elements in that N had a pronounced effect on the uptake of P from a band by increasing the relative feeding power of the corn roots. Bosemark (1954) indicated that N deficiency gives long slender roots, while high banded N levels produce short, stocky roots.

Robinson et al. (1959) showed that at low rather than high temperatures, band placement of P was superior to broadcasting. They concluded that this was due to the relatively greater effectiveness of incorporated P at higher temperatures. This temperature relationship may explain why starter fertilizer is effective in some years at certain locations and not at others. Banding of fertilizer, particularly P has been shown to increase (Welch et al., 1966; Olsen et al., 1961; Duncan and Ohlrogge, 1958; Miller and Ohlrogge, 1958; Robinson et al., 1959; Barber, 1958), or have little effect (Nelson et al., 1949; Truog et al., 1925; Walker and Brooks, 1964; Evans et al., 1970) on yield depending upon factors such as nutrient level in the soil, temperature, moisture, rate of application, crop grown and location. Soil modification will influence some of these factors and contribute to differences in responses to banding or broadcast applications.

Phosphorus plays the major role in growth response to starter fertilizers because large quantities are needed for energy production and transfer in young, rapidly growing plants while release from solid surfaces into the soil solution is slow, making it difficult for the young root system to assimilate sufficient P. Furthermore, because root growth early in the season may be restricted by cool weather, responses to starter fertilizer have sometimes been observed at lower temperatures (Englestad and Allen, 1971; Olsen et al., 1961; Gingrich, 1964) but this is not always the case (Knoll et al., 1964; Power et al., 1964; Wilcox et al., 1962). The cause of such inconsistencies is probably to be found in other unfavorable factors. The reasons why this early growth response to starter P applications is not always transmitted into higher yields are not well understood. More research data are required to satisfactorily explain this phenomenon.

Very little information exists on the placement effects of other nutrient elements on root growth. Haynes and Robins (1948) reported that both B and Ca must be present in all parts of the root zone for normal plant growth. Pearson (1966) reported unpublished data of D. D. Howard and Fred Adams which showed that removal of certain elements, especially N, decreased primary root lengths of cotton (*Gossypium hirsutum* L.) seedlings but increased their dry weight per unit length appreciably.

### 4.4.1.3 Fertigation

Fertigation is the application of soluble fertilizer by injection into irrigation water. Because uniformity of nutrient and water applications go hand in hand good engineering, management and maintenance of the system are required for maximum benefit. With the increased number of sprinkler irrigation systems, especially center pivot systems, fertigation has increased significantly in the Midwest and Southeast of the U.S. in recent years. Although most nutrients readily soluble in water may be applied in this way, certain nutrients may be more effectively applied by other methods. Nutrients which are not sorbed by the soil and tend to move with soil water, are generally considered suitable for fertigation. This would include N, S and probably some of the chelated micro- and secondary elements such as Zn and Mg. Nutrients such as P, K, Ca, and inorganic forms of the micro- and secondary elements may be soluble and compatible under certain conditions but many of these elements react to form insoluble compounds with soil and may accumulate at the soil surface and not be sufficiently distributed throughout the root zone to be taken up by the roots of the growing crop. Hergert and Reuss (1976) found sprinkler-applied Zn moved to a maximum depth of 5 cm on a Haxtum loamy sand (*Pachic Argiustolls*) and a Nunn clay loam (*Aridic Arquistolls*) in Colorado. Except for one year on the loamy sand, P movement was restricted to the surface 4.5 cm. Because of their immobility and fixation in soils, P and Zn are unlikely to move appreciably in soils. This has been discussed in detail by Hemwall (1957) and Mortvedt and Giordano (1967). Hergert (1977) pointed out that sprinkler application of fertilizer nutrients results in relatively little uptake through the leaves of the crop because nutrient concentrations in irrigation water are usually too low for foliar feeding to contribute significantly.

Because nutrients which remain soluble in the soil after application are subject to loss by leaching, the use of fertigation particularly on sandy soils is likely to increase substantially in the forseeable future. In addition, the great convenience in being able to apply fertilizer to a large area at the push of a button is likely to further promote this mode of application. However at this stage, there is a dearth of technical information relating to the frequency and rate of application of soluble fertilizers through the irrigation water particularly in humid areas where sudden downpours can readily leach soluble nutrients recently applied. Many farmers are currently producing high yields under fertigation but it would appear that excessive quantities of soluble nutrients, particularly nitrogen, are being applied. Much research is urgently required before the benefits of fertigation can be maximized.

Foliar applications of fertilizers cannot generally be considered as alleviating nutrient stress by root zone modification. However, under certain alkaline conditions, foliar applied Fe, Zn, Mn, etc. may be more practical than any attempts at modifying the soil to enhance root uptake of the nutrients.

### 4.4.1.4 Subsoiling and Deep Placement

The primary reason for subsoiling and deep plowing is to modify the soil ✗ sufficiently to improve the water transmission properties of the profile and to improve root penetration for increased moisture utilization. When sufficient nutrients and water are available, responses to such modification may not be attained. Under certain conditions, deep placement of fertilizers and liming materials in the subsoil may provide a more favorable environment for plant roots. The interrelationships between fertilizer application and root zone modification will be discussed later in the section dealing with equipment for subsoiling and deep root zone modification.

Nutient stress is often alleviated through modification of the root zone by mechanical placement of liming materials in the subsoil for the purpose of increasing the pH value of acid subsoils and to neutralize the harmful effects of high levels of soluble Al or Mn which may adversely affect root development (Adams and Lund, 1966; Rios and Pearson, 1964; Soileau et al., 1969; Foy and Brown, 1964). In the case of saline-sodic and sodic soils, gypsum may be added in a similar manner as lime to increase the permeability of the argillic (natric) horizon which contains high levels of exchangeable Na and has poor water transmission properties. High rates of gypsum together with subsoiling and/or deep plowing greatly increased crop yields, water-intake rates, and water and root penetration in Oregon (Rasmussen et al., 1972). These aspects received more detailed treatment in Chapter 3.

Effects of subsoil fertilization relative to plant response and yield increase have been conflicting. Duley (1957) summarized the research in the Great Plains and found no evidence of general benefits. Workers in the North Central States (Hobbs et al., 1961; Larson et al., 1960) where frost penetration is deep, indicated that subsoiling and deep placement of lime and fertilizer did not, in most cases, increase yields. In the Southeast where frost action is slight, results have been mixed. In Florida (Robertson et al., 1957) and Louisiana (Patrick et al., 1959) both subsoiling and deep placement of fertilizer increased the yield of corn and cotton. Georgia workers (Boswell et al., 1977; Parker et al., 1975) obtained yield responses to subsoiling but not to deep placed lime or fertilizer for cotton and soybeans, respectively. In Connecticut, De Roo (1961) increased leaf and root growth of tobacco (*Nicotiana tabacum* L.) by subsoiling and fertilization of the lower horizons. In New Jersey, yield of vegetable crops was increased by subsoiling with deep placement of lime and phosphorus (Nissley, 1944). However, in the same state, significant yield increases of cabbage (*Brassica oleracea* L.) tomatoes (*Lycopersicom esculentum,* Mill.), and sweet corn to similar treatments over a 7-year period were not obtained by Brill et al. (1965).

Because numerous factors may influence the effect of subsoiling and deep placement of fertilizers and lime such as specie, moisture, climatic conditions and the soil characteristics, this conflict of results is not surprising. Furthermore, it is doubtful whether yield responses to deep placed lime and fertilizer can be expected unless a definite need was diagnosed at the outset. This has seldom been the case.

### 4.4.2 Crop Selection
#### 4.4.2.1 Root Penetration

Each plant species has a particular type of root growth, but root development within species often varies considerably because of differences in environment. Under various conditions and with various species, root penetration into the soil may vary from a few centimeters (less than 2 or 3) to more than 6 meters.

Byrne (1974) who described the morphology of plant roots very succinctly concluded that there was a "paucity of environmental oriented anatomical investigations". He indicated that plant anatomists generally go to great length to avoid growing root material in soil. This probably indicates why it is often difficult to explain how roots of certain species are able to penetrate very compacted soils. Pearson (1974) and Taylor (1974) who reviewed the significance of rooting patterns to crop production, pointed out that root penetration is dependent upon moisture, pH, nutrients, soil strength, soil porosity and/or voids. Using the results of numerous workers, Taylor (1974) indicated that when the root was exerting its maximum force, root growth pressures usually range from 900 to 1300 kPa which can result in great root penetration.

#### 4.4.2.2 Specie and Cultivar Adaptation

One of the primary ways that man has manipulated crop production or yield in a particular locale is to simply select species adapted to the total environment. This may include physical, chemical and biological modifications, especially those that relate to root zone and nutrient stress effects. It should be noted that different species or cultivars may react very differently at high or low levels of mineral elements in soils. For example, wheat (*Triticum aestivum* L.) and barley (*Hordeum vulgare* L.) cultivars show different Al tolerance in acid soils (Foy et al., 1965). In addition to development of plants that can tolerate large quantities of toxic elements, Stangel (1976) suggested that plant breeders might consider focusing on ways that young plants could accumulate, on a temporary basis, large quantities of nutrients for use at a later stage in plant growth. He suggested that this might be accomplished by promoting extremely rapid root development, supersaturation of nutrients per unit volume of root, or both. Crop selection of species and cultivars that will respond to various management practices such as root zone modification will require cooperative efforts of soil scientists, engineers, plant geneticists, crop physiologists and agronomists.

The variation of elemental uptake among species is influenced by morphological features such as the number, the degree of subdivision, and the rate of growth and extension of the root system. These traits may regulate ion uptake by determining the amount of active surface exposed to the soil solution during root growth. Baker et al. (1964) suggested that differences in mineral accumulation among corn hybrids may be attributed to differences in depth of rooting of the plants.

Because it is not really within the scope of this chapter to discuss crop selection, the interested reader is referred to the material in Chapter 1 and to the proceedings of a workshop on crop selection as related to mineral stress (Wright, 1976).

### 4.4.2.3 Rotations and Multiple Cropping

Because monoculture is often more productive and economic in the short term and because of the present intense economic competition less emphasis has been placed on rotations relative to two or three decades ago. However, if this generation is to be practical, crop selection for high yields and efficient utilization of imput factors must be properly considered. This aspect has been reviewed by Pendleton (1976) to whom the interested reader is referred.

Multiple cropping has received considerable attention recently because of its suitability for use in minimum tillage systems. In addition, the Southeast of the U.S. and many tropical and sub-tropical regions are well adapted to multiple cropping because the environmental conditions are favorable for year-long crop production (Sanford et al., 1973; Bradfield, 1970; Kung, 1969; Gallaher, 1975). Multiple cropping requires very careful selection of crop species as well as cultivars. Root development is very important because the root zone may not be modified to the optimum standard due to limited time between crops. In addition, the total management of the crop including adequate control of pests and soil fertility are critical to the successful cultivation of more than one crop per year. Although multiple cropping is increasing fairly rapidly much research is needed in this area to arrive at best management practices to ensure maximum return. With recent advances in technology and equipment, multiple cropping, especially double cropping is expected to increase as food and fiber demands increase.

### 4.4.3 Equipment

Modern agriculture is totally dependent on various types of mechanical equipment which may range from one-row cultivators to massive land leveling and earth moving machines. There is little doubt that this traffic modifies the root environment appreciably being particularly acute in the case of heavy wheel traffic which results in compaction and degradation of soil physical condition. Because average tractor mass has increased from 3 to 28 metric tons over the past 30 years with attachments adding to this (Voorhees, 1977), one cannot ignore the detrimental modification of the root zone in relation to nutrient uptake or exclusion brought about by equipment traffic. Lowry et al. (1970) demonstrated that when a normally deep-rooted plant encounters mechanical impedence caused by compacted soil layers, it becomes shallow rooted which influences its uptake of nutrients.

Although there are numerous types of equipment involved in land modification, those generally used may be referred to as conventional, subsoiling, no till and minimum tillage, deep modification, injection and barrier placement equipment. Information and data on root zone modification brought about by some of these equipment systems are sparse, especially in relation to alleviating nutrient stress.

#### 4.4.3.1 Conventional

Conventional tillage is defined by the American Society of Agricultural Engineers as "The combined primary and secondary tillage operations normally performed in preparing a seedbed for a given crop and area" (Anon., 1977). Generally the equipment used in conventional tillage consists of a plow to primarily shatter the soil with partial or complete inversion followed by disking or harrowing equipment to pulverize, smooth and pack the soil for seedbed preparation or to aid in controlling weeds.

Earlier workers (Browning and Norton, 1945; Page et al., 1946) showed conventional tillage to be superior to rotatilled, subsurface tillage or disked methods of seedbed preparation for corn production. Current data have been somewhat inconsistent. Virginia and Kentucky workers (Moschler et al., 1972; Thomas et al., 1973) have shown conventionally tilled corn yields to be inferior to no-tillage while in Ohio and Iowa, Van Doren, Jr. et al. (1976) and Mock and Erbach (1977) have shown little difference or conventional tillage to be superior for corn production.

Although crop production for maximum yield is generally the major factor in selecting tillage methods and thus equipment needs, other soil management factors should be considered. The traditional and still most widely used method of seedbed preparation for most row crops is first plowing with a moldboard plow, then utilizing a disk and harrow to prepare a smooth and uniform soil surface. As indicated by Olson and Schoeberl (1970), this method is relatively costly and leaves the soil surface vulnerable to erosion from both wind and water.

Conventional tillage often has a significant impact on the development of the root system for various crops. It has long been recognized that roots tend to proliferate into zones of more favorable nutrient supplies. If conventional tillage is shallow, crop roots may utilize only those nutrients available in the disturbed area. Both laboratory (Taylor and Gardner, 1963; Taylor et al., 1966) and field studies (Carter and Tavernetti, 1968; Dumas et al., 1973; Lowry et al., 1970; Boswell et al., 1977) have shown that compaction increases bulk density and soil strength which prevent development of root systems, resulting in reduced nutrient and moisture utilization. Lowry et al. (1970) found that final plant height and yield of seed cotton were reduced as soil bulk density increased and attributed these reductions to a restricted water supply due to limited rooting volume rather than nutrient stress. As already pointed out, interactions of various parameters are often difficult to separate since the soil-plant system is often complex. There appears to be relatively little data available to suggest the degree to which diagnosed nutrient stress might be alleviated by mechanical root zone modification because of the confounding between moisture and nutrient stress. Improvement of poor conditions in the root zone will result in improved moisture and nutrient utilization and consequently greater yield.

#### ✗ 4.4.3.2 Subsoiling

The tillage operation of subsoiling is defined as any treatment to loosen soil with narrow tools below the depth of normal tillage without inversion and with a minimum mixing of the soil (Anon., 1975). This tillage operation probably has as much effect on the configuration of root systems as any that equipment for soil modification may exert. In discussing the movement of nutrients to roots, Olsen and Kemper (1968) showed that the rate of movement is as important as the amount of available nutrient present while "the resistance offered by the soil to the transfer of nutrients to the roots depends upon the size and shape of the paths along which nutrients must travel," (Barley, 1970). Thus any tillage operation which will improve the conductance of the soil for water and allow increased root proliferation will improve the supply and utilization of nutrients because of the enhanced possibilities for mass flow, diffusion and root extension. All of these phenomena are influenced by root zone modification through various tillage operations with subsoiling exerting a great effect when utilized. Subsoiling and subsequent root penetration are often related to moisture content as well as physical modifications (Olson and Schoeberl, 1970; Taylor and Gardner, 1963; Rios and Pearson, 1964).

Another consideration in root zone modification is alteration of environments such as highly acid subsoils that produce shallow plant rooting and depressed crop yields. These very acid lower horizons may contain high and low amounts of exchangeable Al and bases, respectively, especially on highly weathered soils of the tropics and sub-tropics. After subsoiling, roots penetrating acid subsoils sometimes encounter low concentrations of Al in solution which are detrimental to root growth (Adams and Lund, 1969; Rios and Pearson, 1964). Howard and Adams (1965) and Soileau et al. (1969) have shown that in acid soils primary root elongation of cotton seedlings is not affected by soil solution pH *per se* within the pH range of 4.3 to 6.5 provided Al is absent and Ca and Mg are adequate. The requirement for Ca for the structure and elongation of plant cells is well documented (Soileau et al., 1969; Broyer and Stout, 1959; Jones and Lunt, 1967) and has been shown to affect both root and top growth. However, Boswell et al. (1977) showed that placement of lime in the subsoil trench may not increase yields above those obtained with subsoiling alone. Neither did subsoiling plus lime increase Mg or Ca significantly in cotton tissue when compared to subsoiling alone or conventional tillage even though Ca and Mg levels of the soil were increased at the 18 to 40 cm depths.

In addition to chemically related soil factors, there are mineralogical and physical properties associated with pan horizons that may be influenced by subsoiling. McCracken and Weed (1963) described four types of pan horizons in soils as "plow pan" or "traffic pan", "brittle pan", "organic pan", and "fragipan". Their micromorphological studies indicated that compacted zones in certain soils differ morphologically from adjacent zones which may explain some of the variation obtained with subsoiling studies.

Carlisle and Fiskell (1968) reported that tillage pans in Florida partly or completely prevented corn root penetration into the subsoil, causing root deformities. Subsoiling and deep placement of fertilizer improved corn yields on these Florida loamy fine sands and fine sandy loams containing plow soles, clay zones, or organic hardpans, especially when drought periods were of short duration (Robinson et al., 1959). On a Klej fine sand (*Aquic Quartzipsammets*) which contained no barrier to root penetration, fertilizer on the surface was as good as deep placement and subsoiling alone was of no benefit.

Parker et al. (1975) found a 60 percent increase in soybean yield as a result of subsoiling in the initial year of a 2-year study on a Marlboro sand (*Typic Paleudults*) on the Coastal Plain of Georgia. By including a nematicide (DBCP) in the subsoil trench, they increased yields 88 percent the first year and 95 percent the second year. However, the only fertility treatment that showed an increase was a micronutrient mix but the element(s) responsible was not identified. The above study and studies of Bird et al. (1974) point out the importance of identifying the problem since specific effects, such as nemtode infestation, may be mistaken for nutrient stress.

Root zone modification effects by subsoiling are not restricted to soils of the Southeastern U.S. In California, Stockton et al. (1964) showed that subsoiling 56 to 60 cm directly under the row increased growth, yield, and earliness of cotton on a coarse textured soil that had been artificially compacted. In Indiana, Bertrand and Kohnke (1957) obtained better nutrient uptake when the subsoil was loosened. Subsoil fertilization stimulated root growth more than subsoiling (Kohnke and Bertrand, 1956). In earlier studies in Missouri, Woodruff and Smith (1947) successfully used subsoil shattering, together with lime and P applications, in the improvement of clay pan soil for the growth of sweet clover (*Melilotus alba* Med) and corn.

Stivers et al. (1956) in Virginia showed that subsoiling and deep placement of lime had no real effect on yields of alfalfa, corn, and peanuts (*Arachis hypogaea* L.). Ferrant and Sprague (1940) working in New Jersey with Sassafras loam (*Typic Hapludults*) found that the root development of red clover (*Trifolium pratense* L.) was restricted to the limed zone of the profile and that soil acidity, texture, pore space, and nutrient supply had little effect on root penetration. In Kansas, Hobbs et al. (1961) found that crop yields of wheat, sorghum, barley, oats (*Avena sativa* L.) and corn were seldom increased by chiseling and deep fertilizer placement. This was true despite the fact that chiseling reduced bulk density of the pans and increased permeability. Subsoiling with a chisel-like subsoiler and deep fertilizer placement was ineffective in increasing corn yields in Iowa and Illinois (Larson et al., 1960).

Root zone modification by subsoiling may alleviate nutrient stress but the magnitude of the effect will depend on various interactions with factors such as high Al in the subsoil, nematodes, presence of pans, available moisture and other physical and chemical factors. One should consider the potential impact of these factors prior to subsoiling.

### 4.4.3.3 No-till and Minimum Tillage

A lack of or reduced root zone modification may be associated with no-till or minimum tillage. No-till or zero tillage is a procedure whereby a crop is planted directly into a seedbed not tilled since harvest of the previous crop. Minimum tillage is the minimum soil manipulation necessary for crop production or for meeting tillage requirements under the existing soil and climatic conditions (Anon., 1975). These practices are often referred to as conservation tillage because of their relative effectiveness in relation to soil erosion control, water conservation, soil physical properties, crop production and, more recently, cost reductions resulting from lower energy requirements.

In a recent review, Triplett, Jr. and Van Doren, Jr. (1977) reported that a survey made recently by the Soil Conservation Service indicated that in 1976 some 3 million ha in the U.S. were planted without tillage, that on 21.2

million additional ha, tillage was reduced from the conventional level while conventional tillage accounted for 88 million ha. By the year 2010, greater than 90 percent of the crop area will receive reduced tillage while on more than one-half of the area, some form of no-tillage farming will be practiced.

Data relative to plant nutrient behavior and requirements in relation to minimum tillage operations are somewhat limited except possibly for wheat fallow in semiarid regions. In most tillage comparisons, nutrient levels have not been varied or studied.

The slow diffusion of P in the soil (Olsen and Kemper, 1968) raises a question relative to the availability of surface-applied P that is associated with no-till or minimum tillage since it may be difficult to band and is not incorporated in a tillage operation. Several workers (Belcher and Ragland, 1972; Singh et al., 1966; Triplett, Jr. and Van Doren, Jr., 1969) have shown yields from fertilization studies to be equal to or greater from no-tilled or minimum tillage studies than from conventional tillage methods. Triplett, Jr. and Van Doren, Jr. (1969) reported that P and K concentrations in leaves of corn grown with conventional or without tillage were equal when sampled at tasseling but significantly higher when sampled at the 8 to 10-leaf stage. They also found P and K to be concentrated near the soil surface after a six year cycle of corn grown in monoculture by a no-tillage method. Moschler et al. (1972) found yield and fertilizer efficiency to be higher on no-till corn with surface application than on conventionally tilled corn with an equal disked-in fertilizer application. Larger amounts of residual N, P, K, and organic matter were found under no-tillage than conventional tillage systems. Shear and Moschler (1969) reported higher yields, the accumulation of P in the upper 5 cm of soil, a larger quantity of available P in the plow layer, higher P and K content of leaves and equal or higher N content of leaves sampled at the knee-high stage with no-till than conventional tillage. This accumulation of P at the soil surface may actually be advantageous, as evidenced by greater absorption in the early growth stages (Moschler et al., 1972; Shear and Moschler, 1969; Singh et al., 1966).

Minimum tillage affects N and organic matter contents of soils appreciably. Studies conducted by Beale et al. (1955) on a Cecil sandy loam soil (*Typic Hapludults*) in South Carolina for a 10-year period indicated that the level of plant residue in the surface 15 cm of the soil more than doubled and decreased soil losses significantly under minimum tillage as compared to conventional tillage. Organic matter and total N contents of the soil in the surface 15-cm layer increased much more with mulch tillage than when cover crops were plowed under.

Stanford et al. (1973) adapted data from Black (1973a, 1973b) to show cumulative effects of straw residue at differing rates upon soil properties. Levels of organic C, total N, mineralized N, soluble P, exchangeable K, plant uptake of N and soil water increased with residue rates. These data suggest that minimum tillage, with plant residues present, may greatly influence nutrient availability under conservation tillage conditions. Data for the Great Plains of the U.S. have been reported by McCalla and Army (1961) and Zingg and Whitfield (1957).

One other area related to root zone modification (or lack thereof) by no-till and minimum tillage is the effect of lime on the soil conditions and crop growth. Numerous researchers (Blevins et al., 1977; Blevins et al., 1978;

Moschler et al., 1973) have reported that the soil surface becomes very acid after several years of no-tillage or minimum tillage crop production. This decrease in soil pH is primarily associated with the surface application of N fertilizers (Triplett, Jr. and Van Doren, Jr., 1969; Pierre et al., 1971; Moschler et al., 1973; Blevins et al., 1978). Consequently, lime requirements may be greater under minimum than conventional tillage. Moschler et al. (1973) and Blevins et al. (1978) found significant corn yield increases for broadcast application of lime under no-tillage systems while Estes (1972) and Triplett, Jr. and Van Doren, Jr. (1969) reported increases in surface pH from surface-applied lime but without significant yield response. As expected, ex-changeable Ca and Al increase and decrease respectively as pH approaches a value of 5.6 (Blevins et al., 1978). The above data indicate that roots growing near the surface under no-till and minimum tillage systems are capable of fully supplying the nutrient needs of crops such as corn probably due to the favorable moisture conditions.

**4.4.3.4 Deep Modification**
Deep soil profile modification may be accomplished by various types of equipment but generally large heavy plows designed to disturb soils to depths greater than 30 cm are used in order to break up compacted soil layers, such as tillage, traffic, organic and/or fragipans. Such operations often allow deeper root penetration and water movement, improve drainage and aera-tion, and may reduce runoff and erosion (Brill et al., 1965; Eck and Davis, 1971; Englebert and Truog, 1956; Mech et al., 1967; Rasmussen et al., 1972; Patrick et al., 1959). In many instances deep placement of lime and fertilizer on acid and infertile subsoils is often practiced together with this tillage operation to stimulate deeper rooting.

In much of the deep tillage and profile modification research, conclu-sions regarding root extension have been based on soil water depletion pat-terns (Mech et al., 1967; Hobbs et al., 1961; Eck and Taylor, 1969), rather than monitoring plant nutrient extraction.

Stibbe and Kafkafi (1973) concluded that deep tillage had no significant effect on winter wheat grown on calcareous soils in Israel. Even though the crop responded to P, they found no need to incorporate the fertilizer deeper than 10 cm. Larson et al. (1960) showed that fertilizer placed deep was as ef-fective as fertilizer plowed under for corn production in Iowa and Illinois. Jamison and Thornton (1960) concluded that if the soil surface was ade-quately fertilized, increases from deep fertilization over surface application alone were small and of questionable value.

Other workers have shown yield increases from deep tillage and fertilizer placement. Patrick et al. (1959) found greater root development in the sub-soil and increased yields of corn and cotton from deep fertilizer placement and tillage on soils that possess traffic pans. These soils were the Commerce (Aeric Fluvaquents), Gallion (Typic Hapludalfs), Richland (not classified) and Olivier (Aquic Fragiudalfs) silt loams. These effects were particularly evident in years when the increased root development in the subsoil, resulting from deep placement of fertilizer or tillage, enabled the crop to better with-stand dry periods. Rasmussen et al. (1972) greatly increased crop yields, water intake rates, and water and root penetration by deep plowing, subsoil-ing and amendments (gypsum) on a sodic (slick spot) soil (Nadurargid) in Oregon. Deep profile modification without gypsum was not beneficial.

In most studies on deep tillage or nutrient placement the data reported are limited to soil physical parameters with little information on nutrient uptake. Much more research is required in this area to document the effects of physical and chemical root zone modification on the alleviation of nutrient stress.

### 4.4.3.5 Injection and Barriers

When nutrients are susceptible to loss by leaching or volatilization, equipment is available to prevent or reduce loss by either laying down barriers or injecting the nutrient into the soil at a point where it will be sorbed. This type of equipment can often drastically change the root zone environment.

Although in recent times secondary and micronutrients have been injected into the soil, this discussion will be confined to nitrogen.

Injection of ammonia in sprinkler systems has been shown to be unsatisfactory because losses between the point where the water issued from the sprinkler head to the point of contact with the soil can be as high as 50 percent (Leavitt, 1966). Leavitt (1942) developed the first applicator for direct ammonia injection. Since that time the use of anhydrous ammonia has become a successful agricultural practice. Excellent reviews are contained in the proceedings of a symposium in 1965 (McVicker, 1966). As pointed out in those proceedings, gaseous loss of ammonia increases with increased pH and temperature, and decreased soil moisture and cation exchange capacity.

Injectors may affect soil shattering to some extent and modify root zone environments by allowing plant roots to develop in areas of compaction. On the negative side, large concentrations of ammonia injected into the root zone of young plants may be toxic (Eno and Blue, 1954; Allred and Ohlrogge, 1964).

Plastic barriers at various depths below the surface to conserve moisture have been used to prevent nutrients from being leached by heavy rainfall. These modifications have been evaluated without appreciable success. Parks et al. (1970) did not increase yield, uptake of N by plants or plant recovery of the fertilizer N by utilizing barriers.

Another type of barrier that has shown some promise in sandy soils is a layer of asphalt placed below the plant root zone. Hansen and Erickson (1969) developed the technique by placing an almost impervious 2 mm layer of asphalt at a depth approximately 60 cm below the soil surface to improve the water holding capacity of the soil above the barrier. This technique would aid in preventing movement of mobile nutrients out of the root zone. Some disturbance of the soil occurs but inversion and major shattering does not occur. Erickson et al. (1968a) obtained significant responses to barriers for vegetable crops under both irrigated and natural rainfall conditions on sandy soils. Erickson et al. (1968b) reported spectacular yield responses to asphalt barriers for rice (*Oryza sativa* L.) and sugarcane both under dry land and irrigated conditions in Taiwan. Field trials by Sumner and Gilfillan (1971) with cotton, sugarcane, and alfalfa showed that the productivity of sandy soils in South Africa can be markedly increased by asphalt barriers both under dry land and irrigated conditions.

Root zone modification by mechanical equipment has contributed significantly in reaching yield potentials expected of many agricultural crops in certain locations, especially for newer varieties that contribute to these high yield potentials. Such practices, in compacted soils which can restrict normal root development as well as barrier placements in sandy soils for retention of mobile nutrients, are often invaluable but must be evaluated in economic terms in the light of current energy costs. It is predicted that as additional sources of energy become available to agriculture, many of the above discussed practices will be implemented more fully and additional data will be made available to better relate root zone modifications to nutrient stress.

## References

1   Adams, F. and Z. F. Lund. 1966. Effect of chemical activity of soil solution aluminum on cotton root penetration of acid subsoils. Soil Sci. 101:193-198.

2   Adams, W. E., H. D. Morris, Joel Giddens, R. N. Dawson and G. W. Langdale. 1973. Tillage and fertilization of corn grown on lespedeza sod. Agron. J. 65:653-655.

3   Allred, S. E. and A. J. Ohlrogge. 1964. Principles of nutrient uptake from fertilizer bands: VI. Germination and emergence of corn as affected by ammonia and ammonium phosphate. Agron. J. 56:309-313.

4   Anon. 1975. Glossary of Soil Science terms. Soil Sci. Soc. of Am. Madison, WI p. 34.

5   Anon. 1976. Energy in Agriculture, joint Task Force Report, Southern Regional Agricultural Experiment Stations and ARS, Southern Region, USDA.

6   Anon. 1977. Terminology and definitions for soil tillage and soil-tool relationships. Agricultural Engineers Yearbook. pp. 341-43. ASAE, St. Joseph, MI 49085.

7   Baker, D. E., W. I. Thomas, and G. W. Gorsline. 1964. Differential accumulation of strontium, calcium and other elements by corn (*Zea mays* L.) under greenhouse and field conditions. Agron. J. 56:352-355.

8   Barber, S. A. 1958. Relation of fertilizer placement to nutrient uptake and crop yield. I. Interaction of row phosphate and the soil level of phosphorus. Agron. J. 50:535-539.

9   Barber, S. A. 1962. A diffusion and mass flow concept of soil nutrient availability. Soil Sci. 93:39-49.

10   Barber, S. A. 1966. The role of root interception, mass flow and diffusion in regulating the uptake of ions by plants from soil. Tech. Rep. Ser. Int. Atom. Energy Ag. No. 65:39-45.

11   Barber, S. A. 1968. On the mechanisms governing nutrient supply to plant roots growing in soil. Trans. 9th Int. Congr. Soil Sci. 2:243-250.

12   Barber, S. A. 1974. Influence of the plant root on ion movement in soil. In: E. W. Carson (ed.). The Plant Root and its Environment, University Press of Virginia, Charlottesville, VA. pp. 525-564.

13   Barber, S. A., J. W. Waller and E. H. Nasey. 1963. Mechanisms for the movement of plant nutrients from the soil and fertilizer to the plant root. J. Agr. Food Chem. 11:204-207.

14   Barley, K. P. 1970. The configuration of the root system in relation to nutrient uptake. Advan. Agron. 22:159-201.

15   Beale, O. W., G. B. Nutt, and T. C. Peale. 1955. The effect of mulch tillage on runoff, erosion, soil properties, and crop yields. Soil Sci. Soc. Am. Proc. 19:244-247.

16   Beaufils, E. R. 1957. Research for rational exploitation of *Hevea brasiliensis* using a physiological diagnosis based on the mineral analysis of various parts of the plant. Fertility 3:27-38.

17   Beaufils, E. R. 1959. Physiological diagnosis. III. Method of interpreting data. Rev. Gen. Caoutch. 36:225.

18   Beaufils, E. R. 1971. Physiological diagnosis—a guide for improving maize production based on principles developed for rubber trees. J. Fert. Soc. South Afr. 1:1-3.

19   Beaufils, E. R. 1973. Diagnosis and recommendation integrated system (DRIS). A general scheme for experimentation and calibration based on principles developed from research in plant nutrition. Soil Sci. Bull. 1., University of Natal, South Africa.

20   Beaufils, E. R. and M. E. Sumner. 1976. Application of the DRIS approach for calibrating soil, plant yield and plant quality factors of sugarcane. Proc. South Afr. Sug. Cane Tech. Assoc. 50:118-124.

21   Belcher, C. R. and J. L. Ragland. 1972. Phosphorus absorption by sod-planted corn from surface applied phosphorus. Agron. J. 64:754-756.

22   Bennett, O. L., E. L. Mathias, and P. E. Lundberg. 1973. Crop response to no-till management practices on hilly terrains. Agron. J. 65:488-491.

23   Bertrand, A. R. and H. Kohnke. 1957. Subsoil conditions and their effect on oxygen supply and the growth of corn roots. Soil Sci. Soc. Am. Proc. 21:135-140.

24  Bird, G. W., O. L. Brooks, C. E. Perry, J. G. Futral, T. D. Canerday and F. C. Boswell. 1974. Influence of subsoiling and soil fumigation on the cotton stunt disease complex, *Hoplolaimus colombus* and *Meloidogyne incognita*. Plant Dis. Reptr. 58:541-544.

25  Black, A. L. 1973a. Crop residue, soil water, and soil fertility related to spring wheat production and quality. Soil Sci. Soc. Am. Proc. 37:754-758.

26  Black, A. L. 1973b. Soil property changes associated with crop residue management in a wheat-fallow rotation. Soil Sci. Soc. Am. Proc. 37:943-946.

27  Blevins, R. L., G. W. Thomas, and P. L. Cornelius. 1977. Influence of no-tillage and nitrogen fertilization on certain soil properties after 5 years of continuous corn. Agron. J. 69:383-386.

28  Blevins, R. L., L. W. Murdock, and G. W. Thomas. 1978. Effect of lime application on no-tillage and conventionally tilled corn. Agron. J. 70:322-326.

29  Bosemark, N. A. 1954. The influence of nitrogen on root development. Physiol. Plant 7:497-502.

30  Boswell, F. C., D. A. Ashley, O. L. Brooks, G. W. Bird, T. D. Canerday, J. G. Futral, R. S. Hussey, C. E. Perry, R. W. Roncadori and J. S. Schepers. 1977. Influence of subsoiling, liming, a nematicide, and soil bedding on cotton yield in stunt' areas of Georgia. Ga. Agric. Exp. Stn. Res. Bull. 204. 204. p. 34.

31  Bradfield, R. 1970. Increasing food production in tropics by multiple cropping. In: D. G. Aldrich, Jr. (ed.). Research for the World Food Crisis. Am. Assoc. for Advance. of Sci. Washington, D.C. 92:229-242.

32  Brill, G. D., R. B. Alderfer and W. J. Hanna. 1965. Effects of subsoiling and deep placement of fertilizer on a Coastal Plain Soil and vegetables. Agron. J. 57:201-204.

33  Browning, G. M. and R. A. Norton. 1945. Tillage practices on selected soils in Iowa. Soil Sci. Soc. Am. Proc. 10:461-468.

34  Broyer, T. C. and P. R. Stout. 1959. The macronutrient elements. Ann. Rev. Plant Physiol. 10:277-300.

35  Byrne, J. M. 1974. Root morphology. In: E. W. Carson (ed.) The Plant Root and Its Environment. University Press of Va., Charlottesville, VA. pp. 3-27.

36  Burgess, P. S. 1922. The soil solution extracted by Lipman's direct pressure method compared with 1-5 water extracts. Soil Sci. 14:151-167.

37  Carlisle, V. W. and J. G. A. Fiskell. 1968. Techniques for investigating tillage pans and associated corn root development. Soil Crop Sci. Soc. Fla. Proc. 27th Ann. meeting:159-165.

38  Carter, L. M. and J. R. Tavernetti. 1968. Influence of precision tillage and soil compaction on cotton yields. TRANSACTIONS of the ASAE 11(1):65-73.

39  Chapman, H. D. (ed.) 1966. Diagnostic criteria for plants and soils. University of California, Berkeley.

40  Coe, D. G. 1926. The effects of various methods of applying fertilizers on crop yield. Soil Sci. 21:127-141.

41  De Roo, H. C. 1961. Deep tillage and root growth. Conn. Agric. Exp. Stn. Bull. 644.

42  Duley, F. L. 1957. Subsoiling in the Great Plains. J. Soil and Water Cons. 12:71-74.

43  Dumas, W. T., A. C. Trouse, L. A. Smith, F. A. Kummer and W. R. Gill. 1973. Development and evaluation of tillage and other cultural practices in a controlled traffic system for cotton in the Southern Coastal Plains. TRANSACTIONS of the ASAE 16(5):872-880.

44  Dumenil, L., J. Pesek, J. R. Webb and J. J. Hanway. 1965. P and K fertilizer for corn: How to apply. Iowa Farm Sci. 19:159-162.

45  Duncan, W. G. and A. J. Ohlrogge. 1958. Principles of nutrient uptake from fertilizer bands. II. Root developments in the band. Agron. J. 50:605-608.

46  Eck, H. V. and H. M. Taylor. 1969. Profile modification of a slowly permeable soil. Soil Sci. Soc. Am. Proc. 33:779-783.

47  Eck, H. V. and R. G. Davis. 1971. Profile modification and root yield, distribution and activity. Agron. J. 63:934-937.

48  Engelbert, T. E. and E. Truog. 1956. Crop response to deep tillage with lime and fertilizer. Soil Sci. Soc. Am. Proc. 20:50-54.

49  Englestad, O. P. and S. E. Allen. 1971. Ammonium pyrophosphate and ammonium orthophosphate as phosphorus sources: effects of soil temperature, placement and incubation. Soil Sci. Soc. Am. Proc. 35:1002-1004.

50  Eno, C. F. and W. G. Blue. 1954. The effect of anhydrous ammonia on nitrification and the microbiolgical population in sandy soils. Soil Sci. Soc. Am. Proc. 18:178-181.

51  Erickson, A. E., C. M. Hansen and A. J. M. Smucker. 1968a. The influence of subsurface asphalt barriers on the water properties and the productivity of sand soils. Trans. 9th Int. Cong. Soil Sci. 1:35.

52 Erickson, A. E., C. M. Hansen, A. J. M. Smucker, K. Y. Li, L. C. Hsi, T. S. Wang, and R. L. Cook. 1968b. Subsurface asphalt barriers for the improvement of sugarcans production and the conservation of water on sand soil. Proc. Cong. Int. Soc. Sug. Cane Tech. 13:787-792.

53 Estes, G. O. 1972. Elemental composition of maize grown under no-till and conventional tillage. Agron. J. 64:733-735.

54 Evans, C. E., C. E. Scarsbrook and R. D. Rouse. 1970. Methods of applying nitrogen, phosphorus and potassium for cotton. Ala. Agric. Exp. Stn. Bull. 403. p. 13.

55 Ferrant, N. A. Jr. and H. B. Sprague. 1940. Effect of treating different horizons of Sassafras loam on root development of red clover. Soil Sci. 50:141-161.

56 Food and Agriculture Organization (FAO). 1973. The calibration of soil tests for fertilizer recommendations. Unipub., 345 Park Ave., New York, NY 10010.

57 Foy, C. D. and J. C. Brown. 1964. Toxic factors in acid soils: II. Differential aluminum tolerance of plant species. Soil Sci. Soc. Am. Proc. 28:27-32.

58 Foy, C. D., W. H. Armiger, L. W. Briggle, and D. A. Reid. 1965. Differential tolerance of wheat and barley varieties in acid soils. Agron. J. 57:413-417.

59 Gallaher, R. N. 1975. Triple cropping in the Georgia Piedmont. Ga. Agri. Res. 17:19-25.

60 Gingrich, J. R. 1964. Relationship of soil temperatures, water soluble phosphorus in applied fertilizer, and method of placement to growth of winter wheat. Agron. J. 56:529-532.

61 Hansen, C. M. and A. E. Erickson. 1969. Use of asphalt to increase the water holding capacity of droughty sand soils. I. and Ec. Prod. Res. and Div. 8:256.

62 Haynes, J. L. and W. Robbins. 1948. Calcium and boron as essential factors in the root environment. Agron. J. 40:795-803.

63 Hemwall, J. B. 1957. The fixation of phosphorus by soils. Adv. Agron. 9:95-112.

64 Hergert, G. W. 1977. Sprinkler application of fertilizer nutrients. Fert. Sol. 21:14-20.

65 Hergert, G. W. and J. O. Reuss. 1976. Sprinkler application of P and Zn fertilizer. Agron. J. 68:5-8.

66 Hobbs, J. A., R. B. Herring, D. E. Peaslee, W. W. Harris, and G. E. Fairbanks. 1961. Deep tillage effects on soils and crops. Agron. J. 53:313-316.

67 Howard, D. D. and Fred Adams. 1965. Calcium requirements for penetration of subsoils by primary cotton roots. Soil Sci. Soc. Am. Proc. 29:558-562.

68 Jamison, V. C. and J. F. Thornton. 1960. Results of deep fertilization and subsoiling on a clay pan soil. Agron. J. 52:193-195.

69 Jones, R. G. W. and O. R. Lunt. 1967. The function of calcium in plants. Bor. Rev. 33:407-426.

70 Kenworthy, A. L. 1973. Leaf analysis as an aid in fertilizing orchards. In: L. M. Walsh and J. D. Beaton (eds.) Soil Testing and Plant Analysis. Soil Sci. Soc. of Am., Madison, WI. pp. 381-393.

71 Knoll, H. A., N. C. Brady and D. J. Lathwell. 1964. Effect of soil temperature and phosphorus fertilization on the growth and phosphorus content of corn. Agron. J. 56:145-147.

72 Kohnke, H. and A. R. Bertrand. 1956. Fertilizing the subsoil for better water utilization. Soil Sci. Soc. Am. Proc. 20:581-586.

73 Kung, P. 1969. Multiple cropping in Taiwan. World Crops. 21:128-130.

74 Larson, W. E., W. G. Lovely, J. T. Pesek and R. R. Burwell. 1960. Effect of subsoiling and deep fertilizer placements on yield of corn in Iowa and Illinois. Agron. J. 52:185-189.

75 Leavitt, F. H. 1942. Process of soil fertilization. U. S. Pat. 2, 285,932. June 9 (to Shell Development Co.).

76 Leavitt, F. H. 1966. Agricultural ammonia equipment—development and history. In: M. H. McVickar, W. P. Martin, I. E. Miles and H. H. Tucker (eds.) Agricultural Anhydrous Ammonia. Agric. Ammonia Inst., Am. Soc. of Agron. and Soil Sci. Soc. of Am., Madison, WI. p. 125-142.

77 le Roux, J. and M. E. Sumner. 1967. Studies on the soil solution of various Natal soils. Geoderma 1:125-130.

78 Lowry, F. E., H. M. Taylor and M. G. Huck. 1970. Growth rate and yield of cotton as influenced by depth and bulk density of soil pans. Soil Sci. Soc. Am. Proc. 34:306-309.

79 Lutz, J. A., Jr. and J. H. Lillard. 1973. Effect of fertility treatments on the growth, chemical composition and yield of no-tillage corn on orchard grass sod. Agron. J. 65:733-735.

80 McCalla, T. M. and T. J. Army. 1961. Stubble mulch farming. Advan. Agron. 13:125-196.

81 McCracken, R. J. and S. B. Weed. 1963. Pan horizons in southeastern soils: Micromorphology and associated chemical, mineralogical, and physical properties. Soil Sci. Soc. Am. Proc. 27:330-334.

82   McVickar, M. H., W. P. Martin, I. E. Miles and H. H. Tucker. 1966. Agricultural Anhydrous Ammonia. Agric. Ammonia Inst., Am. Soc. of Agron. and Soil Sci. Soc. Am., Madison, WI. p. 314.

83   Mech, S. J., G. M. Horner, L. M. Cox and E. E. Cary. 1967. Spil profile modification by backhoe mixing and deep plowing. TRANSACTIONS of the ASAE 10(6):775-779.

84   Meldal-Johnsen, A. and M. E. Sumner. 1980. Foliar diagnostic norms for potatoes. J. Plant Nutr. 2:569-576.

85   Melsted, S. W., H. L. Motto and T. R. Peck. 1969. Critical plant nutrient composition values useful in interpreting plant analysis data. Agron. J. 61:17-20.

86   Miller, M. H. and A. J. Ohlrogge. 1958. Principles of nutrient uptake from fertilizer bands. I. Effect of placement of nitrogen fertilizer on the uptake on band-placed phosphorus at different soil phosphorus levels. Agron. J. 50:95-97.

87   Mock, J. J. and D. C. Erbach. 1977. Influence of conservation-tillage environments on growth and productivity of corn. Agron. J. 69:337-340.

88   Mortvedt, J. J. and P. M. Giordano. 1967. Zinc movement in soils from fertilizer granules. Soil Sci. Soc. Am. Proc. 31:608-613.

89   Moschler, W. W., G. M. Shear, D. C. Martens, G. D. Jones and R. R. Wilmouth. 1972. Comparative yield and fertilizer efficiency of no-tillage and conventionally tilled corn. Agron. J. 64:229-231.

90   Moschler, W. W., D. C. Martens, C. I. Rich, and G. M. Shear. 1973. Comparative lime effects on continuous no-tillage and conventionally tilled corn. Agron. J. 65:781-783.

91   Nelson, W. L., B. A. Krantz, C. D. Welch and N. S. Hall. 1949. Utilization of phosphorus as affected by placement: II. Cotton and corn in North Carolina. Soil Sci. 68:137-144.

92   Nissley, C. H. 1944. Tillage and fertilization below plow depth pay big dividends. New Jersey Agriculture. 26:1-2.

93   Olsen, S. R. and W. D. Kemper. 1968. Movement of nutrients to roots. Advan. Agron. 20:91-149.

94   Olsen, S. R., F. S. Watanabe, and R. E. Davidson. 1961. Phosphorus absorption by corn roots as affected by moisture and phosphorus concentration. Soil Sci. Soc. Am. Proc. 25:289-294.

95   Olson, T. C. and L. S. Schoeberl. 1970. Corn yields, soil temperature, and water use with four tillage methods in the western corn belt. Agron. J. 62:229-232.

96   Page, J. B., C. J. Willard and G. W. McCuen. 1946. Progress report on tillage methods in preparing land for corn. Ohio Agric. Exp. Stn., Agron. Mimeo. No. 102.

97   Parker, M. B., N. A. Minton, O. L. Brooks, and C. E. Perry. 1975. Soybean yields and lance nematode populations as affected by subsoiling, fertility, and nematicide treatments. Agron. J. 67:663-666.

98   Parks, C. L., A. W. White and F. C. Boswell. 1970. Effect of a plastic barrier under the nitrate band on nitrogen uptake by plants. Agron. J. 62:437-439.

99   Patrick, W. H., Jr., L. W. Sloane, and S. A. Phillips. 1959. Response of cotton and corn to deep placement of fertilizer and deep tillage. Soil Sci. Soc. Am. Proc. 23:307-310.

100   Pearson, R. W. 1966. Soil environment and root development. In: Plant Environment and Efficient Water Use. Am. Soc. Agron., Madison, WI. pp. 95-126.

101   Pearson, R. W. 1974. Significance of rooting pattern to crop production and some problems of root research. In: E. W. Carson (ed.) The plant root and its evironment. University Press of Va., Charlottesville, VA. pp. 247-270.

102   Pendleton, J. W. 1976. Increasing water use efficiency by crop management. In: M. J. Wright (ed.) Plant adaptation to mineral stress in problem soils. Cornell University, Ithaca, NY. pp. 236-258.

103   Pierre, W. H., J. R. Webb and W. D. Shrader. 1971. Quantitative effects of nitrogen fertilizer on the development and downward movement of soil acidity in relation to level of fertilization and crop removal in a continuous corn cropping system. Agron. J. 63:291-297.

104   Power, J. F., D. L. Grunes, G. A. Reichman and W. O. Willis. 1964. Soil temperature and phosphorus effects upon nutrient absorption by barley. Agron. J. 56:355-359.

105   Rasmussen, W. W., D. P. Moore, and L. A. Alban. 1972. Improvement of a solonetez (slick spot) soil by deep plowing, subsoiling and amendments. Soil Sci. Soc. Am. Proc. 36:137-142.

106   Reid, J. T. 1978. A comparison of the energy input of some tillage tools. ASAE Paper No. 78-1039, ASAE, St. Joseph, MI 49085.

107   Rios, M. A. and R. W. Pearson. 1964. The effect of some chemical environmental factors on cotton root behavior. Soil Sci. Soc. Am. Proc. 28:232-235.

108   Robinson, R. R., V. G. Sprague and C. F. Goss. 1959. The relation of temperature

and phosphate placement to growth of clover. Soil Sci. Soc. Am. Proc. 23:225-235.

109   Robertson, W. K., J. G. A. Fiskell, C. E. Hutton, L. G. Thompson, R. W. Lipscomb and H. W. Lundy. 1957. Results from subsoiling and deep fertilization of corn for 2 years. Soil Sci. Soc. Am. Proc. 21:340-346.

110   Russell, R. S. 1978. Plant root systems: their functions and interaction with the soil. McGraw-Hill Book Co. (UK) Ltd., London.

111   Sanford, J. O., D. L. Myhre, and N. C. Merwine. 1973. Double cropping systems involving no-tillage and conventional tillage. Agron. J. 65:978-982.

112   Shear, G. M. and W. W. Moschler. 1969. Continuous corn by the no-tillage methods: a six-year comparison. Agron. J. 61:524-526.

113   Singh, T. A., G. W. Thomas, W. W. Moschler and D. C. Martens. 1966. Phosphorus uptake by corn (*Zea mays* L.) under no-tillage and conventional practices. Agron. J. 38:147-148.

114   Soileau, J. M., O. P. Engelstad, and J. B. Martin, Jr. 1969. Cotton growth in an acid fragipan subsoil: II. Effect of soluble calcium, magnesium, and aluminum of roots and tops. Soil Sci. Soc. Am. Proc. 33:919-924.

115   Stanford, G., O. L. Bennett and J. F. Power. 1973. Conservation tillage practices and nutrient availability. In: Conservation Tillage Proc. National Conservation Tillage Conference, Des Moines, Iowa. Soil Cons. Soc. of Am., Ankeny, IA.

116   Stangel, P. J. 1976. World fertilizer reserve in relation to future demand. In: M. J. Wright (ed.) Plant adaptation to mineral stress in problem soils. Cornell University, Ithaca, NY. p. 31-46.

117   Stibbe, E. and U. Kafkafi. 1973. Influence of tillage depths and P—fertilizer application rates on the yield of annual cropped winter-grown wheat. Agron. J. 65:617-620.

118   Stivers, R. K., J. H. Lillard, G. D. Jones and A. H. Allison. 1956. The relation of subsoiling, deep lime, and deep phosphate application on the yield of corn, peanuts, and alfalfa in Virginia. Va. Poly. Tech. Res. Inst. Agric. Ext. Serv. Circ. 659.

119   Stockton, J. R., L. M. Carter, and G. Paxman. 1964. Precision tillage for cotton. Calif. Agric. 18(2):8-10.

120   Sumner, M. E. 1977a. Preliminary N, P and K foliar diagnostic norms for soybeans. Agron. J. 69:226-230.

121   Sumner, M. E. 1977b. Preliminary NPK foliar diagnostic norms for wheat. Comm. Soil Sci. Plant Anal. 8:149-167.

122   Sumner, M. E. 1977c. Use of the DRIS system in foliar diagnosis of crops at high yield levels. Comm. Soil Sci. Plant Anal. 8:251-268.

123   Sumner, M. E. 1977d. Effect of corn leaf sampled on N, P, K, Ca and Mg content and calculated DRIS indices. Comm. Soil Sci. Plant Anal. 8:269-280.

124   Sumner, M. E. 1979. Interpretation of foliar analyses for diagnostic purposes. Agron. J. 71(2):343-348.

125   Sumner, M. E. and E. C. Gilfillan. 1971. Asphalt barriers to improve productivity of sandy soils. A preliminary assessment. Proc. S. African Sug. Tech. Assoc. 45:165-168.

126   Syltie, P. W., S. W. Melsted and W. M. Walker. 1972. Rapid tissue test—an indication of yield, plant composition, and soil fertility for corn and soybeans. Comm. Soil Sci. Plant Anal. 3:37-50.

127   Taylor, H. M. 1974. Root behavior as affected by soil structure and strength. In: E. W. Carson (ed.) The plant root and its environment. University Press of Va., Charlottesville, VA. pp. 271-291.

128   Taylor, H. M. and H. R. Gardner. 1963. Penetration of cotton seedling tap roots as influenced by bulk density, moisture content, and strength of soil. Soil Sci. 96:153-156.

129   Taylor, H. M., G. M. Roberson and J. J. Parker, Jr. 1966. Soil strength-root penetration relations for medium-to-coarse-textured soil materials. Soil Sci. 102:18-22.

130   Thomas, G. W., R. L. Belvins, R. E. Phillips and M. A. McMahon. 1973. Effect of a killed sod mulch on nitrate movement and corn yield. Agron. J. 65:736-739.

131   Triplett, G. B., Jr. and D. M. Van Doren, Jr. 1969. Nitrogen, phosphorus, and potassium fertilization of no-tilled maize. Agron. J. 61:637-639.

132   Triplett, G. B., Jr. and D. M. Van Doren, Jr. 1977. Agriculture without tillage. Sci. Am. 236:28-33.

133   Truog, E., H. J. Harper, O. C. Magistad, F. W. Parker and J. Sykora. 1925. Methods of application and effect on germination, early growth, hardiness, root growth, lodging, maturity, quality and yield. Wis. Agric. Exp. Stn. Res. Bull. 65.

134   Van Doren, D. M., Jr., G. B. Triplett, Jr. and J. E. Henry. 1976. Influence of long-term tillage, crop rotation, and soil type combinations on corn yield. Soil Sci. Soc. Am. J. 40:100-105.

135   Voorhees, W. B. 1977. Soil compaction—our newest natural resource. Crops and Soils 29:13-15.

136   Walker, M. E. and O. L. Brooks. 1964. Comparative response of cotton to broadcast application versus row placement of mixed fertilizer. Ga. Agric. Exp. Stn. Mimeo. Series N.S. 193.

137   Walsh, L. M. and J. D. Beaton. 1973. Soil testing and plant analysis. Soil Sci. Soc. of Am., Madison, WI.

138   Welch, L. F., D. L. Mulvaney, L. V. Boone, G. E. McKibben and J. W. Pendleton. 1966. Relative efficiency of broadcast versus banded phosphorus for corn. Agron. J. 58:283-287.

139   Wilcox, G. E., G. C. Martin and R. Lanston. 1962. Root zone temperature and phosphorus treatment effects on tomato seedling growth in soil and nutrient solutions. Am. Sco. Hort. Sci. 80:522-529.

140   Woodruff, C. M. and D. D. Smith. 1947. Subsoil shattering and subsoil liming for crop production on clay pan soils. Soil Sci. Soc. Am. Proc. 11:539-542.

141   Wright, M. J. 1976. Plant adaptation to mineral stress in   problem soils.   Cornell University, Ithaca, NY. p. 420.

142   Zingg, A. W. and C. J. Whitfield. 1957. A summary of research experience with stubble-mulch farming in the western states. USDA Tech. Bull. 1166. 56 pp.

# chapter 5 ▐██████████

## ALLEVIATING AERATION STRESSES

**5**

# 5

# ALLEVIATING AERATION STRESSES

by   R.  Q.  Cannell   and   Michael  B.   Jackson,
Agricultural   Research   Council   Letcombe
Laboratory, Wantage, England

## 5.1 INTRODUCTION

This chapter considers the problems for plant growth and crop production associated with poor soil aeration and assesses various methods for overcoming them. The subject is treated in three stages, by outlining (a) conditions that can give rise to poor soil aeration, (b) responses of plants to such conditions, and (c) various methods of lessening their effects. Readers are likely to have widely differing scientific backgrounds, so relatively few assumptions of previous specialized knowledge of soil aeration problems have been made. Various aspects of this subject have been well reviewed by several authors (eg. in Luthin, 1957; Bergman, 1959; Grable, 1966; van Schilfgaarde, 1974).

## 5.2 THE OCCURRENCE OF POORLY AERATED SOIL CONDITIONS

### 5.2.1 Estimates of the Area of Soils with Excess Water

Poorly drained and badly aerated soils occur in many parts of the world, but reliable estimates of their extent are not readily available. It is difficult to appraise global soil resources because only about one fifth of the world's soils have been surveyed, and the distribution of the surveyed areas is not uniform; the most complete surveys are for European soils (Dudal, 1976). In addition, until recently there was no generally accepted system of soil classification. Efforts of the Food and Agriculture Organization of the United Nations (FAO) and the International Society of Soil Science have produced a Soil Map of the World, from which Dudal (1976) estimated that about 12 percent of the world's soils have excess water.

The main groups of poorly draining soils in the FAO classification are Fluvisols, Gleysols, Vertisols, Planosols and Histosols (Table 5.2-1, where the equivalent terms for these groups in the USDA soil classification are also given). Their main characteristics are: Fluvisols—weakly developed soils from alluvial deposits in active floodplains; Gleysols—soils with excess water as a major factor in their formation; Vertisols—developed from swelling

**Acknowledgement:** We wish to thank most warmly our colleagues and collaborators at the Letcombe Laboratory who commented on and discussed the draft of this chapter, and assisted with the preparation of the bibliography and diagrams.

TABLE 5.2-1. MAJOR WORLD SOILS THAT COMMONLY
CONTAIN EXCESS WATER (from Dudal, 1976)

| Soil associations* dominated by | Area ($10^6$ ha) | Area as percent of world land area |
|---|---|---|
| Fluvisols (Fluvents)† | 316 | 2.40 |
| Gleysols (Aquents, Aquets and Halaquolls)† | 623 | 4.73 |
| Vertisols (Vertisols) | 311 | 2.36 |
| Planosols (Albolls, Albaqualfs and Albaquults) | 120 | 0.91 |
| Histosols (Histosols)† | 240 | 1.82 |
| | 1610 | 12.22 |

*Units of the International Legend (FAO, 1974); the equivalent
terms from the USDA Soil Classification (USDA, 1975) are shown
in parenthesis.
†Large proportions of these soils occur in areas with permafrost
(Dudal, 1978).

clays, showing deep and wide cracks when dry, a turn-over of surface
material by self-mulching and a very narrow range between moisture stress
and excess of water (this latter characteristic results from the very strong
clay-water bonds of the expanding-type clay minerals); Planosols—usually
developed in level or depressed topography, often with sandy surface
horizons, where plants are subjected alternately to seasonal surface waterlogging and severe drought; Histosols—saturated with water for prolonged
periods and with a thick upper horizon of fresh or partly decomposed organic
matter.

The extent of the problems arising from waterlogging also can be partly
gauged from the use of artificial drainage systems. Apart from avoiding
waterlogging, much drainage has been installed to prevent salinity from
building up, and to some extent to improve trafficability of land. Nosenko
and Zonn (1976) have estimated that 155 million ha have been drained. Of
this, 68, 35 and 32 million ha are in North and Central America, Europe (except Russia) and in Asia respectively. Much land has been drained in recent
years. In England and Wales 6 percent of the agricultural land was drained
between 1940 and 1969 (Belding, 1971) and the current rate of work is about
0.1 million ha per annum (Cole, 1976); in Canada the annual rate is believed
to be in excess of 0.06 million ha (Donnan, 1976). The USSR plans to double
the area of drained land to 20 million ha within a few years (Sotnknov, 1970),
and in the Nile delta 0.4 million ha were scheduled to be drained between
1970 and 1978 (Ibrahim, 1976).

## 5.2.2 Factors Influencing Soil Aeration

Anaerobic conditions develop in soil when roots and soil organisms use
oxygen for respiration faster than it can enter the soil by diffusion through interconnected air-filled pores (Currie, 1970; Grable, 1971). Since oxygen diffuses in water about $10^4$ times more slowly than in air, the presence of water-
filled pores is the main restriction to soil aeration. Furthermore, diffusion
out of the soil of gases such as carbon dioxide will likewise be impeded by excess water, and these can then accumulate.

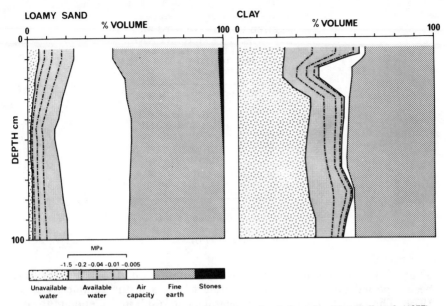

FIG. 5.2-1 Water retention characteristics of loamy sand and clay soils. After Hall et al. (1977).

### 5.2.2.1 The demand for oxygen

The rate of utilization of oxygen by respiration in soil can be large in comparsion with the amount contained within the soil volume that roots usually occupy. For example, in the sand and clay soils illustrated in Fig. 5.2-1, the volumes of oxygen contained after draining freely would be about

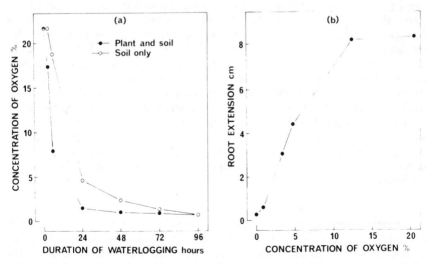

FIG. 5.2-2 (a) The effect of the presence and absence of a tomato plant on the concentration of oxygen dissolved in the soil water during 96 h of waterlogging at 23 °C. (After Jackson and Campbell, 1976b).
(b) The effect of various concentrations of oxygen applied for 3 days to the roots of 3 day old barley seedlings growing in well stirred nutrient solution on root extension after 3 days (day temperature 24 °C, night temperature 14 °C). (Jackson and Dobson, unpublished).

TABLE 5.2-2. THE TIME (DAYS)
NEEDED FOR OXYGEN TO BE
DEPLETED TO 2 PERCENT (v/v)
AT 20 CM DEPTH IN SOIL
WATERLOGGED TO THE
SURFACE

| Soil | Winter | Spring |
|---|---|---|
| | (with mean soil temperature at 20 cm) | |
| | (3 °C) | (13 °C) |
| Sandy loam | 15 | 6 |
| Clay | 10 | 5 |

The experiment was carried out
in lysimeters containing soil mono-
liths and growing winter wheat
(from Cannell (1979))

30 and 5 $L/m^3$ respectively. Rates of consumption up to 17 $L/m^3 \cdot day$ have been reported under crops in temperate climates in the summer (Currie, 1970). Thus sealing the surface of the soil could result in depletion of the oxygen in less than 3 days in the sand and 1 day in the clay, and probably faster in warmer climates. Currie also reported (*loc cit*) that oxygen consumption was slowed to about 1.5 $L/m^3 \cdot day$ in winter, and was nearly halved in the absence of plants.

Waterlogging restricts the supply of oxygen to roots and soil organisms, first by displacing air from the soil, and second by slowing oxygen diffusion. Initially, the water is a source of oxygen for the roots and micro-organisms, but the amount it contains when in equilibrium with air (9.6 mg/L at 20 °C) can be depleted within 2 or 3 days in fallow soil and faster when plants are present (Fig. 5.2-2). R. B. Lee (in unpublished results from Letcombe Laboratory, Wantage, England) has estimated that sterile roots of barley (*Hordeum vulgare* L.) consume about $3.5 \times 10^{-3}$ g $O_2$ g fresh weight/day at 20 °C. The root system of a plant at the beginning of tillering and weighing about 2.5 g may thus require about $9 \times 10^{-3}$ g $O_2$/day, equivalent to 2 to 3 times its own volume or the amount dissolved in about 900 mL of water initially in equilibrium with air. The concentration of oxygen in a waterlogged field soil decreases faster in spring when respiration is more rapid than in winter; it also decreases faster in a clay with smaller drainable pore space than a sandy loam (Table 5.2-2). In a given climate, clay soils are therefore likely to become anaerobic more frequently than sandy ones and remain so for longer periods.

### 5.2.2.2 Pore space and arrangement

In discussing aeration, the soil is conveniently considered as a network of channels (pores) filled with air and water. The sizes, distribution frequency and continuity of soil pores all influence how much soil aeration is inhibited at a given moisture content. The pore space typically accounts for 40 to 60 percent of the total volume of the soil. The individual soil particles vary in size from the smallest clay, to the largest sand and stones, but it is the proportions of the different sizes of aggregates of particles or peds that determines the distribution of pore space within and between them. This in turn influences the potential or force needed to remove water from the pores thereby allowing air to enter.

The pores vary greatly in size, but can be grouped into three categories: first, large pores that are air-filled at 'field capacity'. When a soil drains freely under gravity to its field capacity, the water potential is unlikely to be lower than about -5 to -10 kPa and pores smaller than 60 to 30 $\mu$m will be water-filled (Russell, 1973); in soils containing more than about 30 to 40 percent of clay the water potential at field capacity may be appreciably higher than this (Webster and Becker, 1972) so that only pores larger than 100 to 300 $\mu$m in diameter are air-filled. The diameters of roots (except root hairs) usually exceed these pore diameters; for example in wheat (*Triticum aestivum* L.), and barley, main root axes are 300 to 700 $\mu$m, and secondary laterals are 100 to 300 $\mu$m (Hackett, 1969; Finney and Knight, 1973). Since roots will not enter rigid pores smaller than their own diameter (Wiersum, 1957), it follows that the pores they enter are likely to be air-filled when the soil is at or below field capacity.

The second category are pores that are small enough to hold water against gravity, yet permit water to be withdrawn by roots, down to a potential of about -1500 kPa; this corresponds to diameters between about 60 $\mu$m and 0.2 $\mu$m. If the soil is free-draining and does not shrink on drying pores will be water filled at certain times of the year, depending on rainfall and the extent to which water is extracted by plant roots. In clay soils where there is extensive shrinkage, as water is withdrawn, the pores collapse and air entry is much less (Haines, 1923; Reeve and Hall, 1978).

In a third category are pores so small that plants are unable to remove water from them. In the absence of a generally accepted terminology to relate pore size to function, Greenland (1977) proposed the terms 'transmission', 'storage' and 'residual' pores for the three categories; and for convenience he selected equivalent pore diameters of 50 $\mu$m and 0.5 $\mu$m to separate them.

The proportions of pores in these three categories vary considerably between soils; for example in Fig. 5.2-1 the transmission pores occupy 20 to 30 percent of the soil volume in the loamy sand, but only about 3 percent in the clay soil; in contrast, residual pores occupy a large proportion of the clay.

In addition to pore size distribution, pore continuity can greatly affect the rate of diffusion of gases (Currie, 1970) and drainage of water. The ease of water movement through soil pores is quantified by hydraulic conductivity and soils can be classified for this property. Thomasson et al., (1975) proposed six groups (conductivity in cm/day): very slow, <1; slow, 1 to 10; moderately slow, 10 to 30; moderately rapid, 30 to 100; rapid, 100 to 1000 and very rapid >1000. Knowledge of saturated hydraulic conductivity at different depths in the soil profile is clearly fundamental to the design of drainage systems (see Section 5.4.1.4).

### 5.2.2.3 Soil heterogeneity

Few agriculturally important soils are homogenous, apart from some fine silts and coarse sands. Heterogeneity may be evident from the different proportions of sand and clay down the soil profile, from the degree of aggregation at different depths, and can also be influenced by soil management. The likelihood of poor aeration depends not only on the physical characteristics of the soil profile, but also on soil topography.

The oxygen concentration in a well-drained sandy soil, with interconnected air spaces is similar to that in the atmosphere above. Waterlogging in such soil results mainly from proximity to a groundwater-table, and they are known as 'ground-water soils' (Robson and Thomasson, 1977). By contrast, 'surface-water soils' occur in sites remote from a regional water-table. In these, the downward movement of water is restricted by an impermeable or slowly permeable layer. The degree of waterlogging can vary greatly depending on the depth of the impermeable layer and on the topography of the soil surface, but the topsoil may be partly saturated, (Trafford and Rycroft, 1973) or inundated in patches for several weeks. Such ponding and consequent death of the affected patches of crop are particularly likely on unstable soils prone to slaking after rainfall or frost.

Heterogeneity within a soil profile may also significantly affect aeration. Clay soils are often characterized by large crumbs or peds, and the oxygen concentration from point to point in the soil can then be variable (Greenwood and Goodman, 1967; Dowdell et al., 1979). When shrinkage cracks have formed in clay soils during dry conditions, the planes between the structural units may contribute to localized variation in aeration and it may take several months for the soil to completely swell and for the oxygen concentration to reach its minimum value throughout the soil (Smith and Dowdell, 1974). Greenwood (1975) has estimated the maximum radius of aggregates without being anaerobic in the center (R) from

$$R = \frac{6DSP}{M} \qquad [5.2\text{-}1]$$

where D is the diffusion coefficient of oxygen through the water-saturated soil, S is the solubility of oxygen in the soil water, M is the rate of oxygen utilization, and P is the partial pressure outside the aggregate. From this analysis he has estimated that in British soils, oxygen-free zones could occur if the diameter of the water-saturated aggregates excees 18 mm, as is likely in heavy soils. Both the inability of the roots to enter anaerobic aggregates, and denitrification within anaerobic microsites (see Section 5.2.3), may restrict the availability of nutrients to the plant. An empirical estimate of 10 percent air-filled pore space to avoid anaerobic conditions has been suggested by Grable (1971) and Greenwood (1975), but this depends on the distribution and continuity of the pores, and the demand for oxygen.

Natural variation in the soil physical properties that influence air and water movement may be further modified by soil management procedures. Apart from controlling weeds cultivation has traditionally been considered important for creating a well aerated topsoil, but untimely field operations with heavy equipment can damage weakly structured soils. Compaction of wet soil is particularly likely to diminish the proportion of transmission pores (e.g. Bodman et al., 1958) and their continuity (Blake et al., 1976) and restrict root growth by mechanical impedance (Chapter 2). Compaction by wheels may extend to a depth of 30 cm or more (Soane, 1975), accentuating heterogeneity of porosity in and below the wheel tracks in comparison with untracked areas. During moldboard plowing smearing on the furrow sole and by wheels on the bottom of the furrow can reorientate soil particles in a layer only a few millimeters thick, thereby interrupting the continuity of transmission pores, so that water may accumulate in the top soil and restrict gaseous exchange with the subsoil. This effect is especially likely in clay soils, and may exaggerate the natural tendency for water movement in undrained clays to be predominantly lateral through the cultivated layer. Here the proportion of transmission pores is larger than in the undisturbed subsoil (Trafford and Rycroft, 1973), so tending further to isolate the topsoil from the less aerated subsoil.

### 5.2.3 Effects of Aeration on Soil Conditions

The changes in soil conditions that follow restriction of aeration by excess water have been well discussed in many publications (e.g. Ponnamperuma, 1972; Russell, 1973). In this section we consider the changes likely to affect plant growth.

Roots and soil organisms obtain their energy from the oxidation of organic substances in a series of enzyme-catalyzed reactions involving the transfer of electrons from respirable substrates along a series of organic acceptor molecules. In aerobic respiration the usual terminal acceptor is molecular oxygen from the environment. This is combined with electrons and hydrogen ions to form water; the other end-product is carbon dioxide. When soil is waterlogged, and after free oxygen is exhausted, many microorganisms use other substances from their environment as terminal acceptors of electrons. Roots seem unable to do this but the soil which surrounds them will become increasingly reduced biochemically as anaerobic microorganisms sequentially deplete it of oxidized constituents by donating electrons. This will proceed more quickly at high temperatures and in soils containing readily decomposible organic matter as an electron and energy source. The sequence of the main reduction reactions after the disappearance of molecular oxygen follows the thermodynamically inevitable sequence: nitrate to nitrite, and in turn to nitrous oxide and nitrogen gases; manganic to manganous ions; ferric to ferrous ions; and sulfate to sulfide. This is governed by the relative electron affinities of the acceptors; those with the highest affinity (high redox potential) being chemically reduced before those with a lower affinity (lower redox potential). The possibility that the products of some of these reduction reactions are phytotoxic is discussed later.

In anaerobic conditions, organic substrates are not broken down completely to carbon dioxide. Incompletely oxidized intermediates and end products of attenuated pathways of respiration can therefore accumulate in waterlogged soil. They include lactic acid, ethanol, acetaldehyde and aliphatic acids such as acetic or butyric acid. The possibility that these partly oxidized organic compounds are toxic to waterlogged plants is discussed in Section 5.3.1. The physiologically active gas ethylene is sometimes present in abnormally high concentrations in waterlogged or anaerobic soil (Smith and Jackson, 1974; Ioannou et al., 1977). It seems to be produced at first aerobically, by roots and by soil fungae (Lynch and Harper, 1974), and by aerobic pseudomonad bacteria (Pazout and Wurst, 1977). Production may also occur under anaerobic conditions (Smith and Restall, 1971; Pazout and Wurst, 1977), possibly by extracellular processes with reactants originating from the anoxic lysis of micro-organisms (Smith et al., 1978). These extracellular reactions are thought to require $Fe^{2+}$ ions (Smith et al., 1978). Addition of nitrate can diminish ethylene production by anaerobic soils (Smith and Restall, 1971; Sheard and Leyshon, 1976). The effect may be due to an inhibition of $Fe^{2+}$ production, since the $NO_3^-$ will be reduced rather than $Fe^{3+}$ ions. When ethylene is produced in soil by whatever means, the presence of water will help to retain it, both by impeding its escape to the atmosphere and by maintaining anaerobic conditions that slow its degradation (Cornforth, 1975).

Reduction of nitrate to gaseous nitrous oxide and nitrogen (denitrification) is potentially very important because plant growth can be limited by the nitrate supply. Until recently it has only been possible to estimate indirectly the amount of nitrate lost by denitrification, as the unaccounted fraction of nitrogen budgets; this commonly gave values in the range of 10 to 20 percent of applied fertilizer nitrogen (Broadbent and Clark, 1965; Allison, 1973). An average loss of 19 kg/ha·yr has been estimated by Hauck (1969), but greater losses would be likely in wet conditions that favor denitrification.

Assessment of denitrification losses has been hindered by a lack of techniques to mesure the nitrogen gas component, believed to exceed that of nitrous oxide (readily measured using gas chromatography) by a widely variable factor (Council for Agricultural Science and Technology, 1976). The finding that acetylene can inhibit the reduction of nitrous oxide to nitrogen (Federova et al., 1973; Yoshinari and Knowles, 1976) is potentially of great value; if the inhibition was complete (although it may not be, Yeomans and Beauchamp, (1978)) the total amount of nitrate lost through denitrification could be measured as nitrous oxide. Using this procedure Ryden et al. (1979) found that when acetylene was introduced into an irrigated loam soil, a Haploxeroll, in California, nitrous oxide was the sole product of denitrification and following the application of 335 kg of nitrogen fertilizer/ha, 51 kg N/ha was lost in 123 days. Peak flux densities of nitrous oxide loss from the soil equivalent to 2 kg N/ha·day occurred 1 to 3 days after irrigation ceased, but the rate of loss declined as the soil dried. This trend is in general accord with the finding of Patrick and Wyatt (1964) that repeated aeration and waterlogging of soil (where conditions first favor formation of nitrate from organic matter (mineralization-nitrification) and then its subsequent denitrification), were particularly conducive to large losses of nitrate. Wet soils may contain a mixture of aerobic and anaerobic pockets, and both nitrification and denitrification may occur simultaneously (Greenland, 1962). The acetylene inhibition technique is not yet known to be appropriate on clay soils (e.g. Typic Haplaquepts) where large structural units and a small air-filled pore space are likely to hinder the establishment of effective acetylene concentrations within the peds. Burford et al. (1978) found evidence of denitrification during several years in such clay soils when they were wet and oxygen concentrations small. Even without using acetylene to inhibit nitrous oxide reduction, flux densities of nitrous oxide equivalent to 0.5 kg N/ha·day were measured for short periods (Burford et al., 1979).

The wet conditions favoring denitrification may also contribute to leaching losses of nitrate. For example, Cooke (1976) reported losses of between 0 and 66 kg $NO_3$-N/ha, depending on the wetness of the winter. Thus, waterlogging can slow crop growth both by its direct effects and through the depletion of nitrate by a combination of denitrification and leaching.

## 5.3 PLANT RESPONSES TO POORLY AERATED SOIL CONDITIONS

In this section we ask how waterlogging injury is brought about. The innumerable combinations of soil type, plant species, growth stage, temperature, day length and duration of the stress make it difficult to advance a unifying theory of waterlogging injury. The plasticity of the developmental processes of many plants (by contrast with most animal species) and the varied adaptive mechanisms induced by flooding lie at the heart of the problem. Difficulties in reviewing and interpreting the large literature on waterlogging effects were experienced as long ago as 1921 by Clements who listed over 700 papers in his review of aeration and by Zimmerman (1930), who found the published information 'varied and voluminous'. Despite these problems, a rational basis for exploring the mechanisms of waterlogging injury can be found in the requirements of most plants for oxygen, inorganic nutrients and water from the soil, in the phytotoxicity of substances that may accumulate in waterlogged soil and in the mediating role of plant hormones in many of the morphological changes brought about by waterlogging.

### 5.3.1 Root Growth and Development

Many of the changes in the chemical and physical environment of the roots that follow flooding of soil are detrimental to root growth and some may quickly cause their death. The factors involved are discussed below.

#### 5.3.1.1 Effect of water on the diffusion of gaseous metabolites

Immediately after the roots become surrounded by water, the loss by diffusion of gaseous metabolites such as carbon dioxide and ethylene from the submerged parts is impeded. Ethylene is probably more physiologically active than carbon dioxide and its effects may be amongst the earliest growth responses to inundation. A 0.5 mm layer of water over the roots of white mustard (*Sinapis alba* L.) seedlings is though sufficient to trap growth inhibiting concentrations of the gas (Konings and Jackson, 1979). Ethylene trapping may explain at least in part reports that thin coverings of water can inhibit elongation in other plants (Haberkorn and Sievers, 1977; Larque-Saavedra et al., 1975), promote adventitious rooting (Kawase, 1972), favor the upward growth of previously horizontal stolons (Bendixen and Peterson, 1962), or increase the size of lenticels (Zimmerman, 1930) which can permit oxygen entry into woody plants. Kawase (1974) suggests that enough ethylene gas may accumulate in the roots of waterlogged plants for some to diffuse in physiologically active amounts into the shoot system. Ethylene rarely if ever brings about cell death and its effects as an inhibitor of root elongation are reversible by lowering the concentration (Smith and Robertson, 1971). It is therefore not ethylene that causes the root death commonly found in flooded soils.

#### 5.3.1.2 Damaging effects of small concentrations of oxygen

Root growth and survival require metabolic energy. In aerobic conditions most of this is generated in mitochondria (where electrons from sugars are transferred in energy releasing steps along a series of organic molecules and finally to oxygen). In the absence of oxygen, these electron transfers cannot take place and the energy yield from each sugar molecule is cut back 95 percent (Conn and Stumpf, 1963). Within mitochondria, enzymic catalysis of the transfer of electrons to oxygen is halved by equilibrium concentrations of the gas of between 0.2 and 0.002 percent (equivalent to 1 and 0.01 percent of the oxygen present in air-saturated water) (Longmuir, 1954; Yocum and Hackett, 1957). Concentrations of oxygen outside the root that inhibit growth or respiration are invariably higher than this and have been variously estimated at 14 percent (Eavis et al., 1971), 5 percent (Fig. 5.2-2b), 2.5 percent (Greenwood and Goodman, 1971) or less than 1 percent (Armstrong and Gaynard, 1976). Large differences between the inhibiting concentrations inside and outside the root may reflect both a severe impedance to oxygen diffusion through root cells and a fast rate of oxygen consumption.

Regions of high metabolic rate such as the zones of cell division and elongation are the first to suffer oxygen deficiency (Berry and Norris, 1949), while the more quiescent parts, such as lateral root primordia with their lower demands for energy are more resilient and may survive modest amounts of oxygen stress (Huck, 1970). Problems of energy shortage may be compounded by depletion of respirable reserves arising from the 'Pasteur Effect' (enhanced glycolytic breakdown of sugars under anoxia; Taylor, 1942) and a likely interference with the oxygen requiring phloem transport of respirable substrates from the shoot (Sij and Swanson, 1973). Other substances essential to healthy root growth such as B group vitamins and certain hormones (Luckwill, 1960) may also fail to reach the roots for this reason.

It has been proposed by Crawford (Crawford, 1978, and references therein) that the ethanol produced by anaerobic respiration is harmful to roots and responsible in part at least, for injury to membranes and cell death caused by anoxia. It is questionable, however, whether toxic concentrations of ethanol accumulate in anaerobic roots. Concentrations of 13 mM (Sprent and Gallacher, 1976) and 6.4 mM (Crawford and Bains, 1977) have been measured in anoxic roots but these are well below the 50 mM (Kiyosawa, 1975), 600 mM (Gudjonsdottir and Burstrom, 1962) and 85 mM (Taylor, 1942) variously reported to be required to inhibit root growth severely or affect membranes. On the basis of existing information (eg. Wang and Chuang, 1967; Nashed and Girton, 1958; Lee, 1977) ethanol seems unlikely to do more than modestly depress root growth.

In some species, membrane damage in anoxic roots leads rapidly to the escape of toxic cyanides and phenolics from membrane-bound compartments, (Rowe and Catlin, 1971; Catlin et al., 1977). Injury from such substances will however be secondary to the loss of membrane integrity initially responsible for their release. Acetaldehyde (the penultimate product of glycolysis) is another potential autotoxin but as far as we are aware, toxic concentrations have not been demonstrated in waterlogged plants. Beletskaya (1977) found acetaldehyde in flooded winter wheat (*Triticum aestivum,* L.), but at concentrations 500 times less than those required to inhibit root growth in this species.

### 5.3.1.3 Effects of carbon dioxide

Carbon dioxide produced by aerobic respiration in the early stages of waterlogging and later by anaerobic respiration can accumulate in flooded soil by virtue of its high solubility in water (up to 1.7 g/L at 20 °C) and low rate of escape to the aerial environment imposed by the presence of water. Vine et al., (1942) found 17.5 percent $CO_2$ in a wet silty clay and similar concentrations accumulate in the soil when pea (*Pisum sativum* L.) plants are waterlogged for 2 to 3 days under laboratory conditions (Jackson, 1979). Concentrations of $CO_2$ of this magnitude are usually without much effect (Leonard and Pinckard, 1946; Jackson, 1979) or may even stimulate growth (Talbot and Street, 1968). In highly oxygen deficient conditions, $CO_2$ may be injurious to growth and also depress water uptake (Williamson and Splinter, 1968, 1969; Giesler, 1967). Nitrogen fixation by root nodules on certain leguminous plants has been reported to be much inhibited by 3 percent $CO_2$ (Grobbelar et al., 1971). Reports of extensive injury to plants by applying high concentrations of $CO_2$ (Kramer, 1940; Hagan, 1950) may be more properly related to the exclusion of oxygen brought about by the gassing.

### 5.3.1.4 Effects of ethylene from the soil

As discussed earlier ethylene is unlikely to cause root death since it is a growth regulator rather than a toxin. Its presence in anoxic soils is thus likely to be of little importance to roots unless they are able to obtain oxygen by internal diffusion from the shoot or to survive anoxia by metabolic adaptation (see Section 5.3.4.1). The effects of the gas on such roots could include an inhibition of elongation, cambial activity and stelar differentiation together with a swelling of cortical cells (Radin and Loomis, 1969) and a lower rate of accumulation of dry weight (Jackson and Campbell, 1979). Exposure to the gas may also have long-term consequences for the pattern of root growth and branching (Crossett and Campbell, 1975). The formation by nitrogen fixing bacteria of nodules on the roots of leguminous plants is inhibited by concentrations of ethylene as low as 0.4 ppm and nitrogen fixation by existing nodules is also much impaired by ethylene (Grobbelar et al., 1971; Goodlass and Smith, 1979). The significance of ethylene for growth processes in the shoots of waterlogged plants and for the production of adventitious roots from stem tissue is discussed later (see Section 5.3.3.2).

### 5.3.1.5 The effects of toxic substances from the soil

Most potentially toxic substances are produced in soil under oxygen-free conditions and at low redox potentials (see Section 5.2.3). Such toxins will, therefore, be of importance to roots that have not already been killed by oxygen depletion itself. They have special relevance to rice production, where these conditions commonly exist.

In waterlogged soil, nitrate is an early successor to oxygen as an acceptor of electrons from microbiological respiration. Nitrites produced in this way can inhibit growth at concentrations of 1 mM and above, (Bingham et al., 1954) but only when the pH is sufficiently low to favor the formation of undissociated nitrous acid (Lee, 1979). For this reason at the near neutral pH of most agricultural soils, nitrites will be predominantly dissociated and therefore innocuous. Only in nitrate-rich acid soils is nitrite toxicity likely to be important.

Under severe reducing conditions, $Fe^{3+}$ and $Mn^{4+}$ can be transformed to more soluble $Fe^{2+}$ and $Mn^{2+}$ (see Section 5.2.3). Concentrations of 10 ppm and above are claimed to be injurious (Jones and Etherington, 1970) either directly or by inducing deficiencies in major nutrients (Howeler, 1973). Again, it is in acid soils that the effects seem most severe (Graven et al., 1965; Olomu and Racz, 1974).

After prolonged periods of flooding particularly in warmer climates, hydrogen sulphide can be produced by the bacterial reduction of sulphates and by dissimilation of sulphur-containing amino acids (Ponnamperuma, 1972). At only 2.5 ppm this gas kills roots (Culbert and Ford, 1972) but there is little direct evidence that such concentrations commonly develop in flooded soils. This may be because production is located primarily at the root surface rather than in the bulk soil (Ford, 1973) and because much hydrogen sulphide reacts with iron to give non-toxic insoluble iron sulphides. A soil toxin such as hydrogen sulphide seems to be implicated in the poor growth of certain badly drained citrus groves in California since the absence of oxygen alone is claimed to have little affect on citrus roots (Culbert and Ford, 1972). Like other toxins, hydrogen sulphide is injurious only when undissociated, ie. in acid solutions. Sulphide poisoning can be a serious problem in rice (Vamos and Koves, 1972).

Intermediates and end-products of the anaerobic degradation of soil organic matter may also be phytotoxic. The possible effects on plants of ethanol produced by soil micro-organisms have received little attention (Wang and Chuang, 1967) but those of aliphatic acids such as acetic and butyric have been much studied (references in Lynch, 1976). Wet soils containing large amounts of undegraded organic matter can accumulate millimolar concentrations of these acids, especially at high temperatures (Wang et al., 1967) but in other flooded soils less than 1 mM (Ford, 1965; Jackson, 1979) has been detected. Aliphatic acids such as acetic and butyric remain primarily undissociated in water only at a pH of 4 or less and it is in this form that they exert toxicity (Lee, 1977).

Other potential toxins include phenolic acids such as cinnamic, vanillic and para-coumaric (Chandramohan et al., 1973). Their importance will again be dictated by the soil conditions, particularly pH, and the presence of organic matter as an energy and carbon source for the appropriate micro-organisms. The potential toxicity of any substances present will be irrelevant unless plant roots or seeds have otherwise survived the severely anaerobic conditions which are a prerequisite for toxin production.

### 5.3.2 Seed Germination

Much of the previous discussion concerning the deleterious effects of waterlogging on roots also applies to seeds, which may be especially vulnerable if only because the whole organism rather than just one part is exposed directly to the stress. Features common to many seeds that lessen their chances of survival include a low surface to volume ratio (eg. in peas) which is conducive to particularly steep gradients in oxygen from the soil to the respiring cells of storage tissues. Seeds are also rich in ions and respirable substrates some of which leak out during the imbibition of water (Simon, 1974) and continue to do so in anoxic conditions (Lynch and Pryn, 1977). Inadequate repair of membranes that were imperfect at the beginning of inbibition (Berjak and Villiers, 1972) may explain this. Seed exudates serve as ideal substrates for micro-organisms that compete with the seed for oxygen and may also be a source of phytotoxins (Lynch and Pryn, 1977). In wet soils, mucilagenous exudates can impede the passage of oxygen to the embryo (Heydecker and Chetram, 1971).

A lack of oxygen prevents germination in many species (Taylor, 1942; Morinaga, 1926) reflecting the intensely energy demanding processes of hormone (van Staden and Brown, 1973), protein and nucleic acid synthesis necessary for successful germination (Ching, 1972). It is not immediately apparent why an anoxic environment often kills seeds rather than simply imposing an extension of their 'dormancy'. One possible explanation is that germination is initiated irreversibly by energy derived not from aerobic respiration but from the activation of pre-existing glycolytic enzymes (Ching, 1972). The common occurrence of lactic acid and ethanol in early germination (Sherwin and Simon, 1969; Leblova et al., 1973) support this idea. The anaerobic initiation of metabolic germination would then lead to requirements for energy which cannot be met by glycolysis alone. The seed may thus become fatally trapped between the initiating influence of glycolysis and the absence of the aerobic metabolism required to satisfy the energy needs so created.

Germination may be further inhibited by a tendency for seeds to become water soaked (Grable, 1966) and thus more impermeable to oxygen. A rapid influx of water into the outer cells of the embryo may also cause them to rupture and die in seeds with a mechanically damaged testa (Powell and Matthews, 1978). Like roots, seeds in anoxic conditions are also potentially susceptible to the accumulation of glycolytic end products such as acetaldehyde and ethanol (Holm, 1972; Crawford, 1977) and to aliphatic acids produced by soil micro-organisms. The latter may take on a particular importance when seed is sown in wet soil adjacent to deomposing litter from the previous crop (Lynch, 1977), an increasingly common possibility with modern minimum tillage methods. Waterlogging of the parent plant, especially at the time of pollination may diminish seed size and possibly the content of cytokinin hormones and these factors may in turn depress subsequent germination and seedling vigor (Michael and Sieler-Kelbitsch, 1972).

### 5.3.3 Shoot Growth and Development

The effects of flooding rapidly extend beyond the roots to the shoot system, and can be expressed in one or more of the following gross symptoms; wilting (Kramer and Jackson, 1954), epinastic curvature (Turkova, 1944), chlorosis (Went, 1943) and desiccation (Kramer, 1951), abscission (Lloyd, 1916), slow rates of extension growth and dry matter accumulation (Phillips, 1964; Jackson and Campbell, 1979), hypertrophic swelling at the base of the stem and adventitious rooting (Kawase, 1974) and perhaps complete death. In many sensitive species, when growing rapidly, symptoms may become visible within 24 h or less, reflecting the close inter-relationship between shoot physiology and the responses of roots to soil conditions. We distinguish three classes of effect of roots on shoots; (a) decreased supply of substances essential for ordinary shoot growth and development, (b) increased supply of substances that can modify shoot growth, and (c) accumulations in shoots of substances normally transported down to the roots. Water, hormones, inorganic nutrients and toxins are the principal substances involved. Each will now be considered in more detail.

### 5.3.3.1 Water relationships

Unexpectedly, flooding of the soil can sometimes cause the shoots to wilt within a few hours. This form of physiological drought is thought to be induced by an increase in resistance to water flow in the roots. Metabolic energy seems to be required to maintain the low resistance needed to prevent wilting. Anaerobic conditions (Parsons and Kramer, 1974), may therefore raise resistance to water flow by depressing energy production thus favouring a fall in leaf water potentials to below the minimum required to maintain turgor. Wilting will only be so induced if stomata are open far enouth to allow a rapid rate of water loss. In some species stomata close soon after waterlogging thus preventing wilting (Pereira and Kozlowski, 1977; Jackson et al., 1978). Stomatal closure may be triggered by an initial but short-lived increase in water stress in the leaves resulting from the larger resistance to water movement. This could in turn cause the hormone, abscisic acid (ABA) (a potent promotor of stomatal closure, Raschke, 1975) to accumulate (Hiron and Wright, 1973). Inhibition of nutrient uptake by waterlogged plants (see Section 5.3.3.3) may also increase ABA in the shoots (Mizrahi and Richmond, 1972). Stomata may also be closed more directly be a deficiency of potassium ions (Regehr et al., 1975) or cytokinin and gibberellin hormones from the anaerobic roots (Livne and Vaadia, 1965; Jackson and Campbell, 1979). Any carbon dioxide moving internally into the leaves from the waterlogged soil (Stemmet et al., 1962) may also encourage stomata to close (Raschke, 1975). Further research is needed on the events in flooded plants that often close stomata and on why it fails to happen in all cases.

### 5.3.3.2 Hormonal relationships

Since the work of Went (1938) and Chibnall (1939), the shoots of many plants have been increasingly thought to depend upon healthy roots for at least some of their hormone requirements, with the transpiration stream as their means of transport. Hormones are also synthesized in shoot tissues and some may normally pass into the roots. Waterlogging alters hormone production in both roots and shoots and interferes with their transport. The resulting hormonal imbalance has major repercussions for shoot growth, some of them significant processes of adaptation to waterlogging (see Section 5.3.4.2).

Gibberellins and cytokinins (Crozier and Reid, 1971; Short and Torrey, 1972) are synthesized or transformed from one derivative to another in the apical parts of roots. Since these are the zones most subjected to oxygen stress it is not surprising that the concentration of the hormones in roots or the bleeding sap of detopped plants is very low after waterlogging for 24 h or more (Fig. 5.3-1a; Burrows and Carr, 1969; Reid and Crozier, 1971). If the species is one which produces new, adventitious roots when waterlogged (eg the tomato, *Lycopersicon esculentum* Mill), their emergence coincides with a recovery of hormone levels (Reid and Crozier, 1971). A restricted cytokinin and gibberellin supply to the shoots may be expected to inhibit stem elongation and promote leaf senescence or abscission, (Burrows and Carr, 1969; Reid and Crozier, 1971). The suggestion by Railton and Reid (1973) that leaf epinasty may be promoted by a depletion of cytokinin in waterlogged tomatoes is not supported by observations that removing roots to eliminate this source of cytokinins does not promote epinasty (Jackson and Campbell, 1975a) and thus providing a source of these hormones by inducing the formation of aerobic, adventitious roots that does not diminish epinasty when the original root system below is waterlogged (Jackson and Campbell, 1979, but see W. T. Jackson, 1955).

Unlike the cytokinins and gibberellins, auxin from the roots is not thought to be of great importance for shoot growth, although similar amounts have been measured in both xylem and phloem exudates (Hall and Medlow, 1974). Thus roots both receive and export auxin, but this balance will clearly be upset by any inhibition of auxin synthesis in roots by a deficiency of oxygen (Audus, 1972) and transport to roots from the shoots. There is also the possibility that dying roots will release into the transpiration stream precursors for auxin synthesis (Sheldrake, 1973). Any resulting increase in the auxin content of the shoots of waterlogged plants (Phillips, 1964) may promote adventitious rooting at the base of the stem and stimulate the production of ethylene gas.

The situation regarding ethylene is somewhat paradoxical. The concentration within roots may increase as a consequence of the trapping by water

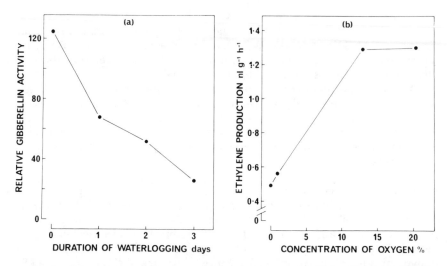

FIG. 5.3-1 Effect of oxygen-stress on two hormones in roots. (a) Gibberellin activity (alpha-amylase test) in the roots of waterlogged tomato plants. (After Reid and Crozier, 1971). (b) Production of ethylene by roots of barley seedings after exposure to different concentrations of oxygen for 3 days. (Jackson and Dobson, unpublished).

(see Section 5.3.1.1). However if waterlogging proceeds sufficiently for oxygen in the soil water to fall to about 2 percent or less, production of ethylene by roots will be severely curtailed (Fig. 5.3-1b; Jackson et al., 1978; Bradford and Dilley, 1978). Reports that the roots of plants waterlogged in soil for extended periods contain unusually large amount of ethylene (Kawase, 1972, 1974) probably result from a combination of an input of ethylene from the soil (see Section 5.2.3) and from the trapping effect of water which may permit a net accumulation despite lowered production of the gas by the roots. In some circumstances physiologically active amounts may diffuse upwards into stem tissue (Kawase, 1974) either dissolved in the transpiration stream or in the gas phase (Jackson and Campbell, 1975b). Extra ethylene can however be produced in the shoots of waterlogged plants in response to oxygen deprivation around the roots (Jackson and Campbell, 1976a; Bradford and Dilley, 1978). This may occur because substances from anaerobic roots pass up into shoot tissues and stimulate ethylene production. The increases in ethylene have been shown to be sufficient to promote the epinastic curvature of leaves of flooded tomatoes (Jackson and Campbell, 1976a). Ethylene may also encourage hypertrophic stem-swellings and adventitious rooting (Kawase, 1974) and the abscission of senescent leaves that are low in auxin (Jackson and Osborne, 1972). Leaf senescence itself may also be promoted although at present there is little evidence that ethylene is highly effective in this regard (eg. Goldney and van Steveninck, 1972). Leaf expansion can however be severely limited by low concentrations of the gas (Abeles et al., 1971).

The production and movement of abscisic acid (ABA) in roots is poorly understood although the root cap is a locally important source of this hormone in the control of root geotropism. More ABA normally moves into the roots from the shoots than *vice versa* (Michniewicz and Galoch, 1972). As with the other hormones, synthesis in the roots and transport in the phloem to the roots will probably be severely depressed by anaerobic soil conditions.

Waterlogging can induce rapid increases in the ABA content of shoot tissue (Hiron and Wright, 1973) possibly brought about by transient water stress in the leaves. In addition to closing stomata (see Section 5.3.3.1) ABA may depress the rate of stem extension and promote leaf senescence (Osborne, 1968) and adventitious rooting (Chin et al., 1969).

Many of the proposed actions of plant hormones in flooded plants described above are speculative. More detailed studies of the sensitivities of growth processes to hormones and further analyses of hormone content are required. It may then be possible to identify more accurately the dominant hormonal influences which reshape the morphology of many species following waterlogging.

### 5.3.3.3 Nutrient relationships

Healthy roots have the ability to absorb selectively inorganic nutrient ions from the soil solution. Much of the uptake is believed to take place across membranes and against gradients of chemical and electrical potential energy. Energy must therefore be expended to move the ions thermodynamically uphill (Baker, 1978). Since the source of this energy is ATP derived mainly from aerobic respiration, it is not surprising that in the absence of oxygen, nutrient absorption is much curtailed (Hopkins et al., 1950; Steward et al., 1936; Rao and Rains, 1976). The rapid *efflux* of ions from anaerobic roots (Lee, 1977) further highlights the loss of metabolic control of ion accumulation that befalls anoxic roots.

The uptake of nitrate remaining in flooded soil and therefore subsequent reductions of nitrate in the root to nitrite and ammonia are also inhibited by anoxia (Lee, 1978). Legumes growing in nitrogen-poor soils are especially vulnerable because low oxygen partial pressures (Cowen, 1978; Pankhurst and Sprent, 1975), $CO_2$ and ethylene (see 5.3.1.3 and 5.3.1.4) each inhibit nodulation by nitrogen fixing bacteria and also nitrogen fixation by existing nodules.

In a study of nitrogen dificiency and chlorosis in flooded barley plants, Drew and Sisworo (1977) reported that after 2 days, nitrogen uptake by flooded plants almost ceased and an internal re-translocation of nitrogen from older to younger leaves then took place. The resulting decline in the nitrogen of the lower leaves preceded chlorosis, suggesting that these symptoms were a result of nitrogen depletion. Their conclusion was supported by the observation that nitrogen fertilizer taken up by a small number of roots that were allowed to grow near to the soil surface in well aerated conditions, prevented chlorosis. Experiments with peas waterlogged for 4 days (Jackson, 1979) have also shown that nitrogen uptake can be inhibited to a greater extent than shoot growth; the resulting drop in the concentration of nitrogen in the rapidly senescing leaves being exacerbated by mobilization into developing fruits.

It has been suggested by Garcia-Novo and Crawford (1973) that nitrate-N may not only serve as a source of nitrogen *per se* for waterlogged plants but also as an accceptor of electrons in plant roots in place of oxygen. No support for such a mechanism could be found by Lee (1978) in a recent study of nitrogen metabolism in anaerobic barley roots, nor by Leonard and Pinckard (1946) with cotton (*Gossypium hirsutum* L.).

### 5.3.3.4 The effect of toxins

In his far-sighted paper of 1951, Kramer deduced that factors emanating from anaerobic roots were responsible for at least some symptons of waterlogging injury to the shoots. This conclusion was based on the observation that removing the roots was less injurious to leaves than waterlogging (see also Burrows and Carr, 1969). However, W. T. Jackson (1956) could not confirm a role for toxins in experiments using approach-grafts and split-root system of tomato. Evidence that ethanol may be tranported in toxic concentrations to shoots from waterlogged roots has been produced by Fulton and Erickson (1964). However, if account is taken of the diluting effect of transpiration on the contents of xylem sap (Meidner and Sheriff, 1976) then the true concentration of ethanol in the plants studied by Fulton and Erickson would be well below that considered by those authors to be growth inhibitory. The significance of ethanol for the growth of shoots thus requires re-examination.

Catlin et al. (1977) have provided strong circumstantial evidence that phenolic compounds released from waterlogged roots of certain walnut (*Juglans*) species enter the transpiration stream and cause severe damage to the aerial shoot. Similarly, hydrogen cyanide derived from the hydrolysis of cyanogenic glucosides in waterlogged roots of peach and plum (Rowe and Catlin, 1971) may also damage shoot tissues. When pea plants are waterlogged, the leaves senesce rapidly and subsequently become desiccated. However, plants grown in anaerobic solution cultures have been found to exhibit these symptoms only slowly (Jackson, 1979). One explanation for this is that toxins derived from the rotting of roots in the soil or from the flooded soil itself are responsible for much of the waterlogging injury.

In strongly reducing conditions (see Section 5.2.3), toxic quantities of iron and manganese from the soil (Jones and Etherington, 1970) or hydrogen sulphide (Vamŏs and Kŏves, 1972) may reach the shoots. Quantitative evidence that the amounts present in the shoots are sufficient to be toxic is, however, difficult to find. Analyses could usefully be concentrated on tissues such as the shoot-tip and young expanding leaves that are possibly the most sensitive to such toxins.

### 5.3.4 Tolerance and Avoidance of Waterlogging Injury

Plants vary widely in their tolerance of submergence and of anaerobic soil conditions. In addition to the more obvious examples of resistant species such as rice ( *Oryza sativa*, L.) and willow (*Salix* spp), differences in suceptability also exist between dry-land plants and between cultivars of various crops (see Section 5.4.3.2). We now discuss some of the physiological and biochemical mechanisms which may confer resistance. These have been placed into two groups, (a) those which may enable plants to tolerate the absence of oxygen (b) those which may enable plants to avoid submergence and anoxia or ameliorate some of their consequences.

### 5.3.4.1 Mechanisms of tolerance

Species for which a waterlogged environment is their normal habitat can grow for limited periods in the absence of oxygen. Seeds of *Peltandra virginica* L. Kunth. (Edwards, 1933) and rice (Taylor, 1942) can germinate readily and grow to a limited extent in atmospheres free of oxygen. The roots of willow may also have this ability (Livingstone and Free, 1917). In rice the leaves and coleoptiles show tolerance rather than the roots. This tolerance appears to be achieved by the possession of a potentially high rate of anaerobic metabolism of respirable substrates to ethanol, a strong Pasteur effect in response to anoxia and the ability to fix $CO_2$ in the dark without oxygen (Taylor, 1942; Avadhami et al., 1978; Wignarajah et al., 1976). These features presumably combine to ensure that sufficient ATP and respirable substrates are made available for some growth and for adequate structural maintenance, especially to membranes. A further adaptive feature of rice is its ability to retain under anaerobic conditions structurally and enzymically intact mitochondria, which permit aerobic respiration and rapid growth to commence at the earliest opportunity (Opik, 1973).

In many dryland species, the property of roots to withstand anoxia rather than to grow under such conditions delineates the more tolerant plants. Extensive investigations by Crawford and co-workers indicate that survival is associated with a low rate of anaerobic respiration, a small Pasteur effect and the formation of malic acid rather than ethanol as a terminal product of glycolysis (Crawford, 1978). These characteristics are the opposite of those in rice and may represent a form of imposed dormancy. The production of malic acid rather than ethanol may assist in maintaining a low metabolic rate since, unlike the ethanol pathway, no ATP is generated when glucose is respired to malate (Davies et al., 1974). However, the absence of ATP production by the malate route could be expected to lead to the death of roots unless energy requiring processes were repressed in synchrony with the limited energy supply. The mechanism of any repression remains to be elucidated. The possibility that malic acid is a less toxic alternative to ethanol has been suggested by Crawford and Bains (1977) as an important part of the tolerance mechanism. Interpretation of this matter awaits more convincing evidence that ethanol is toxic at the concentrations present in anoxic roots (see 5.3.1.2).

### 5.3.4.2 Mechanisms of avoidance

At least four types of avoidance process can be identified: (a) accelerated elongation caused by submergence, (b) the production of replacement roots which grow in the well aerated soil surface, (c) the development of channels (aerenchyma) that may permit the internal aeration of submerged parts and (d) responses by shoots which may minimize injury from waterlogging.

### 5.3.4.2.1 Accelerated elongation.

Submergence stimulates extension growth in rice shoots, especially if the oxygen concentration is also low (optimum is about 3 percent) but not completely absent (Yamada, 1954). This reaction can be related to an increase in growth promoting auxin arising from an inhibition of the oxidative destruction of this hormone (Wada, 1961). Submergence will also increase the internal concentration of ethylene by the trapping effect, (see 5.3.1.1) and ethylene is a powerful stimulator of elongation in rice seedlings (Fig. 5.3-2a; Ku et al., 1970). In other aquatic plants such as *Callitriche platycarpa* Kutz. (Fig. 5.3-2b) and *Ranunculus sceleratus* L. (Musgrave et al., 1972) ethylene has a similar growth promoting effect and will help to ensure that the submerged shoots quickly reach the surface of the water. When rice shoots surface in this way, air may then diffuse internally down to the roots (which have a higher requirement for oxygen than coleoptiles, Kordan, 1975), thus allowing them to grow. The ethylene that rice roots contain by virtue of the

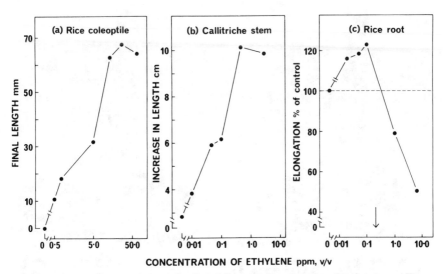

**FIG. 5.3-2 The promotion of elongation by ethylene. (a) In coleoptiles of intact rice seedlings (after Ku et al., 1970).**
**(b) In the shoots of the aquatic plant** *Callitriche platycarpa.* **(After Musgrave et al., 1972).**
**(c) In roots of intact rice seedlings. (After Konings and Jackson, 1979). The arrow indicates the increase in ethylene concentration likely to occur when the roots are submerged in water.**

trapping effect of water can be expected to stimulate root extension rather than inhibit it (Fig. 5.3-2c; Konings and Jackson, 1979).

### 5.3.4.2.2 Production of replacement roots.

The rapid production of new, adventitious roots that grow in the well aerated soil surface is commonly found in species which tolerate waterlogging (Fig. 5.3-3). Herbaceous species such as tobacco (*Nicotiana tabacum* L.) (Kramer, 1951) and peas (Jackson, 1979) that are not able to produce such roots are very rapidly killed or injured by flooding. The emergence of new roots, probably as outgrowths of pre-formed primordia, is favored by the accumulation at the stem base of sucrose (Grinerva and Nechiporenko, 1977) and auxin (Phillips, 1964). Additional ethylene present in the stem will also encourage rooting, in tomato (Jackson and Campbell, 1975b) sunflower (*Helianthus annuus* L.) (Kawase, 1974) and *Zea mays* (Drew, Jackson and Giffard, 1979). The new roots can be expected to replace some of the functions of the original roots (Jackson, 1955; Jackson and Campbell, 1979) but to do so will require oxygen. This can be obtained by the adoption of a horizontal rather than vertical orientation thus maintaining a position near to the soil-air interface. In some species, roots within the waterlogged soil are able to grow upwards and emerge into the well aerated soil surface (Periera and Kozlowski, 1977). The mechanisms controlling the re-orientation of roots are not well understood. Observations that ethylene can promote the upward growth of normally downward growing roots (Crocker, 1948) suggest the gas may be involved in such changes in waterlogged plants.

### 5.3.4.2.3 Development of aerenchyma.

The presence of interconnected airspaces within roots (aerenchyma) is a common feature of flooding-tolerant species or cultivars (Yu et al., 1969;

## WATERLOGGED              CONTROL

FIG. 5.3-3 The promoting effect of waterlogging for 9 days on adventitious rooting by corn plants (cv. Giant Horse Tooth). (Jackson, unpublished).

Arikado, 1955; Fig. 5.3-4). Such roots often have a lower respiratory demand per unit volume (Luxmoore and Stolzy, 1972) and a low resistance to the movement of air from the shoot. In *Zea mays* ethylene appears to be responsible for accelerating aerenchyma formation (Drew et al., 1979). There is considerable experimental and mathematical evidence to support the view that air may diffuse through aerenchymatous roots of some flood tolerant species for up to 50 cm (Coutts and Armstrong, 1976; Philipson and Coutts, 1978; Armstrong, 1964, 1978). Excess oxygen moving out of the root into the anaerobic soil may oxidize substances in the rhizosphere and thus inhibit the formation of toxins (Armstrong, 1964). Internal aeration may be less effective in older plants, possibly as a result of the blockage or collapse of the conducting channels (Arikado, 1955; Yamasaki and Saeki, 1976).

### 5.3.4.2.4 Shoot responses that minimize damage.

Inadequate rates of uptake of water and of inorganic nutrients (see 5.3.3.1 and 5.3.3.3) can contribute to the injurious effects of waterlogging on shoot growth (see 5.3.3.1). The effects may be minimized in some plants by physiological responses such as an overall slowing of growth, the senescence and abscission of the older leaves and the rapid closing of stomata (see 5.3.3.1). Loss of the lower leaves and stomatal closure will reduce the amount of water lost from the plant and hence minimize water stress in the remaining foliage. Epinastic curvature may have a similar effect. If, during senescence and prior to any leaf abscission, nitrogen and other nutrients are transported out of older tissues (Drew and Sisworo, 1977) such recyling may delay the onset of chronic mineral deficiencies in younger parts. In the absence of adaptations by the submerged roots or other organs it is unlikely that such strategies adopted by the aerial tissues will alone enable the plant to survive waterlogging for long (Hook and Brown, 1973).

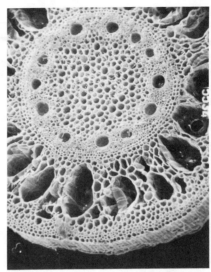

FIG. 5.3-4 Photomicrographs of transverse sections of nodal roots of corn taken by scanning electron microscopy (x 100), showing (a) cortical air spaces (aerenchyma) in roots grown in non-aerated nutrient solution, and
(b) roots in aerated solution. (Unpublished photographs by M. C. Drew, A. Channel, J. P. Garrec and A. Fourcy, Centre d'Etudes Nucleaires, Grenoble, France).

### 5.3.5 Waterlogging at Different Stages of Growth.

The effects of waterlogging discussed in the preceding sections can vary in magnitude between stages of growth. Such differences are important in the design of drainage systems (see 5.4.1.4). Experiments that examine the effects of waterlogging at different growth stages are often difficult to interpret; this is because of the confounding influences of environmental factors, especially temperature, and often inadequate description of the treatments, the growing conditions or stages of plant growth. Furthermore, for most crops the effects of waterlogging have not been examined at all the main stages of growth, and the pattern is thus incomplete. In most experiments, soil has been flooded up to the surface, whereas in practice complete waterlogging of the profile is less common, a difference that may be important for species where adventitious roots do not develop readily (see Section 5.3.4.2.2). Despite these difficulties certain trends are evident for some major groups of crops.

#### 5.3.5.1 Small-grained cereals.

Some experiments suggest that wheat, barley, oats (*Avena sativa* L.) and rye (*Secale cereale* L.) are most sensitive to waterlogging before tillering (about the 3 leaf stage) (Bourget et al., 1966; Watson et al., 1976; Stefanovskii, 1964). But in other experiments the effects have been more pronounced at later stages. Barley plants at the 5 to 8 leaf stage may be especially vulnerable (Leyshon and Sheard, 1978). Russian work on spring barley (Savickaja, 1959; Shazkin, 1960) and wheat (Stefanovskii, 1963, 1965) also indicated greater sensitivity to waterlogging after the formation of spikelet primordia (about the fifth leaf). In the latter work, the most sensitive stage was frequently found to be between stamen initiation and tetrad formation in the anthers when waterlogging may have decreased the viability of pollen (Shazkin, 1960; Shazkin and Federova, 1961; Campbell et al., 1969). The large effect of waterlogging on grain yield of barley noted by Ikeda et al. (1957) during stem elongation, which would include these reproductive stages, is consistent with this hypothesis. However, waterlogging to the soil surface at the time of pollen formation in two experiments under field conditions in the UK had no effect on yield of wheat (Cannell et al., 1980).

Flooding for only a few days (3 to 15 days) during the early stages of growth can cause appreciable plant mortality (Segeta, 1968; Swartz, 1966), but compensatory growth by the surviving plants may cause the loss of yield to be less pronounced. For example, waterlogging to the soil surface for 6 days after germination, but before emergence of winter wheat, decreased the plant population by nearly 90 percent, but grain yield was only 18 percent smaller than in a freely-drained control (Cannell et al., 1980).

After establishment, winter cereals can withstand several weeks of waterlogging during periods of low temperature (Antipov, 1966) with only comparatively small losses in grain yield (Cannell et al., 1980). In the latter work, waterlogging a clay (Aeric Haplaquept) to the soil surface for 6 weeks in January/February and for 25 days in April, depressed yield of winter wheat by up to about 15 percent.

### 5.3.5.2 Corn (Zea mays L.) and Sorghum (Sorghum bicolor (L.) Moench.)

Flooding corn generally seems most harmful at early growth stages (Joshi and Dastane, 1965, 1966; Ritter and Beer, 1969; Chaudhary et al., 1975); the results of the latter workers (Table 5.3-1, experiment c(v)) are representative, and indicate that waterlogging the whole soil profile for more than 48 h is likely to cause a large loss of grain yield. In an Iowa field experiment, corn 30 cm high was killed by 3 to 4 days flooding, and by 6 to 7 days flooding when the plants were 50 to 60 cm high (De Boer and Ritter, 1970). There are exceptions: Ali (1976) found that flooding for 1 to 4 days at the 9 to 10 leaf stage (6 weeks after sowing) had greater effects than at the 4 to 5 or 14 to 15 leaf stages; and in Nigerian work there was no significant effect on yield from 5 days of flooding 3 to 4 weeks after emergence or at flowering (Wien et al., 1979). Possible reasons for the lack of a growth stage effect in the latter experiment include high plant-to-plant genetic variability in the composite used, and that a shallow water-table (at 30 cm) was maintained, except during flooding, so that the root system was adapted to shallow conditions.

Sorghum yield was depressed by up to 30 percent by flooding for 12 days in the early vegetative and boot stages, but not at heading (Howell et al., 1976).

### 5.3.5.3 Grain legumes

Some species in this group, eg Vicia Faba L. are traditionally regarded as waterlogging tolerant, and some autumn sown cultivars may survive long periods of wet soil conditions. However, most grain legumes are very sensitive to short periods of waterlogging, and one of the worst seems to be French bean (Phaseolus vulgaris L.). In high rainfall areas of Latin America this crop is likely to be subjected to repeated short-term flooding. Waterlogging to the soil surface for 12 h in 5 successive weeks decreased yield by 90 percent (Forsythe and Pinchinat, 1971). Even waterlogging for 1 h on 2 to 4 occasions in a 19 day period has significantly depressed yield (Forsythe et al., 1979).

### 5.3.5.3.1 Soybean (*Glycine max.* (L.) Merrill).

Young plants may be more susceptible to flooding, being killed by 3 days of inundation (De Boer and Ritter, 1970), but Wien et al., (1979) found similar effects at two stages of growth; raising the water-table to the surface for 4 to 5 days, either at 3 or 4 weeks after emergence or at flowering reduced seed yield by about 20 percent.

Most experiments have used the extreme of surface waterlogging but the effects of less severe treatments at different stages of growth of soybeans have been studied by Stanley et al. (1980). Water-tables were raised to within 45 or 90 cm from the soil surface for seven days in a rhizotron on three occasions: (a) during vegetative growth (just before flowering), (b) after flowering and (c) after pods had set. The effects on root growth varied substantially. During vegetative growth, the roots ceased to grow while they were inundated, but after draining the soil growth resumed at about the same rate as before the waterlogging. When the watertable was imposed just after flowering, flooded roots appeared normal while the watertable was present but decomposed rapidly on draining the soil; some compensatory root growth occurred in the zone just above the position of the watertable. By the third stage of waterlogging (after pod set), vegetative growth had ceased, and roots died quickly when flooded, and no root growth took place after the treatment. The effects on yield also varied. In comparison with a well-watered control, raising the watertable to 45 cm below the surface before or just after flowering did not affect yield, presumably because the additional water was more advantageous than the effects of flooding on the roots, but during pod-set this high watertable depressed yield by 40 percent. However, raising the watertable to only 90 cm below the surface at all three growth stages depressed yield by similar amounts, about 30 percent, compared to the frequently irrigated but freely drained control.

### 5.3.5.3.2 Cowpea (*Vigna unguiculata* (L.) Walp):

Cowpeas are usually planted during or soon after the tropical rainy season; temporary waterlogging often occurs, and crop failure can result. In controlled environments where 8 day old seedlings were waterlogged, yield was depressed if the treatment lasted 8 or more days (Minchin and Summerfield, 1976). Three cycles of 1 day of waterlogging and 9 days recovery, applied between emergence and flowering, depressed seed yield by about 25 percent, but when the plants were waterlogged for 4 days with 6 days for recovery in each cycle, the losses were 50 percent; after flowering the effects were less pronounced (Minchin et al., 1978). In a Nigerian lysimeter study, 4 or 5 days of flooding 3 to 4 weeks after emergence or at flowering restricted yield of seed by about 20 percent (Wien et al., 1979).

### 5.3.5.3.3 Peas:

This species is also sensitive to poor aeration. In greenhouse experiments, low oxygen content of the soil for twenty-four hours just before and during flowering rapidly caused extensive damage to foliage (Jackson, 1979) and reduced yield by about a third (Erickson and Van Doren, 1960). Even in cooler outdoor conditions waterlogging for 24 hours just before flowering has depressed yield by about 20 percent (Cannell et al., 1979) and the effects were greater with more prolonged waterlogging. At other growth stages effects were less severe, but still statistically significant (Fig. 5.3-5). In lysimeter experiments in field conditions by Belford et al. (1980) when the soil was waterlogged to the surface for 5 days at the 1 to 2, 3 to 4 and 6 to 7

FIG. 5.3-5 The effects of waterlogging to the soil surface at four stages of development on the growth of peas. Treatments were freely drained (c) or waterlogged for 5 days at the 4-5 leaf stage (T1), pre-flowering, 6-7 leaves (T2), flowering, 9-10 leaves (T3) and pod filling (T4). The yields of peas relative to the control were 48, 7, 25, and 29 percent respectively for the waterlogging treatments. Photograph taken 18 days after last treatment (Cannell et al., 1979).

leaf stages (the latter being just before flowering), the losses of yield of seeds (peas) at the freezing stage were 6, 15 and 42 percent respectively. The severity of the effect depended on the duration of the treatment and the height of the watertable; at the 6 to 7 leaf stage either 2 days waterlogging to the surface or 5 days with the watertable at 50 cm below the soil surface had little effect on seed yield.

### 5.3.6 Comparison of Static and Fluctuating Watertables.

Roots are usually most abundant in the upper soil horizons; often three quarters or more are in the top 20 cm and the abundance diminishes linearly or exponentially down the profile. A few roots may reach 1 to 2 m, the maximum depth depending on the soil type and crop species (Gerwitz and Page, 1975; Gregory et al., 1978; Taylor, 1979). The position of the watertable will set the maximum rooting depth of most mesophytic species (Campbell and Seaborn, 1972). In practice the depth of watertable is rarely static (or constant), and the effect of waterlogging will depend on its position relative to that of the root system. The main effects on plants will be derived from the limitation imposed on the maximum depth of rooting, and on the physiology of the root system.

The influence of variation in effective rainfall (rainfall less evaporation) on the depth of the watertable depends on the hydraulic properties of the soil. Furthermore, the effect of a particular depth of watertable on the moisture and aeration status of the soil will vary in relation to soil texture, because of the greater capillary rise in fine textured soils (Kristensen and

Enoch, 1964). In such soils the depth of the watertable is not clearly defined. Nevertheless, information on the effect of watertables at different positions can indicate the range of depths to which watertables should be lowered by drainage, and may be especially useful where crop production is largely dependent on irrigation (see 5.4.2.3).

The abundant published work on this subject has been reviewed by Williamson and Kriz (1970). Shallow watertables can cause large losses of yield, but the reported depth of watertable for the heaviest yield of different crops varies widely, from 15 cm down to 150 cm, partly depending on experimental conditions, eg soil type. Even for the same crop and soil the depth of watertable for maximum yield depends on the wetness of the season (eg Table 5.3-1, experiment (c)), and on the system of watering. In general, where combination watering from the top and bottom is adopted the optimum watertable is deeper than where water is provided only to the deeper parts of the soil profile (Williamson and Kriz, 1970).

From the viewpoint of drainage design (see 5.4.1.4), the effects of fluctuating watertables on crop growth are much more important than those of static watertables (Hiler, 1976), the former being more usual in the field.

TABLE 5.3-1. EFFECTS OF STATIC AND TRANSIENT HIGH
WATER-TABLES ON THE YIELD OF CROPS
The figures in the table have been adapted from (a) Lal and Taylor
(1969), (b) Shalhevet et al., (1969), (c) and (d) Chaudhary et al.,
(1975) and (1974) respectively. The yields are expressed as
percentages of the best treatment in each experiment

| Water-table regime | (a) Corn | (b) Sugar-beet |
|---|---|---|
| Free-drained | 100 | 95 |
| Constant at 90 cm | — | 100 |
| 55 cm | — | 92 |
| 30 cm | 81 | 87 |
| 15 cm | 75 | — |
| Surface flooding in 3 successive | | |
| weeks for 2 days | 63 | — |
| 4 days | 63 | — |
| Fluctuating 90-30 cm every | | |
| 10 days | — | 77 |

| | (c) Corn | | | (d) Wheat | |
|---|---|---|---|---|---|
| | (i) Wet year | (ii) Dry year | | (i) Mean of 2 years | |
| Constant at 60 cm | 59 | 100 | | 91 | |
| 90 cm | 87 | 96 | | 100 | |
| 120 cm | 100 | 85 | | 86 | |
| 150 cm | 78 | 72 | | 74 | |
| Surface flooding (otherwise 60 cm) | every 14 or 18 days (mean) (iii) Wet year | (iv) Dry year | Once at 2 wk (v) | Every 14 or 18 days (ii) | |
| for 1 day | 100 | 100 | 100 | 100 | |
| for 2 days | 70 | 84 | 82 | — | |
| for 3 days | 56 | 61 | 68 | 91 | |
| for 4 days | 44 | 42 | 58 | — | |
| for 5 days | 28 | — | — | 79 | |
| for 7 days | — | — | — | 51* | |
| for 11 days | — | — | — | 26* | |
| | | | | *1 yr only | |

However, the effects of fluctuating watertables on crop growth have not been widely investigated. Few experiments have compared static and fluctuating watertables. Those known to the authors are by Lal and Taylor (1969); Shalhevet et al. (1969) and by Chaudhary et al. (1974, 1975), and are summarized in Table 5.3-1. In the experiments reported by the latter authors the static and fluctuating watertables were not part of the same experiment, but the crops were sown on the same date, and grown in adjacent lysimeters. Owing to the limited amount of other information, comparison of these results is helpful. For each of the four sets of experiments on three crops, while constant shallow watertables have caused some yield depression, except for corn in a dry year (experiment c(ii), Table 5.3-1, transient flooding to the surface of the soil led to a much more pronounced loss of yield. For corn even a single flooding for 2 days or more reduced yield.

Comparison of the effects of less extreme transient waterloggings with static watertables have been rarely studied; that reported by Shalhevet et al. (1969) is the only experiment known to us (Table 5.3-1). The loss in yield by a watertable fluctuating between 90 and 30 cm below the soil surface in comparison with constant watertables at 30 or 90 cm is evident from their results—they conclude that 'the effective' (ie damaging) 'watertable depth for the fluctuating watertable seems to be the high level rather than a time-integrated depth or a weighted-mean depth'.

### 5.3.7 Interactions with Disease

The situation with regard to plant diseases has been reviewed by Bergman (1959). Waterlogging may promote or inhibit disease development, depending upon the ability of a particular pathogen to grow in anaerobic conditions, the presence of other competing micro-organisms, and changes (if any) in the susceptibility of the host to the pathogen. One of the simplest ways in which micro-organisms may be pathogenic is to compete with roots or germinating seeds for available oxygen and for nitrogen (Lynch and Pryn, 1977). As with roots, many soil pathogens will grow less well in poorly aerated soils. Fungal diseases such as 'take-all' of wheat (*Gaeumannomyces graminis*) (Ferraz, 1973) and *Verticillium dahliae* Kleb (Ioannou et al., 1977) are inhibited by low partial pressures of oxygen. Nevertheless a higher percentage of wheat plants growing in lysimeters were infected with 'take-all' after waterlogging than were plants growing on freely drained soil (Cannell et al., 1980). For many potential pathogens, the presence of excess water will provide a suitable medium in which mobile spores may disperse and invade plant roots. *Sclerotinia selerotinium* stem rot in peas (Jones and Gray, 1977), collor rot of apples (*Phytophthora* spp.) and root rot of alfalfa (*Medicago sativa* L.) and avocado (*Persea americana* Mill.) are diseases which are more prevalent in wet soils (Stolzy et al., 1967; Pratt and Mitchell, 1976). In alfalfa, 2-week old seedlings are especially vulnerable to disease at temperatures between 20 to 24 °C.

One factor which may favor infection is the extent to which ethanol is present in the soil. Young et al. (1977) found that zoospores of *Phytophthora cinnamoni* are stimulated to grow towards anaerobic roots which exude ethanol, and Smucker (1971) (PhD dissertation, Michigan State University, Anaerobiosis as it Affects the Exudation and Infection of Pea Roots grown in an Aseptic Mist Chamber) has reported a similar effect on *Fusarium,* a fungus to which peas, beans (Miller and Burke, 1974) and also asparagus

(*Asparagus officinalis* L.) (Gehlker and Scholl, 1974) are susceptible under conditions of low oxygen. Experiments with pea roots (Fig. 5.3-6) confirm the promoting effect of ethanol on growth by micro-organisms at concentrations close to those found in waterlogged soil. Leakage from anaerobic roots of sugars, amino acids and inorganic ions (Grineva, 1961) will also favor micro-biological activity.

The influence of aeration on resistance to infection seems to have been neglected. The susceptiblity of potato tubers to black leg, a bacterial disease may be enhanced by the failure of a protective layer of cork to form in the absence of oxygen (Leach, 1930); in other cases the synthesis of certain phytoalexins that inhibit pathogen attack may be retarded by low partial pressures of oxygen (Griffin, 1969).

## 5.4 ALLEVIATING EFFCTS OF POOR SOIL AERATION

In this section possible ways to minimize effects of inadequate soil aeration are considered under three sub-headings: A. soil improvement, B. soil and crop management, and C. crop improvement. The distinction between soil improvement and management is somewhat arbitrary but the former is mainly concerned with installation of drainage systems, which can be long-lasting in their effect, while the latter is concerned with treatments which may give shorter term benefits. Some procedures for alleviating effects of wet soil, for example the foliar application of growth regulators or the addition of oxidizing agents to the soil are at present highly speculative, and thus discussed only briefly.

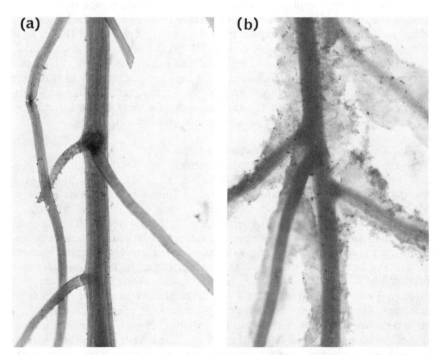

FIG. 5.3-6 The effect of ethanol on growth of micro-organisms around the roots of pea plants grown in (a) nutrient solution only, and (b) nutrient solution containing 3.8 mM ethanol. This concentration is similar to that measured in waterlogged soil. (Jackson, unpublished).

### 5.4.1 Soil Improvement
#### 5.4.1.1 Drainage: general considerations

Drainage has been extensively carried out in many areas of the world, yet not always with clear benefits. Since the purpose is to minimize the effects of excess rainfall, advantages will vary from year to year and also from soil to soil. Trafford (1975) has pointed out the response to drainage will depend considerably on the circumstances prevailing before the installation. He identified four such circumstances (a) the reclamation situation, where without drainage no meaningful farming is possible, (b) the breakdown situation where without drainage the present farming system will have to be abandoned, (c) drainage to enable the farming system to become more intensive and (d) drainage for improved yield in a range of circumstances from those prevailing in the breakdown situation to those where no yield benefit results.

Interpretation and comparison of experiments examining the effects of drainage are complicated by these and several other factors, including rainfall during the experiment, and management of drained and undrained plots in a similar manner (thus minimizing differences which could arise from the opportunity for earlier sowing on drained areas). The importance of this last factor has been shown clearly by Sheard (1978) in Ontario, where on a silt loam soil, much of the increased yield of corn due to drainage could be attributed to the earlier planting that was possible on the drained plots in his experiment.

Very few drainage experiments have been satisfactory from the statistical point of view, often having limited replication.

Drainage often implies the installation of sub-surface pipe systems, but among the first drains to be used were surface drains such as the bedding or ridge and furrow systems of medieval England, (Trafford, 1970). These systems which are simple to provide are still widely used in some high rainfall areas (Forsythe et al., 1979), and especially for sugarcane (Smith, 1976), but may be inconvenient for mechanized cropping. Surface grading to eliminate depressions and to provide larger areas of uniformly graded land may be appropriate in conjunction with sub-surface drains for mechanized agriculture, for example on flat lacustrine soils (Donnan and Schwab, 1974).

#### 5.4.1.2 Drainage: effects on growth and yield

One of the best documented experiments is on a poorly drained silty clay soil (Mollic Haplaquept) in Ohio (Schwab et al., 1974, 1975). In this area about two thirds of the annual precipitation (mean 840 mm) falls between March and September. Undrained plots were compared with surface-drained (by grading to a slope of 0.2 percent), pipe-drained (flat) and a combination of surface and pipe-drained treatments. Irrigation water was applied each year in June to simulate years of above average rainfall. Various crops have been grown since the experiment began in 1959, but corn has predominated. The effects of the different treatments on yield of this crop are given in Table 5.4-1. Apart from a very large average increase in yield from drainage, the coefficient of variation of yield between years was much reduced in pipe-drained plots. The variation in yield from year to year is an indication of the likely economic risk to the farmer. The lower yield on the undrained land was partly due to lower plant populations. Furthermore, in one year, planting was necessarily delayed on plots without pipe drainage because the

TABLE 5.4-1. EFFECT OF DRAINAGE OF A SILT LOAM SOIL IN
OHIO ON GRAIN YIELD AND ITS VARIABILITY IN CORN

| | No drainage (mean yield 3.1 t/ha | Surface drains | Pipe drains | Pipe and surface drains |
|---|---|---|---|---|
| Relative yield (mean of 8 yr) | 100 | 156 | 199 | 210 |
| Coefficient of variation of yield between years (per cent) | 48 | 50 | 21 | 21 |

from Schwab et al. (1975).

soil was too wet. Similarly in some years sowing of oats was sometimes delayed and on one occasion impossible without drainage. The effectiveness of the treatments is further indicated by the positive correlation of corn yields with the flow of drainage water during May, June and July, the period of maximum rainfall and drain flow.

In Iowa, substantial increases in yield of corn due to drainage were also found by Beer et al., (1965), but in New York State over a 15 year period drainage only led to higher yield of corn in three years when rainfall was substantially above average (Black et al., 1976).

Other examples of the effect of drainage on crop yield can be cited in British work. In a clay soil, a Typic Haplaquept, with a hydraulic conductivity of less than 2 cm/day in the subsoil, in an area with a mean annual rainfall of about 630 mm, fairly evenly distributed throughout the year, pipe drains alone at 15 m spacing had limited effect (Trafford and Oliphant, 1977); however, the combined use of pipe and mole drains (unlined circular channels produced with a bullet-shaped tine) at spacings of about 2 m drawn at right angles to and above the pipe drains at a depth of 60 cm, were of considerable benefit, both in removing water and improving crop yield (Table 5.4-2). Smaller increases in yield of wheat due to drainage of other clay soils

TABLE 5.4-2. EFFECT OF DIFFERENT METHODS OF
DRAINAGE IN A CLAY SOIL IN ENGLAND ON
YIELD OF WINTER WHEAT, DEPTH OF WATER-
TABLE AND DRAINAGE OF WATER

| | Yield of wheat* (t ha$^{-1}$) | Mean winter[†] water-table (cm) | Drainage efficiency[†‡] |
|---|---|---|---|
| Undrained | 3.69 (100) | 23 | — |
| Pipe drains | 4.28 (116) | 22 | 42 |
| Pipe drains with subsoiling § | 4.20 (114) | 40 | 41 |
| Pipe drains with mole drains | 4.70 (127) | 40 | 53 |

* Grown in 3 yr.
† Mean of 5 yr.
‡ Total winter drain flow as percentage of rainfall less evaporation.
§ Tine cultivation at 45 cm depth and 1.50 m spacing.
(from Trafford and Oliphant (1977)).

in Britain have been reported by A. C. Armstrong (1978). Information from drainage experiments in several other countries has been summarized by Trafford (1970). Despite the limitations of many drainage experiments, the above examples show that drainage can sometimes have large effect on yields, but also emphasize the need to define the circumstances of the experiment.

The larger crop yields that often result from drainage must be partly due to improved root growth or more efficient functioning of the root system. Widely quoted papers on drainage and soil management (van't Woudt and Hagan, 1957; Russell, 1973) suggest that deeper rooting, leading to greater availability of water during dry weather, is one of the main advantages of draining the soil, but it is difficult to find experimental evidence to support this view. Nevertheless the general desirability of increasing rooting depth to postpone water stress, rather than achieve greater rooting density, has been clearly indicated (at least for soybeans) by Taylor (1979).

Most rice (*Oryza sativa* L.) in Asia is irrigated or rainfed so that water stands on the soil surface for much of the growing season. Nevertheless many years of experience in Japan, Korea and China have shown the benefit of drainage, especially where soil toxins are likely to be present (Nojima, 1963). In Japan installation of drainage has gone on for a long period, and in experiments has been found effective in gley or peat soils where toxic substances tend to accumulate (Section 5.2.3), but not elsewhere; drainage is particularly important where heavy yields are to be achieved. For example Shiroshita et al. (1962) reported yields of rice grain of about 6 t ha$^{-1}$ without drainage, but up to 10 t ha$^{-1}$ with subsurface drainage of an alluvial clay loam. As with field drainage experiments with other crops, the results are specific to the location of the work; with rice the beneficial effect of removing soil toxins may be offset by leaching of nutrients, and there is a lack of information on the optimum percolation rate of water. In south and southeast Asia there has been no critical work on drainage of wet-land rice (personal communication from S. Yoshida, International Rice Research Institute, Philippines). It should be noted, however, that usually it is much more important to conserve water than to remove any excess, though in some areas floods are the major constraint in rice production; solution of the latter problem usually requires construction of dams, large-scale drainage canals and reafforestation. The water and nutrient relations of rice are fully discussed by Yoshida (1981).

### 5.4.1.3 Drainage: effects on soil conditions

Many of the benefits of drainage may arise from 'the direct effects' on crop yield through improved aeration of the soil, but the 'indirect effects' resulting from easier management of the soil may also be important.

In drained soil, higher matric suctions are reached more rapidly and therefore soil strength is greater. The effect of this on reducing the lateral extent and depth of compaction by tractor wheels has been shown by Steinhardt and Trafford (1974). As a result improved trafficability may make possible the earlier application of herbicides and fertilizer to winter cereals, and in the spring may facilitate earlier cultivation and sowing. There is ample evidence that the yields of many spring-sown crops are lower if planting is delayed, eg. barley (Bell and Kirby, 1966), corn (Reeve and Fausey, 1974) and potatoes (Dyke, 1956). Drainage may also increase the number of days in spring when cultivation is possible (Wind and Buitendijk, 1979). The implications of this in British and American conditions have been considered by Smith (1972) and by Reeve and Fausey (1974) respectively. In grassland areas, lowering the watertable lessens poaching (hoofprint) damage by livestock (Massey et al., 1974; Lagocki, 1978).

The thermal capacity of soils increases with moisture content and consequently it is often assumed that drained soils warm up more rapidly in spring. However factual evidence is limited. In Holland, Feddes (1972) found that in spring the mean temperature in the upper 10 cm of well-drained soils was 1 to 2 °C warmer than in soils with shallow watertables; emergence of vegetable crops was also more rapid. In England in a clay soil, a Typic Haplaquept, in a drying period in early spring, the mean soil temperature above 8 cm was often 1 °C warmer on a drained plot than in an undrained area, and the diurnal variation was also greater on the drained areas (Waters, 1977). For crops such as corn that are adversely affected by low soil temperatures, drainage can be especially important.

It is often considered that drainage leads to longer-term improvements in soil structure but few good data on soil physical properties or on drain flow rates and volumes are available. On a clay soil in Holland, high winter watertables in the later years of a drainage experiment led to a decrease in the proportion of transmission pores (van Hoorn, 1958), and a drained clay in England had more transmission pores and greater hydraulic conductivity than undrained areas (Leyton and Yadav, 1960).

In the Ohio drainage experiment discussed earlier (see Section 5.4.1.2) significant effects on several soil properties were evident sixteen years after the start of the experiment (Hundal et al., 1976). The saturated hydraulic conductivity in the top 30 cm of the tile drained treatments was approximately an order of magnitude greater than in undrained land. Since ponding after heavy rain was common on the latter, this seems to be a change of practical importance in a lake-bed soil which is widespread in north-central USA. The greater saturated hydraulic conductivity was apparently due to more transmission pores, although the total porosity in drained and undrained land differed little. However, examination of the patterns of peak flow rates and volumes from the pipe drained treatments did not indicate any significant change in the performance of the drains over a 15 year period (Schwab, 1976).

Evidence that drainage favors the presence of more transmission pores and continuity of these pores is consistent with the general effects of the factors involved. Removal of water by drainage increases the resistance of the soil to compaction by traffic (Steinhardt and Trafford, 1974), and preserves pores for root growth, that should in turn be favored by the improved aeration status of the soil, though the effects of drainage on root growth have been measured rarely (see Section 5.4.1.3). There is good evidence that in general the presence of crops and their roots have themselves led to improvements in soil structure (Low, 1955; Clarke et al., 1967; Cooke and Williams, 1972). However, the mechanisms have not been characterized, but may be associated with the organic residues from the crops. Roots exude large amounts of organic substances, perhaps 10 to 20 percent of the carbon fixed in photosynthesis (Barber and Martin, 1976), on which rhizosphere micro-organisms proliferate. The polysaccharides and other organic materials produced by this biological activity are thought important in forming stable aggregates and therefore soil pores. Goss and Reid (1980) have recently shown that stability of aggregates in an unstable sandy loam was greater after growing ryegrass (*Lolium perenne* L.) than in the absence of roots. Furthermore, although some species of earthworm are able to survive periods of waterlogging, in general most species leave flooded soil for drier sites; earthworm casts also contain more water-stable aggregates than the surrounding soil (Edwards and Lofty, 1977).

Measurements of the stability of soil aggregates might be expected to reflect some of the consequences of draining soil. A higher content of water-stable aggregates was found after five years by Leyton and Yadav (1960). However, Black et al. (1976) found no measureable difference between drained and undrained land over a ten year period, perhaps due to soil heterogeneity; also drainage had little beneficial effect on crop yield in their experiment. However, such changes would not be expected to be important in all soil types or seasons. For fuller discussion of the possible factors involved in the formation and stability of soil pore systems, the reader is referred to E. W. Russell (1973) and R. S. Russell (1977).

Drainage experiments are expensive and therefore limited in number. The need to locate them on major soil types is clear. Furthermore in order to interpret the effects of drainage treatments and to use the results to predict the effects of drainage in other places comprehensive information is needed on the effects on soil conditions and root growth; at present this is lacking.

### 5.4.1.4 Drainage: design

The examples cited above indicate that drainage can sometimes lead to considerable increases in crop production. However drainage design is often empirical. Much has been written about drainage design criteria and equations, and the limitations of different methods. Very useful discussion of this subject, the details of which are outside the scope of this chapter, have been written by Bouwer (1974), Raadsma (1974), Trafford (1975) and Thomasson et al., (1975). The general position is well summarized by Raadsma (1974); despite the evidence from field experiments, and the availability of mathematical models, most drainage design in Europe is based on local practice.

The objectives in draining the soil are to lower the watertable and thereby improve the aeration of the soil to meet the requirements of crop roots, and to increase the trafficability of the soil thereby permitting the maximum benefit to be obtained from earlier sowing and timely field operations, including harvesting. The main variables which the design of drainage systems take into account are the depth and spacing of the drains. The depth determines the maximum height of the watertable when drains cease to flow, while spacing determines the time taken to reach this state. In reviewing the effects of different drain depths and spacings Trafford (1975) pointed out that several experiments have demonstrated benefits from increasing the depth of drains in the range 60 to 140 cm, while most experimenters have failed to find significant differences between various drain spacings. He suggests the following explanation for this. Drain depth will influence the moisture content of the soil profile above the watertable after every rainfall event that gives rise to drainflow; the extent of the effect will depend on the moisture characteristic curve for the particular soil. By contrast all drain spacings will appear equally effective, if the amount of rainfall during the period of the experiment is insufficient to cause high watertables at critical periods for the crop. Where crop yield is used as a measure of the effect, the result may be further obscured, because of the opportunities for compensatory growth between waterlogging and harvest. The latter effect may be most evident in crops such as small-grained cereals where tillering, spikelet number, grain number per spikelet and grain size may each in turn compensate for an earlier stress factor, especially in autumn sown cereals with a long growing period (e.g 5.3.5.1). In crops such as corn with limited tillering ability, plant population is much more important in determining yield. However, in long-term studies where weather patterns have varied appreciably, the advantages of close drain spacing have sometimes been evident (Ericksson, 1979). This emphasizes the need for drainage experiments to be continued for a reasonable period.

Drain spacing (which might be recommended as close as 2 to 3 m in clay (see below) and more than 40 m in sandy soils) is the major variable affecting cost, and is therefore the most important design criterion. In principle, the design of drainage systems can be more rational, but the engineering and soil physical aspects are much better understood than the agronomic aspects. If the hydraulic conductivity of a soil, its drainable porosity and the depth of permeable soil are known it is possible to calculate drain spacing for a given climate (Trafford, 1970). However both Bouwer (1974) and Trafford (1975) emphasize that progress toward more rational drainage design is limited firstly by the cost of obtaining the necessary soil information in relation to the cost of drainage schemes, and secondly by the lack of knowledge of the response of crops to high watertables for short periods at different stages of growth (see 5.3.5). Possible alternative procedures for drainage design avoiding the first problem have been proposed by Bailey (1979). These are, firstly, design based on soil series, using information for a 'standard' soil series profile. Although a practical approach, the method suffers from the extent to which the field to be drained deviates from the standard. And secondly, designs based on values of hydraulic conductivity inferred from other soil physical properties that are more easily determined; the latter method has not been fully evaluated.

Drainage of clay soils presents a special problem. Although mathematical models can be used to assess depth and spacing on soils with low hydraulic conductivity, perhaps less than 10 cm/day, the results are often economically unrealistic. For such soils it is necessary to use very cheap pipes (mole drains) at the calculated spacing, or normal pipes at wider spacing after increasing the hydraulic conductivity of the soil by subsoiling (Trafford, 1975). Loosening the soil to sufficiently increase the hydraulic conductivity can be difficult in humid climates. In the U.K. adequate loosening is rarely achieved below 40 cm (Bailey, 1979). The possible effectiveness of mole drains is indicated in Table 5.4-2. On soils where mole drains are satisfactorily used they usually need to be redrawn at intervals of about 5 years, depending on the soil type and on the soil moisture conditions when the moling operation was carried out; subsequent conditions, eg. deep cracking and freezing may affect their durability. The technique is widely used in Europe on clayey soils (Raadsma, 1974), and increasingly also in New Zealand (Bowler, 1973), but not all soils are suitable for this procedure (Rycroft and Thorburn, 1974).

Drainage can be a relatively high cost investment and therefore must be considered in relation to the likely financial benefit. For high value crops that are very sensitive to waterlogging such as peas (Fig. 5.3-5), intensive drainage systems are justified. For lower value crops also, species can be important. For example on a poorly drained soil in Ontario, an alfalfa-brome (*Bromus inermis* Leyss.) pasture yielded 47 percent less on undrained than drained areas, whereas a trefoil (*Lotus corniculatus* L.)—timothy (*Phleum pratense* L.) pasture was unaffected by drainage (Sheard, 1978, Note on 'four steps to 10.0 t/ha alfalfa', Department of Land Resource Science, University of Guelph, Ontario); the latter species are known to be less sensitive to flooding than the former (Bolton and McKenzie, 1946; Heinrichs, 1970).

### 5.4.1.5 Correcting damage to soil structure.

On soils that are naturally slow draining, artificial drainage will only be useful if continuous transmission pores exist in the topsoil and subsoil. Where these have been lost (see 5.2.3), mechanical creation of fissures by deep tillage or subsoiling may be desirable (Swain, 1975). Subsoiling does not always result in increased yield, and therefore indiscriminate use of the technique should be avoided; careful examination of the soil profile for horizons likely to impede root growth or water movement may help to assess the need. The type of tines and the way in which they are used can greatly influence the effectiveness of the operation. Spoor and Godwin (1978) have shown that there is a critical working depth for rigid tines, below which compaction occurs rather than soil loosening; this critical depth depends on the design and form of the tine, and also on the soil moisture content and density; by attaching wings to the base of the tine, and using shallow tines to loosen the surface layers ahead of the deep tine, the critical depth and soil disturbance was increased.

Other possible means of improving soil structure include addition of organic residues (E. W. Russell, 1977), lime, and soil conditioners of the polyvinyl acetate and polyvinyl alcohol type. Soil conditioners are useful for stabilizing drill rows in soils prone to capping, and Greenland (1977) suggested that they may stabilize fissures created by subsoiling or mole draining.

### 5.4.2 Soil and Crop Management
### 5.4.2.1 Effect of nitrogen supply

The possibility that additional supplies of nitrate or ammonium may help to alleviate the effects of anaerobic soil conditions has been long considered. In solution culture, Woodford and Gregory (1948) showed that the growth of barley in unaerated solutions containing increased concentrations of nitrate was similar to that in aerated solutions with lower nitrate levels. With plants growing in containers of soil, partial or complete compensation for waterlogging damage of corn (Shalhevet and Zwerman, 1958, 1962) and of wheat, barley and oats (Watson et al., 1976) has also been achieved by addition of nitrate, either during or after the waterlogging treatment. For example, Watson et al. (1976) found that when wheat was continuously or intermittently flooded for about 6 weeks after the end of tillering, the grain yields without nitrogen fertilizer were 54 and 64 percent of the non-waterlogged control; with additional nitrogen fertilizer, the corresponding values were 76 and 103 percent of the low nitrogen control. Where the supply of nitrogen prior to waterlogging is restricted, crops may be more susceptible to waterlogging than those well supplied with the nutrient, since additional nitrogen given before anaerobic conditions occur can increase plant nitrogen status during and after the event (Drew and Trought, 1978), and lead to greater yield than in unfertilized plants that are waterlogged (Shazkin and

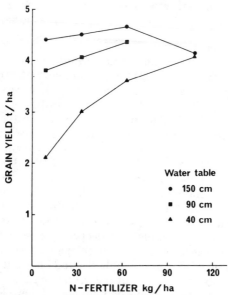

**FIG. 5.4-1 The effect of nitrogen fertilizer on yield of cereals on a clay loam in Holland at various depths of water-table. (From van Hoorn, 1958).**

Federova, 1961). Nitrate fertilizer may favor the formation in the roots of cytokinin hormones which delay leaf senescence (Yoshida and Oritani, 1974).

There are many examples in the literature, of nitrogen fertilizer offsetting the effects of waterlogging in field conditions. In Iowa, flooding of corn for short periods gave a larger decrease in yield at a lower rate of nitrogen application (Ritter and Beer, 1969); in Ontario extra nitrogen fertilizer increased the yield of corn by 70 percent on undrained soil compared with 54 percent on drained plots (Sheard, 1978). In Holland, the adverse effects of high watertables in winter on the yield of small-grained cereals were partially eliminated by the use of additional nitrogen fertilizer in the following spring (van Hoorn, 1958; Sieben, 1964) (Fig. 5.4-1). Sometimes no interaction between drainage and nitrogen fertilizer has been found (Lal and Taylor, 1969; Burke, 1973).

In the field, additional nitrate is normally applied after waterlogging. To raise the redox potential of the soil and to depress the formation of potentially toxic ferric and manganic ions, nitrate would need to be continuously supplied during the waterlogging (Shalhevet and Zwerman, 1958). This is unlikely to be feasible or economic. However, the opportunistic use of extra nitrogen fertilizer after waterlogging can go some way to offset the effects on many crops, but with rising costs of such fertilizers this is inappropriate as a long-term solution.

### 5.4.2.2 Time of planting

If wet soil conditions occur at relatively predictable times then planting to avoid these times might seem advantageous. However if crops in such areas are dependant on the water stored in the soil, any appreciable delay in planting may also lead to loss of yield. In temperate climates the loss of yield due to delayed planting that can be caused by wet soil conditions has already been noted (see 5.4.1.3). Early sowing of winter wheat can lessen the change of pre-emergence waterlogging and associated seedling mortality (see 5.3.5.1 and 5.3.2).

### 5.4.2.3 Interactions between drainage and irrigation

In some places a shallow watertable can complement irrigation by providing water for crop use. In such circumstances it is necessary to strike a careful balance between lowering the watertable sufficiently to avoid problems associated with wet soil conditions, and yet obtain the benefit of the water stored in the soil. If ample water is available for irrigation it might be safer to place drains deeply, but in sandy soils this could lead to over-drainage, risk of drought, and undue reliance on irrigation. In field experiments in North Dakota on sandy soils (Aquic Haplorborolls, Typic Calciaquolls and Typic Haplaquolls) where the watertable declines during the growing season, Follett et al., (1974) and Doering et al., (1976) showed that for alfalfa, corn and sugarbeet (*Beta vulgaris* L.) most irrigation was required where the initial watertable was deepest. Some indication of the minimum watertable was provided in complementary lysimeter studies; yields with a watertable of 45 cm were only about 30 percent of those where the watertable was at about 100 cm, but the irrigation requirement for maximum yield increased as the watertable depth increased beyond about 100 cm (Doering et al., 1976).

In areas with saline ground waters, problems of salinity will be greater with shallow watertables. To maintain a permanent irrigated agriculture and to control salinity, adequate drainage is essential. This subject is well discussed by Bernstein (1974) and by Rhoades (1974).

### 5.4.2.4 Interactions between drainage and cultivation

One of the main functions of cultivation is to control weeds, but following the availability of suitable herbicides, there is world-wide interest in the possibility of simplifying cultivation or even eliminating soil disturbance, except to place seed in the soil; such procedures have been successful in many soils (Cannell et al., 1978; Van Doren et al., 1976). For detailed consideration of this the reader is referred to publications such as Conservation Tillage (1973) and Outlook on Agriculture (1975). In this chapter attention is restricted to aspects which may influence soil aeration; these are most likely to be important in humid regions.

The porosity of soil is generally greater in the cultivated layers of plowed land compared with the same depths in direct-drilled (zero-tilled) land, mainly due to a greater proportion of transmission pores (Finney and Knight, 1973; Pidgeon and Soane, 1977). In heavy textured soils, where artificial drainage is most needed, water mainly moves laterally through the plowed layer if no attempt has been made to modify the subsoil (Trafford and Rycroft, 1973). Thus lower hydraulic conductivity that could restrict lateral movement of water in direct-drilled land could be expected. Aeration may also be restricted, especially on weakly-structured soils that cap (crust).

On the other hand aeration may be promoted in untilled soils by the formation of continuous channels, created by earthworm activity (Baeumer, 1970; Barnes and Ellis, 1979), by cracking in clay soils in dry weather (Ellis et al., 1979) and by decay of roots that are not destroyed each year by plowing. More rapid infiltration of water (Goss et al., 1978), sometimes associated with earthworm channels (Ehlers, 1975), and higher values of unsaturated hydraulic conductivity at potentials greater than -10 kPa have sometimes been found in direct-drilled than plowed soils (Ehlers and van der Ploeg, 1976); these are all evidence that a system of continuous large pores can develop in uncultivated land. However, evidence on the effects of tillage methods on the concentration of oxygen in soils is sparse. Dowdell et al., (1979) found higher concentrations of oxygen in a direct-drilled (zero-tilled) calcareous clay soil (an Aeric Haplaquept) at 15 cm and less frequent occurrence of low oxygen concentrations at 15 and 30 cm than in plowed land during winter when oxygen concentration was at its annual minimum. In other soils containing more clay (Typic Haplaquepts) in wet winters lower values of oxygen concentration have been found after direct drilling (Burford et al., 1979).

Whether the changes outlined in the preceding paragraph are sufficient to affect the drainage requirements of direct-drilled soils is unknown. Field evidence is sparse, but on some experiments on heavy soils in wet years in Germany (Kahnt, 1969), in USA (Griffith et al., 1973; Murdock, 1974) and in Britain (Cannell et al., 1978) direct-drilled crops seem more adversely affected by anaerobiosis than crops grown on plowed soil. It has been suggested that more closely-spaced drains may be needed when such soils are direct-drilled. Few experiments have examined this matter. Chisci (1976) reported results from a clay loam soil in Italy where most of the annual rain (mean 680 mm) falls in autumn and winter, and where storage of water in the soil for use in spring is very important. He found that sub-surface drainage increased the yield of winter wheat in minimally cultivated land by about 44 percent above that of undrained land, compared with only about 5 percent on plowed land, suggesting that more intensive drainage would be required for cropping with minimum cultivation on that soil. Yields of wheat after minimum cultivation were lower than after plowing, possibly due to weeds and greater compaction. On flat land in Ohio, on a silty clay soil (see 5.4.1.2), Schwab et al. (1975) during three years of their experiment, found no interaction between methods of drainage and cultivation on the yield of corn, although zero-tillage was less successful than conventional tillage on this soil due to low plant populations. At times of peak drain flow in June and July, surface run-off was greater from the uncultivated plots, and tile drain flow was greater from the conventionally cultivated plots.

### 5.4.3 Crop Improvement
#### 5.4.3.1 Choice of species

Crop species differ in their sensitivity to waterlogging, and choice of the most tolerant may lead to less variation in yield between years. For example, within small-grained cereals winter wheat is more resistant than winter rye (Bourget et al., 1966; Segata, 1968) and barley is more sensitive than wheat (Tokimasa, 1952; Ikeda et al., 1955; Bourget et al., 1966). The grain legumes, soybeans (De Boer and Ritter, 1970), cowpea (Wien et al., 1979) and French bean (*Phaseolus vulgaris* L.) (Forsythe et al., 1979) are all more affected by waterlogging than corn.

In some areas of the world pastures regularly become flooded (eg. after annual snow melting in Canada), and in other areas intermittent flooding may follow heavy rain. In such conditions, the primary interest is the ability of pasture species to survive flooding. In general, grass species are able to withstand longer periods of flooding than pasture legumes; Bolton and

FIG. 5.4-2 The effect of waterlogging a Vertisol for 2 days on an $F_2$ population of pigeon pea genotypes at the pre-flowering/flowering stage. The plants were photographed 3-4 days after some genotypes had died. (Photograph kindly provided by J. M. Green, International Crops Research Institute for the Semi-Arid Tropics, Hyderabad, India).

McKenzie (1946) found that grasses could survive flooding for 10 to 49 or more days depending on species, and legumes only 9 to 14 days. Comparisons of the sensitivity of different forage species to flooding has been examined in many other experiments; e.g. in grasses by Finn et al., (1961), Rhoades (1964), Anderson (1970), Marshall and Millington (1967) and by Francis and Poole (1973).

### 5.4.3.2 Possibilities of breeding for resistance to waterlogging

Differences in sensitivity to waterlogging between cultivars have been found for several species. These include wheat, barley (Ikeda et al., 1955; Segeta, 1968; Suh, 1973; Yu et al., 1969; Udovenko and Tracher, 1970), rye (Tishkov and Federova, 1964; Udovenko and Lavrova, 1970), lupins (*Lupinus luteus* and *L. angustifolius* L.) (Broue et al., 1976), corn (Marshall et al., 1973), *Bromus mollis* L. (Brown et al., 1976), subterranean clover (*Trifolium subterraneum* L.) (Francis and Devitt, 1969), pea (Jackson, 1978) and bean (*Vicia faba*, L) (Personal communication, Dr. M. A. Hall, Department of Botany, the University College of Wales, Aberystwyth U.K.). However the possibilities of selecting genotypes and breeding for greater resistance to anaerobic conditions seem to have been neglected, although in India genetic differences in the response of pigeon pea (*Cajanus indicus* Spreng) to waterlogging have been observed (Fig. 5.4-2) at the International Crops Research Institute for the Semi-arid Tropics (1976), where work is continuing on selection for waterlogging tolerance. It is believed that this would contribute to the stability of yield over locations and seasons, since in that area the intensity of wetness of soil varies greatly between years.

Tolerance to waterlogging could be due to physiological, biochemical or to morphological differences; for example, in cereals the ability to produce adventitious roots with aerenchyma might be important (see 5.3.4.2.2). Resistance to waterlogging damage may be due in part to resistance to disease. Aldwinkle and Cummins (1974) have identified apple root stocks which are highly susceptible to collar rot, and others with considerable resistance to infection when tested under flooded conditions in the laboratory. Certain selections of alfalfa are also resistant to *Phytophthora* in flooded soils (Lehman, 1972; Pratt and Mitchell, 1976). We are unaware of any estimates of heritability of resistance to waterlogging or of the plant characteristics that are important. Thus the likely success of a breeding program is largely a matter for speculation.

### 5.4.3.3 Application of growth regulators and other chemicals

Several responses of shoot systems to waterlogging, especially in dicotyledenous species, seem to be mediated by changes in their hormone content (see 5.3.3.2). Attempts to minimize these changes by applying growth regulators exogenously have been made at several laboratories in short-term experiments (Railton and Reid, 1973; Selman and Sandanam, 1972; Jackson and Campbell, 1979). Unfortunately, the tomato, which has been used almost exclusively in this work, adapts rapidly to waterlogging by mechanisms such as adventitious rooting, stomatal closure and epinasty, which are probably brought about by the changes in endogenous hormones. *Exogenously* applied hormones interfere with such responses and may thus impair waterlogging tolerance (Jackson and Campbell, 1976b). Growth regulators do not seem to have been tried on species such as *Pisum sativum* or *Phaseolus vulgaris* where adaptive processes under the control of endogenous hormones are less evident.

Other potential remedial measures include the oxidation of the waterlogged environment of the seeds or roots with substances such as barium peroxide (Greilich and Heinze, 1977) or potassium permanganate (Zimmerman, 1930); liming of the soil to reduce manganese toxicity (Graven et al., 1965) and to promote the dissociation of potential phytotoxins such as organic acids (see Section 5.3.1.5), and the application of chlorocholine chloride which may alleviate membrane injury caused by waterlogging (Vorbeikov, 1975). All these suggestions are at present highly speculative and their relevance to practical conditions is unknown.

## 5.5 CONCLUSION

Attempts to improve land by drainage date back at least to Greek and Egyptian civilizations (Donnan, 1976). Although the power and sophistication of present day drainage machinery (eg. laser plant depth control systems) provide the means for rapid installation of drainage systems, much drainage design is empirical. Knowledge of soil science, plant physiology and agronomy is at present inadequate to make full use of the mechanical possibilities. In particular more rational drainage design requires information about soil properties that affect water movement, and of the response of crops to periods of high watertables, in field conditions rather than in artificial environments with atypical conditions of temperature, etc.

The adverse effects of poorly drained soils on crop growth stem from the effects of a limited supply of oxygen for respiration either for root function or for soil organisms. The effects include restricted nutrient and water uptake, and interference with the formation and translocation of plant hormones, and their consequent effects on shoot growth. Indirect effects on plant growth arise from, among other factors, microbial action in anaerobic conditions leading to diminished availability of nitrate and to formation of phytotoxic substances.

Improved knowledge of the physiological mechanisms may lead to alternative means of overcoming the effects of poor soil aeration, but at present drainage seems to be the best available. In addition to directly improving the soil environment for root growth, it can improve crop production indirectly by increasing the trafficability of the soil. This in turn facilitates better soil and crop management by favoring earlier sowing, timely application of herbicides and fertilizer, and less soil compaction. Other techniques such as additional nitrogen fertilizer may improve yields in the year of application. The most neglected aspect seems to be that of breeding and selecting cultivars for increased resistance to anaerobiosis. However, the latter two procedures do not enable advantages to be taken of the indirect benefits on trafficability that drainage can confer.

## References

1    Abeles, F. B., L. E. Forrence and G. R. Leather. 1971. Ethylene and air pollution. Effects of ambient levels of ethylene on the glucanase content of bean leaves. Plant Physiol. 48:504-545.

2    Aldwinkle, H. S. and J. N. Cummins. 1974. Familiar differences in reaction to flood inoculation of young apple seedlings by zoospore suspension of Phytophthora cactorum. Proc. XIX Int. Hort. Congr. 1A Section VII Fruits:328.

3    Ali, M. 1976. Effect of stages and duration of flooding on grain yield of hybrid maize. Ind. J. Agron. 21:4.

4    Allison, F. E. 1973. Soil organic matter and its role in crop production. Elsevier, Amsterdam, London, New York.

5    Anderson, E. R. 1970. Effect of flooding on tropical grasses. Proc. 11th Int. Grassl. Cong. 591-594.

6    Antipov, N. I. 1966. The influence of temperature on the effect of submergence of winter cereals. Fiziologiya Rast. 13:312-319. Seen in Fld. Crop Abst. 19:260, Abst No. 2018.

7    Arikado, H. 1955. Studies on the development of the ventilating system in relation to the tolerance against excess-moisture injury in various crop plants. VI Proc. Crop Sci. Soc. Japan 24:53-58.

8    Armstong, A. C. 1978. The effect of drainage treatments on cereal yields: results from experiments on clay lands. J. Agric. Sci., Camb. 91:229-235.

9    Armstrong, W. 1964. Oxygen diffusion from the roots of some British bog plants. Nature 204:801-802.

10    Armstrong, W. 1978. Root aeration in the wetland condition. p. 269-297. In: D. D. Hook and R. M. M. Crawford (eds). Plant life in anaerobic environments. Ann Arbor Science, Ann Arbor.

11    Armstrong, W. and T. J. Gaynard. 1976. The critical oxygen pressures for respiration in intact plants. Physiol. Plant 37:200-206.

12    Audus, L. J. 1972. Plant growth substances. Chemistry and Physiology: I. Leonard Hill, London.

13    Avadgami, P. N., H. Greenway, R. Lefroy and L. Prior. 1978. Alcoholic fermentation and malate metabolism in rice germinating at low oxygen concentrations. Aust. J. Plant Physiol. 5:15-25.

14    Baeumer, K. 1970. First experiences with direct drilling in Germany. Neth. J. Agric. Sci. 18:283-292.

15    Baker, D. A. 1978. Transport phenomena in plants. Chapman and Hall, London.

16    Bailey, A. D. 1979. Drainage of clay soils in England and Wales. Proc. Int. Drainage Workshop, Int. Inst. Land Reclamation and Improvement, Wageningen, 220-242.

17    Barnes, B. T. and F. B. Ellis. 1979. Effects of different methods of cultivation and direct drilling and disposal of straw residues on population of earthworms. J. Soil Sci. 30:669-679.

18    Barber, D. A. and J. K. Martin. 1976. The release of organic substances by cereal roots into soil. New Phytol. 76:69-80.

19    Beer, C. E., H. P. Johnson and W. D. Shrader. 1965. Yield response of corn in a Planosol soil due to subsurface drainage with variable tile spacings. Iowa Agric. and Home Econ. Stn. Res. Bull. 540.

20    Belding, E. T. 1971. Drainage Survey in England and Wales. Agriculture 78:250-254.

21    Beletskaya, E. K. 1977. Changes in metabolism of winter crops during their adaption to flooding. Soviet Plant Physiol. 24:750-756.

22    Belford, R. K., R. Q. Cannell, R. J. Thomson and C. W. Dennis. 1980. Effects of waterlogging at different stages of development on the growth and yield of peas. J. Sci. Food Agric. 31:857-869.

23    Bell, G. D. H. and E. J. M. Kirby. 1966. Utilization of growth responses in breeding new varieties of cereals. p. 308-319. In: F. L. Milthorpe and J. D. Ivins (eds.) Growth of cereals and grasses, Butterworth Sci. Pub., London.

24    Bendixen, L. E. and M. L. Peterson. 1962. Tropism as a basis for tolerance of strawberry clover to flooding conditions. Crop Sci. 2:223-228.

25    Bergman, H. F. 1959. Oxygen deficiency as a cause of disease in plants. Bot. Rev. 25:418-485.

26    Berjak, P. and T. A. Villiers. 1972. Aging in plant embryos: II Age-induced damage and its repair during early germination. New Phytol. 71:135-144.

27    Bernstein, L. 1974. Crop growth and salinity. p. 39-54 In: J. van Schilfgaarde (ed.) Drainage for agriculture. Am. Soc. Agron., Madison, WI.

28    Berry, L. J. and W. E. Norris. 1949. Studies of onion root respiration: I. Velocity of oxygen consumption in different segments of root and different temperatures as a function of partial pressure of oxygen. Biochem. Biophys. Acta 3:593-606.

29    Bingham, F. T., H. P. Chapman and A. L. Pugh. 1954. Solution culture studies of nitrite toxicity to plants. Soil Sci. Soc. Proc. 18:305-308.

30    Black, R. D., M. F. Walter and F. N. Swader. 1976. Apparent effects of subsurface drainage on soil physical properties. Proc. 3rd Nat. Drainage Symp., ASAE, St. Joseph, MI 49085. p. 123-126.

31    Blake, G. R., W. W. Nelson and R. R. Allmaras. 1976. Persistence of subsoil compaction in a Mollisol. Soil Sci. Soc. Amer. J. 40:943-948.

32    Bodman, G. B., D. E. Johnson and W. H. Kruskal. 1958. Influence of VAMA and of depth of rotary hoeing upon infiltration of irrigation water. Soil Sci. Soc. Am. Proc. 22:463-468.

33    Bolton, J. L. and R. E. McKenzie. 1946. The effect of early spring flooding on certain forage crops. Sci. Agric. 26:99-105.

34    Bourget, S. J., B. J. Finn and B. K. Dow. 1966. Effects of different soil moisture tensions on flax and cereals. Can. J. Soil Sci. 46:213-216.

35    Bouwer, H. 1974. Developing drainage design criteria. p. 67-90. In: J. van Schilfgaarde (ed.) Drainage for agriculture. Am. Soc. Agron., Madison, WI.

36    Bowler, D. G. 1973. The drainage problem and research needs. Inf. Series. New Zeal. Dept. of Sci. and Ind. Res. 96:102-106.

37    Bradford, K. J. and D. R. Dilley. 1978. Effects of root anaerobiosis on ethylene production, epinasty and growth of tomato plants. Plant Physiol. 61:506-509.

38    Broadbent, F. E. and F. E. Clark. 1965. Denitrification. p. 344-359. In: W. V. Bartholomew and F. E. Clark (eds.) Soil nitrogen. Am. Soc. Agron., Madison, WI.

39    Broué, P., D. R. Marshall and J. Munday. 1976. The response of lupins to waterlogging. Aust. J. Exp. Agric. An. Husb. 16:549-554.

40    Brown, A. D. H., D. R. Marshall and J. Munday. 1976. Adaptedness of variants at an alcohol dehydrogenase locus in Bromus mollis L. (soft bromegrass). Aust. J. Biol. Sci. 29:389-396.

41    Burford, J. R., R. J. Dowdell and J. M. Lynch. 1978. Critique of 'Relationships among microbial populations and rates of denitrification in a Hanford soil'. P. 365-377. In: Dr. Nielsen and J. G. MacDonald (eds.) Nitrogen in the environment. Acad. Press. London.

42    Burford, J. R., R. J. Dowdell, R. Crees and K. C. Hall. 1979. Soil aeration and denitrification. Agric. Res. Coun. Letcombe Lab. Ann. Report, 1978:26.

43    Burke, W. 1973. Effect of drainage and nitrogen application on ryegrass yield on blanket peat. Irish J. Agric. Res. 12:159-166.

44    Burrows, W. J. and D. J. Carr. 1969. Effects of flooding the root system of sunflower plants on the cytokinin content in the xylem sap. Physiol. Plant. 22:1105-1112.

45    Campbell, C. A., D. S. McBean and D. G. Green. 1969. Influence of moisture stress,

relative humidity and oxygen diffusion rate on seed set and yield of wheat. Can. J. Pl. Sci. 49:29-37.

46    Campbell, R. B. and G. T. Seaborn. 1972. Yield of flue-cured tobacco and levels of soil oxygen in lysimeters with different water table depths. Agron. J. 64:730-733.

47    Cannell, R. Q. 1979. Effects of soil drainage on root growth and crop production. p. 183-197. In: R. Lal and D. J. Greenland (eds.). Soil physical conditions and crop production in the tropics. J. Wiley & Sons, Chichester.

48    Cannell, R. Q., D. B. Davies, D. Mackney and J. D. Pidgeon. 1978. The suitability of soils for sequential direct drilling of combine-harvested crops in Britain: a provisional classification. Outl. Agric. 9:306-316.

49    Cannell, R. Q., K. Gales, R. W. Snaydon and B. A. Suhail. 1979. Effects of short-term waterlogging on the growth and yield of peas (Pisum sativum L.) Ann. Appl. Biol. 93:327-335.

50    Cannell, R. Q., R. K. Belford, K. Gales, C. W. Dennis and R. D. Prew. 1980. Effects of waterlogging at different stages of development on the growth and yield of winter wheat. J. Sci. Food Agric. 31:117-132.

51    Catlin, P. B., G. C. Martin, and E. A. Olsson. 1977. Differential sensitivity of Juglans hindsii, J. regia, Paradox hybrid and Prerocarya stenoptera to waterlogging. J. Am. Soc. Hort. Sci. 102:101-104.

52    Chandramohan, D., D. Purushothaman and R. Kothandaraman. 1973. Soil phenolics and plant growth inhibition. Plant Soil 39:303-308.

53    Chaudhary, T. N., V. K. Bhatnagar and S. S. Prihar. 1974. Growth response of crops to depth and salinity of ground water and soil submergence: I. Wheat (Triticum aestivum L.) Agron. J. 66:32-35.

54    Chaudhary, T. N., V. K. Bhatnagar and S. S. Prihar. 1975. Corn yield and nutrient uptake as affected by watertable depth and soil submergence. Agron. J. 67:745-749.

55    Chibnall, C. A. 1939. Protein metabolism in plants. Yale Univ. Press, New Haven, CT. 266 pp.

56    Chin, T. Y., M. M. Meyer, Jr., and L. Beevers. 1969. Abscisic acid stimulated rooting of stem cuttings. Planta 88:192-196.

57    Ching, T. M. 1972. Metabolism of germinating seeds. p. 103-218. In: T. T. Kozlowski (ed.) Seed biology. Academic Press, New York.

58    Chisci, G. 1976. Influence of tillage and drainage systems on physical conditions for loam-clay soils cultivation. Proc. 7th Conf. Int. Soil Tillage Res. Org. Uppsala 7:1-6.

59    Clark, A. L., D. J. Greenland and J. P. Quirk. 1967. Changes in some physical properties of the surface of an impoverished red-brown earth under pasture. Aust. J. Soil Res. 5:59-68.

60    Clements, F. E. 1921. Aeration and air content; the role of oxygen in root activity. Carnegie Inst. Washington. Publ. 315.

61    Cole, G. 1976. Land drainage in England and Wales. J. Inst. Water Eng. and Sci. 30:345-361.

62    Conn, E. E. and P. K. Stumpf. 1963. Outlines of biochemistry. Wiley, New York.

62a   Conservation tillage (1973) Soil Cons. Soc. Am., Iowa. 241 pp.

63    Cooke, G. W. 1976. A review of the effects of agriculture on the chemical composition and surface quality of surface and underground waters. In: Agriculture and water quantity, Tech. Bull. 32:5-57. Min. Agric. Fish. Food, H.M.S.O., London.

64    Cooke, G. W. and R. J. B. Williams. 1972. Problems with cultivations and soil structure at Saxmundham. Rothamsted Exp. Stn. Ann. Report, 1971, 2:122-148.

65    Cornforth, I. S. 1975. The persistence of ethylene in aerobic soils. Plant Soil 42:85-96.

66    Council for Agricultural Science and Technology. 1976. Effect of increased nitrogen fixation on stratospheric nitrogen. Report No. 53, Dept. Agron., Iowa State University. 34 pp.

67    Coutts, M. P. and W. Armstrong. 1976. Role of oxygen transport in the tolerance of trees to waterlogging. p. 361-385. In: M. G. R. Cannell and F. T. Last (eds.) Tree physiology and yield improvement. Academic Press, London.

68    Cowan, M. 1978. Influence of nitrate, water potential and oxygen tension in nitrogen fixation in detached pea roots. Qualitas Plantarum 28:65-69.

69    Crawford, R. M. M. 1977. Tolerance of anoxia and ethanol metabolism in germinating seeds. New Phytol. 79:511-517.

70    Crawford, R. M. M. 1978. Biochemical and ecological similarities in marsh plants and diving animals. Naturwiss. 65:194-201.

71    Crawford, R. M. M. and M. A. Baines. 1977. Tolerance of anoxia and the metabolisms of ethanol in tree roots. New Phytol. 79:519-526.

72   Crocker, W. 1948. Growth of plants. Reinhold, New York.
73   Crossett, R. N. and D. J. Campbell. 1975. The effects of ethylene in the root environment upon the development of barley. Plant Soil 42:453-464.
74   Crozier, A. and D. M. Reid. 1971. Do roots synthesize gibberellins? Can. J. Bot. 49:967-975.
75   Currie, J. A. 1970. Movement of gases in soil respiration. In: Sorption and transport processes in soils. Soc. Chem. Ind. Monogr. 37:152-171.
76   Culbert, D. L. and H. W. Ford. 1972. The use of a multi-celled apparatus for anaerobic studies of flooded root systems. Hort. Sci. 71:29-31.
77   Davies, D. D., K. H. Nascimento and K. D. Patil. 1974. The distribution and properties of NADP malic enzyme in flowering plants. Phytochem. 13:2417-2425.
78   DeBoer, D. W. and W. F. Ritter. 1970. Flood damage to crops in depression areas of north-central Iowa. TRANSACTIONS of the ASAE 13:547-549, 553.
79   Doering, E. J., G. A. Reichman, L. C. Benz and R. F. Follet. 1976. Drainage requirement for corn grown on sandy soil. p. 144-147. In: Proc. 3rd Nat. Drainage Symp., ASAE, St. Joseph, MI 49085.
80   Donnan, W. W. 1976. An overview of drainage worldwide. p. 6-9. In: Proc. Third Nat. Drainage Conf., ASAE, St. Joseph, MI 49085.
81   Donnan, W. W. and G. O. Schwab. 1974. Current drainage methods in the U.S.A. p. 93-114. In: J. van Schilfgaarde (ed.) Drainage for agriculture. Am. Soc. Agron., Madison, WI.
82   Dowdell, R. J., R. Crees, J. R. Burford and R. Q. Cannell. 1979. Oxygen concentrations in a clay soil after ploughing or direct drilling. J. Soil Sci. 30:239-245.
83   Drew, M. C. and E. J. Sisworo. 1977. Early effects of flooding on nitrogen deficiency and leaf chlorosis in barley. New Phytol. 79:567-571.
84   Drew, M. C. and M. C. T. Trought. 1978. Effects of the concentration of nitrate on the growth of wheat in anaerobic solution culture. Agric. Res. Council Letcombe Lab. Ann. Report. 1977:63-64.
85   Drew, M. C., M. B. Jackson and S. Giffard. 1979. Ethylene-promoted adventitious rooting and development of cortical air spaces (aerenchyma) in roots may be adaptive responses to flooding in Zea mays L. Planta 147:83-88.
86   Dudal, R. 1976. Inventory of the major soils of the world with special reference to mineral stress hazards. p. 3-13. In: Plant adaptation to mineral stress in problem soils. Spec. Publ., Cornell Univ., Agric. Exp. Stn., Ithaca, New York.
87   Dudal, R. 1978. Land resources for agricultural development. 11th Inter. Congr. Soil Sci. 2:314-340, Edmonton, Alberta, Canada.
88   Dyke, G. V. 1956. The effect of date of planting on the yield of potatoes. J. Agric. Sci. Camb. 47:122-128.
89   Eavis, B. W., H. M. Taylor and M. G. Huck. 1971. Radicle elongation of pea seedlings as affected by oxygen concentration and gradients between shoot and root. Agron. J. 63:770-772.
90   Edwards, C. A. and J. R. Lofty. 1977. p. 133. In: Biology of earthworms. Chapman and Hall, London.
91   Edwards, T. I. 1933. The germination and growth of Peltandra virginica in the absence of oxygen. Bull. Torrey Bot. Club 60:573-581.
92   Ehlers, W. 1975. Observations on earthworm channels and infiltration on tilled and untilled loess soil. Soil Sci. 119:242-249.
93   Ehlers, W. and R. R. van der Ploeg. 1976. Evaporation, drainage and unsaturated hydraulic conductivity of tilled and untilled fallow soil. Z. Pflanzenern Bodenk 3:373-386.
94   Ellis, F. B., J. G. Elliott, F. Pollard, R. Q. Cannell and B. T. Barnes. 1979. Comparison of direct drilling, reduced cultivation and ploughing on the growth of cereals. 3. Winter wheat and spring barley on a calcareous clay. J. Agric. Sci. Camb. 93:391-401.
95   Erickson, A. E. and D. M. Van Doren. 1960. The relation of plant growth and yield to oxygen availability. Proc. 7th Int. Congr. Soil Sci. (Madison, WI.) IV. 54:429-434.
96   Ericksson, J. 1979. Soil Functions and Drainage. Proc. Int. Drainage Workshop, Int. Inst. Land Reclamation and Improvement, Wageningen 180-212.
97   Feddes, R. A. 1972. Effects of water and heat on seedling emergence. J. Hydrol. 16:341-359.
98   Federova, R. I., E. I. Milekhina and N. I. Il'Yukhina. 1973. Evaluation of the method of "gas metabolism" for detecting extraterrestrial life. Identification of nitrogen fixing micro-organisms. Izv. Akad. Nauk. SSSR, Ser. Biol. 6:797-806.
99   Ferraz, J. F. P. 1973. Influence of the soil atmosphere on spread of Ophiobolus

*graminus* along wheat roots. Trans. Br. Mycol. Soc. 61:237-249.

100   Finn, B. J., S. J. Bourget, K. F. Nielsen and B. K. Dow. 1961. Effects of different soil moisture tensions on grass and legume species. Can. J. Soil Sci. 61:16-23.

101   Finney, J. R. and B. A. G. Knight. 1973. The effect of soil physical conditions produced by various cultivation systems on the root development of winter wheat. J. Agric. Sci. Camb. 80:435-442.

102   Follett, R. F., E. J. Doering, G. A. Reichman and L. C. Benz. 1974. Effect of irrigation and water table-depth on crop yields. Agron. J. 66:304-308.

103   Food and Agriculture Organization of the United Nations. 1974. FAO-UNESCO Map of the World, Vol. 1—Legend. Unesco, Paris.

104   Ford, H. W. 1965. Bacterial metabolites that affect citrus root survival in soils subject to flooding. Proc. Am. Soc. Hort. Sci. 86:205-212.

105   Ford, H. W. 1973. Levels of hydrogen sulfide toxic to citrus roots. J. Am. Soc. Hort. Sci. 98:66-68.

106   Forsythe, W. M. and A. M. Pinchinat. 1971. Tolerancia de la variedad de fryol '27-R' a la inundaction. Turrialba 21:228-231.

107   Forsythe, W. M., A. Victor and M. Gomez. 1979. Flood tolerance and surface drainage requirement of *Phaseolus vulgaris* L. p. 205-214. In: R. Lal and D. J. Greenland (eds.) Soil physical conditions and crop production in the tropics. J. Wiley & Sons, Chichester.

108   Francis, C. M. and A. C. Devitt. 1969. The effect of waterlogging on the growth and isoflavone concentration of *Trifolium subterraneum* L. Aust. J. Agric. Res. 20:819-825.

109   Francis, C. M. and M. L. Poole. 1973. Effect of waterlogging on the growth of annual medicago species. Aust. J. Exp. Agric. Anim. Husb. 13:711-713.

110   Fulton, J. M. and A. E. Erickson. 1964. Relation between soil aeration and ethyl alcohol accumulation in xylem exudate of tomatoes. Soil Sci. Soc. Am. Proc. 28:610-614.

111   Garcia-Novo, F. and R. M. M. Crawford. 1973. Soil aeration, nitrate reduction and flooding tolerance in higher plants. New Phytol. 72:1031-1039.

112   Gehlker, H. and W. Scholl. 1974. Ecological factors and cultivation problems in connection with parasitic root rot of asparagus. Zeitschrift fur Pflanzen Krankheiten und Pflanzenschutz. 81:394-406.

113   Geisler, G. 1967. Interactive effects of carbon dioxide and oxygen in soil on root and top growth of barley and peas. Plant Physiol. 42:305-307.

114   Gerwitz, A. and E. R. Page. 1975. An empirical mathematical model to describe plant root systems. J. App. Ecol. 11:773-781.

115   Goldney, D. C. and R. F. M. van Stevenick. 1972. Ethylene production and biochemical changes in detached leaves of *Nymphoides indica*. p. 604-610. In: D. J. Carr (ed.) Plant growth substances 1970. Springer, Berlin.

116   Goodlass, G. and K. A. Smith. 1979. Effects of ethylene on root extension and nodulation of pea (*Pisum sativum* L.) and white clover (*Trifolium repens* L.). Plant Soil 51:387-395.

117   Goss, M. J., K. R. Howse and W. Harris. 1978. Effects of cultivation on soil water retention and water use by cereals in clay soil. J. Soil Sci. 29:475-488.

118   Goss, M. J. and J. B. Reid. 1980. Changes in aggregate stability in a sandy loam effected by growing roots of perennial ryegrass *Lolium perenne* L. J. Sci. Food Agric. 31:325-328.

119   Grable, A. R. 1966. Soil aeration and plant growth. Adv. Agron. 18:57-106.

120   Grable, A. R. 1971. Effects of compaction in content and transmission of air in soils. p. 154-164. In: Compaction of agricultural soils. ASAE. Monograph, ASAE, St. Joseph, MI 49085.

121   Graven, E. H., O. J. Attoe and D. Smith. 1965. Effect of liming and flooding on manganese toxicity in alfalfa. Proc. Soil Sci. Soc. Am. 29:702-706.

122   Greenland, D. J. 1962. Denitrification in some tropical soils. J. Agric. Sci., Camb. 58:227-233.

123   Greenland, D. J. 1977. Soil damage by intensive arable cultivation: temporary or permanent? Phil. Trans. Royal Soc. Lond. B. 281:193-208.

124   Greenwood, D. J. 1975. Measurement of soil aeration. In: Soil physical conditions and crop production. Tech. Bull. 29:261-272. Min. Agric. Fish. Food., H.M.S.O., London.

125   Greenwood, D. J. and D. Goodman. 1967. Direct measurements of the distribution of oxygen in soil aggregates and in columns of fine soil crumbs. J. Soil Sci. 18:182-196.

126   Greenwood, D. J. and D. Goodman. 1971. Studies on the supply of oxygen to the roots of mustard seedlings. New Phytol. 70:85-96.

127   Gregory, P. J., M. McGowan, P. V. Biscoe and B. Hunter. 1978. Water relations of winter wheat: I Growth of the root system. J. Agric. Sci. Camb. 91:91-102.

128   Greilich, J. and G. Heinze. 1977. On the effect of the oxygen releaser barium perox-

ide on seed germination and emergence in seedbeds of different compaction. Archiv. fur Acker und Pflanzbau und Bodenkd., Berlin. 21:127-132.

129   Griffin, D. M. 1969. Soil water in the ecology of fungi. Ann. Rev. Phytopath. 7:289-310.

130   Griffith, D. R., J. V. Mannering, H. M. Galloway, S. D. Parsons and C. B. Richey. 1973. Effect of eight tillage planting systems on soil temperature, percent stand, plant growth and yield of corn on five Indiana soils. Agron. J. 65:321-326.

131   Grineva, G. M. 1961. Excretion by plant roots during brief periods of anaerobiosis. Soviet Plant Physiol. 8:686-691.

132   Grineva, G. M. and G. A. Nechiporenko. 1977. Distribution and conversion of sucrose U-$C^{14}$ in corn plants under conditions of flooding. Soviet Plant Physiol. 24:32-37.

133   Grobbelaar, N., B. Clarke and M. C. Hough. 1971. The nodulation and nitrogen fixation of isolated roots of *Phaseolus vulgaris* L: III The effects of carbon dioxide and ethylene. p. 215-223. In: T. A. Lie and E. G. Mulder (eds.) Plant Soil, special volume "Biological nitrogen fixation in natural and agricultural habitats".

134   Gudjonsdottir, S. and H. Burstrom. 1962. Growth-promoting effects of alcohols on excised wheat roots. Physiol. Plant. 15:498-504.

135   Haberkorn, K. R. and S. A. Sievers. 1977. Response to gravity of roots growing in water. Naturwiss. 64:639-640.

136   Hackett, C. 1969. A study of the root system of barley: II Relationships between root dimensions and nutrient uptake. New Phytol. 68:1023-1030.

137   Hagan, R. 1950. Soil aeration as a factor in water absorption by the roots of transpiring plants. Plant Physiol. 25:748-762.

138   Haines, W. B. 1923. The volume-changes associated with variations of water content in soil. J. Agric. Sci. Camb. 13:296-310.

139   Hall, D. G. M., M. J. Reeve, A. J. Thomasson and V. F. Wright. 1977. Water retention, porosity and density of field soils. Soil Survey Tech. Monograph No. 9. Rothamsted, Harpenden, England.

140   Hall, S. M. and G. C. Medlow. 1974. Identification of IAA in phloem and root pressure saps of *Ricinus communis* L. by mass spectrometry. Planta 119:257-261.

141   Hauck, R. D. 1969. Quantitative estimates of N cycle processes: Review and comments. Conference on recent developments in the use of $^{15}N$ in soil-plant studies. Sponsored by the joint FAO/IAEA Div. of Atomic Energy in Food and Agric., Sofia, Bulgaria.

142   Heinrichs, D. H. 1970. Flooding tolerance of legumes. Can. J. Plant Sci. 50:435-438.

143   Heydecker, W. and R. S. Chetram. 1971. Water relations of beetroot seed germination. II. Effects on the ovary cap on the endogenous inhibitors. Ann. Bot. 35:31-42.

144   Hiler, E. A. 1976. Drainage requirements of crops. Proc. 3rd Nat. Drainage Conf., ASAE, St. Joseph, MI 49085. p. 127-129.

145   Hiron, R. W. P. and S. T. C. Wright. 1973. The role of endogenous abscisic acid in the response of plants to stress. J. Exp. Bot. 24:769-781.

146   Holm, R. E. 1972. Volatile metabolites controlling germination in buried weed seed. Plant Physiol. 50:293-297.

147   Hook, D. D. and C. L. Brown. 1973. Root adaptions and relative flood tolerance of five hardwood species. Forest Sci. 19:225-229.

148   Hoorn, J. W. van. 1958. Results of a ground water level experimental field with arable crops on clay soil. Neth. J. Agric. Sci. 6:1-10.

149   Hopkins, H. T., A. W. Specht and S. B. Hendricks. 1950. Growth and nutrient accumulation as controlled by oxygen supply to plant roots. Plant Physiol. 25:193-209.

150   Hoveland, C. S. and H. L. Webster. 1965. Flooding tolerance of annual clovers. Agron. J. 57:3-4.

151   Howeler, R. H. 1973. Iron-induced oranging disease of rice in relation to physicochemical changes in a flooded oxisol. Soil Sci. Soc. Am. Proc. 37:898-903.

152   Howell, T. A., E. A. Hiler, O. Zolezzi and C. J. Ravelo. 1976. Grain sorghum response to inundation at three growth stages. TRANSACTIONS of the ASAE 19(5):876-880.

153   Huck, M. G. 1970. Variations in tap root elongation rate as influenced by composition of the soil air. Agron. J. 62:815-818.

154   Hundal, S. S., G. O. Schwab and G. S. Taylor. 1976. Drainage system effects on physical properties of a lakebed clay soil. Soil Sci. Soc. Am. J. 40:300-305.

155   Ibrahim, W. R. 1976. Drainage effect on soil characteristics and crop production. p. 117-122. In: Proc. Third Nat. Drainage Conf., ASAE, St. Joseph, MI 49085.

156   Ikeda, T. and others. 1957. Effect of excess moisture in the soil on the growth of barley at various stages. Bull. Div. Pl. Br. Cultur., Tokai-Kinki Nat. Agric. Exp. Station

4:30-37, (seen in Fld. Crop Abstr. 1958, 11:245, Abs. No. 1938).

157   Ikeda, T., S. Higashi, T. Kawaide and S. Saigo. 1955. Studies on the wet-injury resistance of wheat and barley varieties: II. Varietal difference of wet-injury resistance of wheat and barley. Bull. Div. Plant Breed Cultn. Tokai-Kinki Nat. Agric. Exp. Stn. No. 2:11-16.

158   International Crops Research Institute for the Semi-Arid Tropics. 1976. Pigeon pea breeding, ICRISAT Annual Report, 1975-76:93.

159   Ioannou, N., R. W. Schneider and R. G. Grogan. 1977. Effect of flooding on the soil gas composition and the production of microsclerotia by *Verticillium dahliae* in the field. Phytopathology 67:651-656.

160   Jackson, M. B. 1978. Responses of different types of pea to anaerobic soil conditions. Agric. Res. Council Letcombe Lab. Ann. Report 1977:61-63.

161   Jackson, M. B. 1979. Rapid injury to peas by soil waterlogging. J. Sci. Food Agric. 30:143-152.

162   Jackson, M. B. and D. J. Campbell. 1975a. Ethylene and waterlogging effects in tomato. Ann. Appl. Biol. 81:102-105.

163   Jackson, M. B. and D. J. Campbell. 1975b. Movement of ethylene from roots to shoots, a factor in the responses of tomato plants to waterlogged soil conditions. New Phytol. 74:397-406.

164   Jackson, M. B. and D. J. Campbell. 1976a. Waterlogging and petiole epinasty in tomato: the role of ethylene and low oxygen. New Phytol. 76:21-29.

165   Jackson, M. B. and D. J. Campbell. 1976b. Hormones and the responses of tomato plants to waterlogged soil. Agric. Res. Council Letcombe Lab. Ann. Report 1975:42-43.

166   Jackson, M. B., K. Gales and D. J. Campbell. 1978. Effect of waterlogged soil conditions on the production of ethylene and on water relationships in tomato plants. J. Exp. Bot. 29:183-193.

167   Jackson, M. B. and D. J. Campbell. 1979. Effects of benzyladenine and gibberellic acid on the responses of tomato plants to anaerobic root environments and to ethylene. New Phytol. 82:331-340.

168   Jackson, M. B. and D. J. Osborne. 1972. Abscisic acid, auxin and ethylene in explant abscission. J. Exp. Bot. 23:849-862.

169   Jackson, W. T. 1955. The role of adventitious roots in recovery of shoots following flooding of the original root systems. Am. J. Bot. 42:816-819.

170   Jackson, W. T. 1956. The relative importance of factors causing injury to shoots of flooded tomato plants. Am. J. Bot. 43:637-639.

171   Jones, D. and E. G. Gray. 1977. Crop losses due to soil-borne fungus, *sclerotinia sclerotinia*. A.R.C. Res. Rev. 3:79-81.

172   Jones, H. E. and J. R. Etherington. 1970. Comparative studies of plant growth and distribution in relation to waterlogging: I. The survival of *Erica carnea* L. and *E. tetralix* L. and its apparent relationship to iron and manganese uptake in waterlogged soil. J. Ecol. 58:487-496.

173   Joshi, M. S. and N. G. Dastane. 1965. Excess water tolerance of summer cereals. Ind. J. Agron. 10:289-298.

174   Joshi, M. S. and N. G. Dastane. 1966. Studies in excess water tolerance of crop plants: II Effect of different durations of flooding at different stages of growth, under different lay-outs on growth, yield and quality of maize. Ind. J. Agron. 11:70-79.

175   Kahnt, G. 1969. Ergebnisse zweijahriger Direktsaatversuche auf drei Bodentypen. Z. Acker. Pfl. Bau. 129, 227-95.

176   Kawase, M. 1972. Effects of flooding on ethylene concentration in horticultural plants. J. Am. Soc. Hort. Sci. 97:584-588.

177   Kawase, M. 1974. Role of ethylene in induction of flooding damage in sunflower. Physiol. Plant 31:29-38.

178   Kiyosawa, K. 1975. Studies on the effects of alcohols on membrane water permeability of Nitella. Protoplasma 86:243-252.

179   Konings, H. and M. B. Jackson. 1979. A relationship between rates of ethylene production by roots and the promoting or inhibiting effects of exogenous ethylene and water on root elongation. Zeitschrift fur Pflanzenphysiol. 92:385-397.

180   Kordan, H. A. 1975. Relationship between oxygen availability and transverse and vertical shoot geotropisms during germination of submerged rice seedlings. Ann. Bot. 39:249-256.

181   Kramer, P. J. 1940. Causes of decreased absorption of water by plants in poorly aerated media. Am. J. Bot. 27:216-220.

182   Kramer, P. J. 1951. Causes of injury to plants resulting from flooding of the soil. Plant Physiol. 26:722-736.

183    Kramer, P. J. and W. T. Jackson. 1954. Causes of injury to flooded tobacco plants. Plant Physiol. 29:241-245.
184    Kristensen, K. J. and H. Enoch. 1964. Soil air composition and oxygen diffusion rate in soil columns at different heights above a water table. 8th Intern. Congr. Soil Sci., Bucharest, Romania, II:159-170.
185    Ku, H. S., H. Suge, L. Rappaport and H. K. Pratt. 1970. Stimulation of rice coleoptile growth by ethylene. Planta 90:333-339.
186    Lagocki, H. F. R. 1978. Surface soil stability as defined and controlled by a drainage criterion. p. 233-237. In: W. W. Emerson, R. D. Bond and A. R. Dexter (eds.) Modification of Soil Structure. J. Wiley and Sons, Chichester.
187    Lal, R. and G. S. Taylor. 1969. Drainage and nutrient effects in a field lysimeter study: I. Corn yield and soil conditions. Soil Sci. Soc. Am. Proc. 33:937-941.
188    Larque-Saavedra, H. Wilkins and R. L. Wain. 1975. Promotion of cress root elongation in white light by 3, 5-diiodo-4-hydroxy-benzoic acid. Planta 126:269-272.
189    Leach, J. G. 1930. Potato blackleg: The survival of the pathogen in the soil and some factors influencing infection. Phytopath. 20:215-228.
190    Levlova, S., D. Ehlichova and J. Barthova. 1973. The occurrence of ethanol and alcohol dehydrogenase in seed plants. Rostlinna Vyroba (Praha) 19:1209-1220.
191    Lee, R. B. 1977. Effects of organic acids on the loss of ions from barley. J. Exp. Bot. 28:578-587.
192    Lee, R. B. 1978. Inorganic nitrogen metabolism in barley roots under poorly aerated conditions. J. Exp. Bot. 29:693-708.
193    Lee, R. B. 1979. The effect of nitrite on root growth of barley and maize. New Phytol. 83:615-622.
194    Lehman, W. F. 1972. Phytophthora root rot. Rhizoctonia root canker and flooding injury of alfalfa in southwestern United States. p. 15. In: Rep. of 23rd Alfalfa Improvement Conf.
195    Leonard, O. A. and J. A. Pinckard. 1946. Effect of various oxygen and carbon dioxide concentrations on cotton root development. Plant Physiol. 2:18-36.
196    Leyshon, A. J. and R. W. Sheard. 1978. Growth and yield of barley in flooded soil: Ethylene generation and ER relationships. Can. J. Soil Sci. 58:347-355.
197    Leyton, L. and J. S. P. Yadav. 1960. Effect of drainage on certain physical properties of a heavy clay soil. J. Soil Sci. 11:305-312.
198    Livingstone, B. E. and E. E. Free. 1917. The effect of deficient soil oxygen on the roots of higher plants. John Hopkins, Univ. Circ. 3:182-185.
199    Livne, A. and Y. Vaadia. 1965. Stimulation of transpiration rate in barley leaves by kinetin and gibberellic acid. Physiol. Plant 18:658-664.
200    Lloyd, F. E. 1916. The abscission of flowerbuds and fruits in Gossypium and its relation to environmental changes. Trans. Royal Soc. Canada 10:55-61.
201    Longmuir, I. S. 1954. Respiration of bacteria as a function of oxygen concentration. Biochem. 57:81-87.
202    Low, A. J. 1955. Improvements in the structural state of soils under leys. J. Soil Sci. 6:179-199.
203    Luckwill, L. C. 1960. The physiological relationships of root and shoot. Sci. Hort. 14:22-26.
204    Luthin, J. N. 1957. Drainage of agricultural lands. Am. Soc. Agron., Madison, WI.
205    Luxmoore, R. J. and L. H. Stolzy. 1972. Oxygen consumption rates predicted from respiration, permeability and porosity measurements on excised wheat root segments. Crop Sci. 12:442-445.
206    Lynch, J. M. 1976. Products of soil micro-organisms in relation to plant growth. C. R. C. Crit. Rev. Microbiol. 5:67-107.
207    Lynch, J. M. 1977. Phytotoxicity of acetic acid produced in the anaerobic decomposition of straw. J. Appl. Bacteriol. 42:81-87.
208    Lynch, J. M. and S. H. T. Harper. 1974. Formation of ethylene by a soil fungus. J. Gen. Microbiol. 80:187-195.
209    Lynch, J. M. and S. J. Pryn. 1977. Interaction between a soil fungus and barley seed. J. Gen Microbiol. 103:193-196.
210    Marshall, D. R., P. Broue and A. J. Pryor. 1973. Adaptive significance of alcohol dehydrogenase isoenzymes in maize. Nature New Biol. 244:16-18,
211    Marshall, T. and R. J. Millington. 1967. Flooding tolerance of some Western Australian pasture legumes. Aust. J. Exp. Agric. Anim. Husb. 7:367-371.
212    Massey, W., D. W. Rycroft, A. Thorburn and T. S. Threadgold. 1974. An investiga-

tion into poaching of grassland in Lancashire. Tech. Bull. 74/6, Min. Agric. Fish. Food, London.

213  Mees, G. C. and P. E. Weatherley. 1957. The mechanism of water absorption by roots: II. The role of hydrostatic pressure gradients across the cortex. Proc. Royal Soc. B. 147:381-391.

214  Meidner, H. and D. W. Sheriff. 1976. Water and plants. Blackie, Glasgow.

215  Michael, G. and H. Seiler-Kelbitsch. 1972. Cytokinin content and kernel size of barley grain as affected by environment and genetic factors. Crop Sci. 12:162-165.

216  Michniewicz, M. and E. Galoch. 1972. Dynamics of endogenous inhibitor of abscisic acid properties in willow cuttings. Bull. Acad. Pol. Sci. 20:333-337.

217  Miller, D. E. and D. W. Burke. 1974. Influence of soil bulk density and water potential on fusarium root rot of beans. Phytopathology 64:526-529.

218  Minchin, F. R. and R. J. Summerfield. 1976. Symbiotic nitrogen fixation and vegetative growth of cowpea (*Vigna unguiculata* L.) in waterlogged conditions. Plant Soil 45:113-127.

219  Minchin, F. R., R. J. Summerfield, A. R. J. Eaglesham and K. A. Stewart. 1978. Effects of short-term waterlogging on growth and yield of cowpea (*Vigna unguiculata* L.) J. Agric. Sci., Camb. 10:355-366.

220  Mizrahi, Y. and A. E. Richmond. 1972. Abscisic acid in relation to mineral deprivation. Plant Physiol. 50:667-670.

221  Morinaga, T. I. 1926. The favorable effects of reduced oxygen supply upon the germination of certain seeds. Am. J. Bot. 13:159-166

222  Murdock, L. 1974. No-tillage on soil with restricted drainage. p. 16-19. In: R. E. Phillips (ed.) Proc. No-Tillage Res. Conf., Univ. Kentucky, Lexington.

223  Musgrave, A., M. B. Jackson and E. Ling. 1972. Callitriche stem elongation is controlled by ethylene and gibberellin. Nature New Biol. 238:93-96.

224  Nashed, R. B. and R. E. Girton. 1958. Inhibition by ethanol of the growth and respiration of maize roots and coleoptiles. Am. J. Bot. 45:190-193.

224a  Nojima, K. 1963. Irrigation and drainage. p. 399-425. In M. Matsubuyashi, R. Ito, T. Takasi, T. Nomoto and N. Yamada (eds.) Theory and practice of growing rice. Fuji Publishing Co., Ltd., Tokyo.

225  Nosenko, P. P. and I. S. Zonn. 1976. Land drainage in the world. Int. Commission of Irrigation and Drainage. I.C.I.D. Bull. Jan. 1976:65-70.

226  Okajima, H. 1973. The correlation between the activity of rice roots and flooded soils. Tech. Bull. No. 19, p. 28, Food and Fertilizer Technology Center, PO Box 3387, Taipei City, Taiwan.

227  Olomu, M. O. and G. J. Racz. 1974. Effect of soil water and aeration on iron and maganese utilization by flax. Agron. J. 68:523-526.

228  Opik, H. 1973. Effect of anaerobiosis on respiratory rate, cytochrome oxidase activity and mitochondrial structures in coleoptiles of rice (*Oryza sativa* L.) J. Cell Sci. 12:725-739.

229  Osborne, D. J 1968. Hormonal mechanisms relating senescence and abscission. p. 815-840. In: F. Wightman and G. Setterfield (eds.) Biochemistry and physiology of plant growth substances. Runge Press, Ottawa.

230  Outlook on Agriculture. 1975. 8, Special Number, 211-260.

231  Pankhurst, C. E. and J. I. Sprent. 1975. Surface features of soy bean root nodules. Protoplasma 85:85-98.

232  Parsons, L. R. and P. J. Kramer. 1974. Diurnal cycling in root resistance to water movement. Physiol. Plant 30:19-23.

233  Patrick, W. H. Jr. and R. Wyatt. 1964. Soil nitrogen loss as a result of alternate submergence and drying. Soil Sci. Soc. Am. Proc. 28:647-653.

234  Pazout, J. and M. Wurst. 1977. Effect of aeration on ethylene production by soil micro-organisms. Folia Microbiol. 22:458-459.

235  Pereira, J. S. and T. T. Kozlowski. 1977. Variation among woody angiosperms in response to flooding. Physiol. Plant 41:184-192.

236  Philipson, J. J. and M. P. Coutts. 1978. The tolerance of tree roots to waterlogging: III. Oxygen transport in lodge pole pine and sitka spruce roots of primary structure. New Phytol. 80:341-349.

237  Phillips, I. D. J. 1964. Root-shoot hormone relations: II Changes in endogenous auxin concentration produced by flooding of the root system in *Helianthus annuus*. Ann. Bot. 28:37-45.

238  Pidgeon, J. D. and B. D. Soane. 1977. Effects of tillage and direct drilling on soil properties during the growing season in a long-term barley mono-culture system. J. Agric. Sci., Camb. 88(2):431-442.

239   Ponnamperuma, F. N. 1972. The chemistry of submerged soils. Adv. Agron. 24:29-96.
240   Powell, A. A. and S. Matthews. 1978. The damaging effect of water on dry pea embryos during inhibition. J. Exp. Bot. 29:1215-1229.
241   Pratt, R. G. and J. E. Mitchell. 1976. Interrelationships of seedling age, inoculum, soil moisture level, temperature, and host and pathogen genotype in phytophthora root rot of alfalfa. Phytopathology 66:81-85.
242   Raadsma, S. 1974. Current drainage practices in flat areas of humid regions in Europe. p. 115-143. In: J. van Schilfgaarde (ed.) Drainage for agriculture. Am. Soc. Agron., Madison, WI.
243   Radin, J. W. and R. S. Loomis. 1969. Ethylene and carbon dioxide in the growth and development of cultured radish roots. Plant Physiol. 44:1584-1589.
244   Railton, I. D. and D. M. Reid. 1973. Effects of benzyladenine on the growth of waterlogged tomato plants. Planta 111:261-266.
245   Rao, K. P. and D. W. Rains. 1976. Nitrate absorption by barley. Plant Physiol. 57:55-58.
246   Raschke, K. 1975. Simultaneous requirement of carbon dioxide and abscisic acid for stomatal closing in Xanthium strumarium L. Planta 125:243-259.
247   Reeve, M. J. and D. G. M. Hall. 1978. Shrinkage in clayey subsoils of contrasting structure. J. Soil. Sci. 29:315-323.
248   Reeve, R. C. and N. R. Fausey. 1974. Drainage and timeliness of farming operations. p. 55-66. In: J. van Schilfgaarde (ed.) Drainage for agriculture. Am. Soc. Agron., Madison, WI.
249   Regehr, D. L., F. A. Bazzaz and W. R. Boggess. 1975. Photosynthesis, transpiration and leaf conductance of Populus deltoides in relation to flooding and drought. Photosynthetica 9:52-61.
250   Reid, D. M. and A. Crozier. 1971. Effects of waterlogging on the gibberellin content and growth of tomato plants. J. Exp. Bot. 22:39-48.
251   Rhoades, E. D. 1964. Inundation tolerance of grasses in flooded areas. TRANSACTIONS of the ASAE 7(2):164-169.
252   Rhoades, J. D. 1974. Drainage for salinity control p. 433-468. In: J. van Schilfgaarde (ed.) Drainage for agriculture. Am. Soc. Agron., Madison, WI.
253   Ritter, W. F. and C. E. Beer. 1969. Yield reduction by controlled flooding of corn. TRANSACTIONS of the ASAE 12(1):46-47, 50.
254   Robson, J. D. and A. J. Thomasson. 1977. Soil water regimes. Soil Survey Tech. Monogr. 11. Soil Survey of England and Wales, Harpenden, England.
255   Rowe, R. N. and P. B. Catlin. 1971. Differential sensitivity to waterlogging and cyanogenesis by peach, apricot and plum roots. Am. Soc. Hort. Sci. 96:305-308.
256   Russell, E. W. 1973. p. 849. In: Soil conditions and plant growth. Longmans, London.
257   Russell, E. W. 1977. The role of organic matter in soil fertility. Phil. Trans. Royal Soc. Lond. B 281:209-219.
258   Russell, R. S. 1977. p. 298. In: Plant root systems. McGraw-Hill Book Co (U.K.) Ltd., London.
259   Rycroft, D. W. and A. A. Thorburn. 1974. Water stability tests on clay soils in relation to mole draining. Soil Sci. 117:306-310.
260   Ryden, J. C., L. J. Lund, J. Letey and D. D. Focht. 1979. Direct measurement of denitrification loss from soils: II Development and application of field methods. Soil Sci. Soc. Am. J. 43:110-118.
261   Savickaja, N. N. 1959. The effect of abundant soil moisture on barley plants at different periods of their development. Doklady Akad. Nauk SSSR 128:850-852. Seen in Fld. Crop Abst. (1960), 13:19, Abs. 71.
262   Schilfgaarde, J. van. 1974. Drainage for agriculture. Am. Soc. Agron., Madison, WI.
263   Schwab, G. O. 1976. Age effects on surface and sub-surface drain flow. p. 10-13. In: Proc. 3rd Nat. Drainage Symp., Chicago, Dec. 1976. ASAE, St. Joseph, MI 49085.
264   Schwab, G. O., N. R. Fausey and D. W. Michener. 1974. Comparison of drainage methods in a heavy-textured soil. TRANSACTIONS of the ASAE 17(2):424-425, 428.
265   Schwab, G. O., N. R. Fausey and C. R. Weaver. 1975. Tile and surface drainage of clay soils: II.Hydrologic performance with field crops (1962-1972). III Corn, oat and soybean yields (1962-1972). Res. Bull. 1081, Ohio Agric. Res. Devel. Center:37.
266   Segeta, V. 1968. Resistance of winter wheats and ryes to direct effects of flooding. Ved. Pr. Ustr. vyzk. Ust. rostl. Vyroby Praze-Ruzyni 13:229-237. Seen in Field Crop Abst. (1970) 23:293, Abs. No. 1966.
267   Selman, I. W. and S. Sandanam. 1972. Growth responses of tomato plants in non-

aerated water culture to foliar sprays of gibberellic acid and benzyladenine. Am. J. Bot. 36:837-848.

268 Shalhevet, J., H. Enoch and S. Dasberg. 1969. Response of sugar beet to soil drainage and aeration. Israel J. Agric. Res. 19:161-170.

269 Shalhevet, J. and P. J. Zwerman. 1958. Nitrogen response of corn under variable conditions of drainage—a preliminary greenhouse study. Soil Sci. 85:255-260.

270 Shalhevet, J. and P. J. Zwerman. 1962. Nitrogen response of corn under variable conditions of drainage—a lysimeter study. Soil Sci. 93:172-182.

271 Shazkin, F. D. 1960. The effect of surplus moisture on plants at different development periods. Fiziol. Rast. 7:269-275.

272 Shazkin, F. D. and Yu N. Federova. 1961. The effect of surplus soil moisture and nitrogen on some of the physiological processes and the yield of barley in connection with its development stages. Dokl. Akad. Nauk. 139:151-153.

273 Sheard, R. W. 1978. Tile drainage benefits corn production. Information for industry personnel. Ontario Ministry of Agric. and Food, Univ. of Guelph.

274 Sheard, R. W. and R. J. Leyshon. 1976. Short term flooding of soil: Its effect on the composition of gas and water phases of soil and on phosphorous uptake of corn. Can. J. Soil Sci. 56:9-20.

275 Sheldrake, A. R. 1973. The production of hormones in higher plants. Biol. Rev. 48:509-559.

276 Sherwin, T. and E. W. Simon. 1969. The appearance of lactic acid in Phaseolus seeds germinating under wet conditions. J. Exp. Bot. 20:776-785.

276a Shiroshita, T., K. Ishii, J. Kaneko and S. Kitajima. 1962. Studies in the high productivity of paddy rice by increasing the manurial effects. J. Cent. Agric. Expt. Sta., 1:47-108.

277 Short, K. C. and J. G. Torrey. 1972. Cytokinins in seedling roots of pea. Plant Physiol. 49:155-160.

278 Sieben, W. H. 1964. The effect of drainage conditions on nitrogen supply and yield (In Dutch). Landb. Tijdschr. 76:784-802.

279 Sij, J. W. and C. A. C. Swanson. 1973. Effect of petiole anoxia on phloem transport in squash. Plant Physiol. 51:368-371.

280 Simon, E. W. 1974. Phospholipids and plant membrane permeability. New Phytol. 73:377-420.

281 Smith, A. M., P. J. Milham and W. L. Morrison. 1978. p. 329-336. In: M. W. Loutit and J. A. R. Miles (eds.) Microbial Ecology, Springer, Berlin.

282 Smith, K. A. and R. J. Dowdell. 1974. Field studies of the soil atmosphere: I Relationship between ethylene, oxygen, soil moisture content and temperature. J. Soil Sci. 25:217-230.

283 Smith, K. A. and M. B. Jackson. 1974. Ethylene, waterlogging and plant growth. Agric. Res. Council Letcombe Lab. Ann. Rep. 1973. p. 60-75.

284 Smith, K. A. and S. W. F. Restall. 1971. The occurence of ethylene in anaerobic soil. J. Soil Sci. 22:430-443.

285 Smith, K. A. and P. D. Robertson. 1971. Effect of ethylene on the root extension of cereals. Nature 234:148-149.

286 Smith, L. P. 1972. The effect of weather, drainage efficiency and duration of spring cultivations on barley yields in England. Outl. Agric. 7:79-83.

287 Smith, W. 1976. Aspects of drainage related to cultivation and harvesting of sugar cane. Proc. 3rd Nat. Drainage Conf., ASAE, St. Joseph, MI 49085. p. 139-143.

288 Soane, B. D. 1975. Studies on some soil physical properties in relation to cultivations and traffic. In: Tech. Bull. 29:160-182. Min. Agric. Fish. Food, H.M.S.O., London.

289 Sotnkov, V. P. 1970. Soil and climatic conditions of Soviet agriculture. p. 54-64. In: Agriculture of the Soviet Union. M.I.R. Publ., Moscow.

290 Spoor, G. and R. J. Godwin. 1978. An experimental investigation into the deep loosening of soil by rigid tines. J. Agric. Engng. Res. 23:243-258.

291 Spent, J. I. and A. Gallacher. 1976. Anaerobiosis in soybean root nodules under water stress. Soil Biol. Biochem. 8:317-320.

292 Staden, J. van and N. A. C. Brown. 1973. The effect of oxygen on engogenous cytokinin levels and germination of Leucadendron daphnoides seed. Physiol. Plant 29:108-111.

293 Stanley, C. D., T. C. Kaspar and H. M. Taylor. 1980. Soybean top and root response to temporary water-tables imposed at different stages of growth. Agron. J. 72:341-346.

294 Stefanovskii, I. A. 1963. Resistance of spring wheat to excess moisture at various stages of growth. Agrobiologiya 5:778-779.

295  Stefanovskii. I. A. 1964. Effect of mineral fertilizers on yield of spring wheat under waterlogged conditions. Agrokhimiya, Moscow 9:89-92.

296  Stefanovskii, I. A. 1965. Resistance of grain crops to waterlogging. Vest. Sel-kholz., Nauki, Moscow 3:56-60.

297  Steinhardt, R. and B. D. Trafford. 1974. Some effects of sub-surface drainage and ploughing on the structure and compactability of a clay soil. J. Soil Sci. 25:138-152.

298  Stemmet, M. C., J. A. de Bruyn and P. B. Zeeman. 1962. The uptake of carbon dioxide by plant roots. Plant Soil 17:357-364.

299  Steward, F. C., W. E. Berry and T. C. Broyer. 1936. The absorption and accumulation of solutes by living cells: VIII. The effect of oxygen upon respiration and salt accumulation. Ann. Bot. 50:345-366.

300  Stolzy, L. H., G. A. Zentmyer, L. J. Kholz and C. K. Labanauskas. 1967. Oxygen diffusion, water and *Phytophthora cinnamomi* in root decay and nutrition of avocados. Proc. Am. Soc. Hort. Sci. 90:67-76.

301  Suh, H. S. 1973. Studies on the wet-injury resistance of wheat and barley varieties: II Relation between wet-injury resistance and root growth in wheat and barley. Korean J. Breeding (Yik Jong Hak Joi Ji) 5:91-97.

302  Swain, R. W. 1975. Subsoiling. In: Soil physical conditions and crop production. Tech. Bull. 29, p. 189-204. Min. Agric. Fish Food, H.M.S.O., London.

303  Swartz, G. L. 1966. Flood tolerance of winter crops in southern Queensland. Queensland J. Agric. An. Sci. 23:271-277.

304  Talbot, B. and H. E. Street. 1968. Studies of the growth in culture of excised wheat roots. V. Influence of carbon dioxide on growth and branching. Physiol. Plant. 21:800-805.

305  Taylor, D. L. 1942. Influence of oxygen tension on respiration, fermentation and growth in wheat and rice. Am. J. Bot. 29:721-738.

306  Taylor, H. M. 1979. Postponement of severe water stress in soybean. World Soybean Res. Conf. II, Raleigh, NC. p. 161-178.

307  Thomasson, A. J., D. Mackney, B. D. Trafford and R. A. Walpole. 1975. Soils and field drainage. Soil Survey Tech. Monog. 7:80, Rothamstad,Harpenden, England.

308  Tishkov, E. N. and L. I. Federova. 1964. The resistance of winter rye cultivars to flooding, snow mould and *Fusarium* rot of seed. Trudy Mosk. Otd. vses. Inst. Rasteniev. I:164-174, Seen in Field Crop Abst. 19:263, Abs. 2046.

309  Tokimasa, F. 1952. Studies in the harm of the excessive moisture in the soil to the growth of barley and wheat. Proc. Crop Sci. Soc. Japan 20:266-267.

310  Trafford, B. D. 1970. Field drainage. J. Royal Agric. Soc. 131:129-152.

311  Trafford, B. D. 1975. Improving the design of practical field drainage. Soil Sci. 119:334-338.

312  Trafford, B. D. and J. M. Oliphant. 1977. The effect of different drainage systems on soil conditions and crop yield on a heavy clay soil. Expl. Husb. 32:75-85.

313  Trafford, B. D. and D. W. Rycroft. 1973. Observations on the soil-water regimes in a drained clay soil. J. Soil Sci. 24:380-391.

314  Turkova, N. S. 1944. Growth reactions in plants under excessive watering. Compt. Rend. (Doklady) Academy Sci., U.R.S.S. 42:87-90.

315  Udovenko, G. V. and M. A. Lavrova. 1970. Resistance of winter wheat cultivars to early-spring flooding in relation to their physiological characters. Trudy prikl. Bot. Genet. Selek. 43:67-78. (Seen in Fld. Crop Abst. (1971) 24:415, Abst. No. 3055).

316  Udovenko, G. V. and M. V. Tracher. 1970. Resistance of winter rye cultivars to early-spring flooding in relation to their physiological characters. Trudy prikl. Bot. Genet. Selek. 43:79-87, Seen in Fld. Crop Abst. 1971, 24:427, Abs. No. 3144.

317  United States Department of Agriculture. 1975. Soil taxonomy, a basic system of soil classification for making and interpreting soil surveys. Soil Survey Staff, Soil Conservation Service. Agriculture Handbook No. 486, U.S. Government Printing Office, Washington, DC.

318  Vamŏs, R. and E. Kŏves. 1972. Role of light in the prevention of the poisoning action of hydrogen sulphide in the rice plant. J. Appl. Ecol. 9:519-525.

319  Van Doren, D. M., G. B. Triplett and J. E. Henry. 1976. Influence of long term tillage, crop rotation and soil type combinations on corn yield. Soil Sci. Soc. Am. J. 40:100-105.

320  Vine, H., H. A. Thompson and F. Hardy. 1942. Studies on aeration of cacao soils in Trinidad: II. Soil-air composition in certain cacao soil-types in Trinidad. Trinidad Trop. Agric. 19:215-223.

321  Vorobeikov, G. A. 1975. Effect of chlorocholine chloride (C.C.C.) on resistance of wheat to excessive and inadequate water supply during the critical period. Fiziol. Rast. 23:573-578.

322   Wada, S. 1961. IAA-oxide inhibitors contained in rice coleoptiles. Sci. Rep. Tohoku Univ. Ser, IV (Biol). 27:237-249.

323   Wang, T. S. C., and T. T. Chuang. 1967. Soil alcohols, their dynamics and their effect upon plant growth. Soil Sci. 104:40-45.

324   Wang, T. S. C., S-Y. Cheng and H. Tung. 1967. Dynamics of soil organic acids. Soil Sci. 104:138-144.

325   Waters, P. 1977. The effect of drainage on soil temperature at Drayton EHF. Tech. Bull. 77/6 Fld. Drainage Exptl. Unit, Ministry of Agric. Fish. and Food, London p. 16.

326   Watson, E. R., P. Lappins and R. J. W. Barrow. 1976. Effect of waterlogging on the growth, grain and straw yield of wheat, barley and oats. Aus. J. Exp. Agric. An. Husb. 16:114-122.

327   Webster, R. and P. H. T. Becket. 1972. Suctions to which soils in south central England drain. J. Agric. Sci. Camb. 78:379-387.

328   Went, F. W. 1938. Transplantation experiments with peas. Am. J. Bot. 25:44-55.

329   Went, F. W. 1943. Effect of the root system on tomato stem growth. Plant Physiol. 18:51-65.

330   Wien, C., R. Lal and E. L. Pulver. 1979. Effects of transient flooding on growth and yield of some tropical crops. p. 235-245. In: R. Lal and D. J. Greenland (Eds.) Soil physical conditions and crop production in the tropics. J. Wiley and Sons, Chichester.

331   Wiersum, L. K. 1957. The relationship of the size and structural rigidity of pores to their penetration by roots. Plant Soil 9:75-85.

332   Wignarajah, K., H. Greenway and C. D. John. 1976. Effect of waterlogging on growth and activity of alcohol dehydrogenase in barley and rice. New Phytol. 77:585-592.

333   Williamson, R. E. and G. J. Kriz. 1970. Response of agricultural crops to flooding, depth of water table and soil gaseous composition. TRANSACTIONS of the ASAE 13(2):216-220.

334   Williamson, R. E. and W. E. Splinter. 1968. Effect of gaseous composition of root environment upon root development and growth of Nicotiana tabacum L. Agron. J. 60:365-368.

335   Williamson, R. E. and W. E. Splinter. 1969. Effects of light intensity, temperature and root gaseous environment on growth of Nicotiana tabacum L. Agron. J. 61:285-289.

336   Wind, G. P. and J. Buitendijk. 1979. Simulation over 35 years of the moisture content of a top soil with an electronic analog for three drain depths and three drain spacings. Proc. Int. Drainage Workshop, Int. Inst. Land Reclamation and Improvement, Wageningen, 214-219.

337   Woodford, E. K. and F. G. Gregory. 1948. Preliminary results obtained with an apparatus for the study of salt uptake and root respiration of whole plants. Ann. Bot. 12:335-370.

338   Woudt, B. D. vant' and R. M. Hagan. 1957. Crop responses at excessively high soil moisture levels. p. 514-611. In: J. N. Luthin (ed.) Drainage of agricultural lands. Am. Soc. Agron. Madison, WI.

339   Yamada, N. 1954. Auxin relationships of the rice coleoptile. Plant Physiol. 29:92-96.

340   Yamasaki, S. and T. Saeki. 1976. Oxygen supply from shoots to roots relative to the total oxygen consumption in rice shoots. Bot. Mag. Tokyo 89:309-319.

341   Yeomans, J. C. and E. G. Beauchamp. 1978. Limited inhibition of nitrous oxide reduction in soil in the presence of acetylene. Soil Biol. Biochem 10:517-519.

342   Yocum, C. S. and D. P. Hackett. 1957. Participation of cytochromes in the respiration of the aroid spadix. Plant Physiol. 32:186-191.

343   Yoshida, R. and T. Oritani. 1974. Studies on nitrogen metabolism in crop plants: XIII. Effects of nitrogen top-dressing on cytokinin content in the root exudate of rice plants. Proc. Crop Sci. Soc. Japan 43:47-51.

344   Yoshida, S. 1981. Fundamentals of rice crop science. International Rice Research Institute, Los Banos, Philippines. (in press).

345   Yoshinari, T. and R. Knowles. 1976. Acetylene inhibition of nitrous oxide reduction by denitrifying bacteria. Biochem. Biophys. Res. Comm. 69:705-710.

346   Young, B. R., F. J. Newhook and R. N. Allen. 1977. Ethanol in the rhizosphere of seedlings of Lupinus angustifolius L. New Zealand J. Bot. 15:189-191.

347   Yu, P. T., L. H. Stolzy and J. Letey. 1969. Survival of plants under prolonged flooded conditions. Agron. J. 61:844-847.

348   Zimmerman, P. W. 1930. Oxygen requirements for root growth of cuttings in water. Am. J. Bot. 17:842-861.

# chapter 6

ALLEVIATING PATHOGEN STRESS

**6**

**6**

# ALLEVIATING PATHOGEN STRESS

by    Stuart D. Lyda, Professor, Department of Plant
      Sciences, Texas A&M University

## 6.1 INTRODUCTION

Soil is one of the most complex environments that agricultural scientists encounter. It is teeming with a myriad of organisms living in a state of dynamic equilibrium, very dependent upon environmental and edaphic conditions. The chemical and physical properties of soil influence the qualitative and quantitative composition of its biomass.

Fluctuations in biomass occur with cropping sequences, with variations in temperature, moisture, fertility, and even with the growth or reproductive stages of plants supported by this seemingly inert material. Fortunately, most organisms improve the soil as a medium for plant growth.

The soil-borne organisms that concern the plant pathologist include actinomycetes, bacteria, fungi and nematodes. There are several soil-borne virus pathogens, often vectored by another type of organism, i.e. fungi or nematodes, but they will not be discussed in this chapter.

Each group of pathogens will be considered along with methods of managing plant diseases. Modification of the root zone by chemical, physical or biological means to alleviate stress on the plant by diseases of parasitic or non-parasitic origin will be discussed. A prelude to such a discussion necessitates a general section on concepts of plant disease and their causes as well as defining some basic terminology peculiar to this field.

No attempt will be made to include all of the literature in this subject. There are several good review articles touching on various aspects of root zone manipulations and/or modifications by biological, chemical or physical means. Emphasis has been placed upon research conducted since the International Symposium on Factors Determining the Behavior of Plant Pathogens in Soil, Berkeley, CA, April 7-13, 1963.

## 6.2. DYNAMIC ASPECTS OF THE SOIL

Microorganisms living in the soil utilize many of the same nutrients required by higher plants for growth and reproduction. Some microorganisms utilize metabolites produced by higher plants and other organisms, and many secrete metabolites that influence the ambient environment. Microorganisms that inhabit the same ecological niche, but do not interact with other organisms, are neutral. They have a commensalistic relationship, i.e. they feed at the same table without adversely affecting one another. This is a rare phenomenon in soil, because most microorganisms require similar types of nutrients for sustenance, thus they are competitors for available nutrients. These populations of organisms are influenced by the competitive ability of a particular species as opposed to another. Some organisms are held to certain levels by predacious organisms usually from the soil fauna.

There are four major groups of soil borne microorganisms of interest to plant and soil scientists: (a) bacteria, (b) actinomycetes, (c) fungi, and (d) nematodes. Some salient features of each group are presented below.

### 6.2.1 Bacteria

Bacteria are single-celled organisms in the kingdom Monera. These primitive organisms along with the blue-green algae, are procaryotic. They lack a nuclear membrane, and endoplasmic reticulum, Golgi apparatus and lysomes. They contain both DNA and RNA, ribosomes, proteins and they can generate ATP for the synthesis of organic compounds.

Bacteria, like plant cells, are enclosed within a cell wall. Whereas the cell wall of higher plants is chiefly cellulose, the bacterial cell wall is murein, a polymer of amino acids and amino sugars (Keeton, 1972).

In soil, the bacterial population fluctuates in response to nutritional, environmental and biological factors. A decrease in bacterial numbers is noted with increasing soil depths (Waksman, 1952). They occur in the greatest numbers as compared to other soil microbes, but because of their small size they do not contribute more than about one-third to the biomass. The mass of one bacillus (rod-shaped) bacterium ($0.75 \times 2.5$ $\mu$m) is 1.02 pg.

TABLE 1. THE COMPARATIVE RANGE IN
NUMBERS OF REPRESENTATIVE GROUPS
OF MICROBIAL AND MACROBIAL
ORGANISMS IN SOIL

|  |  | Numbers |
|---|---|---|
| 1. | Flora* | /g |
|  | Actinomycetes | $10^6$-$10^7$ |
|  | Algae | $10^3$-$10^5$ |
|  | Bacterial | $10^6$-$10^8$ |
|  | Fungi | $10^4$-$10^5$ |
| 2. | Fauna | /m$^2$ |
|  | Earthworms | 30-3,000 |
|  | Pot-worms | 200-20,000 |
|  | Slugs and snakes | 100-8,500 |
|  | Millipedes and centipedes | 900-1,700 |
|  | Woodlice | 100-400 |
|  | Spiders | 180-840 |
|  | Beetles and larvae | 500-1,000 |
|  | Fly maggots | $\cong$ 1,000 |
|  | Ants | 200-500 |
|  | Springtails | 10,000-40,000 |
|  | Mites | 20,000-120,000 |
|  | Nematodes | 1.8-120 $\times$ $10^6$ |

*Waksman, S. A. 1927. Principles of soil micro-
biology. 897 p. The Williams and Wilkins Co.,
Baltimore.
†McEkevan, D. K. 1965. The soil fauna—Its
nature and biology, p. 38
In: K. F. Baker and W. C. Snyder (eds.) Ecology
of Soil-Borne Plant Pathogens, Univ. of Calif.
Press, 571 p.

The minuteness of bacteria is evident when it takes one trillion bacteria to weigh one gram (Pelczar et al., 1977). Clark used this value to calculate the contribution bacteria makes to the soil biomass. Two billion cells/g of soil add approximately 4500 kg in the upper 15 cm of a hectare of land (Clark, 1967). One gram of fertile soil may contain from $2 \times 10^6$ to $2 \times 10^8$ bacteria (Waksman, 1927). Their numbers are influenced by soil properties, reaction, moisture, temperature and nutrition.

## 6.2.2 Fungi

Fungi are among the largest of the categories of microorganisms. There are more than 4,000 genera and 100,000 species of fungi (Benjamin et al., 1964). Fortunately, only about 200 genera of fungi are pathogenic to plants.

Fungi are placed in a separate kingdom in classification of the animate world. They are placed in the kingdom Fungi, because of their absorptive mode of nutrition as opposed to a photosynthetic nutrition for green plants and an ingestive mode of nutrition by animals (Whittaker, 1969). Fungi possess a true nucleus, mitochondria, ribosomes and Golgi bodies characteristic of eukaryotic organisms. Most plant pathogenic fungi have a filamentous body (thallus). Collectively, the thread-like filaments are called mycelium and individually the threads are called hyphae. The diameter of the mycelium varies from $< 2$ $\mu$m to $> 20$ $\mu$m.

Many fungi produce vegetative and reproductive stages. Reproduction may occur through asexual or sexual processes. The diverse spore size, shape and manner of germination are criteria used to classify fungi into taxanomic groups. There are four classes: (a) phycomycetes, (b) ascomycetes, (c) basidiomycetes and (d) deuteromycetes (fungi imperfecti). The phycomycetes possess a mycelium, without crosswalls, so the thallus is one long tube with many nuclei (coenocytic). The sexual stage in the phycomycetes results from the union of an antheridium and oogonium which gives rise to an oospore. The oospore is a thick-walled, diploid spore. Meiosis occurs upon germination restoring the haploid nuclear condition. Asexual spores (sporangia or conidia) are borne on fruiting branches called conidiophores or sporangiophores. The spores germinate by releasing motile zoospores into the soil solution or some may germinate directly by forming a germ tube. The class phycomycetes contains genera of some omnipresent and very serious soil-borne pathogens (*Aphanomyces, Phytophthora,* and *Pythium*). These pathogens are favored by wet, poorly drained soils.

The ascomycetes are sometimes called sack fungi. Their sexual spores are produced in sacks called asci. Usually, there are eight spores per sack and on release they serve to initiate new infections. The sacks may be enclosed in a protective structure (apothecium, cleistothecium, perithecium), produced in stromatic host tissue, or they may be naked. Asexual spores are produced in large numbers on conidiophores and serve to give secondary spread to the pathogen from the point of initial infection. Examples of soil-borne pathogens in this class include *Ceratocystis, Hypomyces, Hymenula* and *Gaeumannomyces (Ophiobolus), Whetzelinia (Sclerotinia).*

Basidiomycetes produce sexual spores on club-shaped fruiting structures called basidia. Basidia may occur separate or they may be produced in protective structures termed basidiocarps. Fungi in this class are among the highest evolved. They include the very host-specific rusts and smuts;

however, they also include some very serious soil-borne pathogens, *Armillaria mellea, Clitocybe tabescens, Corticium solani, (Sclerotium rolfsii), Thanatephorus cucumeris (Rhizoctonia solani)*.

Fungi in the class deuteromycetes lack a sexual stage. They produce asexual spores or they are composed of sterile mycelium. Many soil-borne pathogens are members of this class: *Verticillium dahliae, Phymatotrichum omnivorum, Fusarium oxysporum*.

### 6.2.3 Actinomycetes

These microscopic organisms possess morphological traits of bacteria and fungi. The name actinomycete means "ray fungi". They resemble bacteria in that they are unicellular, procaryotic organisms. Their cell wall does not contain cellulose or chitin and they are sensitive to chemicals and phage active upon bacteria. They resemble fungi in their filamentous nature, except the mycelium is very narrow (1 $\mu$m) (Waksman, 1967).

Actinomycetes occur in most soils. They are readily isolated from neutral to alkaline soils, with the optimum pH between 7.0 to 8.0 (Waksman and Curtis, 1916).

In total numbers the actinomycetes are second to bacteria, however in biomass they are low because of their slow growth rate and extremely small mycelium. The size of the population ranges from $10^5$ to $10^8$/g of soil in temperate soils. The numbers are reduced in acidic soil, water-logged soils, peat and tundra. They thrive in dry alkaline soils (Alexander, 1961).

Actinomycetes are probably best known for their ability to synthesize antibiotics. The well-known antibiotics streptomycin, aureomycin and terramycin are metabolic by-products of these organisms (Miller, 1961).

### 6.2.4 Microfauna

The living soil contains numerous organisms more closely related to animals than plants. They lack a cell wall, they are motile and they feed by ingesting food. The phylum Protozoa is estimated to contain nearly 30,000 species (Garrett, 1963). Free-living members of the phylum are classified into three groups: Sarcodina, Mastigophora, and Ciliophora.

*Amoeba* spp-Creeping represent forms of Protozoa in the group Sarcodina. They move over surfaces by extending finger-like projections of protoplasm (pseudopodia). Bacteria and other food particles are engulfed by the pseudopodia and digested in a cytoplasmic vacuole. *Paramecium* spp. have hair-like appendages on the cell surface known as cilia. The cilia propel the cell through the soil moisture, and cilia lining the mouth serve to move microbial cells into the cell. Since these types of organisms ingest their food and destroy the prey, they are called predators.

The phylum Nematoda comprises about 10,000 species of microscopic nematodes. Along with the phylum Protozoa, they make up the bulk of the soil microfauna (Garrett, 1963).

These invertebrate animals occur in soil as free-living (saprophagous) or plant parasites. They are small, yet visible to the naked eye when suspended in water and properly lighted. Nematodes inhabiting the soil may be classified in three groups: (a) the saprophagous species that derive their sustenance from decaying organic matter, (b) the predacious species that feed on small animals and (c) the parasitic species that feed on algae, fungi, and higher plants (Christie, 1959). In addition to feeding damage, they may also create avenues for entry of other soil-borne pathogens. Some nematodes are vectors of soil-borne viruses.

Nematodes in general constitute a small portion of the soil biomass yet the damage they inflict on cultivated crops, in some cases, may be devastating.

## 6.3 CONCEPTS AND TERMINOLOGY USED IN THE STUDY OF PLANT DISEASES

Plant disease results from a continuous irritation in the metabolism of the plant (Horsfall and Dimond, 1959; Wood et al., 1940). This is contrasted to injury, which is a transient condition such as might be inflicted by a cultivator or by chewing insects. Diseases may be separated into two broad classes based upon their cause: (a) parasitic and (b) nonparasitic or physiogenic. Nonparasitic diseases are caused by nutritional imbalances (deficiencies, excesses), environmental pollutants, unfavorable light quantity or quality, unfavorable temperature, moisture irregularities and improper aeration.

Parasitic diseases are caused by microorganisms, viruses and phanerogamic parasites. The organism or agent causing disease is a pathogen. It may be specific for a certain cultivar within a species or it may be nonspecific and parasitize many different species or genera of plants, i.e., it has a broad host range.

Manifestation of disease is variable. Early stages of disease may produce physiological changes in the host that are invisible to man, and as these changes continue the plant responds by producing morphological changes that are visible. The physiological and morphological changes differ with each type of disease. Most plant diseases can be categorized according to symptom expression: (a) abnormal coloration, (b) wilting, (c) necroses (dead tissue), (d) abnormal growth increase, (e) stunting.

In the first category (abnormal coloration), plants may appear light green to yellow in color, which is termed chlorosis. There are numerous environmental and parasitic organisms that will produce chlorosis in plants. A nitrogen or sulfur deficiency may cause a general chlorosis, and some vascular and root parasites will inflict this condition. Not all chlorotic symptoms are visible over the entire plant. Some pathogens produce yellowing between the leaf veins and this is called interveinal chlorosis. In some diseases only the veins become chlorotic (veinclearing). In other diseases the tissue has irregular light and dark green areas spread over the leaf surface. This is termed mosaic, and it is especially prevalent with diseases caused by some of the viral pathogens.

Plants with wilt-type diseases (Fusarium wilt, Verticillium wilt, Cephalosporium wilt, etc.) possesses a vascular system inhabited by the pathogen. There are several theories of the mechanisms involved in wilting (Vascular occlusion, toxin secretion, combination of the two), but the results are the same. The leaves do not secure adequate water during periods of rapid transpiration, thus they become flaccid and may die if the wilting period is prolonged. The temperature of plants with vascular parasites is usually 2 to 3 deg above that of healthy plants. Plants so affected often wilt during the heat of day and regain their turgidity as the transpiration rate subsides.

Necrosis or death of host tissue is the symptom manifest by plants parasitized by root rotting organisms, or it may result from toxic chemicals elaborated by decomposing organic matter, or metabolic by-products of soil organisms. Pre-emergence and post-emergence damping-off of seedlings typifies a necrotic type disease, however, in no way is this type of disease restricted to seedlings. Mature plants have root rots, cankers, mummified fruit and necrotic, localized foliar lesions, all of which are necrotic responses.

Plants showing abnormal growth are diseased. Some pathogens stimulate excessive cell division (hyperplasia) and some induce excessive cell size (hypertrophy). Crown gall (*Agrobacterium tumefaciens*), club root of crucifers (*Plasmodiophora brassicae*), and host response to root knot nematodes (*Meloidogyne* spp.) are examples of diseases in this category.

Stunting of plants is prevalent when cells fail to divide normally (hypoplasia), and when they are smaller than those of healthy cells (hypotrophy). Diseases of this type may be physiogenic or parasitic. Alfalfa dwarf, a disease caused by a gram-negative bacterium transmitted by leafhoppers falls in this category. A very common physiogenic disease in some areas of the country is zinc deficiency, which causes a stunting effect in some plants. Zinc is required for the synthesis of indole acetic acid and in the absence of this growth regulating compound, plant cells do not elongate normally.

Morphological changes in the host as a response to diseases are not transmitted, the pathogen is transmitted. Disease is a condition of the plant. Diseases caused by nutritional imbalances, etc. are non-parasitic so they are non-transmissible and noninfectious. The pathogen is transmissible and it is infectious. Visible manifestations of the pathogen upon or in the host are called signs.

## 6.4 PRINCIPLES OF PLANT DISEASE CONTROL

In seeking a method for controlling plant diseases several principles may apply: (a) exclusion, (b) eradication, (c) host resistance, and (d) protection. Obviously, if a pathogen is not present in a given area, quarantines or embargoes might be an effective means of disease control by exclusion. This is applicable to only certain types of pathogens, since quarantines would be ineffective in restricting pathogens disseminated by air currents. Exclusion would be suitable for pathogens transmitted by seed, soil and propagating material. In these examples, man is usually the vector of the pathogen.

Once a pathogen is introduced into a new sphere, eradication might be an effective control measure. This is usually a drastic and expensive approach, but satisfactory results have been achieved with certain types of pathogens, especially viruses. Some bacterial diseases of apple and citrus also have been controlled by eradication.

The two principles most applied to agronomic crops are host resistance and protection. Host resistance is a desirable method of control since once a resistant variety is developed the remaining costs for disease control are minimal. This sounds good in principle, but there are not many examples of plant diseases that have been completely controlled by the use of resistant varieties. Microorganisms reproduce rapidly and therefore new genotypes evolve that are capable of parasitizing the newly developed, resistant variety. Breeding for resistance is a perpetual task, as neither the host nor the pathogen is static. Walker (1959) stated, "By and large the development of resistant varieties must be looked upon as a continuing program. The potential variability of most pathogens (including viruses) will not permit any currently successful variety to remain so for an indefinite period."

Protection is a very broad principle that includes cultural practices, environmental manipulation, and direct prophylactic measures through the use of chemicals. Some authors include biologocal control as a separate category for plant disease control (Agrios, 1978; Baker and Cook, 1974).

Since this chapter is devoted to root zone modification to alleviate pathogen affects, the remaining terminology will be slanted towards soilborne pathogens and diseases they cause.

As mentioned earlier, there are many kinds of organisms in the soil, but not all of them have the same survival potential. Organisms that persist in the soil indefinitely are termed soil inhabitants, whereas those that are introduced are known as soil invaders. Some microorganisms are confined to root tissue and are called root invaders. Generally the soil invaders and root invaders do not persist for long periods.

Most of the soil inhabitants produce specialized survival propagules that permit them to pass through adverse conditions such as an inadequate nutrient supply, insufficent water, unfavorable temperatures and flushes of antagonistic organisms. Numerous fungi produce single-celled, double-walled spores called chlamydospores for this purpose. The fusaria may produce chlamydospores from mycelium or they may form from macroconidia. Other fungi produce compact masses of mycelium, usually surrounded by a melanized rind, that are called sclerotia. They may be regular or irregular in shape and size depending upon the organism forming them. If they are very small, the sclerotia are called microsclerotia, eg. propagules produced by *Verticillium dahliae*. Chlamydospores and sclerotia are produced asexually. Some of the lower fungi (phycomycetes) produce survival spores called oospores. These are produced as a result of sexual reproduction, but unlike the asexual spores they serve to perpetuate the species through adverse conditions.

Survival propagules reside in the soil for indefinite periods. When conditions are favorable, they germinate and infect the host. These favorable conditions might be induced by substances diffusing from roots of host plants, or they might just be environmental. Sclerotia and chlamydospores germinate to form mycelium (the thread-like body of the fungus) which grows through the soil. If a susceptible root is contacted, the organism establishes a parasitic relationship and gains sustenance for additional growth. If nutrients are not secured before the energy reserve of the spore is utilized, the fungus responds by forming a smaller survival propagule or it dies. In the phycomycetous fungi, osspores may germinate directly in some species, or they may germinate and release motile spores called zoospores. These single-celled spores are propelled through the soil solution by flagella, appendages attached to the spore. Motile inoculum is also produced by some nematodes. Cysts or eggs germinate and release larvae into the soil solution, which migrate to the root surface of susceptible hosts.

## 6.5 METHODS OF MODIFYING ROOT ZONE
## TO ALLEVIATE PATHOGEN EFFECTS

Information on methods used to modify the root zone to alleviate pathogen effects falls in several broad categories: (a) biological, (b) chemical, (c) cultural, (d) physical, and (e) a combination of all. The paucity of absolute data can be understood when one realizes the root inhabits an invisible, three dimensional environment. To extricate the interacting variables from this complex system and study them individually is perhaps a waste of time. Models are very difficult to conceive, and at best they would probably only be suited for one situation.

### 6.5.1 Biological Methods of Modifying Root Zone to Alleviate Pathogen Effects

In biological methods to alleviate pathogen effects in the root zone, scientists are breeding plants that harbor particular microorganisms in or on the root. These organisms appear to be under genetic control of the host plant. Selection of proper crops in the rotation sequence may provide a form of biological control. Certain crop residues are colonized by microorganisms that produced substances that are antagonistic to root rotting fungi, or they may induce spore germination and lysis. Any method that reduces the activity or survival of the pathogen through the action of other living organisms is biological control.

Chemical methods of modifying the root zone have probably been more widely used than any other method. They can be applied to seed to protect the soil zone around the emerging primary root, and some seed protectant chemicals move into the seedling and provide protection of the hypocotyl. Others are absorbed and give a chemotherapeutic action within the seed against internal infection. Chemicals may be mixed into soil that covers the seed and provide a region of the soil protected from organisms that feed upon plant tissue. Some chemicals are injected into the soil and provide a fumigant action. These types of chemicals usually diffuse throughout the treated area and kill many types of soil organisms, beneficial as well as harmful.

Physical methods of root zone modification to alleviate pathogen effects are widely used. The value of improved soil aeration and reduced soil strength is well documented. With improved soil aeration, beneficial microorganisms flourish and some of the root rotting pathogens lose their competitive, saprophytic ability. Deep plowing serves to break up the plow sole and it enables roots to penetrate deeper into the profile. A better developed root system gives the plant an advantage when growing in infested soils. Deep plowing incorporates spores that have fallen on the soil surface into a region where they are less efficient in causing disease. Deep plowing removes surface plant debris which is often colonized and used as an energy source for root pathogens that infect plants near the soil surface. Often these organisms produce toxic substances that kill root cells in advance of their mycelium.

These types of root zones modifications do not provide absolute values to be used in developing computer models. The voluminous literature on total numbers of actinomycetes, bacteria, fungi and nematodes is rather meaningless. The method used in collecting the data is probably the most serious objection, as the media used may select for certain groups and exclude others. High sporulating organisms may appear to predominate, but in actuality they may be quiescent in the soil. The data vary considerably from one soil type to the next, from one crop to the next, and even with respect to distance from the root. Relative values are often used to ascertain responses to a particular chemical, biological or physical treatment, i.e., do certain crops enhance fluctuations of the microbial populations?, do chemicals reduce populations?

Quantitative values of individual genera or species may be important. Knowledge of the numbers of infective units per volume of soil can be used to predict disease severity, if other variables are known.

## 6.5.2 Chemical Methods of Modifying
## the Root Zone to Alleviate Pathogen Effects
### 6.5.2.1 Fungicides

Fungicides play an essential role in modifying the root zone to alleviate pathogen effects. Many soil-borne organisms, causing pre-emergence, post-emergence damping-off, root rots and root diseases caused by nematodes have a broad host range. It is necessary to protect from parasites the young seedlings until they become established. A week or two of protection can make the difference between a healthy productive plant, a debilitated plant, or a dead plant.

Adding chemicals to the soil for pest control of cultivated crops has been practiced for over 100 years. Attempts to eliminate the root aphid, *Phylloxera vastatrix*, which was damaging the vineyards in Europe, prompted the injection of carbon disulfide (Tietz, 1970). Inorganic compounds, such as sodium carbonate, sodium chloride and copper sulfate were used as seed protectants prior to the use of soil fumigants. Copper and sulfur fungicides were used in the latter part of the 19th century and continue to be used today in large quantities to control foliar pathogens.

Organic fungicides have (nearly) replaced the inorganic fungicides to control soil-borne pathogens. The modern era of organic fungicides is less than 50 years, yet some of the most exciting fungicides, the systemics, have been released in the last decade. In 1934, Tisdale and Williams patented dithiocarbamic acid to control and prevent growth of fungi and microbes. The chemical was not used for this purpose until 1941-43, when the copper supply became scarce due to the demands of World War II (McCallan, 1967).

Thiram was the first derivative of dithiocarbamic acid to be widely used and even today it is one of the most widely used seed protectants. The chemical was not sold as a commercial fungicide until 1951, some 17 years after its discovery (Sharvelle, 1969).

The dicarboximides, another class of organic fungicides, evolved about the same period as the dithiocarbamates. Captan, trichloromethyl-thiodicarboximide, was introduced as a broad spectrum fungicide in 1951 (Sharvelle, 1969). It is widely used today as a seed protectant and soil fungicide. Both classes of fungicides have persisted for many years because of their nonspecific mode of action. They react with functional groups on organic molecules, thus they are capable of affecting microbial metabolism in many places.

Knowledge of the physiology of plant pathogens has aided in disease control. Combinations of fungicides are used today to control a complex array of organisms that feed on developing roots. Fungicides are formulated to meet this purpose.

Several new fungicides have been marketed in the last decade that become systemic in the plant. This has been especially valuable for treating seed and killing internal pathogens. The oxathiins and benzimidazoles represent two classes of systemic fungicides. Carboxin, dihydrocarboxanilidomethyloxathiin, is used as a seed treatment of grains to control loose smut (Von Schmeling and Kulka, 1966). In these types of diseases, the pathogen

### TABLE 2. COMMON CHEMICALS USED TO ALLEVIATE SOIL-BORNE DISEASE CAUSED BY FUNGI AND NEMATODES
(Modified from list of Chemical and Common Names Suggested for Use by AR Scientists Reporting Research Results on Pesticides, Second Edition, 1978)

| Trade name | Chemical name | Class | Use |
|---|---|---|---|
| Aldicarb | 2-methyl-2(methylthio)propionaldehyde-0-(methylcarbamoyl)oxime | I, N* | SN† |
| Benomyl | methyl 1-(butylcarbamoyl)-2-benzimidazole carbamate | F, PGR, N | SF |
| Captan | N-(trichloromethyl)thio-4-cyclohexene-1,2-dicarboximide | F, PGR | SP |
| Carbofuran | 2,3-dihydro-2,2-dimethyl-7-benzofuranyl methylcarbamate | I, N | SN |
| Carboxin | 5,6-dihydro-2-methyl-1,4-oxathiin-3-carboxanilide | F | SSP |
| Chloranil | tetrachloro-p-benzoquinone | F | IFF, SP |
| Chloroneb | 1,4-dichloro-2,5-dimethoxybenzene | F | IFF, SSP |
| Chloropicrin | trichloronitromethane | I, F, N | Fμ |
| Chlorothalonil | tetrachloroisophthalonitrile | F | SF |
| 1,3-D | 1,3-dichloropropene and related chlorinated $C_3$ hydrocarbons | F, N | Fμ |
| Dazomet | tetrahydro-3,5-dimethyl-2H-1,3,5-thiadiazine-2-thione | F, N | SP |
| DBCP | 1,2-dibromo-3-chloropropane | F, N | Fμ |
| Demeton | 0,0-diethyl O(and S)-(2-ethylthio)ethyl) phosphorothioates | I, N | SN |
| Diazinon | 0,0-diethyl 0(2-iopropyl-6-methyl-4-pyrimidinyl) phosphorothioate | I, N | N |
| Dichlofenthion | 0-2,4-dichlorophenyl), 0,0-diethyl phosphorothioate | I, N | N |
| Dichlone | 2,3-dichloro-1,4-naphthoquinone | F | SP |
| Dimethoate | 0,0-dimethyl S-(N-methylcarbamoylmethyl) phosphorodithioate | I, N | SN |
| Disulfoton | 0,0-diethyl S-(ethylthio)ethyl)phosphorodithioate | I, N | SN |
| Ethylene dibromide | 1,2-dibromoethane | I, N | Fμ |
| ETMT | 5-ethoxy-3-(trichloromethyl)-1,2,4-thiadizaine | F | IFF |
| Fenaminosulf | sodium p-(dimethylamino) benzene-diazosulfonate | F | IFF |
| Fensulfothion | 0,0-diethyl-0-(p-(methylsulfinyl)phenyl) phosphorothioate | I, N | SN |
| Fonofos | 0-ethyl S-phenyl ethylphosphonodithioate | I, N | N |
| Hexachlorobenzene | Hexachlorobenzene | F | SP |
| Metham | Sodium methyldithiocarbamate | F, I, N | Fμ |
| Methomyl | S-methyl N-(methycarbamoyl)oxy)thio-acetimidate | I, N | N |
| Methyl bromide | bromomethane | I, F, N | Fμ |
| Methyl isothiocyanate | methyl isothiocyanate | N | Fμ |
| Nabam | disodium ethylenebis(dithiocarbamate) | F | IFF |
| Oxamyl | methyl N',N'-dimethyl-N-((methylcarbamoyl) oxy)-1-trioxaminidate | I, N | SN |
| Oxycarboxin | 5,6-dihydro-2-methyl-1,4-oxathiin-3-carboxanilide 4, 4-dioxide | F | SF |
| PCNB | pentachloronitrobenzene | F | IFF |
| Phenamiphos | ethyl 4-(methylthio)-m-tolyl isopropyl-phosphoroamidate | N | SN |
| Phorate | 0,0-diethyl S-((ethylthio)methyl) phosphorodithioate | I, N | SN |
| TCMTB | 2-((thiocyanomethyl)thio)benzothizole | F, N | IFF |
| Thiabendazole | 2-(4-thiazolyl) benzimidazole | F | IFF |
| Thionazin | 0,0 diethyl 0-pyrazinyl phosphorothioate | I, N | N |
| Thiophanate-methyl | dimethyl ((1,2-phenylene)bis(iminocarbo-nothioyl))-bis(carbamate) | F | IFF |
| Thiram | bis(dimethylthiocarbamoyl)disulfide | F | SP |
| Carbon disulfide | carbon disulfide | F, N, I | Fμ |

*F = fungicide, I = insecticide, PGR = plant growth regulator, N = nematicide

†Fμ = fumigant, IFF = infurrow fungicide, N = nematicide, SF = systemic fungicide, SN = systemic nematicide, SP = seed protectant, SSP = systemic seed protectant.

survives in the embryonic tissue of the seed and must be killed to prevent its moving systemically in the meristematic tissues. Hoffman (1971) reported that thiabendazole would control common and dwarf bunt of wheat when it was used as a seed treatment. This fungicide would protect the young seedling from infection by soil-borne chlamydospores germinating and penetrating the young seedling.

Some herbicides and growth regulators have an effect upon soil-borne pathogens. The growth retardant, Pydanon, was found by Buchenauer and Erwin (1976) to reduce the foliar symptons of Verticillium wilt of tomato and cotton, when it was applied as a soil drench. They also noted there was a retardant growth of the plant and a reduced number of *V. dahliae* propagules in the soil (Erwin et al., 1976). Many of the new organic chemicals have activity as fungicides and nematicides. Some also have plant growth regulator effects (Table 2).

### 6.5.2.2 Fertilizers used to control diseases.

Many diseases cannot be controlled without the use of chemicals. It was previously mentioned that some diseases are caused by the absence of essential minerals required by the plant. An anemic person, suffering from iron deficiency, is unhealthy and can only be restored to a healthy condition by supplementing the diet with the required element. A plant can also become diseased by iron deficiency and it responds by reduced chlorophyll synthesis manifested by interveinal chlorosis and poor growth. Physiogenic diseases caused by nutritional inadequacies may usually be corrected by supplying the needed element in a form that the plant can absorb and metabolize. Often the essential nutrient is present in the soil but it is not available to the plant. Some elements are oxidized and become insoluble at high soil pH (alkaline condition) such as boron, copper, iron, manganese and zinc. A low pH (acidic condition) may induce molydbenum deficiency or excessive manganese and aluminum.

Fertilization practices affect the soil microorganisms. Some microorganisms are unable to utilize nitrate nitrogen and must receive a reduced form. In general, the conditions that favor maximum plant growth are also favorable for microbial growth. Application of chemicals to fallow soil, that would be toxic to a growing plant, would be a method of reducing pathogen populations. Rush and Lyda (1978) reported that anhydrous ammonia, applied deep in soils infested with *Phymatotrichum omnivorum*, reduced the incidence of cotton root rot the following season. Rush et al. (1979) later showed that anhydrous ammonia is very toxic to the sclerotia of this pathogen, disrupting cell membranes and killing the multicellular sclerotia upon contact. Smiley et al., (1970) studied anhydrous ammonia as a fungicide for fusarial pathogens in Washington. They observed that the fungal populations dropped to zero within the ammonia retention zone. Subsequently, Smiley and Cook, (1973) studied the effects of anhydrous ammonia on *Gaeumannomyces graminis*, a fungus causing take-all of wheat. A nitrification inhibitor, 2-chloro-6-(trichloromethyl) pyridine was used with the anhydrous ammonia to keep it from being oxidized. The pH of rhizosphere soil dropped and under low pH conditions the growth of the fungus was inhibited. Conversely, pH of rhizosphere soil became more alkaline if a nitrate form of nitrogen was used.

The type of fertilizer can influence disease severity. Take-all of wheat was suppressed by maintaining a substantial proportion of the rhizosphere nitrogen in the ammonium form (Smiley, 1974). Some authors propose that ammonia is the natural inhibitor in soils that causes soil fungistasis (Ko et al., 1974; Hora et al., 1977). In our experiences, ammonia is very toxic to survival propagules. If it is applied during the proper time and at the proper depth, it will probably replace several soil fumigants.

Certain fertilizer and soil chemical amendment practices to reduce disease severity may have side effects. In some regions of the United States, lime is used to change soil pH and reduce soil-borne diseases. Jones and Wolf (1970) added hydrated lime plus ground limestone to the soil to control Fusarium wilt of tomato. It reduced the incidence and severity of the disease, however, the results could be reversed by adding manganese plus zinc or iron plus zinc in the lignosulfonate form. Micronutrients are essential for growth of the pathogen. The authors proposed that control by liming the soil was not due to a change in pH, but it made the micronutrients unavailable for the pathogen.

Many of the nonparasitic diseases are caused by improper plant nutrition. Table 3 lists a number of disease symptoms caused by the deficiency of one or more essential elements. Plants respond to these adversities in many ways, such as characteristic discolorations, leaf curling, abnormal growth, and breakdown of internal tissues. Mineral deficiencies can usually be corrected by supplying the nutrient in a form available to the plant, sometimes as chelates applied to the foliage. These types of diseases are nonparasitic, noninfectious, and endemic.

### 6.5.3 Biological Control

In the evolution of agricultural crops, root disease must be the exception rather than the rule, and generally biological control of root pathogens must be effective. Crop yields continue to improve in these times of highly mechanized and sophisticated production systems. Knowingly or not, the farmer has learned to produce higher yielding crops through a sometimes complex decision making process. But in most instances the farmer is doing so by practicing some form of biological control.

The term biological control has different connotations depending upon who uses it. Entomologists often use this term to imply maintaining the population of harmful insects by the introduction of predators, parasites or pathogens, or at least using one organism to control another.

A somewhat broader definition was used by Beirne (1967). He states, "Any living organism that can be manipulated by man for pest control purposes are biological control agents". Many plant pathologists use this broader concept of biological control. They include changing environmental conditions to favor multiplication of antagonistic organisms through crop rotation, cultural practices, crop management and host plant resistance. Baker and Cook (1974) write, "Biological control is the reduction of inoculum density or disease-producing activities of a pathogen or parasite in its active or dormant state, by one or more organisms, accomplished naturally or through manipulation of the environment, host, or antagonist, or by mass introduction of one or more antagonists". This seems to be an all inclusive definition, but it does apply to those situations where an organism is manipulated or held in abeyance by the action of another organism or organisms.

TABLE 3. SYMPTOMS OF SEVERAL NONPARASITIC DISEASES
IN PLANTS CAUSED BY THE DEFICIENCY OF AN
ESSENTIAL ELEMENT

| Deficient nutrient | Symptom and disease |
|---|---|
| Boron | Leaves on terminal buds become light green and break down. Plants are stunted with distorted stems and leaves. Fruits and roots may crack and rot in the center. Brown heart of crucifers, heart rot of sugar beets and tip blight of tobacco result from boron deficiency. |
| Calcium | Young leaves are distorted with tips hooked back and curled margins. Terminal buds die, and plants have a poorly developed root system. Black heart of celery, blossom end rot of tomato and bitter pit of apple result from calcium deficiency. |
| Copper | Tips of leaves wither and margins become chlorotic. Leaves of cereals may not unroll and appear wilted. Heads are dwarfed and distorted. Reclamation disease of oats and withertip of apple are caused by copper deficiency. |
| Iron | Young leaves become severely chlorotic with green veins. There may be some necrotic spots develop in leaves, and in severe cases leaves will abcise. |
| Magnesium | Leaves show a general chlorosis starting with the lower leaves. In corn leaves may become white as chlorophyll synthesis is reduced. Leaves tend to roll upward, and entire plants are stunted. Sand drown disease of tobacco is caused by magnesium deficiency. |
| Molybdenum | Malformed plant growth and poor fruit set. Whiptail of cauliflower is caused by molybdenum deficiency. Alfalfa, beans, corn, flax, lettuce, oats and peas have been affected. |
| Nitrogen | Plants grow poorly and show a light green color. The lower leaves turn yellow first and scorch with a prolonged deficiency. |
| Phosphorus | Plants grow poorly with short shoots. Leaves are bluish-green with purplish tints as carbohydrates accumulate in the leaves. Shows first on older leaves. |
| Potassium | Plants have thin shoots and in severe cases they die back. Older leaves become chlorotic and die at the tips and margins. |
| Sulfur | Young leaves are pale yellow, without necrotic spots. Cotton appears yellow and dwarfed as a result of sulfur deficiency. |
| Zinc | Leaves show interveinal chlorosis similar to iron deficiency. Internodes are short and often leaves appear to emerge from the same location on the shoot. Apple and pecan rosette, little leaf of stone fruits, and white bud of corn are caused by zinc deficiency. |

### 6.5.3.1 Host resistance

Host resistance is a form of biological control if one uses the broad definition of biological control to include reduction of inoculum density or disease-producing activities of a pathogen by manipulation of the host. Baker and Cook (1974) refer to direct and indirect host resistance, but imply there is good evidence that all plants have a general resistance to pathogens. Direct resistance limits the pathogen by chemical and morphological barriers. Such mechanical barriers include thick cell walls or cuticle, suberized cells or lignified cells. Some plants possess fungitoxic constituents such as chlorogenic acid, phenolics and glycosides that impede growth and development of the pathogen. Physiological incompatability in which the pathogen is unable to secure essental nutrients from the host would also be classified as a form of direct resistance. These factors all are genetically controlled and are a characteristic of the host.

Indirect resistance includes those conditions which stimulate the host to the pathogen. These organisms may be under rhizosphere influence or they may be in the rhizoplane. It is conceivable that plants can be developed that support a microbial population in their immediate sphere to repress growth of root pathogens. Roots exude organic substances into the soil that influence that quality of soil microorganisms (Rovira, 1965; Rovira, 1969). The region so influenced by the root is called the rhizosphere. This term was first used by Hiltner (1904), who observed this region of intense microbial activity close to the root. Most compounds present in plants may also be detected in the rhizosphere. Sugars, organic acids and amino acids were reported to be detected in the rhizosphere of various crops by Ivarson et al. (1970), Katnelson et al. (1955), Schroth and Hendrix (1962), Sulochana (1962a). Later more complex, organic compounds were discovered. Terpenoids have been detected as exudations from cotton roots (Hunter et al., 1978). Basipetal translocation and root exudation of foliar applied herbicides, dicamba and picloram, were reported by Hurtt and Foy (1964). Vitamins were detected in low concentrations from root exudates which could meet nutritional requirements for some rhizosphere microorganisms (Sulochana, 1962b; Sulochana, 1962c).

Since so many interacting variables affect the quality and quantity of the soil microbial population, it is necessary to compare the microorganisms in the rhizosphere with the microorganisms in soil not under root influence. The ratio (R/S) between rhizosphere microorganisms (R) and microorganisms in soil devoid of roots (S) termed "rhizosphere effect", (Katznelson, 1965). This effect with actively growing crops is more pronounced with bacteria, with values ranging from 10 to 20 (Katznelson, 1965). Progress in breeding plants with a rhizosphere unfavorable to plant pathogens has been reported. Bird et al. (1978) reported that, in cotton, exudates from seeds, seedlings and older plants can be genetically controlled. The exudates provide a nutritional base selectively favorable for actinomycetes and bacterial populations antagonistic to and competitive with soil-borne pathogens. Bird et al. (1979) reported that exudates from seeds varied significantly among multi-adversity resistant (MAR) cultivars in electrical resistance, concentration of sodium, calcium, potassium, magnesium and carbohydrates. In addition there were differences in rhizoplane populations. Through statistical analyses, they concluded the exudate components and the microorganism data were significant in explaining variability in resistance to Phymatotrichum root rot, Verticillium wilt and seedling diseases.

Kraft and Roberts (1970) found that populations of *Fusarium solani* f. sp. *pisi* increased more in rhizosphere soil from roots of a susceptible variety than in soil from the roots of resistant lines. Exudates from the roots of resistant pea varieties, reduced the spore germination of *F. solani* f. sp. *pisi* and mycelial growth of *Pythium ultimum.*

### 6.5.3.2 Crop rotations and organic amendments

The importance of modifying the root zone by breeding or selecting plants with rhizosphere populations antagonistic to soil-borne pathogens was discussed. Rhizosphere organisms are also amenable to other kinds of manipulations. Scientists have noticed that disease incidence and severity tend to decline after certain crops or organic amendments are included in the rotation.

There are many examples about the benefits of crop residues or preceding crops in reducing disease incidence. The reasons for the beneficial responses are debatable. Lewis and Papavizas (1977) reported that decomposing mature residues of rye, oat, soybean, sorghum, barley, buckwheat, timothy and corn were generally more effective than immature residues in reducing bean root rot caused by *Fusarium solani* (Mart.) Appel and Wr. f. sp. *phaseoli* (Burk.) Sndy. and Hans. The residues were inhibitory to chlamydospore germination *in vitro* and in soil. They attributed this inhibition to nutrient deficiency in the soil, as well as to the formation of inhibitory substances from the decomposing residue.

When organic matter decomposes in the soil an enviroment is created that may reduce the germinability of fungal spores. This phenomenon, which they called fungistasis, was first described by Dobbs and Hinson (1953). The nature of fungistasis has not been resolved. Some authors believe that exogenous nutrients are depleted from the soil solution by microbial activity and they are not available for spore germination (Ko and Lockwood, 1967), others are proponents of an inhibitor hypothesis (Hora and Baker, 1979; Ko et al., 1974; Romine and Baker, 1973) and some believe both mechanisms may exist (Hora et al., 1977).

## 6.5.4 PHYSICAL METHODS OF ALLEVIATING PATHOGEN STRESS

There are four physical factors that have a profound affect of soil-borne pathogens, (a) moisture, (b) temperature, (c) aeration and (d) mechanical strength of the soil. Methods of regulating or manipulating these factors to reduce the incidence of plant disease will be discussed.

### 6.5.4.1 Soil Moisture

Adequate soil moisture is essential for maximizing potential crop yields. It also influences the incidence and severity of soil-borne diseases. In many regions soil moisture is maintained through irrigation, but in other regions the farmer must depend upon natural precipitation.

Soils with excessive soil moisture may predispose plants to infection by phycomycetous fungi, i.e. those that produce motile asexual spores. These fungi are very sensitive to changes in the matric potential. Sterne et al. (1977) found *Phytophthora cinnamomi* infection on roots of *Persea indica* seedlings to be less than 10 percent at -25 kPa (0.25 bars). At 0 kPa root infection was between 80 to 100 percent.

Potato scab, on the contrary, is more severe under dry soil conditions. This disease is caused by the Actinomycete, Streptomyces scabies, and it can be controlled by irrigation during the period of tuber formation.

Plant populations affect the soil moisture availability. Though soil surface temperatures may be lower with dense plant populations, soil moisture may be rapidly depleted through transpiration. All of the parameters interact.

### 6.5.4.2 Soil Temperature

Several cultural practices may be used to regulate soil temperature under field conditions.

The time of planting is one of the most widely used means of achieving soil temperature favorable for plant growth, and unfavorable for certain soil-borne pathogens.

Row spacing and plant density affect the radiant energy reaching the soil surface. Soil temperatures are reduced with narrow row spacing where the plant canopy covers the soil.

Raised beds are used to improve aeration and increase soil temperatures. This practice is used to reduce the severity of soil-borne diseases caused by pathogens thriving under cool conditions.

Pullman et al. (1978, 1979) used plastic tarping to increase soil temperatures and reduce pathogen propagules in the soil. Polyethylene plastic tarps (1 and 4 mil) were compared in field tests as a method of increaseing soil temperautre through the "greenhouse effect". Soil temperatures at varying soil depths under the tarps increased 13 to 14 °C at 5 cm and 8 to 9 °C at 46 cm. The increased soil temperature reduced propagules of *Verticillium dahliae* from 274 to 3 per g soil in the 30 to 46 cm zone.

Propagules of *Pythium* spp. were reduced from 254 to 95 per g soil in the 30 to 46 cm zone.

Acala SJ 2 cotton was planted in the tarped and nontarped plots to evaluate disease control. The percentage of plants manifesting disease symptoms varied from 77 in the nontarped soils to 3 in the soils tarped for 8 wks with 4 mil film at Shafter, CA.

Elevation of soil temperature affects growth and survival of the flora and fauna, but costs of implementation often preclude its use in managing soil-borne diseases.

### 6.5.4.3 Aeration

There are many examples attesting to the value of plowing the soil deep to improve plant growth and decrease the severity of root diseases. Shear and Miles (1927) noted the organism, causing cotton root rot in Texas, grows best where soil aeration is the poorest. They observed that plowing soil (18 to 23 cm deep) in the fall reduced the disease to 12 percent dead plants as compared to 96 percent dead plants in soils prepared in the normal manner, i.e., bedded with a "middle buster" in the spring. They also found that deep fall plowing was superior to deep plowing in the spring. Rogers (1942) noted that subsoiling (38 to 76 cm deep) consistently reduced the incidence of root rot, but the increased yields were insufficient to pay for the costs of the operation. He also recognized the value of plowing early in the fall to reduce root rot. Benefits from deep tillage are believed to be derived from improved soil aeration.

**Phymatotrichum omnivorum,** the cotton root rot pathogen, thrives in a soil environment with elevated carbon dioxide content (Lyda, 1975; Lyda and Burnett, 1975). The fungus grew through soil columns flushed with 0.5 to 5 percent $CO_2$, but it did not grow in soil columns flushed with air of 50 percent $CO_2$. Under field conditions the soil $CO_2$ content was observed to increase each season until the soils began to dry and crack. The increased soil $CO_2$ content was proposed as a competitive advantage for the pathogen over other soil microorganisms. Highly aerated soils would not be conducive for growth of *P. omnivorum*.

Deep plowing and burial of organic matter is used to control southern blight caused by *Sclerotium rolfsii* (Aycock, 1966). This pathogen grown on dead organic matter and secretes oxalic acid which diffuses and kills root cells in advance of mycelial growth (Higgins, 1927; Bateman and Beer, 1965).

In a comparative study of tillage methods to reduce Phymatotrichum root rot, Lyda (1975) reported rototilling and ditching to a 60-cm depth were superior to surface mulching or disk plowing. Improvement of soil aeration and redistribution of sclerotia were suggested as explanations for the differences in disease. After a period of three years, the level of root rot was high in all treatments, indicating the fungus had reestablished in the deep tilled soils.

### 6.5.4.4 Mechanical Strength

Burke et al. (1970) observed *Aphanomyces euteiches* was prevalent in the plowed layer, and the number of propagules was markedly reduced in the subsoil. They proposed the disease problem was enhanced by plant roots unable to penetrate the compacted layer, whereas in fields having soils more easily penetrated by roots there was no disease problem. Root distribution has been used as a criterion for selecting plants that will escape high levels of inoculum. Phillips and Wilhelm (1971) noted that Waukena White, a cotton cultivar from *Gossypium barbadense,* produced few lateral roots in the upper 45-cm of soil where the highest numbers of *Verticillium dahliae* microsclerotia occur. A susceptible cultivar, Acala SJ-1 from *Gossypium hirsutum,* produced most of its lateral roots in the upper 45-cm of soil. The tap roots of both cultivars penetrated vertically to a depth of 120 cm until they reached the claypan. Burnett and Tackett (1969) studied the effects of soil profile modification on root development of cotton and sorghum. They observed a more extensive root branching on plants grown in soils that had been rotatilled or ditched, as compared to the conventional tillage treatment (disking). In addition to enhanced root development in the rotatilled and ditched  soil, aeration was improved and was much more uniform than in the shallow disked soils.

### SUMMARY

Agricultural scientists continually seek methods of improving crop productivity. Pests (insects, pathogens and weeds) reduce potential yields and in many regions they are deterrents for planting crops. Soil that becomes infested with long surviving pathogens is often abandoned or planted to less desirable, nonsusceptible crops. In the United States, the successful farmer uses several methods to manage pest problems. The successful farmer has long practiced integrated pest management.

Roberts (1979) states, "Every plant pest control practice, whether chemical, cultural, or biological, exemplifies a principle, which principle is the scientific basis for the success of the practice. First principles, to which most would agree, are: (a) reduce the initial population of noxious organisms, and (b) slow their rates of increase." Nearly every control strategy today is based on these two principles. The methods of reducing inoculum potential and slowing the rate of inoculum increase vary with the type of pathogen, the host and the environment.

Means for alleviating pathogen effects are also influenced by these factors. It may be justifiable to employ expensive chemicals to control root rot on perennial plants, but not an annual plants. Economics usually dictate the method of choice.

Host resistance is a desirable control measure for pathogens with a narrow host range, i.e. the more specialized pathogens. Resistance has not proven effective in controlling diseases caused by indiscriminate pathogens such as *Phymatotrichum omnivorum, Rhizoctonia solani,* and *Sclerotium rolfsii.*

Nearly every choice of modifying the root zone to alleviate pathogen effects is expensive or nonpersistent. Physical disruption of the soil profile requires considerable expenditures of energy and time, especially in heavy soils. These methods are usually very effective in abating soil-borne diseases for one or two years, but the costs of implementation are greater than the returns from fewer diseased plants.

Biological control through the use of green manure crops are also effective in reducing the severity of some soil-borne diseases, but this is also an expensive venture. Often the farmer loses one crop in addition to the costs of seeding and incorporating the green manure crop.

In essence there do not appear to be any inexpensive, consistent, effective means of root zone modification for alleviating pathogen effects. The challenge scientists have is to discover new information about and ways of manipulating this invisible, three-dimensional environment.

**References**

1  Agrios, G. N. 1978. Plant Pathology. Academic Press, New York. 703 p.

2  Alexander, M. 1961. Introduction to soil microbiology. John Wiley and Sons, Inc. New York. 472 p.

3  Aycock, R. 1966. Stem rot and other diseases caused by Sclerotium rolfsii. North Carolina Agric. Exp. Sta. Tech. Bull. 174. 202 p.

4  Baker, K. F. and R. J. Cook. 1974. Biological control of plant pathogens. W. H. Freeman and Co., San Francisco. 433 p.

5  Beirne, B. P. 1967. Biological control and its potential. World Rev. Pest. Control 6:7-20.

5a  Benjamin, C. R., W. C. Haynes and C. W. Hesseltine. 1964. Micro-organisms: what they are, where they grow, what they do. USDA Misc. Pub. No. 955.

6  Bird, L. S., D. L. Bush, R. G. Percy, and F. M. Bourland. 1978. Genetic research improves disease resistance in cotton. Texas Agr. Progress 24:22-23.

7  Bird, L. S., C. Liverman, R. G. Percy, and D. L. Bush. 1979. The mechanism of multi-adversity resistance in cotton: theory and results. Proc. Beltwide Cotton Prod. Res. Conf. 1979:226-230.

8  Buchenauer, H. and D. C. Erwin. 1976. Effect of the plant growth retardant Pydanon on Verticillium wilt of cotton and tomato. Phytopathology 66:1140-1143.

9  Burke, D. W., D. J. Hagedorn, and J. E. Mitchell. 1970. Soil conditions and distribution of pathogens in relation to pea root rot in Wisconsin soils. Phytopathology 60:403-406.

10  Burnett, E. and J. L. Tackett. Effect of soil profile modification on plant root development. p. 329-337. 9th Int. Congr. Soil Sci. Trans.

11  Christie, J. R. 1959. Plant Nematodes—their bionomics and control, H & W. B. Drew Co., Jacksonville, FL. 256 p.

12  Clark, F. E. 1967. Bacteria in soil. In: A. Burgess and F. Raw (eds.). Soil biology. Academic Press, New York. 535 p.

13  Dobbs, G. G. and W. H. Hinson. 1953. A widespread fungistasis in soils. Nature 172:197-199.

14  Erwin, D. C., S. D. Tsai, and R. A. Khan. 1976. Reduction of the severity of Verticillium wilt of cotton by the growth retardant, tributyl ([5-chloro-2-thienyl] methyl) phosphonium chloride. Phytopathology 66:106-110.

15  Garrett, S. D. 1963. Soil fungi and soil fertility. The McMillian Co., New York. 165 p.

15a  Higgins, B. B. 1927. Physiology and parasitism of *Sclerotium rolfsii.* Phytopathology 17:417-448.

15b   Hi, L. 1904. Uber newer Erfahrungen und probleme auf dem giebiet der Boden-bacteriologie und unter Besanderer Beruchsichtigung der grundungung und brache. Arb Deut. Landwirsch Ges 98:59-78.

16   Hoffman, J. A. 1971. Control of common and dwarf bunt of wheat by a seed treatment with thiabendazole. Phytopathology 61:1071-1074.

17   Hora, T. S. and R. Baker. 1979. Volatile factor in soil fungistasis. Nature 225:1071-1072.

18   Hora, T. S., R. Baker, and G. J. Griffin. 1977. Experimental evaluation of hypothesis explaining the nature of soil fungistasis. Phytopathology 67:373-379.

19   Horsfall, J. G. and A. E. Dimond. 1959. The diseased plant. p. 6-7. In: J. G. Horsfall and A. E. Dimond (eds) Plant pathology, Vol. I. Academic Press, New York. 674 p.

20   Hunter, R. E., J. M. Halloin and J. A. Veech. 1978. Exudation of terpenoids by cotton roots. Plant and Soil 50:237-240.

21   Hurtt, W. and C. L. Foy. 1965. Excretion of foliarly applied dicamba and picloram from root of black valentine beans grown in soil, sand, and culture solution. Proc. Northeastern Weed Control Conf. 19:302.

22   Ivarson, K. C., F. J. Sowden and H. R. Mack. 1970. Amino acid composition of rhizosphere as affected by soil temperature, fertility, and growth stage. Can. J. Soil Sci. 50:183-189.

23   Katznelson, H. 1965. Nature and importance of the rhizosphere. p. 187-209. In: K. F. Baker and W. C. Snyder (eds.). Ecology of soil-borne plant pathogens. University of California Press, Berkeley. 571 p.

24   Katznelson, H., J. W. Rouatt and T. M. B. Payne. 1955. The liberation of amino acids and reducing compounds by plant roots. Plant and Soil 7:35-48.

25   Keeton, W. T. 1972. Biological science. W. W. Norton & Co., Inc., New York. 888 p.

26   Ko, W. H. and J. L. Lockwood. 1967. Soil fungistasis: relation to fungal spore nutrition. Phytopathology 57:894-901.

27   Ko, W. H., F. K. Hora and E. Herlicska. 1974. Isolation and identification of a volatile fungistatic substance from alkaline soil. Phytopathology 64:1398-1400.

28   Kraft, J. M. and D. D. Roberts. 1970. Resistance in peas to Fusarium and Phythium root rots. Phytopathology 60:1814-1817.

29   Lewis, J. A. and G. C. Papavizas. 1977. Effect of plant residues on chlamydospore germination of Fusarium solani f. sp. phaseoli and on Fusarium root rot of beans. Phytopathology 67:925-929.

30   Lyda, S. D. 1975. Studies on Phymatotrichum omnivorum and Phymatotrichum root rot. In: G. C. Papvizas (ed.) The Relation of Soil Microorganisms to Soilborne Plant Pathogens. So. Coop. Ser. Bull. Virginia Polytech. Inst. and State Univ. Press, Blacksburg, VA. 183 p.

31   McCallan, S. E. A. 1967. History of fungicides. p. 1-37. In: D. C. Torgenson (ed.) Fungicides, Vol. I. Academic Press. New York. 607 p.

32   Miller, M. W. 1961. The Pfizer handbook of microbial metabolites. McGraw-Hill Book Co., Inc. New York. 772 p.

33   Pelczar, M. J., Jr., R. D. Reid and E. C. S. Chan. 1977. Microbiology. McGraw-Hill Book Co., New York. 952 p.

34   Philips, D. J. and S. Wilhelm. 1971. Root distribution as a factor influencing symptom expression of Verticillium wilt of cotton. Phytopathology 61:1312-1313.

35   Pullman, G. S., J. E. DeVay, and R. H. Garber. 1978. Effects of plastic tarping on soil temperatures and survival of soil-borne propagules of Verticillium dahliae and Pythium spp. Proc. Beltwide Cotton Prod. Res. Cong. p. 28-29.

36   Pullman, G. S., J. E. DeVay, and R. H. Garber. 1979. Plastic tarping: An effective control for Verticillium wilt in cotton. Proc. Beltwide Cotton Prod. Res. Cong. p. 25.

37   Roberts, D. A. 1979. The terminology of plant-pest control Phytopathology 69:1043 (Abstract).

38   Rogers, C. J. 1942. Cotton root rot studies with special reference to sclerotia, cover crops, rotations, tillage, seeding rates, soil fungicides, and effects on seed quality. Texas Agri. Exp. Sta. Bull. 614. 45 p.

39   Romaine, M., and R. Baker. 1973. Soil fungistasis: evidence for an inhibitory factor. Phytopathology 63:756-759.

40   Rovira, A. D. 1965. Plant root exudates and their influence upon soil microorganisms. p. 170-186. In: K. F. Baker and W. C. Snyder (eds.). Ecology of soil-borne plant pathogens. Univ. of Calif. Press, Berkeley. 571 p.

41   Rovira, A. D. 1969. Plant root exudates. Bot. Rev. 35:35-53.

42   Rush, C. M. and S. D. Lyda. 1978. Anhydrous ammonia fumigation of Phymatotrichum-infested montmorillonitic clay. Proc. Beltwide Cotton Prod. Res. Cong.

1978:23-25.
43  Rush, C. M., C. M. McClung, and S. D. Lyda. 1979. Cellular effects of anhydrous ammonia on Phymatotrichum omnivorum sclerotia. Phytopathology 69:1044 (Abstract).
44  Schroth, M. N. and F. E. Hendrix, Jr. 1962. Influence of nonsusceptible plants on the survival of Fusarium solani f. phaseoli in soil. Phytopathology 52:906-909.
45  Sharvelle, E. G. 1979. Chemical control of plant diseases. University Publishing College Station, TX. 340 p.
46  Shear, C. L. and G. F. Miles. 1907. The control of Texas root-rot of cotton. U.S. Dept. Agric. Bur. Indus. Bull. 261:39-42.
47  Smiley, R. W. 1974. Take-all of wheat as influenced by organic amendments and nitrogen fertilizers. Phytopahtology 64:822-825.
48  Smiley, R. W. and R. J. Cook. 1973. Relationship between take-all of wheat and rhizosphere pH in soils treated with ammonium vs. nitrate-nitrogen. Phytopathology 63:882-889.
49  Smiley, R. W., R. J. Cook and R. I. Papendick. 1970. Anhydrous ammonia as a soil fungicide against Fusarium and fungicidal activity in the retention zone. Phytopathology 60:1227-1232.
50  Sterne, R. E., G. A. Zentmeyer and M. R. Kaufman. 1977. The influence of matrix potential, soil texture, and soil amendment on root disease caused by Phytophthora cinnamoni. Phytopathology 67:1490-1500.
51  Sulochana, C. B. 1962a. Amino acids in root exudates of cotton. Plant & Soil 16:327-334.
52  Sulochana, C. B. 1962b. B vitamins in root exudates of cotton. Plant & Soil 16:327-334.
53  Sulochana, C. B. 1962c. Cotton roots and vitamin requiring and amino acid requiring bacteria. Plant & Soil 16:335-346.
54  Tietz, H. 1970. One centennium of soil fumigation: Its first years. p. 203-207. In: T. A. Tousson, R. V. Bega, and P. E. Nelson (eds.). Root diseases and soil-borne pathogens. University of California Press, Berkeley. 252 p.
55  Von Schmeling, B. and M. Kulka. 1966. Systemic fungicidal activity of 1,4-oxathiin derivatives. Science 152:659-660.
56  Waksman, S. A. 1927. Principles of soil microbiology. The Williams & Wilkins Co., Baltimore. 897 p.
57  Waksman, S. A. 1952. Soil microbiology. John Wiley & Sons, Inc. New York. 356 p.
58  Waksman, S. A. 1967. The actinomycetes. A summary of current knowledge. The Ronald Press Co., New York. 280 p.
59  Waksman, S. A. and R. E. Curtis. 1916. The actinomycetes of the soil. Soil Sci. 1:99-134.
60  Walker, J. C. 1959. Progress and problems in controlling plant diseases by host resistance. p. 32-41. In: C. S. Holton, G. W. Fisher, R. W. Fulton, Helen Hart and S. E. A. McCallan (eds.). Plant pathology—problems and progress 1908-1958, University of Wisconsin Press, Madison. 588 p.
61  Whittaker, R. H. 1979. New concepts of kingdoms of organisms. Science 163:150-160.
62  Wood, J. I., N. E. Stevens and D. Reddick. 1940. Report of the committee on technical words. Phytopathology 30:361-368.

# chapter 7

**ALLEVIATING TEMPERATURE STRESS**

**7**

# 7
# ALLEVIATING TEMPERATURE STRESS

by    Ward B. Voorhees, USDA-SEA-AR, North Central
Soil Conservation Research Laboratory, Morris,
MN; R. R. Allmaras, USDA-SEA-AR, Columbia
Plateau Conservation Research Center, Pendleton,
OR; Clarence E. Johnson, Department of
Agricultural Engineering, Auburn University,
Auburn, AL.

## 7.1 INTRODUCTION

Agricultural scientists and engineers have long recognized, even as early as 1900 (Richards et al., 1952), that soil and air temperatures are important and often critical environmental factors for plant growth and productivity. There is, therefore, a large body of scientific literature concerning plant response to management of soil temperature. However, cause and effect were often not clearly defined until the development of equipment in the early 1950's for accurate and continuous sensing and recording of temperature. Scientists had long suspected that heat and water flux in soils were closely related, but the mechanisms and importance were not recognized until the work of Philip and de Vries (1957). Other new research since 1950 that has helped our understanding of plant response to soil temperature is measured plant response to temperature under controlled conditions, description of plant root growth in the field, theory and measurement of heat and water flux in soil, physiological modeling of plant response to stress, and description of heterogeneity of field environment. Technology has improved to the stage where we can systematically change soil temperature in the field, and also understand the temperature mediated responses to traditional soil management practices. Gigantic advances in farm machinery design, also since the early 1950's, have provided us with tools to implement specific soil management objectives.

Successful modification of soil temperature requires a knowledge of:

1    mechanisms controlling soil temperature,

2    relationships between soil temperature, root growth, and plant component of interest, and

3    effects of cultural practices on soil temperature.

A systematic approach is illustrated schematically in its simplest form by Fig. 7.1-1.

The farmer, manager, or producer, is most interested in the route marked by the broken line in Fig. 7.1-1. The marketable product may be grain or other reproductive part of the plant, total biomass, or some quality consideration such as flower brilliance. One problem the producer will have is to recognize when a new or modified cultural practice is necessary to improve

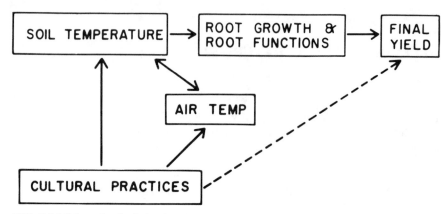

FIG. 7.1-1 Schematic of relationships between soil temperature modifications and final yield.

the final yield (or product). The researcher, in his quest to more fully understand the whole system and to provide guidance to the user, needs to consider the processes along both the solid vertical and the horizontal arrows. This latter pathway involves causes and effects which can and must be quantified.

Our order of discussion will be to first consider the mechanisms of soil temperature change in the root zone. Many factors interact to make the system complex; however, the discussion will be developed by emphasizing sources containing more detailed evaluations and descriptions. The next section will deal with root responses to soil temperature, but, since the plant is a complex feedback system, we must consider root and shoot responses together. Finally we discuss what soil management practices can be used to control soil temperature.

## 7.2 MECHANISMS AND PRINCIPLES
## OF SOIL TEMPERATURE CONTROL

Soil temperature depends on radiation balance at the soil surface, soil heat flux, and soil water flux. Radiation and thermal energy balances at the soil surface are discussed first because of their dominant influence. Subsequent discussions about the flux of heat within the soil emphasize its secondary, but still important, role in determining soil temperature. Both can be controlled by the producer. Water fluxes are associated with soil heat flux. Thus, we will discuss water flux only as it influences heat flux (chapter 10 will consider water flux and associated available water effects on root environment). For detailed discussions about radiant energy balance refer to Gates (1965), and van Wijk (1966). Soil heat flux and soil temperature are discussed in detail by van Wijk (1965, 1966). Fluxes of heat, vapor, and momentum in the lowest two meters of the atmosphere interact with both radiation balance at the earth's surface and soil heat flux (van Wijk (1966), Webb (1965), Tanner (1968) and Lemon et al. (1973)). Monteith (1973) addresses all of these subjects from the viewpoint of environmental physics.

### 7.2.1 Radiation Balance at the Earth's Surface

Radiation from the sun is the primary source of heat influencing soil temperature. The mean total solar radiation impinging on the outer extremity of the earth's atmosphere (the solar constant) is about 1.39 kW/m²; it varies less than ±5 percent because of earth-to-sun distance and solar activity. Not all of this direct beam extraterrestrial solar radiation reaches the earth's surface because much of it is reflected, absorbed and reradiated as thermal radiation, or scattered and diffused in the earth's atmosphere. Fig. 7.2-1 gives a schematic estimate of these radiation components at the earth's surface, where they can be managed to alter soil temperature.

### 7.2.1.1 Direct Beam Radiation

The intensity of direct solar radiation received on a horizontal area at the earth's surface is affected by solar altitude, terrestial latitude, solar declination, solar hour angle, radius vector of sun-to-earth distance, and transmission coefficient of the atmosphere (Gates, 1965; List, 1966). During passage of solar radiation through the atmosphere, the ultraviolet, and infrared to some extent, is selectively blocked. Thus, direct beam at the earth's surface has characteristic wavelength range from 0.3 to 3.0 μm, while longwave radiation has a characteristic 3 to 30 μm wavelength.

A 10-deg change in latitude can increase direct beam radiation on a horizontal surface as much as 30 percent (Fig. 7.2-2); season can change the radiation intensity by a factor of at least 3. Solar hour produces the diurnal influence. Direct insolation on slopes, i.e., receipt of radiation on a surface normal to the slope, can be estimated (using the cosine law) by comparing radiation expected on a horizontal surface (Gates, 1965). It is possible then to compare slope and latitude influences on the receipt of radiation at one instant or integrated over a day. Thus, Shul'gin (1965) noted that a field with a 1-deg north facing slope will have heat intervals similar to a level field

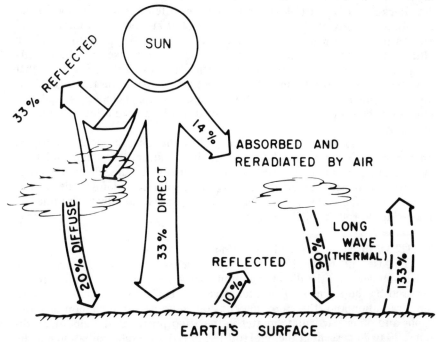

FIG. 7.2-1 Schematic of radiation balance at earth's surface (annual global average).

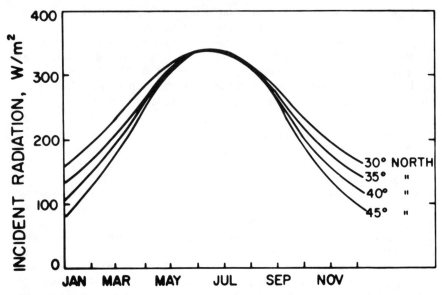

**FIG. 7.2-2 Daily total direct and sky radiation incident upon a horizontal surface at various north latitudes during cloudless days (after Becker and Boyd, 1961).**

situated about 100 km farther north. Smith et al. (1964) reported that mean temperatures on north-facing slopes are almost 3 °C cooler than south-facing slopes. Macyk et al. (1978) concluded that the growing season was 1 week longer on a south than a north facing slope at about 54 deg N latitude.

Becker and Boyd (1961) reported little difference in daily total solar radiation intensity at altitudes up to 300 m, but at 1,500 m, it increased 12 and 18 percent in winter and summer, respectively. Additional increase in altitude to 3,000 m resulted in another 6 percent increase in radiation. At high altitudes, the minimum soil temperature on a south slope may be higher than the maximum soil temperature on a north slope (Shreve, 1924). These altitude influences are more difficult to assess because of cloudiness and dust effects on the transmission coefficient for direct beam radiation through the atmosphere.

Direct solar radiation can be changed by changing the slope and aspect on a microscale, as with bedding or ridging. Diurnal, seasonal, and solar elevation (azimuth angle) effects dominate climate and the selection of crops and crop sequences. Later in the discussion, solar elevation will be seen as important in the understanding of diurnal and seasonal effects on radiation scattering on rough soil surfaces.

### 7.2.1.2 Diffuse and Thermal Radiation

Diffuse and thermal radiation components at the earth's surface are much smaller than the direct radiation component, but they should also be assessed by the manager of soil temperature. Diffuse sky radiation is scattered radiation and, therefore, depends on cloudiness, the amount of water vapor and other particulate matter in the atmosphere. Under cloudy conditions, 20 to 80 percent of the total (diffuse plus direct) solar radiation may be diffuse (Gates, 1965). Our inability to manage soil temperature by changing the incidence of diffuse radiation is illustrated by the work of Anderson and Denmead (1968) showing that leaf orientation, by describing both aspect and inclination from horizontal, has a large effect on the hourly receipt of direct solar radiation but no significant effect on receipt of diffuse radiation.

Thermal or long wave rediation arises because all materials above absolute zero radiate energy at wavelengths depending on temperature of the body (Wiens displacement law), and amounts depending on the fourth power of the absolute body temperature (Stefan-Boltzman law). The balance is shown in Fig. 7.2-1 along with the short-wave components. A good black body absorber is also a good black body emitter; plants, soils, sand, snow, clouds are characteristically black body. Clouds at 0 °C at their base can radiate about 0.279 $kW/m^2$ of thermal radiation downward to the soil surface. Gates (1965) gives an account of the long wave absorption and reradiation patterns of the atmosphere. Compared to short wave radiation, the diurnal variation in long wave radiation is small. There is a net upward flux of long wave radiation. Idso and Jackson (1969) showed that incoming thermal radiation can be estimated based on air temperatures; however, the manager of soil temperature cannot control this component.

### 7.2.2 Energy Balance at the Soil Surface

Once the incoming radiation has reached the earth's surface, it is then available for heating the soil, evaporating water, and heating the air. The energy balance equations at the soil surface may be written (van Wijk, 1966) as follows:

$$(1 - \alpha_s) R_s + R_1{}^{net} = R_N = H_{soil} + H_a + H_e, \quad \dots\dots\dots\dots\dots \quad [7.2-1]$$

where $\alpha_s$ is the reflection coefficient for the incoming short-wave radiation, $R_s$ is incoming short wave radiation and $R_1{}^{net}$ is net long wave radiation. $R_N$ is net radiation which is available for heat flow into the soil, ($H_{soil}$), heat flow into the air, ($H_a$), and evaporation of water, ($H_e$). Usually the R and H terms in equation [7.2-1] are expressed as energy per unit of horizontal surface per unit time. The reflection coefficient, $\alpha_s$, is dimensionless. The heat and energy involved in respiration and photosynthesis are relatively small and are ignored in this equation.

$R_N$ is commonly measured with a net radiometer and is the sum of unreflected short wave plus net long wave radiation. $R_N$ can also be estimated from using: (a) upright and inverted pyranometers to measure unreflected short wave, and (b) estimates of incoming and outgoing long wave (Monteith, 1973). Monteith (1973) and Gates (1965) both show that $R_1{}^{net}$ is small, with outgoing larger than incoming. $R_1{}^{net}$ can make a significant contribution to $R_N$ (Allmaras et al., 1977), thus the manager of soil temperature must be aware of management effects of $R_1{}^{net}$.

Terms on the right hand side of equation [7.2-1] are interactive, so that $H_{soil}$ will depend on $H_a$ and $H_e$. $H_{soil}$ also depends on plant cover. Brown (1964) found only 6 percent of $R_N$ impinging on a corn crop was partitioned to $H_{soil}$; on a bare soil, $H_{soil}$ ranged from 6 to 29 percent of $R_N$ (Allmaras et al., 1977). Thus about 6 to 90 percent of $R_N$ normally partitions to $H_a + H_e$. $R_N$ is relatively easily measured; we shall show in our later discussion that $H_{soil}$ is more difficult to measure. Either $H_a$ or $H_e$ are difficult to measure directly. Tanner (1968) and Webb (1965) give the details for measurement of

TABLE 7.2-1. SHORT-WAVE REFLECTANCE
OF NATURAL SURFACES
(From van Wijk, 1966)

| Surface | Short-wave reflectance, % |
|---|---|
| Fresh snow | 80-85 |
| Lime | 45 |
| Quartz sand | 35 |
| Dark clay, wet | 2-8 |
| Dark clay, dry | 16 |
| Sand, wet | 9 |
| Sand, dry | 18 |
| Bare fields | 15-25 |
| Wet plowed fields | 5-14 |
| Grass, green | 16-27 |
| Grass, dried | 16-19 |
| Stubble fields | 15-17 |
| Grain crops | 10-25 |
| Pine, spruce wood | 10-14 |
| Deciduous wood | 16-37 |
| Desert, midday | 15 |
| Water* | 2 |

*Water with sun angle from 0 to 30 deg, 0 deg
is perpendicular to the water surface.

these two heat balance terms. $H_e$ is most easily measured by lysimetry or soil loss of water. There is no easy way to estimate $H_a$ because of the complex patterns and short time frame changes in turbulence, momentum transfer, and heat transfer in the air layer, especially near the air-soil interface.

### 7.2.2.1 Reflection Coefficient

The reflection coefficient, $\alpha_s$ in equation [7.2-1], is defined as the ratio of radiant energy reflected to the total radiant energy incident upon the surface. The short-wave reflectance (Table 7.2-1) may vary from 85 percent for fresh snow to 2 percent for wet, dark clay soil. Note that a wet surface, either soil or vegetation, reflects less than a dry surface, and that the reflectance of plant residues is somewhat greater than for soil surfaces. Herein lies the greatest opportunity for management of soil temperature (Van Doren and Allmaras, 1978).

Micro- and random-roughness produced by soil aggregates, or larger scale roughness produced by tillage operations, both cause changes in reflectivity of bare soil (Coulson and Reynolds, 1971; Allmaras et al., 1972; Idso et al., 1975). Bowers and Hanks (1965) showed that, as aggregate diameter increases, reflectivity of a 1-$\mu$m monochromatic beam decreases. Thus soil roughness is also a means to manage soil temperature when there is not a dominating plant cover. When plant growth produces a leaf area index (total one-side leaf area per unit ground area) of about 3, the receiving and radiating surface plane is somewhere within the plant canopy, not at the soil surface.

The reflectance of soils and plant residues changes as wavelength of the incident radiation changes. Short-wave radiation was earlier defined as having wavelength ranging from 0.3 to 3.0 $\mu$m. Fig. 7.2-3a shows that the 0.75 to 1.3 $\mu$m region is apparently better than either the <0.75 or >1.3 $\mu$m wavelength for distinguishing among reflectances of soil-tillage-straw treatments. The condition of the residue (standing versus matted) also affects

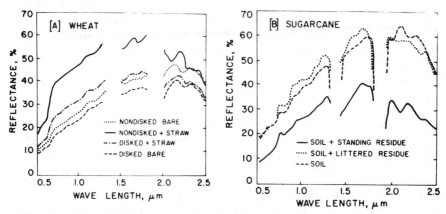

FIG. 7.2-3 Effects of tillage and residue on soil reflectance at various wavelengths for (A) wheat residues (Gausman et al., 1977) and (B) sugar cane residue (Gausman et al., 1975).

reflectance throughout the 0.5 μm to 2.5 μm wavelength range (Fig. 7.2-3b). Bower and Hanks (1965) showed that soil water content affected reflectance over the whole 0.5 to 2.5 μm wavelength range. Condit (1972) and Montgomery and Baumgardner (1974) produced curves of reflectance versus wavelength of incident radiation for dry and wet soils. Idso et al. (1975) were able to show that albedo, normalized to remove solar zenith angle effects, was a linear function of surface soil water content over the range 0 to 0.2 m³/m³. Thus, failure to recognize the effects of water content and varying wavelength on reflectance can produce biases when comparing various tillage-residue combinations.

### 7.2.2.2 Soil Heat Flux

Vertical heat flux downward through a very thin soil layer may be expressed as:

$$H(z,t) = -\lambda \frac{dT(z,t)}{dz}, \dots\dots\dots\dots\dots\dots\dots\dots\dots\dots\dots [7.2-2]$$

where H is a generalized soil heat flux density (heat passing through a unit area per unit time) for some given depth (z) and time (t), λ is the thermal conductivity, dT/dz is the temperature gradient, and a negative sign denotes flow from warmer to cooler regions. A common unit for H is W/m² and degree/m for dT/dz. λ then has units of W/m °C.

Soil heat flux density at the soil surface, H (o, t), is the most important quantity determining the regime of temperature in the soil; unfortunately, it is extremely difficult to measure at the soil surface because of abrupt changes in λ and dT (o, t)/dz near the surface.

Four well recognized methods to measure H (z, t) are described by Kimball and Jackson (1975). In the thermal gradients method, equation [7.2-2] is applied directly, but λ is difficult to measure (see a later discussion). The method cannot be used to measure H (o, t). Heat flux plates, a thermopile incorporated into a thin plate, may be used to determine heat flux (Fuchs and Tanner, 1968); however, a heat flux plate may converge or diverge soil heat flux if it has a thermal conductivity different than soil;

heat flux plates may also interfere with vapor and liquid water flow. For this reason heat flux plates are to be used at depths deeper than 10 cm. A third method to estimate soil heat flux is the calorimetric method, in which the change in soil heat storage in a very thin soil layer is computed over a small time period:

$$-\frac{\partial H(z,t)}{\partial z} = C \frac{\partial T}{\partial t} , \dotfill [7.2-3]$$

where C is a volumetric heat content of the soil to be discussed later. Although this method, when expanded to a finite depth and time interval, can be used to estimate H (o, t), it suffers from inaccurate temperature measurements at depths greater than 30 cm. A fourth and most useful method, the combination method, utilizes calorimetry (as in equation [7.2-3]) at depths of 30 cm or less and the heat flux plate or a thermal gradient estimate (method 1) to estimate heat flux past a selected depth in the soil (10- to 30-cm range). H (o, t) is most accurately measured using this combination method.

A formal statement for heat flow in soils can be obtained by examining the change in heat flux density in a very small layer of soil and the corresponding heat content changes in that layer:

$$\frac{\partial H(z,t)}{\partial z} = \frac{\partial}{\partial z} (-\lambda \frac{\partial T}{\partial z}) = -(\Delta z) C \frac{\partial T}{\partial t} \quad \dotfill [7.2-4]$$

If both $\lambda$ and C are unchanged with time and depth the classical heat flow continuity equation is formed:

$$\frac{\partial T}{\partial t} = K \frac{\partial^2 T}{\partial z^2}, \dotfill [7.2-5]$$

where $K = \lambda/C$ is the thermal diffusivity with units of m²/sec. When various initial and boundary conditions are imposed, depending on the heat flow system (van Wijk, 1966; Kirkham and Powers, 1972), solutions for T (z, t) can be obtained. Most notable of these solutions is the description of T (z, t) for soils exposed to H (o, t) that is a periodic function of time. When H (o, t) is a periodic function of time, the surface temperature, T (o, t), will also be periodic:

$$T(o,t) = \overline{T} + A(o) \sin \omega t \dotfill [7.2-6]$$

where $\overline{T}$ is the mean surface temperature, A(o) is the amplitude of the oscillation, and $\omega$ is the period of oscillation $= (2\pi/24)/hr$ for diurnal or $(2\pi/365)/day$ for annual cycle. This periodic function corresponds approximately to either the diurnal or annual heat waves. This solution is treated in detail by van Wijk (1965, 1966). Examples of diurnal and annual soil temperature variations are shown in Figs. 7.2-4 and 7.2-5, respectively.

The daily sum of surface heat flux density ($H_{soil}$ in equation [7.2-1] and H(o, t) in later definitions) does not reflect the potential soil temperature variation that can be produced by diurnal fluctuations of soil heat flux density. There are instances when temperature variation may be more important than mean temperature. For a given insolation, or constant $R_N$ and $H_e$, the amplitude of the heat flux density is proportional to the thermal admittance, $(\lambda C \omega)^{1/2}$. van Wijk (1965) shows that the comparative soil temperature

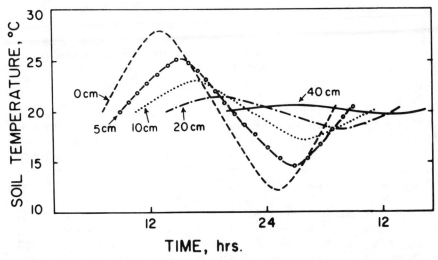

FIG. 7.2-4 Diurnal soil temperature variation with depth in a sand soil (van Wijk, 1966).

amplitude of two soils with different λ and C can be different because of different thermal admittance.

Thus, with constant $R_N$ and $H_e$, there must be a different $H_a$ if $H_{soil}$ is changed. The ration of amplitudes of $H_a$ over the soils with different thermal admittance is treated in the same way as the amplitudes of $H_{soil}$ (van Wijk, 1965) to give a first approximation of soil thermal properties effects on air temperature fluctuations in a plant canopy or within 2 meters above the soil

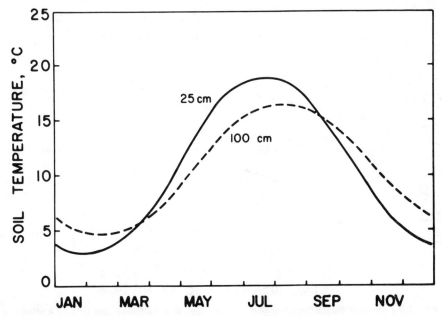

FIG. 7.2-5 Average annual soil temperature at two depths at Wageningen, The Netherlands (van Wijk, 1966).

surface. An application of this problem is the prevention of night frosts by mulch layers.

### 7.2.2.3 Thermal Diffusivity, Thermal Conductivity, and Heat Capacity.

In equation [7.2.5], thermal diffusivity, K, was defined as:

$$K = \frac{\lambda}{C}, \dots\dots\dots\dots\dots\dots\dots\dots\dots\dots\dots\dots\dots\dots\dots\dots \quad [7.2\text{-}7]$$

K has units m²/sec, and can be estimated directly from estimates of λ and C (Fig. 7.2-6). K was assumed to be constant with depth and time in equation [7.2-5]. The change in K with soil water content for three soils of different texture (Fig. 7.2-6) immediately suggests that K is not constant in soil. One of the difficulties experienced in estimating K by harmonic analyses or Fourier techniques applied to T(z, t) measured over at least one cycle of oscillation is that K is not constant with depth (Allmaras et al., 1977; Carlson, 1963; van Wijk, 1966). Consequently, numerical techniques along with best guesses of K can be used to compute heat fluxes that match measured T (z, t), given a known functional form of H (o, t). Examples are the work of Hanks et al., (1971) and Wierenga et al. (1969). Both the harmonic and cited numerical techniques assume that equation [7.2-5] will not be invalidated by air turbulence in the soil, or by generation and loss of heat, such as in vaporization and condensation. Allmaras et al. (1977) showed that

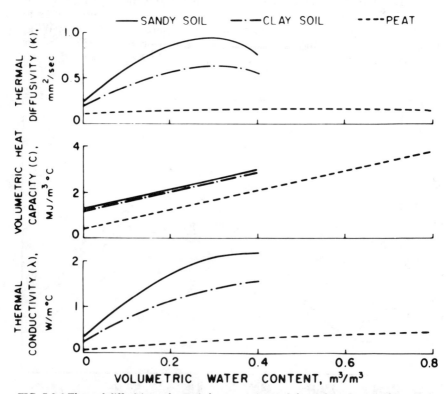

FIG. 7.2-6 Thermal diffusivity, volumetric heat capacity, and thermal conductivity for a sandy soil, clay soil, and peat at various water contents. Volume fraction of pore space was 0.4 for the sandy and clay soils, and 0.8 for peat (data from van Wijk, 1966).

turbulence affects K in soils of low density produced by tillage. The more serious problem of vapor transfer in response to thermal gradients and isothermal vapor-pressure gradients was first discussed comprehensively by Philip and de Vries (1957). Because direct measurements of vapor transfer in soils have not been obtained in the field, there is still controversy about their importance in heat flux (Kimball et al., 1976; Hadas, 1977a).

The volumetric heat capacity of a soil, C, can be estimated (van Wijk, 1966) from readily measured soil properties:

$$C = X_s\, C_s + X_w\, C_w + X_a\, C_a \quad \dots\dots\dots\dots\dots\dots\dots\dots\dots\dots\dots\dots \quad [7.2\text{-}8]$$

where $X_s$, $X_w$, and $X_a$ are the volume fractions of solid, water, and air, and the C's are the respective volumetric heat capacities. Volumetric heat capacity of a substance, C, is the product of the density and specific heat (amount of heat required to raise the temperature of one gram by 1 °C). $C_a$ is sufficiently low that the term $X_a\, C_a$ can be neglected.

The volumetric heat capacity for soil mineral matter, soil organic matter and water is about 1.9, 2.5, and 4.2 MJ/m³ °C, respectively (de Vries, 1966). Equation [7.2-8] becomes

$$C = 1.9\, x_m + 2.5\, x_o + 4.2\, x_w, \quad \dots\dots\dots\dots\dots\dots\dots\dots\dots\dots\dots \quad [7.2\text{-}9]$$

where C is the volumetric heat capacity of the soil and $x_m$, $x_o$, and $x_w$ are the volume fractions of mineral matter, organic matter, and water, respectively, in the soil. Equation [7.2-9] estimates C within 5 percent; thus, there is little research effort currently to estimate C from another independent means, such as by calorimetry. The thermal conductivity of soil is also estimated from easily measured soil constitutents (van Wijk, 1966) as follows:

$$\lambda = \sum_{i=1}^{N} k_i\, X_i\, \lambda_i \Big/ \sum_{i=1}^{N} k_i\, X_i, \quad \dots\dots\dots\dots\dots\dots\dots\dots\dots\dots\dots \quad [7.2\text{-}10]$$

where $X_i$ is the volume fraction of the $i^{th}$ soil component, $k_i$ are shape factors affecting heat flow, and $\lambda_i$ are thermal conductivities. One of the soil components is the continuous medium. Examples of computation by van Wijk (1966), Kimball et al. (1976), and Allmaras et al. (1977), showed that $\lambda$ estimates by equation [7.2-10] were within 10 percent of those values obtained by direct measurement in the field and laboratory.

Direct measurements of $\lambda$ can be made by harmonic analyses, alignment of null temperature gradient and heat flow (Kimball and Jackson, 1975), or cylindrical heat probe (de Vries and Peck, 1958). As mentioned earlier, harmonic analysis makes rigid assumptions about the applicability of equation [7.2-5]; cylindrical heat probes have a serious contact impedance when used in loose soil (Hadas, 1974), and may underestimate $\lambda$ of relatively moist soils (Sepaskhah and Boersma, 1979). The null-alignment method gives good estimates of $\lambda$ at the reference depth because it does not have the restrictive assumptions that are inherent in harmonic analyses or cylindrical heat probes.

### 7.2.2.4 Damping Depth

The depth to which heat moves from the soil surface is described by the damping depth:

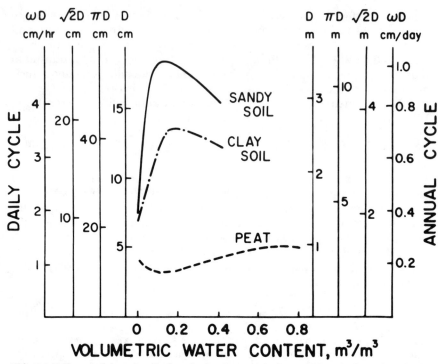

**FIG. 7.2-7 Change of damping depth and related quantities for a sandy soil, clay soil, and peat at various water contents. Left- and right-hand axes refer to daily and annual cycles, respectively, (from Monteith, 1973).**

$$D = (2 \lambda / C\omega)^{1/2}, \quad \dots\dots\dots\dots\dots\dots\dots\dots\dots\dots\dots\dots \quad [7.2\text{-}11]$$

where at some depth $z = D$, the amplitude of the temperature wave is only 0.37 times that at the surface. To describe the diurnal or annual damping depth $\omega = (2\pi/24)/\text{hr}$ or $\omega = (2\pi/365)/\text{day}$, respectively (see Fig. 7.2-7 for estimates for a sandy, clay, or peat soil).

Fig. 7.2-7 also shows related quantities of D that characterize soil heat flow based on the generalized soil temperature:

$$T(z,t) = T + A(z) \sin (\omega t - z/D), \quad \dots\dots\dots\dots\dots\dots\dots \quad [7.2\text{-}12]$$

where $A(z)$ is the amplitude of the temperature wave and $(\omega t - z/D)$ is the phase angle. At depth $z = \pi D$, the phase angle for the temperature wave is lagged exactly 1/2 of the $2\pi$ cycle, i.e., when the temperature is at a maximum at the surface, the temperature will have reached its minimum at depth $z = \pi D$ (Monteith, 1973). The quantity $\sqrt{2} D$ is an expression of the effective depth for heat flow; it is based on differentiation of equation [7.2-12] with respect to z and setting $z = 0$. Another useful multiple of D is $\omega D$ which describes the velocity of the temperature maximum through the soil.

### 7.2.2.5 Energy Balance Management—A Summary

The thermal properties of soil can be modified five ways: (a) changing reflectance, (b) changing thermal conductivity, $\lambda$, (c) altering heat capacity, C, (d) modifying sensible heat loss, and (e) changing the shape, slope, or aspect orientation of the soil surface. Reflectivity of the soil can be changed by changing the color. Surface applications of various materials range from light colored sand or chalk to plant residues to coal dust. Varying the water content also affects soil color. Changes of roughness also affect reflectivity. The thermal conductivity and heat capacity are probably most easily changed by changing the soil water content and soil porosity (or soil bulk density). Texture also affects conductivity and heat capacity, but is less easily altered under field conditions. One exception would be mixing of soil horizons of different texture by deep tillage. Modification of evaporation by mulches, tillage, etc. dictates to a large extent the amount of radiation available to heat the soil. Changing the shape of the soil surface can affect the angle of radiation incidence and water drainage. The magnitude of the above factors will be discussed in a following section.

### 7.2.3 Soil Heterogeneity

Mathematical treatment of heat flow in soils is more easily accomplished when homogeneity (over time and depth) of soil thermal properties is assumed. This is seldom the case, especially with cultivated soils which normally have a relatively loose surface layer overlying one or more layers of subsoil that may differ quite drastically in texture, bulk density, and water content. Even noncultivated soils are not vertically uniform. Virgin prairies may have a thin surface layer with a relatively high organic matter content which affects reflectivity, conductivity, and heat capacity. Nevertheless, the vertical heterogeneity of soil properties can be mathematically treated by assuming unique thermal properties for each distinct soil layer (van Wijk, 1966).

Horizontal heterogeneity may be more difficult to manage. On a practical field basis, the farm operator must often compromise and use one management system to modify a range of temperature problems in the same field. Using remote sensing techniques, Myers et al. (1970) reported soil temperature difference of several degrees within one field due to soil water differences. Thus, the magnitude of spatial variability in soil temperature may be larger than the change produced by some management system. This may be true for smaller-scaled variability also, such as wheel traffic-induced soil compaction. Voorhees (1976) and Voorhees et al. (1978) reported significant changes in soil bulk density and soil temperatures in wheel tracks. Similar differences in tillage-induced physical properties between the row and interrow may also produce horizontal variability in soil temperature (Larson, 1964; Cassel and Nelson, 1979). Allmaras and Nelson (1971) showed that these heterogenieties become diffuse as far as temperature is concerned at depths of about 0.4 m.

## 7.3 PLANT RESPONSE TO SOIL TEMPERATURE

Plant response to temperature is typified in Fig. 7.3-1 showing relative root dry weight as a function of the controlled temperature around the roots. This response curve has an optimum temperature where root dry weight is a maximum; it shows a decreasing root growth when root temperatures are either above or below the optimum. The optimum temperature range may be sufficiently wide to suggest no sensitivity to temperature.

Temperature influences processes at the cellular level such as osmotic

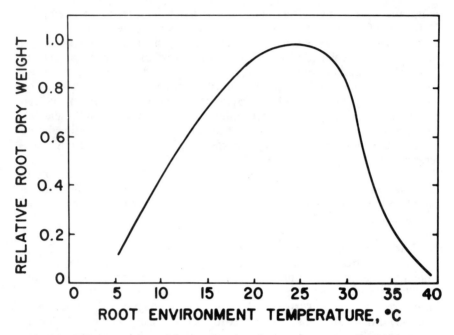

FIG. 7.3-1 Relative root dry weight as a function of temperature (generalized from data by Brouwer, 1962).

potential, hydration of ions, Gibbs free energy available for work, stomatal activity and transpiration, membrane permeability, solute solubilities, diffusion, and enzymatic activity. Each of these processes or reactions has its own velocity response to temperature, commonly designated as a $Q_{10}$ which is defined as the number of times that the process rate increases with a 10 °C rise in temperature. Mathematically, it can be expressed as:

$$Q_{10} = \left(\frac{V_2}{V_1}\right)^{\frac{10}{T_2-T_1}}, \quad \dotfill \quad [7.3\text{-}1]$$

where $V_2$ and $V_1$ are the reaction rate at temperatures $T_2$ and $T_1$, respectively; temperature $T_2$ is greater than $T_1$. $Q_{10}$ may range from 1.0, for photochemical reactions unaffected by temperature, to values as great as 6.0 for high temperature denaturation of enzymes. The coupling of reactions and processes in plants even further complicates plant response to temperature and makes response difficult to predict.

Many plant growth response curves are produced using a constant temperature input such as shown in Fig. 7.3-1. In the field, however, soil temperature is characterized by a diurnal temperature wave with amplitudes as great as ± 10 °C. If plant response does follow a single reaction as described in equation [7.3-1], the mean reaction may differ from the reaction produced by a constant temperature.

An optimum reaction curve was synthesized in Fig. 7.3-2 to simulate an observed growth curve such as in Fig. 7.3-1. This optimum reaction-curve (A-B in Fig. 7.3-2) was synthesized from a series coupling of reaction curves A and B. The $Q_{10} = 2$ for curve A typifies chemical reaction rates in enzymes

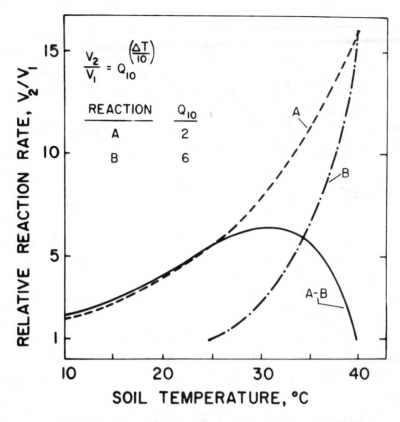

**FIG. 7.3-2** A possible optimum-type root response curve to soil temperature produced by series coupling of two hypothetical reactions, each with a different $Q_{10}$ (after Salisbury and Ross, 1969).

in response to temperature, and the $Q_{10} = 6$ for curve B characterizes high temperature denaturization (deactiviation) of an enzyme. Fig. 7.3-3 illustrates how the analysis described by Currie (1972) can be applied to Fig. 7.3-2 to compare reaction rates at a constant temperature with the mean reaction rate over an alternating temperature cycle. A harmonic cycle of temperature with amplitude of $\pm$ 10 °C and a mean of 25 °C was assumed. This temperature cycle is plotted in the lower left of Fig. 7.3-3 with time, t, progressing down the vertical axis. The $Q_{10}$ relationship for reaction A is plotted in the upper left and shows a doubling of reaction rate with a 10 °C rise in temperature. These two portions of the graph are used to obtain the relative reaction rate $V_2/V_1$, over the harmonic temperature cycle as shown in the upper right, with time progressing to the right on the horizontal axis. If the reaction rate at the mean temperature is set at 1.0, the mean relative reaction rate, $\overline{V}$, integrated over the complete cycle in the upper right, is 1.13 times greater than expected from a constant temperature. This mean relative reaction rate would be the same as that produced had the temperature been a constant 26.8 °C (obtained by projecting $\overline{V}$ back to the $\Delta T$ axis in lower left). Similarly, the harmonic cycle projected a relative reaction rate of 2.65 for the reaction of curve B; a constant temperature of 25 °C would have seriously

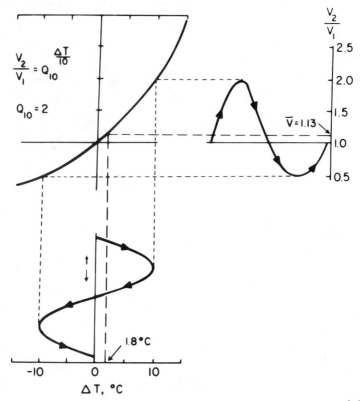

**FIG. 7.3-3 The effect of a harmonic temperature variation on a biological reaction with $Q_{10}$ of 2.**

underestimated the importance of the denaturing reaction in a diurnal field environment. If the denaturing reaction were irreversible, the diurnal cycle would have had an even greater impact. When applied to the optimum shaped curve A-B, a harmonic temperature cycle with a mean of 25 °C projected a mean relative reaction rate only 0.83 as great as that for a constant temperature of 25 °C. This mean relative reaction rate is the same as that produced if held at a constant temperature of 22.4 °C. It is possible then that, even with an optimum type response curve, information obtained under constant temperature can give 30 percent errors in predicted reaction under diurnally varying soil conditions.

Walker (1970) compared corn (*Zea mays* L.) response to constant soil temperatures with those obtained under diurnally varying soil temperatures—both under carefully controlled conditions (Walker, 1969). Growth was greater under diurnally varying (amplitude of ±3 °C) temperatures both at 23 and 26 °C mean soil temperatures. Walker (1970) concluded that the coincident light and positive temperature increment during the daytime explained the advantage of alternating temperature; when the light and negative temperature increment coincided during the daytime, a negative response occurred from alternating compared to constant temperature. Had the tests been performed at a higher average temperature, a different result might have been expected after the optimum temperature had been reached during the light period.

Comparisons of responses to different constant temperatures are a useful method to study plant response to soil temperature, but care must be used during field application of plant response information obtained under constant soil temperature. A change of reaction rate is the least that may be expected. Comparisons of field responses at different average temperatures are useful because soil temperature managements may change average temperatures as much as 3 °C with less than a 20 percent change in amplitude. As will be discussed later, the measurement of soil temperature as indicative of root environment is more difficult in the field. The difficulty relates to the horizontal nonuniformity of soil temperature produced by canopy cover, mulch concentrations, and other factors discussed later. Besides the errors discussed above with respect to diurnally alternating versus constant temperature, the horizontal heterogeneity of soil temperature complicates field application of response data obtained under constant soil temperatures (Allmaras et al., 1973). This aspect of soil temperature control will also be discussed later.

Plant response to soil temperature entails the effect on germination (and emergence) and effects after emergence. Germination response to soil temperature is generally more marked than in later growth. Furthermore, the cause-effect or dose-response relations are more easily traced to a known soil temperature regime at the site of the germinating and emerging seedling.

During germination and emergence, the root system has just begun development. This early growth and expansion of the root system depends on soil temperature environment; optimum and excessively non-optimum temperatures will be considered. Once there is a root of specific size and distribution in the soil, the activity of a root continues to depend on soil temperature. These activities are water and nutrient absorption. Because plants respond to environmental stimuli as a system, the interaction between shoot and root growth must be considered. A distinction must be made here between crops in which the shoot is harvested versus root crops. The field manager of soil temperature finds it informative to evaluate shoot or root growth (whichever portion is harvested) relationships to measured soil temperature mainly in the 5- to 20-cm depth.

Theory or simulation techniques are not adequate to project field crop responses to soil temperature changes following emergence. The interested reader may wish to verify that some of the most advanced plant growth simulations, such as the following: alfalfa (*Medicago sativa* L.) (Holt et al., 1975), general (de Wit and Brouwer, 1970), cotton (*Gossypium hirsutum* L.) (McKinion et al., 1975), have not yet incorporated soil temperature. Recent progress has been made toward incorporation of soil temperature into plant growth simulations (Lambert et al., 1976).

### 7.3.1 Germination and Emergence Response to Soil Temperature

Germination of an intact seed is completed when the radicle (embryonic root) emerges from the seed covering structure; emergence is then completed when the young shoot emerges through the soil surface. Germination is sensitive to soil temperature because imbibition of water by the seed requires enzymatic and respiratory activity that has a $Q_{10}$ indicative of chemical and biological processes. Substances must be mobilized and translocated to the growing points. Growth during the germination-to-emergence period also has a high $Q_{10}$ value, possibly because of enzymatic activity required to mobilize food reserves in the seed. Root development during this phase is critical; shoot-to-root ratios are lower during this period and the early vegetative period than at the end of the vegetative growth phase.

Richards et al. (1952) reviewed the older literature on germination and emergence responses to soil temperature. Much of this literature lists the cardinal (minimum, optimum, and maximum) soil temperatures for germination. More recent studies on some plants have distinguished between rate of germination and ultimate percentage germination. While these two responses may be frequently correlated (Gordon, 1972) for both fast (cereals) and slow germinating seeds (Sitka spruce (*Picea sitchensis*)), rate becomes increasingly important for reduction of disease and insect injury. Hegarty (1972) found that the percentage germination of carrot (*Daucus* L.) seed was not sensitive to either constant or fluctuating soil temperatures as long as the average temperature was above 10 °C or there were no severe chilling or heat stresses when the radicle emerged from the seed. Meanwhile, Hegarty (1972) found that the rate of germination and emergence (reciprocal of the time when one-half of the total germinated and emerged had occurred) was a linear function of the average soil temperature between seeding and one-half emergence. He further observed that diurnally varying soil temperatures hastened the rate of germination.

Wanjura and Buxton (1972a, 1972b) determined a germination and emergence simulation for cotton in two stages. First, they determined the time for radicle emergence as a function of soil temperature. Between 15 and 32 °C the relationship followed the Van't Hoff-Arrhenius relation; above 32 °C radicle emergence time leveled off and became insensitive to soil temperature. Various moisture contents in the seed-bed were parametrically shown in the relationship. Later Wanjura and Buxton (1972b) measured the time for the hypocotyl to elongate as a function of imposed soil temperatures, moistures, and mechanical impedances. Finally Wanjura et al. (1973) incorporated the germination and hypocotyl elongation phases into the simulation model that took into account four factors: seeding depth, soil moisture, soil impedance to hypocotyl elongation, and soil temperature.

Blacklow (1973) developed a germination and emergence simulation for corn from first following the time course of imbibition (Blacklow, 1972a). Imbibition was assumed completed by extrapolation of the imbibition curve to coincide with the end of the exponential phase. The mathematical relation is shown in differential form:

$$\frac{dw}{dt} = b + k\,(\alpha - W_o^1)\,\exp\,(-kt), \quad \dotfill [7.3\text{-}2]$$

where w is water content of the seed, t is time, and $W_o^1$ is estimated water content at beginning of imbibition. The parameters k, b, and $\alpha$ are soil temperature dependent; they relate to seed permeability during the linear and exponential phases of imbibition. Separate evaluations were made (Blacklow, 1972b) to describe radicle and coleoptile lengths as a linear function of time. Blacklow (1973) developed incremental forms of these relations to simulate germination and emergence response to varying field soil temperatures and moisture conditions.

Soybean (*Glycine max* L.) germination and emergence is soil temperature sensitive, but hypocotyl elongation is especially sensitive to soil mechanical strength at low soil temperatures (Knittle et al., 1979). Based on hypocotyl elongation and associated soil temperature, moisture, and penetrometer resistance measurements in the field on a silty clay loam, Knittle et al. (1979) developed a hypocotyl elongation rate separately for the germination and hypocotyl elongation phases. As with the carrot, cotton, or corn simulations discussed above, the equation of Knittle et al. (1979) can be incremented for simulation of soybean germination and emergence response to soil temperature.

Numerous other studies have measured germination and emergence responses to temperature and have shown strong interactions with moisture. Lindstrom et al. (1976) developed a linear relation between rate of emergence (days⁻¹) and reciprocal of the absolute soil temperature; they further developed interactions with soil water potential in the seedbed. Other studies on temperature-moisture interactions in germination are: grasses (McGinnies, 1960), grain sorghum (*Sorghum vulgare*) (Evans and Stickler, 1961; Monk, 1977).

### 7.3.2 Root Growth and Optimal Soil Temperature Response

Root branching and elongation of individual root axes are both involved in determining the final shape of a root system. Several recent reviews indicate that most plant species exhibit an optimum temperature for maximum root elongation rate (Cooper, 1973; Nielsen and Humphries, 1966). These temperatures were reported as 20 °C for sunflower (*Helianthus sp.*) (Galligar, 1938), 25 °C for loblolly pine (*Pinus taeda* L.) (Barney, 1951), 30 °C for pecan (*Carya illinoensis*) (Woodroof, 1934), tomato (*Lycopersicon esculentum* Mill.) (White, 1937), and maize (Anderson and Kemper, 1964), and 33 °C for cotton (Arndt, 1945). Root branching showed a similar response (Cooper, 1973). Thus, the optimum temperatures for maximum production of root material for several plant species ranges from 20 to 30 °C as generalized in Fig. 7.3-1. Some notable exceptions are oats (*Avena sativa* L.) at 5 °C at maturity (Nielsen et al., 1960), Kentucky bluegrass (*Poa pratensis* L.) at 15 °C (Brown, 1939), and strawberry (*Fragaria sp.*) at 10 to 20 °C (Brouwer, 1962).

Just as there are differences in root elongation and branching rate response to temperature among different genera of plants, there are also differences among species within a genus (Heinrichs and Nielsen, 1966), and among cultivars within a species (Johnson and Hartman, 1919).

Within the range from 10 to 30 °C, root diameter generally decreases and color darkness increases as temperature increases (Cooper, 1973). The color change may indicate advancing maturity or suberization of the tissue as temperature increases. Suberized cell walls contain suberins and lignins, which are fatty and waxy materials.

Most of the measured growth responses to soil temperature have been obtained in relatively short time exposures of single roots to constant temperature. In the field, roots are exposed to temperature environments that fluctuate over the entire growing season. Earlier we noted that a commonplace optimum temperature curve could be produced by coupling of two processes operating in the same enzyme system. A long term growth period in the field is likely to involve simultaneous and parallel coupled reactions, as well as sequential coupled reactions. Thus, the temperature range for optimum root elongation rate and branching in the field will likely be wider than that shown in Fig. 7.3-1. Within root-crop species the optimum temperature for root dry matter decreases with age or maturity in sugar beets (*Beta vulgaris*) (Radke and Bauer, 1969) and in turnips (*Brassica napus* L.) (Army and Miller, 1959). It is not known if the same trend exists for root elongation and branching, because as the growing season progresses rooting depth in crops like wheat (*Tritium aestivum* L.) and corn increases as new roots develop at the deeper depths.

Later in this chapter we shall show that root function, or activity, is also affected by soil temperature. Thus, the manager of soil temperature must utilize indices of response that integrate root growth rate and function (or activity).

### 7.3.3 Root Growth and Extreme Soil Temperatures

Many crops grown in temperate zones have a growing season bracketed on both ends by freezing temperatures. Tropically grown crops are often exposed to excessively high soil temperatures at least at the beginning or end of the growing season. Stress physiology may be useful to assist with identification of plant injury and to anticipate what types of injury can be produced by adverse temperatures (Levitt, 1972, 1978). A zero stress condition exists when soil temperatures are optimal. A chilling stress would then occur at soil temperatures below the optimum. An analogous situation can be produced by elevated temperatures or heat stress. As an aid to understanding the observed stress, one should consider what plant mechanism (tolerance or avoidance) is used to reduce the adverse effect. If a plant is unaffected by exposure of the living cell to stress, it has tolerance; if the plant avoids exposure of the living cell, it has an avoidance mechanism. Stress physiology perhaps is most useful for plant breeding and selection.

Soil temperatures at shallow depths often exceed 50 °C, which can kill even woody seedlings by killing a narrow strip of bark around the stem at the soil level. Julander (1945) reported definite injury to under-ground stolens of range grasses at 48 °C. Proebsting (1943) concluded that high soil temperature accounted for the reduced root growth in the surface 30 cm of California orchards. Suggested mechanisms for injury at higher-than-optimum soil temperatures are (a) that because of the higher reaction rate the mass action may be curtailed by exhaustion of reactants or accumulation of reaction products, (b) the accelerated breakdown and inactivation of metabolites, enzymes, and other growth regulating substances, and (c) the imbalance of reaction rates for various processes (Langridge, 1963). Proteins in enzymes are especially sensitive to high temperature destruction; respiration may starve a plant at high temperatures unless photosynthesis also accelerates (Christiansen, 1978).

The mechanism limiting root and plant growth by chilling soil temperature appears to be mainly decreased water absorption (Kramer, 1969; Crookston et al., 1974). Because most water uptake is supplied by passive absorption, low soil temperature reduces water absorption by increasing the viscosity of water and decreasing cell membrane permeability. Other effects of low temperature are reduced metabolic activity and decreased root growth (dry matter production), as shown in Fig. 7.3-1. The addition of fertilizer, particularly phosphorus, may compensate to some extent for the reduced growth in cold soils (Nielsen and Humphries, 1966). In general, plant species that normally grow in warm soil are affected more by chilling soil temperatures than are species that normally grow in cool soil. This is

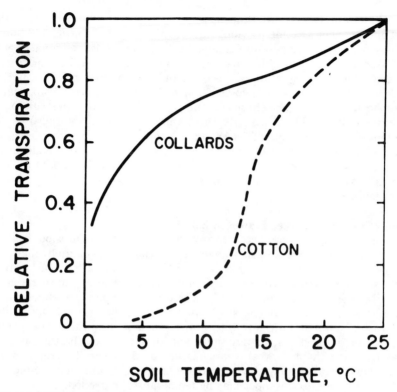

**FIG. 7.3-4 Effects of low soil temperatures on relative transpiration of cotton and collard** (redrawn from Kramer, 1942).

shown in Fig. 7.3-4 for collards (*Brassica oleracea acephala* Dc.) (a cool-season crop) and cotton (a warm-season crop). Besides cool air temperature effects on soil temperature (see section 7.2.2), cold irrigation water can also lower soil temperature (Schroeder, 1939; Raney and Mihara, 1967).

### 7.3.4 Soil Temperature Mediated Activity of Roots

Soil temperature's influence on root activity is not limited to the growth features discussed in relation to optimum and extreme temperatures. Root functions are also temperature mediated. These root functions control the supply of water and nutrients for shoot growth, and mediate the sink strength for photosynthate produced in the shoot.

### 7.3.4.1 Water Absorption

Temperature affects water uptake by several mechanisms. The viscosity of water at 25 °C is about half that at 0 °C. In addition, increased temperature increases cell membrane permeability and metabolic activity. Thus, both passive and active uptake of water is generally increased by increasing temperature. Takeshima (1964) reported twice as much water uptake by rice (*Oryza sativa* L.) at 28 °C as compared with 13 °C. Brouwer and van Vliet (1960) reported a sevenfold increase in water uptake by pea (*Pisum sativum* L.) roots when temperature was increased from 5 to 22 °C.

Water uptake at a given time is also affected by the temperature of the root environment at some previous time (Kuiper, 1964; Tagawa, 1937) and also by the rate of temperature change (Bohning and Lusanandana, 1952).

Water movement within the plant and subsequent transpiration are also affected by root temperature. Typical examples are shown in Fig. 7.3-5 for loblolly pine (Barney, 1951), coffee (*Coffea arabica*) (Franco, 1958), maize (Grobbelaar, 1963), pea (Brouwer and van Vliet, 1960), and sunflower (Whitfield, 1932). There are differences among species in the range of the optimum temperature band; for example, sunflower is apparently less sensitive than loblolly pine. One might infer that soil temperature management is more crucial for loblolly pine than for sunflower, but that would be true only if maximum transpiration were the only goal of soil temperature management.

### 7.3.4.2 Nutrient Uptake, Translocation, Assimilation

Before a nutrient can become incorporated into the plant shoot it must exist outside the root in a form available to the root, be taken up by the root, and be assimilated into some growth metabolite before or after translocation to the shoot. Soil temperature strongly influences nutrient transformations in the soil (see Stanford et al., 1973, for a resume of temperature effects on N cycles in soil); it also influences root absorption of the available nutrients, and assimilation of nutrients into growth metabolites within the plant. Photosynthetic products, mainly sugar and starches, must be translocated away from the shoot (leaves) to a sink such as the roots. Several growth-regulating substances may also be produced in the root, and their production and translocation to the shoot may be temperature controlled.

Thornley (1977) modeled the partition of dry matter between tomato roots and shoot based on two simple assumptions: (a) dry matter utilization

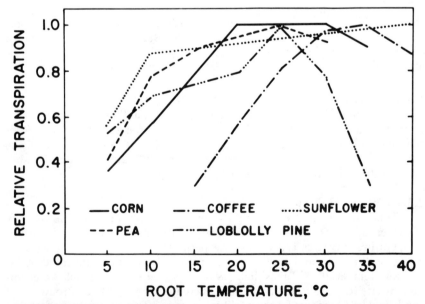

FIG. 7.3-5 Influence of root temperature on relative transpiration of several plant species (redrawn from Cooper, 1973).

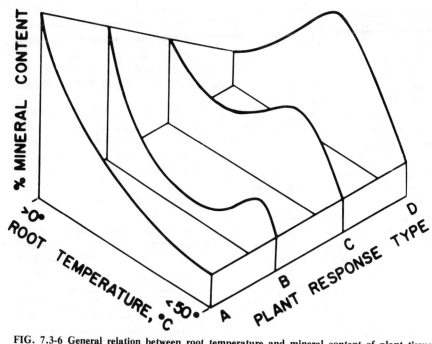

FIG. 7.3-6 General relation between root temperature and mineral content of plant tissue (redrawn from Cooper, 1973).

for growth was proportional to the product of the C and N substrate concentrations, and (b) the rate of movement between shoot and root was proportional to substrate concentration differences. The only effect of root temperature was to change the specific root uptake rate. Elaboration upon these two assumptions is discussed by Thornley (1976). The reader is referred to reviews by Street (1966), Nielsen (1974), Nielsen and Humphries (1965), Cooper (1973), and Commonwealth Agricultural Bureau (1976) for more details on soil temperature affects on nutrient uptake by specific plants.

Measurements of the nutrient content of the whole plant may provide an integration of soil temperature effects on nutrient availability, nutrient absorption, assimilation, and translocation. Cooper (1973) plotted the nitrogen, phosphrorus, and potassium content (product of concentration and dry matter production) as a function of root temperature for several plant species. From this he developed the generalized relationships shown in Fig. 7.3-6 which depict the range of root temperature effects on nutrient uptake. Response curve A shows decreasing mineral content with increasing temperature; apparently, the optimum temperature ranges for dry matter production and nutrient uptake are strongly divergent. Curve D shows an optimum temperature at which mineral content of the plant tissue is at a maximum; most species follow curve D perhaps, because both nutrient uptake and dry matter production have the same optimum temperature range. Response curves B and C incorporate features of the extreme response curves A and D.

For a given species different nutrients may follow a different mineral uptake response curve. Nitrogen uptake by corn may follow curve C, while phosphorus uptake may follow curve D. There is no apparent tendency for a nutrient to follow one curve regardless of species. Nitrogen uptake often follows curve A in cabbage (*Brassica oleracea capitata* L.), curve B in apple (*Malus* sp.), curve C in corn, and curve D in coffee. Phosphorus uptake may follow curve B in apple and curve D in corn; potassium uptake may follow curve A in snap beans (*Phaseolus vulgaris* L.) and curve D in tomato and barley (*Hordeum vulgare* L.).

Root temperature also affects absorption of foliar applied nutrients. Phillips and Bukovac (1967) concluded that increased absorption at higher soil temperatures (18 to 24 °C) was due to increased respiratory and metabolic activity in the root, causing the roots to act as a sink for phosphorus compounds. This flow of carbohydrates was substantiated by Brouwer and Levi (1969), who also showed that the history of root temperature affected leaf size at the time of foliar nutrient application.

Root morphology and concentration are partially a result of differing soil temperature in the row and in between rows (Allmaras and Nelson, 1973). The root length density (m roots/m$^3$ soil) may be especially important in the case of phosphorus and potassium, which are relatively immobile in the soil and may rely on extensive root exploration for maximum uptake.

Allmaras and Nelson (1971, 1973) concluded that the optimum temperature for corn root initiation was not the same as for root branching and elongation. Williams (1968) reported similar effects for perennial grass. The angle at which secondary roots grow out from the primary root is also temperature dependent (Mosher and Miller, 1972). The angle of corn root growth was most horizontal at 17 °C (Onderdonk and Ketcheson, 1973). Above or below this temperature, root growth was more vertical.

### 7.3.5 Interactions of Root Growth, Shoot Growth and Crop Yield

Growth has often been defined as a permanent increase in weight or volume of the whole plant. This definition is too vague to define cause and effect. It may also be a meaningless definition for the grower who is generally interested in marketing just a portion of the plant. It does, however, infer relationships between roots and shoots and temperature effects on root-growth factors affecting yield. Photosynthates, such as carbohydrates, are produced in the shoot while mineral nutrients and water are absorbed by the roots. Source-sink relationships somewhat control the partition and transport of photosynthate to the roots, and nutrients plus water to the shoots. Some control on the ultimate partition of these growth substances is exerted by plant hormones, some of which are produced in the root. The details of these source-sink responses between shoot and root are not well enough developed to be used in simulating influences produced by soil temperature control in the field (Nielsen, 1974).

### 7.3.5.1 Foliar and Grain Crops

Grobbelaar (1963) and Watts (1973) concluded that leaf area per plant was the main determinant of differences in relative growth rate of corn as influenced by root temperature. Watson (1956) showed that differences in leaf area of winter wheat, spring barley, potatoes (*Solanum tuberosum* L.) and sugar beet was primarily responsible for growth and yield differences. Evans (1972) cites studies showing a close relation between final growth and leaf area duration, which is the area under a curve of leaf area versus time. Thus, if an indicator of growth that relates soil temperature to biomass yield is sought, some leaf area measurement should be considered (leaf area per plant, leaf area per unit of soil surface area, leaf area duration). Milthorpe (1956) concluded that adverse environmental conditions in the early growth stages seriously retarded leaf development; this observation puts added emphasis on optimum soil temperature for early seedling growth. Seed germination is also known to be temperature sensitive. From a management standpoint, this is fortunate because soil temperatures are most easily modified at shallow depths where most early seedling root growth occurs. Soil temperature modification is also most effective before the plant develops a canopy to shade the soil surface.

The optimum soil temperature for plant yield is affected by water and nutrient supplies, stage of growth, plant species, rate and duration of soil temperature change (either daily or seasonally), and other environmental factors. The optimum soil temperatures for germination and root growth are generally higher and lower, respectively, than the optimum soil temperature for maximum shoot growth. Optimum soil temperature generally decreases with advancing maturity. Thus, it is difficult, if not impossible, to determine an optimum soil temperature for yield. Extensive reviews by Richards et al. (1952), Nielsen and Humphries (1965), and Willis and Amemiya (1973) provided data for optimum soil temperature for maximum yield for representative plant species (Table 7.3-1).

### 7.3.5.2 Root Crops

Root crops, such as potato and sugar beet, deserve additional attention because the underground portion, rather than some component of the shoot, is the harvestable product. Potato tubers initiated faster and in greater numbers at relatively cool soil temperatures of 10 to 15 °C, (Epstein, 1966; Yamaguchi et al., 1964). At higher soil temperature, 15 to 20 °C, fewer tubers were initiated, but they grew faster (Borah and Milthorpe, 1963). Yield was highest and sugar content was lowest at 15 to 24 °C while russeting, a desirable characteristic for some varieties, was highest at 30 °C (Yamaguchi et al., 1964). Thus, in addition to yield, other factors must be considered when managing soil temperature.

Ito and Takeda (1963) showed that although sugar beet root yield was highest at 20 °C, percent sugar was highest at 15 °C. This shift in optimum temperature occurred in the last 30 days of growth and emphasizes the importance of cooler air temperature, and, therefore, cooler soil temperature, during later stages of growth. Radke and Bauer (1969) confirmed this shift in optimum temperature.

### 7.3.6 Interaction with Other Factors of Soil Environment

Many other soil environmental factors can interact with temperature to affect root growth and even mask the temperature effects. Soil microoorganisms often have optimum temperatures similar to that for root growth and thus compete with roots for nutrients and oxygen. Harmful disease organisms may mask an expected favorable temperature response in root growth. They may also compound unfavorable temperature responses. For example, *Gibberella saubinetii* can cause seedling blight on corn at 8 to 16 °C, which is cooler than optimum for corn growth, but on wheat the organism is effective at 16 to 28 °C, warmer than optimum for wheat growth

TABLE 7.3-1. OPTIMUM SOIL TEMPERATURE FOR
MAXIMUM CROP YIELD FOR REPRESENTATIVE
PLANT SPECIES*

| Plant species | | Tempera- |
|---|---|---|
| Common name | Scientific name | ture, °C† |
| Alfalfa | Medicago sativa L. | 21-27 |
| Apple | Malus sp. | 25 |
| Barley | Hordeum vulgare L. | 18 |
| Beans | Phaseolus vulgaris | 28 |
| Brome grass | Bromus inermis Leyss. | 19-27 |
| Coffee | Coffea arabica | 20-(night)-26(day) |
| Corn | Zea mays L. | 25-30 |
| Cotton | Gossypium hirsutum L. | 28-30 |
| Gardenia | Gardenia gradiflora | 23 |
| Grape | Vitis sp. | 28 |
| Guayule | Parthenium argentatum | 28 |
| Jack Pine | Pinus banksiana | 27 |
| Kentucky bluegrass | Poa pratensis L. | 15 |
| Lucerne | Medicago sativa L. | 28 |
| Oats | Avena sativa | 15-20 |
| Onion | Allium sp. | 18-22 |
| Orange | Citrus sp. | 25 |
| Orchard grass | Dactylis glomerata L. | 20 |
| Pea | Pisum sativum L. | 18-22 |
| Potato | Solanum tuberosum L. | 20-23 |
| Rice | Oryza sativa L. | 25-30 |
| Rye grass | Lolium perenne L. | 20 |
| Snapdragon | Antirrhinum L. | 20 |
| Soybean | Glycine max L. | 22-27 |
| Soybean (innoculated) | Glycine max L. | 19 |
| Soybean (not innoculated) | Glycine max L. | 30 |
| Squash | Cucurbita sp. | 27 |
| Strawberry | Fragaria sp. | 18-24 |
| Sugar beet | Beta vulgaris | 20-24 |
| Sugar cane | Saccharum officinarum L. | 25-30 |
| Sunflower | Helianthus sp. | 23 |
| Tobacco | Nicotiana tabacum sp. | 22-26 |
| Tomatoes | Lycopersicon esculentum Mill. | 26-34 |
| Wheat | Tritium aestivam L. | 20 |

*Sources: Commonwealth Agricultural Bureau (1976), Cooper
(1973), Nielsen and Humphries (1965), Nielsen (1974), Richards
et al. (1952), and Willis and Amemiya (1973).
†A range of optimum soil temperatures may indicate lack of refinement in defining cause and effect, or may be due to more than one
data source from slightly different sets of experimental conditions.
Generally, the reported temperature values are measured early in the
growing season and at depths < 30 cm. Crop yield generally refers to
the harvested portion of the plant.

(Sewell, 1970). In general, warm-season crops, such as cotton and corn, are
more susceptible to pathogenic damage at low soil temperatures, and cool-
season crops such as peas, spinach (Spinacia sp.), and wheat, are more
susceptible at high soil temperatures (Richards et al., 1952).

The optimum soil temperature for rhizobia bacteria growth and symbiotic nitrogen fixation generally coincides with the optimum temperature
for the host legume plant. Nitrification, the process whereby soil nitrogen is
changed to a form available for plant uptake, also proceeds well at this optimum temperature. Soil compaction, on the other hand, exerts its own in-

fluence on nodule development and may mask temperature effects (Voorhees et al., 1976). Furthermore, soil compaction may alter the branching response of roots (Voorhees et al., 1975) to temperature and potentially change the uptake of nutrients. Thus, the interaction of soil temperature with other soil environment factors complicates the delineation of cause and effect under field conditions.

### 7.3.7 Guidelines for Anticipating Response to Soil Temperature

Plant response to soil temperature is sufficiently complex to rule out theoretical or simulation approaches for anticipating response to either modeled or observed soil temperature changes. A stepwise analysis is required utilizing measured responses to controlled changes in soil temperature.

Before selecting a crop species it should be recognized that soil temperature is one factor that makes corn most adapted to the Corn Belt or cotton most adapted to the Cotton Belt and not vice versa. Thus, if any new crop is to be selected, the manager should observe what plant species generally grow in the test area; if a new crop is contemplated, a careful check is in order to determine temperature optima and susceptibility to chilling, or freezing, and heat injury. Table 7.3-1 provides summary soil temperature optima. Stress physiology literature provides the best background information on the heat or chilling injury produced by short term soil temperature fluctuations (Levitt, 1972, 1978; Christiansen, 1978). Much of that literature is crop specific; the injuries can often be surmised based on known plant reactions to the chilling or heat stress.

After the desired soil temperature (optimum temperature) is determined, a daily average soil temperature (continuous, average of 1300 and 0700 hour, or average of maximum and minimum) should be measured to determine whether the average daily soil temperature is above or below the stated soil temperature optimum. A later section discusses what changes can be made.

The following measurements, in order of most likely success, can be made to assess plant growth response to an experimental temperature-management treatment: leaf area, dry matter production, nutrient uptake, plant shape. For the soil temperature manager interested in a production factor such as green biomass, nutrient uptake, or quality of a flower or fruit, reference should be made to the reviews of soil temperature and crop production cited earlier. These reviews often produce dose-response relationships without theoretical analysis suitable for simulation.

It should also be recognized that pests, such as pathogens, insects, and weeds respond to soil or air temperature, whichever is their environment. With the advent of integrated pest managements, simulations are being utilized for prediction of environmental factors, such as soil temperature in disease development (Waggoner et al., 1972; Waggoner and Horsfall, 1969).

## 7.4 MODIFICATION OF SOIL TEMPERATURE ENVIRONMENT

The producer or researcher interested in modifying the soil temperature environment must determine: (a) when soil temperature is limiting crop production, and (b) how soil temperature can be modified. Section 7.3.7 summarized guidelines for assessing plant response to soil temperature. Briefly, the optimum soil temperature depends on the plant response of interest and the stage of growth. The data in Table 7.3-1, along with some measurements of average soil temperature at about the 10-cm depth, can serve as a first evaluation of the need to modify soil temperature. Optima in Table 7.3-1 are roughly 2 to 5 °C lower than optima for seed germination. Field research has repeatedly shown a close relation between soil temperature and germination, leaf development, and early growth. Sometimes a poor relation between early growth and harvested yield may then develop. This response reversal often occurs in seed crops like corn, soybeans, and cereals. Many soil managements produce a joint response of soil temperature and water; the ultimate plant response then relates to the more severely limiting factor. Soil water influences were treated in chapter 3.

### 7.4.1 Field Criteria for Modification

Soil temperature is most easily managed by treatments applied before planting or shortly thereafter; soil temperature responses to treatments generally are immediate. Allmaras et al. (1977) showed that soil temperature changes at 50-cm depth were produced within 3 days after tillage; the test soil had an approximate damping depth of 18-cm.

Field research, over a geographic area to provide a wide variation of soil temperature, shows that changes of the average soil temperature at a 10-cm depth produce corn growth responses consistent with an optimum growth curve, such as in Fig. 7.3-1. When the average 10-cm soil temperature at planting time was below optimum in Ontario, Canada or Ames, Iowa, corn yields increased when soils were heated using heating cables (Ketcheson, 1970; Willis et al., 1957). Negative corn responses from soil cooling with a straw mulch were noted when the 10-cm soil temperature was below optimum in northern Corn Belt locations, and little or no responses from soil cooling when soil temperatures were at or near optimum in Georgia (van Wijk et al., 1959; Allmaras et al., 1964; Moody et al., 1963). Consistent with this pattern is the expectation that soil cooling will increase corn growth when average soil temperature at the 10-cm depth is above the optimum. Similar responses of wheat (Black, 1970) and soybeans led Allmaras et al. (1973) to suggest that at latitudes greater than 40 deg N, it may be desirable to increase soil temperature, at least for some crops. It is interesting to note that 40 deg N latitude corresponds closely to the isoline for frost penetration to 50-cm (USDA Yearbook, 1941), and to the isoline for optimum corn planting date of May 1 (Larson and Hanway, 1977). The latter isoline is based on soil temperature. Similar approaches have been used with soil temperature as an input to determine geographical distribution of wheat (Williams, 1969) and grasslands (Coupland, 1961).

The studies of Olson and Horton (1975) in South Dakota provide an interesting contrast to the above growth suppression with straw mulch applications at or near planting time in northern latitudes in the U.S. Mulches applied at planting time or after the corn was 30-cm tall in 1-m spaced rows reduced soil temperature, but the late applied mulch always increased corn growth, possibly because of the lower temperature optima for older plants and enhanced water supply provided by less evaporation under the mulch. Unger (1978) reported slower early sorghum growth due to reduced soil temperature with high (> 8 t/ha) mulch rates but more rapid growth later in the season because of higher soil water content.

Another factor that affects interpretation of temperature modification is the horizontal nonuniformity of soil temperature caused by plant canopy cover. Denmead et al. (1962) showed that net radiation, part of which is available for soil heating, equation [7.2-1], is spatially nonuniform in a corn field even under a LAI of 2.5 in rows spaced 1-m apart. Wind turbulence was spatially affected by soybean row geometry (Perrier et al., 1972). Soil temperatures in the row and between rows were not different at depths greater than 40-cm (Allmaras and Nelson, 1971). Thus, it is commonplace for the field environment under the plant canopy to differ from exposed adjacent areas; analagous differences may occur between the row and interrow positions. These environmental differences determine where soil temperature measurements should be made and the locus of rooting activity.

### 7.4.2 Model or Simulation Criteria

Soil temperature control can be attained by more than one method, in which case it may be desirable to estimate the change or evaluate cause and effect before making an expensive field change. There are a number of soil temperature simulations (as a function of depth and time) given the heat input or temperature at the soil surface (Wierenga et al., 1969; Wierenga and de Wit, 1970; Hanks et al., 1971).These simulations can be used to predict an average daily temperature change at the soil depth of interest; with information in Section 7.3.7, a decision can be made about modification.

A recent simulation by Cruse et al. (1980) gives guidelines for using weather station information to estimate the 5-cm depth soil temperature in a tilled soil that may or may not have received treatments to modify radiant energy input. The effects of tillage-induced soil surface roughness on radiation reflection and subsequent soil temperature is considered in detail as a subroutine of that simulation (Linden, 1979). Other parts of the simulation will be discussed in later sections.

### 7.4.3 Soil Temperature Modification

The theory and general classifications of methods, radiative and thermal, for changing soil temperature were summarized in Section 7.2.2.5. Methods of achieving these modifications follow with a discussion of expected magnitude and duration. Field studies will be cited when possible.

### 7.4.3.1 Mulch

**Plant Residues on Soil Surface:** A surface mulch is probably the most effective means for modifying soil temperature, with plant residues being the most common mulching material. The amount of residue available after harvest depends on a number of factors but will typically vary from over 7 t/ha for rice to less than 1 t/ha from cotton (Table 7.4-1). These plant residues are typically subjected to at least one tillage operation prior to planting. Depending on the tillage system, various amounts of residue may be left on or near the soil surface. This can vary from 90 percent with an operation that merely undercuts the dead plant to less than 10 percent with a moldboard plow (Table 7.4-2).

Surface placed residues alter soil temperature primarily by changing albedo, or reflection coefficient, thereby altering the net radiation balance at the soil surface. Thus, the proportion of the soil surface covered by residue becomes important. Several equations have been developed to relate residue weight to fraction of soil surface covered by residue. These include:

**TABLE 7.4-1. TYPICAL AMOUNTS OF PLANT RESIDUE AT MATURITY FOR SELECTED CROPS**

| Crop | Ratio (straw/grain)* | Grain yield† tonnes/ha | (bu/acre) | Residues tonnes/ha | (T/a) |
|------|------|------|------|------|------|
| Barley | 1.5 | 2.15 | (40) | 3.23 | (1.44) |
| Corn | 1.0 | 5.65 | (90) | 5.65 | (2.52) |
| Cotton | 1.0 | 0.58 | (520)‡ | 0.58 | (0.26) |
| Oats | 2.0 | 1.72 | (48) | 3.44 | (1.54) |
| Rice | 1.5 | 4.79 | (4,274)‡ | 7.18 | (3.21) |
| Rye | 1.5 | 1.57 | (25) | 2.35 | (1.05) |
| Sorghum | 1.0 | 3.70 | (59) | 3.70 | (1.65) |
| Soybean | 1.5 | 1.88 | (28) | 2.82 | (1.26) |
| Wheat - | | | | | |
| winter | 1.7 | 2.22 | (33) | 3.77 | (1.68) |
| spring | 1.3 | 1.88 | (28) | 2.45 | (1.09) |

*Ratios from Larson et al. (1978).
†Estimated from Agricultural Statistics (USDA, 1975).
‡Pounds per acre.

$$\ln (1-Y) = -0.157 - 0.59 \ X \quad \dots\dots\dots\dots\dots\dots\dots\dots \quad [7.4\text{-}1]$$

for wheat (Van Doren and Allmaras, 1978), where $Y$ = fraction of soil surface covered by residue and $X$ is dry residue weight (t/ha); and

$$Y = 100 \ (1\text{-}e^{-ax}) \quad \dots\dots\dots\dots\dots\dots\dots\dots\dots\dots\dots \quad [7.4\text{-}2]$$

where $Y$ = percent cover, $X$ is dry residue weight (t/ha) and $a$ has values of 0.392, 0.742, and 1.334 for corn, soybeans, and oats, respectively (L. L. Sloneker, Personal Communication, 1979. USDA-SEA-SR, Morris, MN.).

Van Doren and Allmaras (1978) developed a relationship between soil temperature reduction and fraction of soil surface covered by residue. Their semi-empirical equation was:

$$\Delta T = T_S - T_R = K_1 (R_{S1} - R_{S \ MIN}) \times [MFRAC \times (P_R - P_S)], \dots \quad [7.4\text{-}3]$$

where

| | |
|---|---|
| $T_s$ | = average temperature of bare soil |
| $T_R$ | = average temperature of residue covered soil |
| $K_1$ | = constant |
| $R_{S1}$ | = daily incoming short wave radiation with an atmospheric transmission coefficient estimated for same time span as temperature measurements |
| $R_{S \ MIN}$ | = daily incoming short wave radiation at the winter solstice when $\Delta T$ is assumed to be zero |
| MFRAC | = fraction of soil surface covered by plant residue |
| $P_R$ | = reflection coefficient of residue |
| $P_S$ | = reflection coefficient of soil. |

If radiation data is not available, $K_2$, $\overline{T}_1$ and $\overline{T}_{MIN}$ can be substituted for $K_1$, $R_{S1}$, and $R_{S \ MIN}$, respectively, where

| | |
|---|---|
| $K_2$ | = constant accounting for energy sinks other than soil heating, such as heating of air or evaporation of water |

**TABLE 7.4-2. PLANT RESIDUES REMAINING ON SOIL
SURFACE AFTER VARIOUS COMBINATIONS OF FALL
AND SPRING TILLAGE OPERATIONS\***

| Crop | Tillage method | | Percent of original plant residues left on soil surface after spring tillage operation |
| | Fall | Spring | |
|---|---|---|---|
| Oat and | None | None | 85 |
| Wheat | None | Sweep | 88 |
| | None | Rodweeder | 90 |
| | None | Heavy duty cultivator or chisel plow | 76 |
| | None | Tandem disk | 53 |
| | None | One-way disk- 7.5 cm deep | 60 |
| | None | Plow and disk | 12 |
| | None | Plow | 10 |
| | Chisel | None | 66 |
| | Sweeps | None | 51 |
| | Disk | Tandem disk | 39 |
| | Sweeps | Tandem disk | 32 |
| | Chisel | Tandem disk | 32 |
| | Plow | Tandem disk | 12 |
| Sorghum | None | 2 sweep passes | 52 |
| | None | One-way disk and sweep | 26 |
| | None | Tandem disk and sweep | 24 |
| | None | 2 tandem disk passes | 13 |
| | Sweep | 2 sweep passes | 43 |
| | One-way disk | 2 sweep passes | 36 |
| | Tandem disk | 2 sweep passes | 44 |
| | Tandem disk | Tandem disk and sweep | 9 |
| Corn | None | No-till | 84 |
| | None | Tandem disking | 45 |
| | None | Plow and tandem disk | 10 |
| | Tandem disk | Tandem disk | 41 |
| | Chisel | Tandem disk | 32 |
| | Plow | Tandem disk | 8 |
| Soybean | None | No-till | 63 |
| | None | Plow and tandem disk | 5 |
| | None | Chisel | 25 |
| | Chisel | Field cultivator | 25 |
| | Chisel | None | 20 |
| | Plow | Disk | 4 |

\*Source of data: Fenster (1977), Greb and Black (1962), Johnson
(1977), Sloneker and Moldenhauer (1977), and Unger et al. (1977).

$\overline{T}_1$ = mean monthly air temperature

$\overline{T}_{MIN}$ = lowest mean monthly air temperature when $\Delta T$ is assumed to be zero.

Temperature reductions for a soil in Iowa and a soil in Alberta, Canada, calculated from equation [7.4-3], are shown in Fig. 7.4-1 as a function of residue cover. Complete residue cover could be expected to decrease soil temperature at the 2.5-cm depth by 4 °C.

$P_R$ and $P_S$ are the most difficult parameters in equation [7.4-3] to estimate. The reflection coefficient for soil, $P_S$, depends on soil color, soil wetness, and soil roughness as discussed in Section 7.2. Van Doren and Allmaras (1978) describe some of the comparative plant residue and soil reflectances (wet or dry). After a reflection coefficient for wet or dry soil (residue) is determined, the reflection coefficient in equation [7.4-3] is fur-

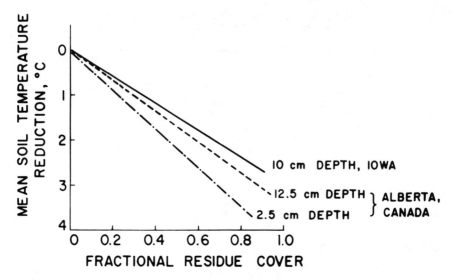

FIG. 7.4-1 Calculated temperature reduction caused by surface residue of chopped corn stalks in Iowa, and wheat straw in Canada (after Van Doren and Allmaras, 1978).

ther refined depending on the proportion of the time when wet or dry coefficients occur. Linden (1979) further refined $P_s$ to account for soil roughness influences. Cruse et al. (1980) added a multiplier to equation [7.4-3], which is the quotient of the soil thermal admittances for mulched and unmulched soils. Thermal admittance is based on heat conduction properties of the soil (Section 7.2.2.2); soil-water content increases under the mulch would affect the thermal admittance.

For every t/ha of small grain residue, soil temperatures decrease about 0.15 to 0.30 °C at the 10-cm depth (Allmaras et al., 1973). Under warmer and probably drier climates, Lal (1974) reported a decrease of up to 8 °C at a 5-cm soil depth with 2 t/ha of residue, and Smith (1936) recorded a 5 °C decrease at a depth of 30 cm with a surface straw mulch (quantity unknown). A 5-cm thick layer of straw mulch reduced excessively high daytime soil temperatures in Israel sufficiently to allow two crops a year instead of one (Neumann, 1953).

Band placement of mulch may offer a means of avoiding some undesirable temperature depressions associated with uniform mulching. Allmaras and Nelson (1971) showed reduced mean daily soil temperatures up to 2 °C at the 5-cm depth in response to 4.5 t/ha of residue. This mulch, when placed over the row, reduced corn yield. If the mulch was placed only between the row, then yields were increased because of a gain in soil water and no temperature reduction over the row. Similar results were reported by Lal (1976) under tropical climatic conditions.

The effectiveness of mulch decreases with time for several reasons. The color of plant-residues generally darkens with age, thus decreasing the reflection coefficient. Von Hoyningen-Huene (1971) determined that reflection from fresh straw was double that of a bare loamy sand; after 6 months, there was only a 6 percent difference. Hanks et al. (1961) reported an initial 15 percent decrease in net radiation on surfaces covered with fresh straw (compared to bare soil), but no differences after 3 months. Albedo data were not reported, but temperature differences at a depth of 5-cm in response to 2 t/ha of residue were slight after 50 days (Lal, 1974).

Plant residues decay with time, the rate of which depends on climatic conditions and type of residue. Within 4 months after spring application in Iowa, 50 percent of surface applied corn residue had decomposed (Parker, 1962), while in Montana only 31 percent of surface applied wheat straw decomposed after an 18-month period (Brown and Dickey, 1970). Cruse et al. (1979) assumed exponential residue decay using a coefficient controlled by decrease in the reflection coefficient.

Another factor that decreases mulch effectiveness with time is plant canopy development which reduces radiant energy at the soil surface (Adams and Arkin, 1977; Adams et al., 1976). Also, soils commonly become drier as the growing season progresses and thus have a higher reflectance which reduces the reflection coefficient differences between a bare and residue covered soil.

Surface residues can increase soil temperature during the winter mainly through their insulating properties. Gradwell (1955) reported the diurnal temperature range of a bare soil at a depth of 2.5 cm was twice that of a soil covered with or shaded by plant material when air temperatures were near freezing. When the sun is at a low angle, the reflection coefficient of surface residues increases (Allmaras et al., 1972). A higher wintertime reflection coefficient occurs on bare and residue-covered soil; thus, with lower net radiation in wintertime, residues have much less effect on average soil temperature in winter. In fact, equation [7.4-3] assumes no effect on soil temperature at the winter solstice in the northern hemisphere because $R_N$ is so small; snow cover dominates the radiant exchange because of its high reflection coefficient (Table 7.2-1).

When there are only small amounts of crop residues available, best water conservation may result from concentrating the residue to provide alternating bare soil and strips of residue. Uniform placement of mulch is best for reducing evaporation up to 7 days after a rainfall, while strip placement of mulch is more efficient for longer time periods (Bond and Willis, 1969). The mechanism could be the tendency for mulches to offer resistance to water vapor movement (Van Doren and Allmaras, 1978). Compared with wheat straw mixed uniformly in the soil, straw placement in a layer just below the soil surface or entirely on the soil surface reduced evaporation by 19 and 57 percent, respectively (Unger and Parker, 1968). Finally, mulch can increase water infiltration, thus requiring more of the radiant energy to heat the water rather than merely heating the soil.

**Other Mulches:** Plastic mulches can modify soil temperature by changing the reflection coefficient and the amount of heat used for evaporation. Clear plastic raised soil temperature by 5 to 7 °C, which was injurious to sweet potatoes (*Ipomoea batatas* Lam.) (Paterson et al., 1970) while similar soil temperature increases were beneficial for corn growth (Free and Bay, 1965). Ekern (1967) reported increased pineapple (*Ananas comosus*) growth due to a 1.6 °C increase under plastic and paper film mulch. Black plastic may not deteriorate as fast as clear plastic under field conditions, while clear plastic may cause larger temperature increases. Plastic film is generally economical only for truck and specialty crops.

A mulch of sand 13-cm thick on top of peat soil can reduce the temperature amplitude to the extent that it may be an economically feasible means of decreasing the risk of night time frost damage in nurseries or truck crops (van Wijk and Derksen, 1961).

### 7.4.3.2 Tillage

Tillage may influence soil temperatures by one or more of the following mechanisms. Surface roughness may influence the reflection of radiant energy and turbulent exchanges at the surface. Aggregation, porosity, and moisture content, (drying commonly produced by tillage) all have an influence on soil thermal properties such as thermal conductivity and heat capacity. There is often a difference in tillage (and wheel traffic) in the row area versus the interrow; seedbed shaping may also be used to facilitate planting or harvest operations. Soil temperatures in these surface configurations are affected by exposure to direct beam radiant energy. Each type of tillage often produces a characteristic surface residue placement, which especially affects the reflection of short wave radiation.

Tillage effects on soil temperature are often confounded by the amount of plant residues left on the soil surface by each tillage method (Table 7.4-2). The effects of these residues on soil temperature can be estimated using equation [7.4-3]. Griffith et al. (1973) tested eight different tillage systems in Indiana and concluded that systems that left the most residues on the soil surface resulted in the lowest soil temperature. Compared with conventional fall plowing, a no-till system reduced the soil temperature 3.8 °C at the 10-cm depth. This difference persisted for 8 weeks after planting. Plant residues on noncultivated soil in the United Kingdom insulated the soil from temperature fluctuations and resulted in less freezing compared to a plowed soil during mild winters (Hay, 1977). This noncultivated soil had lower heat sums (degree hours above 5 °C per day) than the plowed soil for 20 days after planting because of greater radiation reflectance and a higher thermal diffusivity (Hay et al., 1978).

Surface roughness, especially random roughness, differs depending on the tillage implement and soil properties. Surface roughness is difficult to measure, but fortunately three classes of random roughness magnitudes are sufficient to estimate the effect on the reflection coefficient (Linden, 1979; Cruse et al., 1980). Linden (1979) developed a special model to estimate the reflection coefficient; the input to soil temperature is then provided by a modification of $P_s$ in equation [7.4-3]. Field evidence to quantify the contribution of random roughness to soil temperature change is confounded by the usual positive correlation of random roughness with total porosity and diameter of pores in the tilled layer. Model projections (Cruse et al., 1980) suggest that a random roughness change from 0.2 to 2.5 cm may produce a 3 °C increase at 5-cm depth. Random roughness is partially a function of the sequence of previous tillage operations (Table 7.4-3). Random roughness influence on the reflection coefficient and on net radiation was the factor that caused fall-plowed soil to be 2 °C warmer at planting than soil that had not been plowed until several days before spring planting (Allmaras et al., 1972). This soil temperature difference began immediately after fall plowing and extended to a depth of 1.5 m at planting time. The random roughness influence on radiant energy balance was much greater at low sun angle of 23 degrees. Snow cover or large amounts of residue will obliterate the random roughness influence on the surface energy balance.

Tillage readily changes many soil properties within the tilled layer that can be shown to affect soil thermal properties. These properties are moisture content, total porosity, and diameters of aggregates (clods) and interaggregate (interclod) voids, as discussed earlier in Section 7.2. Depending on depth of tillage and degree of residue incorporation, the mineral and organic constituents of soil may be changed. These properties are taken into account for computation of thermal conductivity and heat capacity. These thermal properties may then be used to project soil temperature changes using models discussed earlier in Section 7.4.2. van Wijk (1966) and others (see Section 7.2) discuss simpler models involving layered soils and projected influence on amplitudes and phase shifts of soil temperature.

The diameter distribution of aggregates (or clods) and associated interaggregate porosity generally reflects tillage, while the porosity of aggregates reflects wheel traffic compaction (Voorhees et al., 1978) and long term management of organic matter (Larson and Allmaras, 1971). The average soil aggregate diameter under row crop tillage in a sub-humid climate is approximately 1 cm (Allmaras et al., 1965; Voorhees et al., 1978), depending on tillage method. Under drier, semi-arid climates, the average diameter for a range of tillage is 2 to 3 cm (Hadas et al., 1978). The air-filled porosity within aggregates increases as aggregate diameter increases over a range of aggregate water contents (Voorhees et al., 1966). Thermal conductivity of beds of dry aggregates decreased as the aggregate size increased (Hadas, 1977b), with values ranging from 0.063 to 0.252 W/m°C. Allmaras et al. (1977) reported thermal conductivity values of moist aggregates of similar internal porosity almost an order of magnitude higher. This was likely due to the higher thermal conductivity of water compared to air.

Total porosity of the soil can also be affected by tillage systems (Table 7.4-3). Secondary tillage generally, but not always, decreases plow layer porosity, while post-plant cultivation shows no consistent trends (Allmaras et

**TABLE 7.4-3. RANDOM ROUGHNESS AND FRACTIONAL TOTAL POROSITY OF TILLED LAYER AS AFFECTED BY TILLAGE SYSTEMS***

| Tillage method | | Random roughness measured after spring tillage, cm | | Total fractional porosity measured after spring tillage | |
|---|---|---|---|---|---|
| Fall | Spring | Range | Mean[†] | Range | Mean[†] |
| Plow[‡] | — | 1.96 - 7.26 | 3.24 | 0.54 - 0.96 | 0.72 |
| Tandem disk[‡] | — | 1.75 - 1.83 | 1.79 | 0.54 - 0.69 | 0.61 |
| Chisel[‡] | — | 1.35 - 1.68 | 1.52 | 0.68 - 0.69 | 0.68 |
| Plow | —[§] | 1.30 - 3.63 | 2.22 | 0.48 - 0.68 | 0.60 |
| Plow | Disk and harrow | 0.89 - 1.52 | 1.27 | 0.58 - 0.66 | 0.61 |
| None | Plow | 1.17 - 3.61 | 2.28 | 0.55 - 0.84 | 0.66 |
| None | Plow, disk and harrow | 0.69 - 2.16 | 1.15 | 0.53 - 0.82 | 0.67 |
| None | Plow and wheel tracked | 0.93 - 1.15 | 1.06 | 0.52 - 0.56 | 0.54 |
| None | Rotary till | 1.55 - 1.75 | 1.65 | 0.64 - 0.67 | 0.66 |
| None | None | 0.46 - 0.86 | 0.59 | 0.52 - 0.53 | 0.52 |

*Source of data: Allmaras et al., 1966, 1972, 1977; Voorhees, 1973; Voorhees et al., 1979. Soil textures ranged from loam to clay.
[†]Average coefficient of variation for all treatments was 6 and 4 percent for random roughness and total fractional porosity, respectively.
[‡] Measured after fall tillage, before over wintering.
[§]Measured in spring before secondary tillage operations.

TABLE 7.4-4. TILLAGE EFFECTS ON THE ENERGY BALANCE
COMPONENTS AT THE SURFACE OF A CLAY LOAM
(After Allmaras et al., 1977)

| Date and treatment | Components of energy balance,* $kW/m^2$ | | | | Percent volumetric water in 0- to 15-cm layer |
|---|---|---|---|---|---|
| | $R_N$ | $H_{soil}$ | $H_a$ | $H_e$ | |
| 23 June 1975 | | | | | |
| Plow | 370 | 24 | 128 | 218 (5.2)[†] | 23.3 |
| PDH | 335 | 45 | 108 | 182 (4.4) | 25.1 |
| PP | 347 | 46 | 111 | 190 (4.6) | 25.2 |
| Plow + M | 280 | 13 | 96 | 172 (4.1) | 28.2 |
| 30 June 1975 | | | | | |
| Plow | 329 | 54 | 159 | 116 (2.8) | 20.6 |
| PDH | 314 | 77 | 135 | 103 (2.5) | 24.3 |
| PP | 303 | 86 | 135 | 82 (2.0) | 25.0 |
| Plow + M | 272 | 42 | 133 | 97 (2.3) | 26.6 |

*Components (see equation [7.2-1] are $R_N$ = net radiation, $H_{soil}$ =
soil heat flux, $H_a$ = sensible heat to the air, and $H_e$ = evaporation.
Components are daily sums for the positive net radiation period.
[†]Parentheses values are evaporation in mm/day.

al., 1966). Voorhees (1973) reported a 10 to 15 percent decrease in plow layer porosity within the growing season just due to raindrop impact, natural settling, and drying and shrinkage.

Soil water content and initial porosity at tillage time both affect the soil structure produced by a given tillage operation (Allmaras et al., 1967; Ojeniyi and Dexter, 1979a). Generally, total porosity and random roughness are increased more by tillage when the soil water content is drier than the lower plastic limit (LPL). Ojeniyi and Dexter (1979a) found the lowest porosity when tilled at 0.9 LPL. Finer particles and aggregates sift to the bottom of the tilled layer and most aggregate breakup occurs with the first secondary tillage (Ojeniyi and Dexter, 1979b); both effects were intensified by tillage when the soil was wetter than the LPL. Thus, by timing the tillage operation to occur at a certain soil water content, porosity and roughness can be somewhat controlled. Also, when initial porosity is high, tillage-induced random roughness will not increase as much as when initial porosity is low (Allmaras et al., 1967).

Few field experiments have the relevant microclimatic, soil temperature, and soil structural data necessary for comparing tillage effects on soil temperature. One such experiment (Allmaras et al., 1977) was conducted at three location-years in west central and southwestern Minnesota on clay loam to silt loam soils. Tillage treatments (all performed in spring) were: (a) plow—moldboard plowing with no subsequent disturbance, (b) plow-disk-harrow (PDH)—moldboard plowing followed immediately by one tandem disking and one spike-tooth harrow operation, and (c) plow-pack (PP)—moldboard plowing followed by both a uniform packing of the surface with a tractor and one harrow operation. The resulting soil structures and surfaces are thus tillage induced with no confounding surface placement of residue. A fourth treatment, plow + mulch (M), is included to characterize surface residue influence for a known underlying tillage.

Table 7.4-4 typifies the effects of tillage on components of the energy balance at the soil surface. Net radiation was higher on the plow and PDH

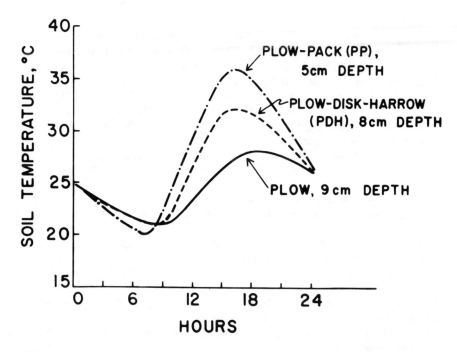

**FIG. 7.4-2** Diurnal soil temperature at indicated depth for various tillage treatments (from Allmaras et al., 1977).

treatments than on the PP treatment—because a more randomly rough surface increased radiation scatter within the rough surface layer. Among the bare surface treatments, the increased net radiation on the plow treatment was distributed to heating the air and evaporation; meanwhile the plow treatment had less soil heat flow. These distributions of net radiation are consistent with turbulent air exchanges in the plow treatment. Comparison of the clean and residue covered plow treatment shows that residues did indeed reduce net radiation and soil heat flow; evaporation was reduced by the residue only on the first day following a heavy rain.

Computed heat capacity for the PP treatment was 2.09 MJ/m³°C, while that for the other two bare treatments in Table 7.4-4 were less than 1.88; this difference in heat capacity was primarily caused by the 20 percent less fractional total porosity in the PP treatment compared to the other two treatments.

Thermal conductivity calculated from measured interaggregate void, intraaggregate void, volumetric water, and other nontillage related soil properties (see Section 7.2) varied from 0.38 to 1.05 W/m°C; thermal conductivity, measured from harmonic analyses of measured soil temperatures, was within 10 percent of those calculated except for the plow treatment, where turbulent exchanges increased thermal conductivity as much as 100 percent over that calculated, assuming only a combination of heat conduction and vapor diffusion. Calculated thermal conductivity usually ranked in the order: PP > PDH > Plow, an order reversed from that for total porosity. Measured thermal conductivity of the plow treatment was usually the greatest. Fig. 7.4-2 shows typical diurnal soil temperatures on 30 June resulting from the three tillage methods. While daily minimum temperatures were similar, the maximum for PP was about 5 and 8 °C higher than for PDH and Plow, respectively. The soil temperature on the PP treatment was measured at a depth of 5 cm, and at a depth of about 8-cm on the other two treatments, which could account for some of the difference in maximum temperature. These soil temperature differences were established within a few days after tillage was performed and lasted for at least 6 weeks. Like mulch effects, tillage effects will probably decrease with time due to drying and weathering of the surface structure and due to increasing plant canopy cover.

Shaping of the soil surface into beds or ridges affects soil temperature. At a 10-cm depth, soil temperature at planting time on a 30-cm high ridge was about 1.5 °C warmer than on level soil and about 3 °C warmer than at the bottom of a 30-cm deep furrow between ridges (Burrows, 1963). By mid-summer, the differences were slight. In drier soils, small but insignificant soil temperature increases occurred in 18-cm high ridges, which suggests that a bed or ridge configuration might be beneficial only in areas where seedbed soil is cold and wet at planting time (Adams, 1967).

Kouwenhoven (1978) reported that increasing ridge size reduced daily mean soil temperature, that increasing ridge flatness increased mean soil temperature early in the season but decreased temperature after complete canopy cover, and that soil temperature was increased with increasing size of soil structural units. All of these observations were closely related to the water regimes.

An important aspect of tillage, and one that is frequently ignored, is the accompanying effects of tractor wheel traffic on soil structure. Willis and Raney (1971) reviewed the relationship between compaction and soil temperature, and Voorhees et al. (1978) reported soil structural changes mediated by wheel traffic. Thermal conductivity and heat capacity of the soil, as well as the energy balance at the soil surface, can be expected to change in response to changes in total porosity, aggregate size and density, and soil surface roughness brought about by wheel traffic. The 5-cm depth soil temperature in the wheel track of a clay loam soil was as much as 3 °C cooler than the nontracked soil through much of the growing season (Voorhees, 1976), presumably because of the effect of increased volumetric water content on heat capacity. A wheel-tracked soil may accumulate up to 10 percent more degree hours favorable for germination and early growth than a nontracked soil (Voorhees, 1976). Wheel traffic can modify the relative surface elevation between the row and the interrow by up to 15 cm, thus creating a ridge or valley depending on relative placement of wheel traffic with respect to the rows (Fig. 7.4-3). Wheel traffic, then, is another variable that must be considered in temperature management.

Soil compacted to a density of 1200 kg/m³ emitted up to 0.84 MJ/m² more heat during the night than soil at a density of 900 kg/m³ (Gradwell, 1963). Thus, a compacted soil with a bare surface (which allows soil heat to be emitted to the air during the night) can be a significant factor in frost protection (Gradwell, 1968).

### 7.4.3.3 Irrigation and Drainage

Irrigation and drainage are generally management practices to control soil water, but they also can have a significant effect on soil temperature. Since water has a higher heat capacity and thermal conductivity than air, a

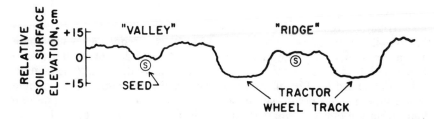

FIG. 7.4-3 Relative soil surface elevation as affected by tractor wheel traffic of row-crop planting operation (unpublished work conducted by W. B. Voorhees at Lamberton, Minnesota, 1973).

wet soil will generally be cooler than a dry soil. Much of the radiant energy impinging on a wet soil goes into heating or into evaporating the water, thus leaving the wet soil cooler than a dry soil. Draining a wet soil will generally result in increased soil temperature, which can be desirable. This is evident on a microscale with ridging or bedding-type tillage practices (Burrows, 1963; Adams, 1967) and also on a larger scale where the surface layer of a well-drained soil may be 3 to 7 °C warmer than in a comparable poorly-drained soil (Brady, 1974). Arkhipova (1955) reported maximum 10-cm depth soil temperature of 25.8 °C on dry land compared with only 15.8 °C in a bog area.

Irrigation can lower the soil temperature. In Arizona, irrigation reduced soil temperature up to 1.1 °C at the 30-cm depth. This reduction was still evident after 5 days, but not at 14 days after irrigation. Irrigating at 4- to 6-week intervals kept the soil in the top 60 cm about 5 °C cooler than nonirrigated soil, which was beneficial for citrus orchards (Smith et al., 1931). Continuous irrigation with 21 to 26 °C water during hot weather in Arizona provided enough evaporative cooling to improve lettuce (*Lactuca* sp.) stands (Wharton and Hobart, 1931).

The original source of surface irrigation water in the western U.S. is often snow melt. When the distance to the storage reservoir is relatively short, the water temperature in the reservoir can be lowered considerably. River water temperature below the Shasta Dam in California dropped from 16 to 7 °C due to the discharge of cold water from the bottom of the reservoir (Anonymous, 1959). Underground water, another major source of irrigation water, is usually within a few degrees of the annual mean air temperature (Raney and Mihara, 1967), which means it is usually colder than the soil on which it is being applied. The effects of irrigation water temperature, however, are small and of short duration (Wierenga et al., 1970, 1971). Soil temperature differences resulting from irrigating with cool (4.1 °C) or warm (21.6 °C) water lasted for only 14 hours at the 1-cm depth and 42 hours at the 20-cm depth (Wierenga et al., 1970). The main soil cooling effect of irrigation water results from evaporative cooling rather than the direct effect of adding cool water. One instance where irrigation water temperature can be important is in rice culture for which irrigation water is continuously added, thus resulting in lower temperature and reduced or delayed growth (Raney and Mihara, 1967). Partial solutions to this problem include: (a) taking warmer water from the top layer of a reservoir rather than from the cooler bottom, (b) providing temporary storage or basins where water can warm up, or (c) using broad shallow canals rather than deep narrow ones for water transport.

Irrigation frequency has an effect on soil temperature. Kohl (1973) compared daily sprinkler irrigations with a 10 to 14-day frequency on a silt loam. The mean daily soil temperature at the 10-cm depth 7 days after an irrigation was as much as 2 °C warmer than for the daily irrigated soil. This relationship can be used to more efficiently manage soil temperature. For example, to encourage rapid potato sprout growth, it may be desirable to maintain long irrigation intervals which would result in less cooling of the soil. Later, when tuber initiation and growth respond favorably to a cooler soil, frequent irrigations can be used to keep the soil temperature at a lower level.

Another soil temperature aspect of irigation involves off-season application of water. By applying water during the winter, growing season soil temperature did not exceed 26 °C in an area of the USSR, where soil temperature (15 to 20-cm depth) normally is as high as 30 °C, which inhibits potato tuber growth (Shubin, 1960). Cost of off-season irrigation may be less because of labor and equipment supply and technical aspects of applying proper amounts of water to suit the plant's demands would be lessened. Willis et al. (1977) reported that fall-applied water decreased soil temperature at the 30- to 120-cm depth as much as 2 °C as compared with the normal temperature during the next year's growing season.

### 7.4.3.4 Crop Canopy Effects

As discussed in previous sections, many soil temperature differences induced early in the season are negated by complete crop canopy cover because a considerable amount of incoming radiation may be retained in the canopy. Increased fertility resulted in increased leaf growth and canopy shading, thus lowering soil temperature (Hay et al., 1978). Soil temperatures under unshaded sod was over 7 °C warmer at midafternoon than under a pine forest canopy (Hall et al., 1958). This temperature difference can be an important factor when interseeding low growing crops between taller crops in a multiple-cropping system.

Rosenberg (1966) reported generally higher soil temperature during the day in a field of soybeans sheltered by a snow fence windbreak than in the open field. The windbreak apparently altered the advective transfer of heat to and from the soil. Windbreaks can also modify soil temperature by catching snow and increasing the soil water content (attended by higher heat capacity of the soil and more potential evaporative cooling).

### 7.4.3.5 Waste Management

Methods of modifying soil temperature that may become more common in the future involve waste materials. Applications of sewage sludge, with about 24 percent organic matter, decreased the thermal conductivity of soil but resulted in higher specific heat because of increased water content (Gupta et al., 1977). This combination resulted in the maximum soil temperature (15-cm depth) being 3 °C lower and the minimum soil temperature being 4 °C higher on plots receiving 900 t/ha sludge compared to no sludge.

Waste heat in the form of hot water from power generating plants offers a potential method of modifying soil temperatures. By circulating the waste hot water in pipes through the soil, temperatures were increased by 5 to 7 °C at the soil surface, and by 8 to 11 °C at a depth of 20 cm (Skaggs et al.,

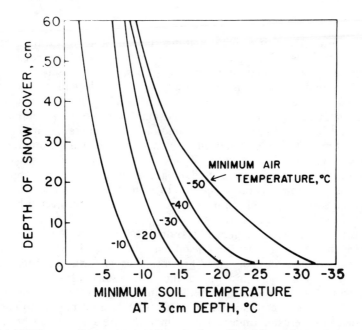

FIG. 7.4-4 Minimum soil temperature at 3-cm depth as affected by minimum air temperature and depth of snow cover (after Shul'gin, 1967).

1976). This increase can significantly hasten germination and increase early growth. Yield increases of 30 to 40 percent were predicted by using this method (Boersma et al., 1972). Such use of waste heat may be feasible for specialty crops, truck gardens, greenhouses, or nurseries.

### 7.4.3.6 Snow Cover

Snow cover can be the dominant factor controlling winter soil temperature. The depth of frost penetration influences spring soil temperature at planting time, and can also affect over-winter survival of fall-planted crops. Windbreaks of trees, standing plant residues, or other materials, can be effective means of trapping snow in the northern Great Plains (Willis, 1977).

Snow cover reduces heat conduction to the soil surface (Legget and Crawford, 1952), and serves as a sink for heat fluxes at both the upper and lower surfaces of the cover (Granger et al., 1977). The insulation effectiveness of the snow cover depends on its density and thickness. Shul'gin (1967) reported snow density to range from 40 kg/m³ for fresh fallen snow to 700 kg/m³ for consolidated snow; thus, thermal conductivity of snow is commonly an order of magnitude greater and less than for air and mineral soil, respectively. Relationships between air temperature and snow depth, such as shown in Fig. 7.4-4, can be used to predict minimum soil temperatures. Frost penetration depth can be reduced by 30 to 60 cm for each 30 cm of snow cover (Willis, 1977).

## 7.5 SUMMARY

The problem of modifying soil to alleviate a temperature stress is somewhat unique. The plant's "optimum" temperature requirement changes with stage of growth and plant species, and is closely interrelated with other factors. From a research standpoint, it is difficult to definitively measure a yield response to soil temperature. Unlike other soil parameters, such as water or nutrient content, a soil sample can not be taken from the field for subsequent determination of amount of heat contained in the soil at sampling time. The theory has been developed for measuring amount of heat in the soil with depth and time. Simulations are available to project soil temperature, but too few sets of data are available for a comprehensive evaluation of the projection adequacy. In spite of these deficiencies in soil temperature research, soil temperature changes can be projected based on changes in field management.

From a practical management standpoint, it appears that crop yield can be expected to benefit most from a soil temperature regime that encourages fast germination and good early growth. This is fortunate because soil temperature is most easily modified at the shallow soil depths (20 cm) early in the growing season; mulch management and tillage are the two most common methods for accomplishing this. If mulch is used mainly to conserve water or control erosion under cooler climates, there are methods of managing the mulch without imposing a low soil temperature stress.

Irrigation and drainage are mainly water control practices, but, at the same time, modify soil temperature. Irrigation of dry soils in hot climates may result in beneficial lowering of soil temperature, while draining wet soils in cool climates often increases soil temperature sufficiently to hasten germination and early growth. Use of plastic film mulch and waste heat from power generating plants are limited to smaller field areas of high cash value crops.

## References

1   Adams, John E. 1967. Effect of mulches and bed configuration. I. Early-season soil temperature and emergence of grain sorghum and corn. Agron. J. 59:595-599.

2   Adams, John E., and G. F. Arkin. 1977. A light interception method for measuring row crop ground cover. Soil Sci. Soc. Am. J. 41:789-792.

3   Adams, John E., G. F. Arkin, and J. T. Ritchie. 1976. Influence of row spacing and straw mulch on first stage drying. Soil Sci. Soc. Am. J. 40:436-442.

4   Allmaras, R. R., A. L. Black, and R. W. Rickman. 1973. Tillage soil environment, and root growth. Proc. Natl. Conf. on Conservation Tillage, Des Moines, Iowa. pp. 62-86.

5   Allmaras, R. R., W. C. Burrows, and W. E. Larson. 1964. Early growth of corn as affected by soil temperature. Soil Sci. Soc. Am. Proc. 28:271-275.

6   Allmaras, R. R., R. E. Burwell, and R. F. Holt. 1967. Plow-layer porosity and surface roughness from tillage as affeted by initial porosity and soil moisture at tillage time. Soil Sci. Soc. Am. Proc. 31:550-556.

7   Allmaras, R. R., R. E. Burwell, W. E. Larson, and R. F. Holt. 1966. Total porosity and random roughness of the interrow zone as influenced by tillage. USDA Conservation Research Report No. 7. 22 pp.

8   Allmaras, R. R., R. E. Burwell, W. B. Voorhees, and W. E. Larson. 1965. Aggregate size distribution in the row zone of tillage experiments. Soil Sci. Soc. Am. Proc. 29:645-650.

9   Allmaras, R. R., E. A. Hallauer, W. W. Nelson, and S. D. Evans. 1977. Surface energy balance and soil thermal property modifications by tillage-induced soil structure. Minn. Agr. Exp. Sta. Tech. Bull. 306.

10   Allmaras, R. R., and W. W. Nelson. 1971. Corn (*Zea mays* L.) root configuration as influenced by some row-interrow variants of tillage and straw mulch managements. Soil Sci.

Soc. Am. Proc. 35:974-980.

11  Allmaras, R. R., and W. W. Nelson. 1973. Corn root-configuration response to soil temperature and matric suction. Agron. J. 65:725-730.

12  Allmaras, R. R., W. W. Nelson, and E. A. Hallauer. 1972. Fall versus spring plowing and related soil heat balance in the western Corn Belt. Minn. Agr. Exp. Sta. Bull. 283.

13  Anderson, Margaret C., and O. T. Denmead. 1969. Short wave radiation on inclined surface in model plant communities. Agron. J. 61:867-872.

14  Anderson, W. B. and W. D. Kemper. 1964. Corn growth as affected by aggregate stability, soil temperature and soil moisture. Agron. J. 56:453-456.

15  Anonymous. 1959. United States Department of Interior, Bureau of Reclamation, Division of Irrigation Operation. Annual Report p. 10-17.

16  Arkhipova. 1955. Thermal regime of soil on a reclaimed bog. Trudy Glovnoi Geofizicheskoi Observatorii. 49:30-41. (Translated from Russian 1PST Cat. No. 1979).

17  Army, T. J. and E. V. Miller. 1959. Effect of lime, soil type and soil temperature on phosphorus nutrition of turnips grown on phosphorus-deficient soils. Agron. J. 51:376-378.

18  Arndt, C. H. 1945. Temperature-growth relations of the root and hypocotyls of cotton seedlings. Plant Physiol. 20:200-220.

19  Barney, C. W. 1951. Effects of soil temperature and light intensity on root growth of loblolly pine seedlings. Plant Physiol. 26:146-163.

20  Becker, C. F. and J. S. Boyd. 1961. Availability of solar energy. AGRICULTURAL ENGINEERING 42(6):302-305.

21  Black, A. L. 1970. Soil water and soil temperature influences on dryland winter wheat. Agron. J. 62:797-801.

22  Blacklow, W. M. 1972a. Mathematical description of the influence of temperature and seed quality on imbibition of seeds by corn (*Zea mays* L.). Crop Sci. 12:643-646.

23  Blacklow, W. M. 1972b. Influence of temperature on germination and elongation of the radicle and shoot of corn (*Zea mays* L.). Crop Sci. 12:647-650.

24  Blacklow, W. M. 1973. Simulation model to predict germination and emergence of corn (*Zea mays* L.) in an environment of changing temperature. Crop Sci. 13:604-608.

25  Boersma, L. E., W. R. Barlow, and K. A. Rykbost. 1972. Use of reactor cooling water from nuclear power plants for irrigation of agricultural crops. WRRI-12. Water Resources Research Institute, Oregon State University, Corvallis, Oregon.

26  Bohning, R. H. and B. Lusanandana. 1952. A comparative study of gradual and abrupt changes in root temperature on water absorption. Plant Physiol. 27:475-488.

27  Bond, J. J. and W. O. Willis. 1969. Soil water evaporation: surface residue rate and placement effects. Soil Sci. Soc. Am. Proc. 33:445-448.

28  Borah, M. N. and F. L. Milthorpe. 1963. Growth of the potato as affected by temperature. Indian J. Plant Physiol. 5:53-72.

29  Bowers, S. A. and R. J. Hanks. 1965. Reflection of radiant energy from soils. Soil Sci. 100:130-138.

30  Brady, Nyle C. 1974. The Nature and Properties of Soils. 8th ed. MacMillan Publishing Co., Inc., New York.

31  Brouwer, R. 1962. Influence of temperature of the root medium on the growth of seedlings of various crop plants. Jaarb. Inst. Biol. Scheik. Onderz. Landb Gewass. pp. 11-18.

32  Brouwer, R., and E. Levi. 1969. Response of bean plants to root temperatures. IV. Translocation of Na$^{22}$ applied to the leaves. Acta Bot. Neerl. 18:58-66.

33  Brouwer, R., and G. van Vliet. 1960. The influence of root temperature on growth and uptake of peas. Jaarb. Inst. Biol. Scheik. Onderz. Landb Gewass. pp. 23-26.

34  Brown, E. M. 1939. Some effects of temperature on the growth and chemical composition of certain pasture grasses. Mo. Agr. Exp. Sta. Res. Bull. 299.

35  Brown, K. W. 1964. The energy budget at the earth's surface. Vertical fluxes within the vegetative canopy of a corn field. 1962. Interim Rpt. 64-1. U.S. Dept. Agr., Ithaca, New York.

36  Brown, P. L. and D. T. Dickey. 1970. Loses of wheat straw residue under simulated conditions. Soil Sci. Soc. Am. Proc. 34:118-121.

37  Burrows, W. C. 1963. Characterization of soil temperature distribution from various tillage-induced microreliefs. Soil Sci. Soc. Am. Proc. 27:350-353.

38  Carlson, J. E. 1963. Analyses of soil and air temperature by Fourier techniques. J. Geophys. Res. 68:2217-2232.

39  Cassel, D. K. and L. A. Nelson. 1979. Variability of mechanical impedance in a tilled one-hectare field of Norfolk sandy loam. Soil Sci. Soc. Am. J. 43:450-455.

40  Christiansen, M. N. 1978. The physiology of plant tolerance to temperature extremes.

pp. 173-191. In, G. A. Jung (ed) Crop Tolerance to Suboptimal Land Conditions. ASA Spec. Pub. 32. Am. Soc. Agron., Madison, Wisc.

41    Commonwealth Agricultural Bureau. 1976. Plant nutrition and growth as affected by soil and root temperature. Annotated Bibliography No. SB 1831. Rothamsted Exp. Sta., Harpenden, U.K.

42    Condit, H. R. 1972. Application of characteristic vector analysis to spectral energy distribuion of daylight and spectral reflectance of American Soils. Appl. Opt. 11:74-86.

43    Cooper, A. J. 1973. Root temperature and plant growth. Research Review No. 4. Commonwealth Bureau of Horticulture and Plantation Crops. East Malling, Maidstone, Kent. 73 pp.

44    Coulson, K. L. and D. W. Reynolds. 1971. The spectral reflectance of natural surfaces. J. Appl. Met. 10:1285-1295.

45    Coupland, R. T. 1961. A reconsideration of grassland classification in the Northern Great Plains of North America. J. Ecol. 49:135-167.

46    Crookston, R. Kent, J. O'Toole, Rita Lee, J. L. Ozbun, and D. H. Wallace. 1974. Photosynthetic depression in beans after exposure to cold for one night. Crop Sci. 14:457-464.

47    Cruse, R. M., D. R. Linden, J. K. Radke, W. E. Larson, and K. Larntz. 1980. A model to predict tillage effects on soil temperature. Soil Sci. Soc. Am. J. 44:378-383.

48    Currie, J. A. 1972. The seed-soil system. pp. 463-480. In: W. Heydecker (ed) Seed ecology. Proc. 19th Easter School. Pennsylvania State Univ. Press.

49    Denmead, O. T., L. J. Fritschen, and R. H. Shaw. 1962. Spatial distribution of net radiation in a corn field. Agron. J. 54:505-510.

50    de Vries, D. A. 1966. Thermal properties of soils. pp. 210-235. In: W. R. van Wijk (ed) Physics of plant environment. John Wiley & Sons, New York.

51    de Vries, D. A. and A. J. Peck. 1958. On the cylindrical probe method of measuring thermal conductivity with special reference to soil. I. Extension of theory and discussion of probe characteristics. Aust. J. Physics. 11:255-271.

52    de Wit, C. T. and R. Brouwer. 1970. The simulation of photosynthetic systems. In prediction and measurement of photosynthetic productivity. Proc. IBP/PP Technical Meeting. Trebon. Centre Agr. Publ. Doc., Wageningen.

53    Ekern, P. C. 1967. Soil moisture and soil temperature changes with the use of black vapor-barrier mulch and their influence on pineapple growth in Hawaii. Soil Sci. Soc. Am. Proc. 31:270-275.

54    Epstein, E. 1966. Effect of soil temperature at different growth stages on growth and development of potato plants. Agron. J. 58:169-171.

55    Evans, G. C. 1972. The quantitative analysis of plant growth. Univ. California Press, Berkeley.

56    Evans, W. F. and F. C. Stickler. 1961. Grain sorghum seed germination under moisture and temperature stresses. Agron. J. 53:369-372.

57    Fenster, C. R. 1977. Conservation tillage in the Northern Plains. J. Soil Water Conserv. 32:37-42.

58    Franco, C. M. 1958. Influence of temperature on growth of coffee plant. Bull. 1BEC Res. Inst. No. 16, 21 pp.

59    Free, G. R. and Clyde Bay. 1965. Effects of plastic mulch on the growth, maturity, and yields of corn. Soil Sci. Soc. Am. Proc. 29:461-464.

60    Fuchs, M. and C. B. Tanner. 1968. Calibration and field test of soil heat flux plates. Soil Sci. Soc. Am. Proc. 32:326-328.

61    Galligar, G. C. 1938. Temperature effects upon the growth of excised root tips. Plant Physiol. 13:835-844.

62    Gates, D. M. 1965. Radiant energy, its receipt and disposal. pp. 1-26. In: Agricultural meteorology. Meteorology Monograph 6. Am. Meteorol. Soc., Boston, Mass.

63    Gausman, H. W., A. H. Gerbermann, C. L. Wiegand, R. W. Leamer, R. R. Rodriguez, and J. R. Noriega. 1975. Reflectance difference between crop residues and bare soils. Soil Sci. Soc. Am. Proc. 39:752-755.

64    Gausman, H. W., R. W. Leamer, J. R. Noriega, R. R. Rodrigues, and C. L. Wiegand. 1977. Field-measured reflectances of disked and nondisked soil with and without wheat straw. Soil Sci. Soc. Am. J. 41:793-769.

65    Gordon, A. G. 1972. The rate of germination. pp. 391-410. In: W. Heydecker (ed) Seed ecology. Pennsylvania State Univ. Press.

66    Gradwell, M. W. 1955. Soil frost studies at a high country station—II. N. Z. J. Sci. Technol. 37B:267-275.

67    Gradwell, M. W. 1963. Overnight heat losses from soil in relation to its density. N. Z.

J. Sci. 6:463-473.

68  Gradwell, M. W. 1968. The effect of grass cover on overnight heat losses from soil. N. Z. J. Sci. 11:284-300.

69  Granger, F. J., D. S. Chanasyk, D. H. Male and D. I. Norum. 1977. Thermal regime of a prairie snow cover. Soil Sci. Soc. Am. J. 41:839-842.

70  Greb, B. W. and A. L. Black. 1962. Sorghum residue reduction in a stubble mulch fallow system. Agron. J. 54:116-119.

71  Griffith, D. R., J. V. Mannering, H. M. Galloway, S. D. Parsons, and C. B. Richey. 1973. Effect of eight tillage-planting systems on soil temperature, percent stand, plant growth, and yield of corn on five Indiana soils. Agron. J. 65:321-326.

72  Grobbelaar, W. P. 1963. Responses of young maize plants to root temperatures. Meded. Landbhogesch. Wageningen. 63:1-71.

73  Gupta, S. C., R. H. Dowdy, and W. E. Larson. 1977. Hydraulic and thermal properties of a sandy soil as influenced by incorporation of sewage sludge. Soil Sci. Soc. Am. J. 41:601-605.

74  Hadas, A. 1974. Problems involved in measuring the soil thermal conductivity and diffusivity in a moist soil. Agric. Meteor. 13:105-113.

75  Hadas, A. 1977a. Evaluation of theoretically predicted thermal conductivities of soils under field and laboratory conditions. Soil Sci. Soc. Am. J. 41:460-466.

76  Hadas, A. 1977b. Heat transfer in dry aggregated soil: I. Heat conduction. Soil Sci. Soc. Am. J. 41:1055-1059.

77  Hadas, A., D. Wolf and I. Meirson. 1978. Tillage implements-soil structure relationships and their effects on crop stands. Soil Sci. Soc. Am. J. 42:632-637.

78  Hall, N., V. S. Russell, and C. D. Hamilton. 1958. Trees and microclimate. Climatology and microclimatology. Arid Zone Res. Vol. II:175-177. UNESCO.

79  Hanks, R. J., D. D. Austin, and W. T. Ondrechen. 1971. Soil temperature estimation by a numerical method. Soil Sci. Soc. Am. Proc. 35:665-667.

80  Hanks, R. J., S. A. Bowers, and L. D. Bark. 1961. Influences of soil surface conditions on net radiation, soil temperature, and evaporation. Soil Sci. 91:233-238.

81  Hay, R. K. M. 1977. Effects of tillage and direct drilling on soil temperature in winter. J. Soil Sci. 28:403-409.

82  Hay, R. K. M., J. C. Holmes, and E. A. Hunter. 1978. The effects of tillage, direct drilling and nitrogen fertilizer on soil temperature under a barley crop. J. Soil Sci. 29:174-183.

83  Hegarty, T. W. 1972. Temperature relations of germination in the field. pp. 411-432. In: W. Heydecker (ed) Seed ecology. Pennsylvania State Univ. Press.

84  Heinrichs, D. H. and K. F. Nielsen. 1966. Growth responses of alfalfa varieties of diverse genetic origin to different root zone temperatures. Can. J. Plant. Sci. 46:291-298.

85  Holt, D. A., R. J. Bula, G. E. Miles, M. W. Schreiber, and R. M. Peart. 1975. Environmental physiology, modeling, and simulation of alfalfa growth: I. Conceptual development of SIMED. Res. Bull. 907. Purdue Agricultural Experiment Station.

86  Idso, S. B. and R. D. Jackson. 1969. Thermal radiation from the atmosphere. J. Geophys. Res. 74:5397-5403.

87  Idso, S. B., R. D. Jackson, R. J. Reginato, B. A. Kimball, and F. S. Nakayama. 1975. The dependence of bare soil albedo on soil water content. J. Appl. Meteorol 14(1):109-113.

88  Ito, K., and T. Takeda. 1963. Studies on the effect of temperature on the sugar production in sugar beet. I. The effect of temperature on the growth. Crop Sci. Soc. Jap. Pro 31:272-276.

89  Johnson, J. and R. E. Hartman. 1919. Influence of soil environment on the root-rot of tobacco. J. Agr. Res. 17:41-86.

90  Johnson, W. E. 1977. Conservation tillage in western Canada. J. Soil Water Conserv. 32:61-65.

91  Julander, O. 1945. Drought resistance in range and pasture grasses. Plant Physiol. 20:573-599.

92  Ketcheson, J. W. 1970. Effects of heating and insulating soil on corn growth. Can. J. Soil Sci. 50:379-384.

93  Kimball, B. A. and R. D. Jackson. 1975. Soil heat flux determination; a null-alignment method. Agr. Meteorol. 15:1-9.

94  Kimball, B. A., R. D. Jackson, R. J. Reginato, F. S. Nakayama, and S. B. Idso. 1976. Comparison of field measured and calculated soil heat fluxes. Soil Sci. Soc. Am. J. 40:18-25.

95  Kirkham, D. and W. L. Powers. 1972. Advanced soil physics. Wiley-Interscience,

New York, NY.
    96  Knittle, K. H., J. S. Burris, and D. C. Erbach. 1979. Regression equations for rate of soybean hypocotyl elongation by using field data. Crop Sci. 19:41-46.
    97  Kohl, R. A. 1973. Sprinkler-induced soil temperature changes under plant cover. Agron. J. 65:962-964.
    98  Kouwenhoven, J. K. 1978. Ridge quality and potato growth. Neth. J. Agr. Sci. 26:288-303.
    99  Kramer, P. J. 1942. Species differences with respect to water absorption at low soil temperatures. Am. J. Bot. 29:828-832.
    100  Kramer, P. J. 1969. Plant and soil water relationships: A modern systhesis. McGraw-Hill Book Company, New York.
    101  Kuiper, P. J. C. 1964. Water uptake of higher plants as affected by root temperature. Meded. Landb Hogesch. Wageningen. 64:1-11.
    102  Lal, R. 1974. Soil temperature, soil moisture and maize yield from mulched and unmulched tropical soils. Plant Soil 40:129-143.
    103  Lal, R. 1978. Influence of within-and between-row mulching on soil temperature, soil moisture, root development and yield of maize in a tropical soil. Field Crops Research 1:127-139.
    104  Lambert, J. R., D. N Baker, C. J. Phene. 1976. Dynamic simulation of processes in soil under growing crops—RHIZOS. In: Computers applied to the management of large scale agricultural enterprises. Proc. of a US-USSR Seminar, Moscow, Rig. Kishinev.
    105  Langridge, J. 1963. Biochemical aspects of temperature response. Annual Review of Plant Physiol. 13:411-462.
    106  Larson, W. E. 1964. Soil parameters for evaluating tillage needs and operations. Soil Sci. Soc. Am. Proc. 28:118-122.
    107  Larson, W. E. and R. R. Allmaras. 1971. Management factors and natural forces as related to compaction. pp. 367-428. In: K. K. Barnes et al. (ed) Compaction of agricultural soils. ASAE, St. Joseph, MI 49085.
    108  Larson, W. E. and J. J. Hanway. 1977. Corn production. In: G. F. Sprague (ed) Corn and corn improvement. Agronomy 18:625-670. Am. Soc. Agron., Madison, Wisc.
    109  Larson, W. E., R. F. Holt, and C. W. Carlson. 1978. Residues for soil conservation. pp. 1-16. In: W. R. Oschwald (ed) Crop residue management systems. ASA Special Pub. No. 31. Am. Soc. of Agron., Madison, Wisc.
    110  Lemon, E. R., D. W. Stewart, R. W. Shawcroft, and S. E. Jensen. 1973. Experiments in predicting evapotranspiration by simulation with a soil-plant-atmosphere model (SPAM). pp. 57-76. In: R. Russell Bruce, K. W. Flack, and H. M. Taylor (eds) Field soil water regime. ASA Spec. Pub. No. 5. Am. Soc. of Agron., Madison, Wisc.
    111  Legget, R. F. and C. B. Crawford. 1952. Soil temperatures in water works practice. J. Am. Waterworks Assoc. 44:923-939.
    112  Levitt, J. 1972. Responses of plants to environmental stresses. Academic Press, New York. 697 pp.
    113  Levitt, J. 1978. Crop tolerance to suboptimal land conditions—a historical overview. pp. 161-171. In: G. A. Jung (ed) Crop tolerance to suboptimal land conditions. ASA Spec. Pub. No. 32. Am. Soc. Agron., Madison, Wisc.
    114  Linden, D. R. 1979. A model to predict soil water storage as affected by tillage practices. Ph.D, thesis University of Minnesota, St. Paul.
    115  Lindstrom, M. J., R. I. Papendick, and F. E. Koehler. 1976. A model to predict winter wheat emergence as affected by soil temperature, water potential, and depth of planting. Agron. J. 68:137-141.
    116  List, R. J. 1966. Smithsonian Meterological Tables. 6th revised edition. Smithsonian Instit. Press.
    117  McGinnies, W. J. 1960. Effects of moisture stress and temperature on six range grasses. Agron. J. 52:159-162.
    118  McKinion, J. M., D. N. Baker, J. D. Hesketh, and J. W. Jones. 1975. Part 4.—SIMCOT II: a simulation of cotton growth and yield. pp. 27-82. In: Computer simulation of a cotton production system: Users manual. ARS-S-52 USDA.
    119  Macyk, T. M., S. Pawlak, and J. D. Lindsay. 1978. Relief and microclimate as related to soil properties. Can J. Soil Sci. 58:421-438.
    120  Milthorpe, F. L. 1956. The growth of leaves. Proc. of 3rd. Easter School. Butterworths Scientific Publications, London.
    121  Monk, R. L. 1977. Characterization of grain sorghum emergence. M. S. Thesis, Texas A&M University, College Station, TX.

122   Moody, J. E., J. N. Jones, Jr., and J. H. Lillard. 1963. Influence of straw mulch on soil moisture, soil temperature, and growth of corn. Soil Sci. Soc. Am. Proc. 27:700-703.

123   Monteith, J. L. 1973. Principles of environmental physics. American Elsevier Publ. Co., Inc., New York, NY.

124   Montgomery, O. L. and M. F. Baumgardner. 1974. The effects of physical and chemical properties of soil on the spectral reflectance of soils. LARS Information Note 112674. Purdue Univ., West Lafayette, Ind.

125   Mosher, P. N. and M. H. Miller. 1972. Influence of soil temperature on the geotropic response of corn roots (*Zea mays* L.). Agron. J. 64:459-462.

126   Myers, V. K., M. D. Heilman, R. J. P. Lyon, L. N. Namken, D. Simonett, J. R. Thomas, C. L. Wiegand, and J. T. Woolley. 1970. Soil, water, and plant relations. pp. 253-297. In: Remote sensing, with special reference to agriculture and forestry. Natl. Academy of Sciences, Washington, DC.

127   Neumann, J. 1953. Some microclimatological measurements in a potato field. Israel Meteorological Service. Misc. Paper No. 6, Series C.

128   Nielsen, K. F. 1974. Roots and root temperature. pp. 293-333. In: E. W. Carson (ed) The plant root and its environment. University Press of Virginia, Charlottesville.

129   Nielsen, K. F., R. L. Halstead, A. J. MacLean, R. M. Holmes, and S. J. Bourget. 1960. The influence of soil temperature on the growth and mineral composition of oats. Can. J. Soil Sci. 40:255-264.

130   Nielsen, K. F., and E. C. Humphries. 1965. Soil temperatures and plant growth (an annotated bibliography, 1951-1964). S.C. 108—1965. Rothamsted Exp. Sta., Harpenden, U.K.

131   Nielsen, K. F. and E. C. Humphries. 1966. Effects of root temperature on plant growth. Soils Fert. 29:1-7.

132   Ojeniyi, S. O. and A. ·R. Dexter. 1979a. Soil factors affecting the macro-structures produced by tillage. TRANSACTIONS of the ASAE 22(2):339-343.

133   Ojeniyi, S. O. and A. R. Dexter. 1979b. Soil structural changes during multiple pass tillage. TRANSACTIONS of the ASAE 22(5):1068-1072.

134   Olson, T. C. and M. L. Horton. 1975. Influences of early, delayed, and no-mulch residue management on corn production. Soil Sci. Soc. Am. Proc. 39:353-356.

135   Onderdonk, J. J. and J. W. Ketcheson. 1973. Effect of soil temperature on direction of corn root growth. Plant Soil. 39:177-186.

136   Parker, D. T. 1962. Decomposition in the field of buried and surface-applied corn-stalk residue. Soil Sci. Soc. Am. Proc. 26:559-562.

137   Paterson, D. R., D. E. Speights, and J. E. Larsen. 1970. Some effects of soil moisture and various mulch treatments on the growth and metabolism of sweet potato roots. J. Am. Soc. Hort. Sci. 95:42-45.

138   Perrier, E. R., J. M. Robertson, R. J. Millington, and D. B. Peters. 1972. Spatial and temporal variation of wind above and within a soybean canopy. Agr. Meteorol. 10:421-442.

139   Philip, J. R. and D. A. de Vries. 1957. Moisture movement in porous materials under temperature gradients. Trans. Am. Geophys. Union. 38:222-232.

140   Phillips, R. L. and M. J. Bukovac. 1967. Influence of root temperature on absorption of foliar applied radiophosphorus and radiocalcium. Proc. Am. Soc. Hort. Sci. 90:555-560.

141   Proebsting, E. L. 1943. Root distribution of some deciduous fruit trees in a California orchard. Proc. Am. Soc. Hort. Sci. 43:1-4.

142   Radke, J. K. and R. E. Bauer. 1969. Growth of sugar beets as affected by root temperature. Part I. Greenhouse studies. Agron. J. 61:860-863.

143   Raney, F. C. ȧnd Yoshiaki Mihara. 1967. Water and soil temperature. In: R. M. Hagan, H. R. Haise, and T. W. Edminster (eds) Irrigation of agricultural lands. Agronomy 11:1024-1936. Am. Soc. of Agron., Madison, Wisc.

144   Richards, S. J., R. M. Hagan, and T. M. McCalla. 1952. Soil temperature and plant growth. pp. 303-480. In: B. T. Shaw (ed) Soil physical conditions and plant growth. Agron. Monograph 2. Academic Press, New York, NY.

145   Rosenberg, N. J. 1966. Microclimate, air mixing and physiological regulation of transpiration as influenced by wind shelter in an irrigated bean field. Agr. Meteorol. 3:197-224.

146   Salisbury, F. B. and C. Ross. 1969. Plant physiology. Wadsworth Publishing Co., Belmond, CA. pp. 761.

147   Schroeder, R. A. 1939. The effect of root temperature upon the absorption of water by the cucumber. Univ. Missouri Res. Bull. 309:1-27.

148   Sepaskhah, A. R. and L. Boersma. 1979. Thermal conductivity of soils as a function of temperature and water content. Soil Sci. Soc. Am. J. 43:439-444.

149    Sewell, G. W. F. 1970. The effect of altered physical condition of soil on biological control. pp. 479-494. In: K. F. Baker and W. C. Snyder (eds) Ecology of soil-borne plant pathogens. Prelude to biological control. University of California Press, Berkeley, CA.
150    Shreve, F. 1924. Soil temperature as influenced by altitude and slope exposure. Ecol. 5:128-136.
151    Shubin. 1960. Winter irrigation to control drought and soil overheating. 7th Int. Congress of Soil Sci. 1:616-621.
152    Shul'gin, A. M. 1965. The temperature regime of soils. Translated from Russian. TT-65-50083. Office of Tech. Serv., U.S. Dept. of Comm., Washignton, DC.
153    Shul'gin, A. M. 1967. Soil climate and its control. Translated from Russian. TT-72-51048. National Tech. Information Serv., U.S. Dept. of Comm., Springfield, VA.
154    Skaggs, R. W., D. C. Sanders, and C. R. Willey. 1976. Use of waste heat for soil warming in North Carolina. TRANSACTIONS of the ASAE 19:159-167.
155    Sloneker, L. L. and W. C. Moldenhauer. 1977. Measuring the amounts of crop residue remaining after tillage. J. Soil Water Conserv. 32:231-236.
156    Smith, C. D., F. Newhall, L. H. Robinson, and D. Swanson. 1964. Soil—temperature regimes—their characteristics and predictability. SCS-TP-144. USDA, Washington, DC.
157    Smith, G. E., A. F. Kinnon, and A. G. Cairns. 1931. Irrigation investigations in young grapefruit orchards on the Yuma Mesa. Arizona Agr. Exp. Sta. Tech. Bull. 37:554-589.
158    Smith, G. E. P. 1936. Control of high soil temperature. AGRICULTURAL ENGINEERING 17(9):383-385.
159    Stanford, G., M. H. Frere, and D. H. Schwaninger. 1973. Temperature coefficient of soil nitrogen mineralization. Soil Sci. 115:321-323.
160    Street, H. E. 1966. The physiology of root growth. Annual Review of Plant Physiol. 17:315-344.
161    Tagawa, T. 1937. The influence of the temperature of the culture water on the water absorption by the root and on the stomatal aperture. J. Fac. Agr. Hokkaido Imp. Univ. 39:271-296.
162    Takeshima, H. 1964. Studies on the effects of soil temperature on rice plant growth. 3. Effects of root temperature upon water and nutrient absorption at different stages and in alternating temperature. Proc. Crop Sci. Soc. Japan. 32:319-324.
163    Tanner, C. B. 1968. Evaporation of water from plants and soil. pp. 73-106. In: T. T. Kozlowski (ed) Water deficits and plant growth, Vol. 1 Development, control and measurement. Academic Press, New York, NY.
164    Thornley, J. H. M. 1976. Mathematical models in plant physiology. Academic Press, New York, NY.
165    Thornley, J. H. M. 1977. Modeling as a tool in plant physiological research. pp 339-350. In: J. J. Landsberg and C. V. Cutting (eds) Environmental effects on crop physiology. Academic Press, New York, NY.
166    United States Department of Agriculture Yearbook. 1941. Climate and man. United States Government Printing Office, Washington, DC.
167    United States Department of Agriculture. 1975. Agricultural Statistics. United States Government Printing Office, Washignton, DC.
168    Unger, P. W. 1978. Straw effects on soil temperatures and sorghum germination and growth. Agron. J. 70:858-864.
169    Unger, P. W. and J. J. Parker. 1968. Residue placement effects on decomposition, evaporation, and soil moisture distribution. Agron. J. 60:469-472.
170    Unger, P. W., Allen F. Wiese, and Ronald R. Allen. 1977. Conservation tillage in the Southern Plains. J. Soil Water Conserv. 32:43-48.
171    van Doren, D. M., Jr., and R. R. Allmaras. 1978. Effect of residue management practices on the soil physical environment, microclimate, and plant growth. pp. 49-83. In: W. R. Oschwald (ed) Crop residue management systems. ASA Spec. Pub. No. 31. Am. Soc. of Agron., Madison, Wisc.
172    van Wijk, W. R. 1965. Soil microclimate, its creation, observation, and modification. pp. 59-73. In Agr. Meteorol. Monograph 6. Am. Meteorol. Soc., Boston, Mass.
173    van Wijk, W. R. 1966. Physics of plant environment. Second edition. John Wiley & Sons, New York.
174    van Wijk, W. R. and W. J. Derksen. 1961. Surface temperature on a soil consisting of two layers. Agron. J. 53:245-246.
175    van Wijk, W. R., W. E. Larson, and W. C. Burrows. 1959. Soil temperature and the early growth of corn from mulched and unmulched soil. Soil Sci. Soc. Am. Proc. 23:428-434.
176    Von Hoyningen-Huene, S. 1971. Einfluss einer Strohdicke auf den Strahlungshausalt

des Endbodens. (The influence of straw cover on the radiation balance of the soil.) Agr. Meteorol. 9:63-75 (English summary).

177  Voorhees, W. B. 1973. Traffic patterns for controlled soil compaction. North Central Soil Conserv. Res. Center Annual Report. Morris, MN, pp. 69-82.

178  Voorhees, W. B. 1976. Plant response to wheel traffic-induced soil compaction in the Northern Corn Belt of the United States. Proc. 7th Conf. Int. Soil Tillage Res. Organ., Uppsala, Sweden. 44:1-44:6.

179  Voorhees, W. B., R. R. Allmaras, and W. E. Larson. 1966. Porosity of surface soil aggregates at various moisture contents. Soil Sci. Soc. Am. Proc. 30:163-167.

180  Voorhees, W. B., V. A. Carlson, and C. G. Senst. 1976. Soybean nodulation as affected by wheel traffic. Agron. J. 68:976-979.

181  Voorhees, W. B., D. A. Farrell, and W. E. Larson. 1975. Soil strength and aeration effects on root elongation. Soil Sci. Soc. Am. Proc. 39:948-953.

182  Voorhees, W. B., C. G. Senst, and W. W. Nelson. 1978. Compaction and soil structure modification by wheel traffic in the Northern Corn Belt. Soil Sci. Soc. Am. J. 42:344-349.

183  Voorhees, W. B., R. A. Young, and Leon Lyles. 1979. Wheel traffic considerations in erosion research. TRANSACTIONS of the ASAE 22(4):786-790.

184  Waggoner, P. E., J. G. Horsfall, and R. J. Lukens. 1972. EPIMAY. A simulator of southern corn leaf blight. Connecticut Agr. Exp. Sta. Bull. 729. 84 pp.

185  Waggoner, P. E., and J. G. Horsfall. 1969. EPIDEM. A simulator for plant disease written for a computer. Connecticut Agr. Exp. Sta. Bull. 698. 80 pp.

186  Walker, J. M. 1969. One-degree increments in soil temperatures affect maize seedling behavior. Soil Sci. Soc. Am. Proc. 33:729-736.

187  Walker, J. M. 1970. Effects of alternating versus constant soil temperatures on maize seedling growth. Soil Sci. Soc. Am. Proc. 34:889-892.

188  Wanjura, D. F. and D. R. Buxton. 1972a. Water uptake and radicle emergence of cottonseed as affected by soil moisture and temperature. Agron. J. 64:427-431.

189  Wanjura, D. F. and D. R. Buxton. 1972b. Hypocotyl and radicle elongation of cotton as affected by soil environment. Agron. J. 64:431-434.

190  Wanjura, D. F., D. R. Buxton, and H. N. Stapleton. 1973. A model for describing cotton growth during emergence. TRANSACTIONS of the ASAE 16(2):227-231.

191  Watson, D. J. 1956. Leaf growth in relation to crop yield. pp. 178-194. In: F. L. Milthorpe (ed) The growth of leaves. Proc. of 3rd Easter School. Butterworths Scientific Publications, London.

192  Watts, W. R. 1973. Soil temperature and leaf expansion in *Zea mays*. Expl. Agr. 9:1-8.

193  Webb, E. K. 1965. Aerial microclimate. pp. 27-58. In: Agr. Meteorol. Monograph 6. Am. Meteorol. Soc., Boston, Mass.

194  Wharton, M. F., and C. Hobart. 1931. Studies in lettuce seedbed irrigation under high temperature conditions. Arizona Agr. Exp. Sta. Tech. Bull. 33:283-303.

195  White, P. R. 1937. Seasonal fluctuations in growth rates of excised tomato root tips. Plant Physiol. 12:183-190.

196  Whitfield, C. J. 1932. Ecological aspects of transpiration. 2. Pikes Peak and Santa Barbara regions: edaphic and climatic aspects. Bot. Gaz. 94:183-196.

197  Wierenga, P. J., and C. T. de Wit. 1970. Simulation of heat transfer in soils. Soil Sci. Soc. Am. Proc. 34:845-848.

198  Wierenga, P. J., Robert M. Hagan, and E. J. Gregory. 1971. Effects of irrigation water temperature on soil temperature. Agron. J. 63:33-36.

199  Wierenga, P. J., R. M. Hagan, and D. R. Nielsen. 1970. Soil temperature profiles during infiltration and redistribution of cool and warm irrigation water. Water Resour. Res. 6:230-238.

200  Wierenga, P. J., D. R. Nielsen, R. M. Hagan. 1969. Thermal properties of a soil based upon field and laboratory measurements. Soil Sci. Soc. Am. Proc. 33:354-360.

201  Williams, G. D. V. 1969. Applying estimated temperature normals to the zonation of the Canadian Great Plains for wheat. Can. J. Soil Sci. 49:263-276.

202  Williams, T. E. 1968. Root activity of perennial grass swards. pp. 270-279. In: W. J. Whittington (ed) Root growth. Proc of 15th Easter School. Butterworths Scientific Publications, London.

203  Willis, W. O. 1977. Soil temperature and tillage. Research progress and needs, conservation tillage. ARS-NC-57:19-22.

204  Willis, W. O. and M. Amemiya. 1973. Tillage management principles: Soil temperature effects. Proc. of Nat. Conf. on Conservation Tillage, Des Moines, Iowa. pp. 22-41.

205  Willis, W. O., W. E. Larson, and D. Kirkham. 1957. Corn growth as affected by soil

temperature and mulch. Agron. J. 49:323-328.

206    Willis, W. O., and W. A. Raney. 1971. Effects of compaction on content and transmission of heat in soils. pp. 165-177. In: K. K. Barnes et al (ed) Compaction of agricultural soils. ASAE, St. Joseph, MI 49085.

207    Willis, W. O., P. J. Wierenga, and R. T. Vredenburg. 1977. Fall soil water: Effect on summer soil temperature. Soil Sci. Soc. Am. J. 41:615-617.

208    Woodroof, J. G. 1934. Pecan root growth and development. J. Agr. Res. 49:511-530.

209    Yamaguchi, M., H. Timm, and A. R. Spurr. 1964. Effects of soil temperature on growth and nutrition of potato plants and tuberization, composition and periderm structure of tubers. Proc. Am. Soc. Hort. Sci. 84:412-423.

# chapter 8

## ALLEVIATING CHEMICAL TOXICITIES: LIMING ACID SOILS

**8**

# 8

# ALLEVIATING CHEMICAL TOXICITIES: LIMING ACID SOILS

by    Fred Adams, Professor, Department of Agronomy and Soils, Auburn University, Auburn, AL

## 8.1 INTRODUCTION

Although liming of acid soils is an ancient agricultural practice, it was well into the twentieth century before American farmers began to recognize its merit as a practical and economical part of successful farming. Even today, farmers are believed to be using far less lime than that needed for maximum crop production.

Early use of lime was limited to small agricultural areas that were adjacent to convenient lime sources, because transportation limited the distance that lime could be transported, and spreading was accomplished by hand tools. Lime sources were limited to materials that were fine enough to react with the soil. They were the naturally fine, calcareous powders of marl and chalk or limestones that were made powdery by calcination. Advancing technology completely revolutionized the availability of agricultural lime in the early 1900's, and lime is now accessible to practically all agricultural lands in the world.

With the advent of machinery for crushing of crystalline limestones, a great deal of definitive work was done between 1920 and 1950 in determining the relative effectiveness of various sizes of limestone particles (Barber, 1967). Out of these experiments came recommendations for the most practical mix of different size separates, based on the cost of crushing, on reaction rates with soil, and on the economic return from crops. These same experiments provided the basis for the state laws and regulations that have been promulgated to control the sale of agricultural lime in much of the USA.

Methods for accessing soil acidity and identifying soils that needed lime were generally ineffective until the adoption of the pH concept in soils. Soil pH is probably the most commonly measured soil chemical property today, and it is related to many aspects of soil fertility and plant growth. In diagnosing fields with production problems, soil pH should be one of the first measurements made. Soil pH in a water slurry is the basis on which most soil test laboratories identify soils that need to be limed; soil pH in a buffered solution often serves as a measure of the amount of lime that is needed.

Wherever annual rainfall exceeds evapotranspiration, conditions are favorable for the leaching of bases and salts from soil profiles and for the accumulation of soil acidity. Some agricultural areas consist of soils that have formed from exposed, highly weathered geologic deposits that are quite acid and low in bases. Such areas occupy enormous expanses in the tropics and subtropics, where liming agricultural soils presents a major problem in soil management. In much of the temperate-zone agriculture, it is quite likely that soil acidity is accumulating in proportion to the ammoniacal N being used and to the N that is symbiotically fixed by legume crops.

In spite of the fact that several aspects of acid-soil infertility have been clearly delineated, the general relationship involving soil pH, lime rates, and crop response in the field is still largely empirical (Pearson, 1975; Pearson and Adams, 1967). However, acid soils are believed to be infertile because one or more of the following is affecting plant growth: H toxicity, Al toxicity, Mn toxicity, Ca deficiency, Mg deficiency, or Mo deficiency. Soil chemists have studied the chemistry of Al, Mn, Ca, Mg, and Mo (Coleman and Thomas, 1967); soil microbiologists have studied the functioning of rhizobia and legume nodules (Munns, 1978); mineral nutritionists have studied the effects of toxic and deficient levels of ions and the selective mechanisms by which plants are protected from extreme concentrations of certain ions (Jackson, 1967).

The process of natural selection has provided a wide variation in the genetic tolerance of plant species for the various factors of acid-soil infertility. Species that evolved on acid, base-poor soils are generally more tolerant of high Al and Mn and low Ca, Mg, and Mo than are species that evolved on neutral or alkaline soils. Nitrogen-fixing rhizobia species follow a similar selection pattern.

The enormity of the problems associated with low soil pH is attested to by the annual use of tens of millions of tons of agricultural lime. Still, agronomists and horticulturists contend that only a fraction of that needed is actually used. Although crop production is often limited by an inadequate liming program, the actual magnitude of the problem is not readily ascertainable.

## 8.2 SOIL pH AND CROP YIELD

Although soil pH is used to identify a soil that needs to be limed, it is often a highly changeable quantity because of the dynamic nature of various soil processes and the interaction of these processes with plants and microorganisms. The general relationship between soil pH and crop yield can be understood only by understanding its separate components.

### 8.2.1 Development of Soil-Profile Acidity

Wherever rainfall is sufficient to leach through the soil to the groundwater, the soil profile loses bases and tends toward increasing acidity. If a biological $H^+$-producing mechanism is also present in the soil, the acidification process will be greatly accelerated. Major examples of $H^+$ producers in soils are: (i) S or $S^{2-}$ oxidation to $SO_4^{2-}$, (ii) $NH_4^+$ oxidation to $NO_3^-$, (iii) differential ion uptake by legumes; the cations absorbed from soil by legumes greatly exceed the anions absorbed because most of the N in the plants come from N-fixing bacteria.

The amount of acidity produced by the oxidation of sulfur and sulfides is highly predictable from such equations as

$$S + \tfrac{3}{2} O_2 + H_2O = SO_4^{2-} + 2H^+ \dots\dots\dots\dots\dots\dots\dots \text{[8.2-1]}$$

and

$$2FeS_2 + 7H_2O + 7\tfrac{1}{2} O_2 = 4SO_4^{2-} + 8H^+ + 2Fe(OH)_3 \dots\dots\dots \text{[8.2-2]}$$

The acidity from N fertilizers is less predictable, but amounts can be expected to fall within well-defined limits. For example, $(NH_4)_2SO_4$ oxidizes according to the equation

$$(NH_4)_2 SO_4 + 4O_2 = 2NO_3^- + SO_4^{2-} + 4H^+ + 2H_2O \dots\dots\dots \text{[8.2-3]}$$

and the maximum acidity possible from 28 g of N is equivalent to 200 g of $CaCO_3$. If nitrification is followed by complete denitrification, according to the equation

$$2NO_3^- + 2(CH_2O) = N_2O + 2OH^- + 2CO_2 + H_2O \dots\dots\dots\dots \text{[8.2-4]}$$

then one-half of the $H^+$ ion produced by nitrification in equation [8.2-3] is neutralized by the $OH^-$ subsequently produced by denitrification (Hiltbold and Adams, 1960). Consequently, the minimum acidity possible from 28 g of N from $(NH_4)_2SO_4$ is equivalent to 100 g of $CaCO_3$.

In normal arable soils, the acidity produced by nitrification will also be reduced, i.e., partially neutralized, by the absorption action of plant roots. Since roots absorb ions independently of one another, a process called differential ion uptake, there is a tendency for ion absorption to unbalance the electroneutrality of the ambient solution. A fundamental law of chemistry, however, requires that a solution's electroneutrality be maintained, and roots are capable of doing that while absorbing an excess of cations or anions. Roots apparently have a mechanism whereby they exchange $H^+$ ions for absorbed solution cations and $HCO_3^-$ ions for absorbed solution anions. For example, if cation absorption exceeds anion absorption by 10 meq, roots will contribute 10 meq of $H^+$ to the solution. Similarly, $H^+$ ions in solution would be reduced by this amount if anion uptake exceeded cation uptake by 10 meq.

Since $NO_3^-$ can be absorbed in relatively large amounts by plants, high $NO_3^-$ contents in the soil solution can be expected to increase soil pH because of the preferential uptake of $NO_3^-$ relative to $Ca^{2+}$, $Mg^{2+}$, and $K^+$ (Pierre et al., 1970; Pierre and Banwart, 1973). Pierre (1928) applied this concept to his early experimental finding that soil acidity from N fertilizers was approximately midway between the calculated maximum and minimum values (see the above example for $(NH_4)_2SO_4$). Pierre's values for various N fertilizers are imprecise, but they have served as very useful guidelines for many years in predicting the approximate amount of lime needed to neutralize acidity from N fertilizers.

The $H^+$-producing action of N-fixing rhizobium results from the fact that legumes absorb more cations than anions from the soil when N is obtained almost entirely from $N_2$. Based on data from a pot experiment, Nyatsanga and Pierre (1973) calculated that 10 metric tons of alfalfa would produce acidity equivalent to 600 kg of $CaCO_3$. They found good agreement between developed soil acidity and the excess-base contents of legumes, i.e., $(Ca + Mg + K + Na) - (Cl + SO_4 + H_2PO_4 + NO_3)$ of the plant tissue. The greater the $N_2$ fixation, the greater the amount of acidity produced.

Except for soils that develop from highly weathered geologic materials, soil-profile acidity develops first in the surface soil, where biological activity is greatest. As surface-soil acidity intensifies, $H^+$ and $Al^{3+}$ move downward in the profile in conjunction with the leaching of $NO_3^-$, $Cl^-$, and $SO_4^{2-}$. Exchangeable bases in the subsoil are displaced by $Al^{3+}$ and subsoils become increasingly acid. The movement of $Al^{3+}$ from the surface soil to the subsoil will not be significant unless surface-soil pH is about 5.0 or less because of the relationship between pH and solution Al. The acidity that accumulates in subsoils is not readily correctable by conventional means and presents a special problem when it is too great for satisfactory root growth.

### 8.2.2 Soil pH

Soil pH is probably the most commonly measured, as well as one of the most useful, chemical soil property. It predicts the need for lime, the likelihood of excess phytotoxic ions, the activity of microorganism, and the relative availability of most inorganic nutrients. In spite of the practical implications to be derived from knowing soil pH, the value *per se* is neither a quantitative measure of $H^+$ activity in soil solution nor total titratable acidity. Its usefulness lies in the fact that empirical relationships have been established between soil pH and other soil properties.

The typical recommendation from a soil testing laboratory is to lime a soil if its pH is less than some specified value, e.g., pH 5.5. Unfortunately, soil pH is not static, even where total soil acidity might not change, and it can vary considerably within the same field at any sampling date. Most causes of soil pH variation are reasonably well known, while others are still somewhat obscure.

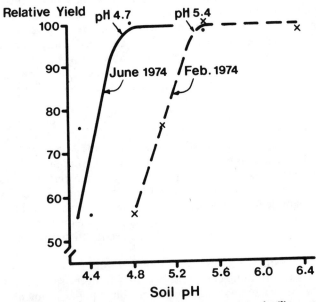

FIG. 8.2-1 Effect of sampling date, as influenced by fertilizer, on critical soil pH; Feb. sample preceded fertilizer application; June sample was at harvest time (Adams, 1978).

The most immediate and striking change in soil pH occurs from the "salt effect" that occurs with the addition of fertilizer. Fertilizer effects have been fairly well documented, as shown by the data in Fig. 8.2-1. In poorly buffered soils, such as the one in Fig. 8.2-1, soil pH will usually decrease temporarily about 0.5 to 1.0 pH unit with the addition of normal fertilizer rates. Although total soil acidity may not change with the fertilizer addition (e.g., in the absence of nitrification), soil pH definitely does because additional $Al^{3+}$ and $H^+$ ions move from exchange sites to soil solution as a consequence of exchanging with the fertilizer cations. Soil pH is primarily an expression of $H^+$ in soil solution. As fertilizer salts are removed from soil by plant uptake or by leaching, soil pH returns toward its original value. Other factors, such as the junction potential at the soil-electrode interface, also affect measured pH.

Effects of fertilizer salts on soil pH will be lessened by leaching samples prior to pH measurements (Hester and Shelton, 1933) or by measuring pH in 1 $N$ KCl solutions (Puri and Ashgar, 1938) or 0.01 $M$ $CaCl_2$ solutions (Schofield and Taylor, 1955). Although there are clear advantages for measuring soil pH by one of these methods, the advantages have apparently not outweighed the convenience of measuring soil pH in a water slurry. The majority of published soil pH's continue to be reported as that measured in a water slurry, and most soil testing laboratories recommend liming based on soil-water pH's. Furthermore, Davies (1971) showed no advantage for measuring pH in 0.01 $M$ $CaCl_2$ with several hundred British soil samples.

The observed phenomenon of seasonal variation in pH can often be explained in terms of fertilizer-salt and nitrification effects. However, year-to-year variations are somewhat more baffling. For example, Moschler et al. (1962) failed to alleviate annual variations in soil pH from the same test plots by measuring pH in salt solutions. This is not an unusual observation, and no completely satisfactory explanation has been offered as yet.

In trouble-shooting nutritional problems in the field, it is often convenient to measure soil pH while the sample is still field moist. This will usually give a pH that is somewhat higher than that measured in laboratories, where soils are air dried first. The air-drying process generally lowers soil pH by 0.2 to 0.3 unit, although some decrease much more and others show no change (Volk and Bell, 1944). Since most yield-response data are calibrated against soil pH on air-dried samples, the effect of drying on soil pH affect interpretations when pH is near a critical value.

Water-logging a soil for an extended period will increase soil pH significantly, particularly when decomposable organic matter is present. The induced oxygen deficiency causes $Fe^{3+}$-oxide to be reduced to $Fe^{2+}$-oxide with a concomitant increase in pH according to the reaction

$$Fe_2O_3 + 3H_2O + 2e^- = Fe(OH)_2 + 2OH^- \quad\dots\dots\dots\dots \text{[8.2-5]}$$

Soil pH may be increased by as much as two pH units by this reaction (Redman and Patrick, 1965). This reaction explains why liming is an ineffective amendment on flooded soils that are used for growing rice (*Oryza sativa* L.).

Even when great care is taken to obtain a soil sample that represents a particular field, the inherent inconstancy of soil pH severely strains the concept that a soil needs to be limed if its pH falls below a pre-determined value.

## 8.2.3 Critical Soil pH

In spite of the quantitative uncertainties of soil pH and problems associated with sampling, soil pH remains the most useful index available to predict the need for lime, as long as measuring techniques are adequately standardized. Probably the most useful soil pH is the minimum pH above which liming will not increase crop yield. This is conveniently called the "critical" pH. When soil pH falls below this value, liming can be expected to improve yield. Although the concept of critical pH is an oversimplification of complex soil and plant factors, it is the single, most-used criterion for predicting the need for lime. Its simplicity offsets its uncertainty, and it is highly suited to the screening of thousands of samples by soil testing laboratories. The concept of critical pH led Spurway (1941) to compile soil pH preferences for some 1700 plant species. Such cataloging must be used with great care, however, because critical pH will vary among soil types (Adams and Pearson, 1967) and among cultivars (Foy, 1976). An example of critical-pH variation among soil types is shown in Fig. 8.2-2.

Critical pH varies among soil types because of differences in acid-soil infertility factors. Acid soils are believed to be infertile because plant growth is limited by one or more of the following factors: H toxicity, Al toxicity, Mn toxicity, Ca deficiency, Mo deficiency. Soils at the same pH may have different factors limiting plant growth. Furthermore, the intensity of the same factor may differ. For example, Al toxicity has been found to differ significantly among soils at the same pH (Adams and Lund, 1966; Richburg and Adams, 1970), probably because soil-solution Al levels differ.

Critical pH varies among species and cultivars because of differences in tolerances to phytotoxic ions and differences in efficiencies with which plants absorb and utilize essential inorganic ions. Although many studies have been made on plant adaptation to various mineral stresses (Wright, 1976), the mechanisms responsible for differences in tolerance and efficiency are poorly understood.

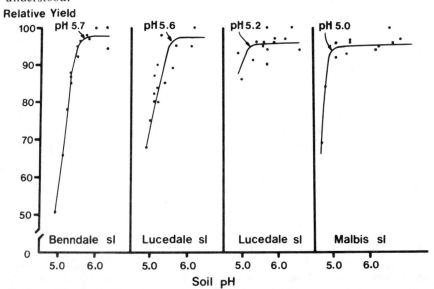

FIG. 8.2-2 "Critical" soil pH for maximum yield of soybeans (*Glycine max* L.) in field experiments on different soil types (Rogers et al., 1973).

TABLE 8.2-1. RELATIVE TOLERANCES OF
LOW SOIL pH OF SOME MAJOR
AGRONOMIC CROPS IN THE USA (Adams
and Pearson, 1967; Woodruff, 1967)

Least acid-tolerant: critical pH = 6.0-6.5

Alfalfa (*Medicago sativa* L.)
Ball clover (*Trifolium nigrescens* L.)
Red clover (*Trifolium pratense* L.)
Sugar beet (*Beta vulgaris* L.)
Sweet clover (*Melilotus indica* L.)

Medium acid-tolerant: critical pH = 5.5-6.0

Barley (*Hordeum vulgare* L.)
Birdsfoot trefoil (*Lotus corniculatus* L.)
Cotton (*Gossypium hirsutum* L.)
Crimson Clover (*Trifolium incarnatum*)
Fescue (*Festuca arundinacea* Schreb.)
Lespedeza (*Lespedeza striata* Thumb.)
Millet (*Setaria italica*)
Sorghum (*Sorghum vulgare* Pers.)
Soybeans (*Glycine max* L.)*
Wheat (*Triticum aestivum* L.)*

Most-acid-tolerant: critical pH = 5.0-5.5

Bahiagrass (*Paspalum notatum* Flugge)
Barley (*Hordeum vulgare* L.)*
Bermudagrass (*Cynodon dactylon* (L.) Pers.)
Corn (*Zea mays* L.)
Oats (*Avena sativa* L.)
Pangolagrass (*Digitaria decumbens* Stent.)
Peanuts (*Arachis hypogaea* L.)
Potato (*Solanum tuberosum* L.)
Rye (*Secale cereale* L.)
Soybeans (*Glycine max* L.)*
Vetch (*Vicia* sp.)
Wheat (*Triticum aestivum* L.)*

*Some crops are listed in two categories because
important cultivars differ significantly in their
pH tolerance.

Because of differences in soils at the same pH and differences in cultivars of the same species, a compilation of species according to critical soil pH is fraught with uncertainties. However, itemizing into broad categories can serve as a useful guide (Table 8.2-1). Alfalfa (*Medicago sativa* L.), for example, is very intolerant of low pH whereas bermudagrass (*Cynodon dactylon* Pers.) is highly tolerant. On the other hand, some cultivars of barley (*Hordeum vulgare* L.) are quite tolerant of low pH while others are not (Reid, 1976).

For critical soil pH to be interpreted properly, obtaining and analyzing soil samples must be accomplished by a standard procedure so that pH variation caused by electrolyte levels in the soil solution is minimized. Data in Fig. 8.2-1 illustrate what a drastic effect fertilizer can have on defining critical soil pH.

### 8.2.4 Lime Requirement Methods

Soil pH may alert the grower to the need for lime, but it gives no measure of the amount needed. The amount needed is called the "lime re-

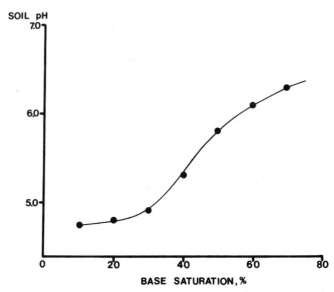

FIG. 8.2-3 Average relationship between base saturation and soil pH of 348 separate soil samples from Ultisols in Alabama (Adams and Evans, 1962).

quirement", and it is defined as the amount of agricultural lime needed to raise soil pH to the desired level for a particular crop. Lime requirement is a simple concept, but its quantitative determination is based on integrating different soil chemical measurements with the characteristic response expected for a particular crop. The major soil properties associated with lime requirement are pH, base saturation, and cation-exchange capacity (CEC).

### 8.2.4.1 Base Saturation and Soil pH

Base saturation is defined as the fraction of CEC that is satisfied by the basic cations of Ca, Mg, K, and Na, with all values expressed as meq/100 g. The remaining exchangeable cations in an acid soil are normally Al and Mn and traces of H. For a particular soil, there is a well-defined, positive correlation between base saturation and soil pH. If soils of similar characteristics are grouped together, the average relationship can be graphed, as in Fig. 8.2-3, and soil pH can be used to estimate base saturation. Since base saturation is defined in terms of CEC, it is one of the variables used to determine lime requirement.

### 8.2.4.2 Cation-Exchange Capacity and Soil pH

The CEC of a soil is not a constant entity but increases with increasing pH. The increase in CEC that occurs at higher pH values results from two basic phenomena: (a) H ions from organic ligands, such as COOH, increasingly dissociate as pH is increased, thereby creating additional negative exchange sites; (b) the positive charge of Al- and Fe-hydroxy compounds or polymers is lowered by increased pH, thereby increasing the net negative charge of the clay-mineral surface. Thus, soils high in organic matter have high pH-dependent CECs; those high in Al- and Fe-hydroxy polymers also have significant pH-dependent CECs but not nearly as great as organic soils.

Because base saturation is defined as $\Sigma$ bases/CEC, the calculated base saturation will depend upon the pH at which CEC is measured. Common extracting solutions may vary considerably in pH, and CEC may differ by several fold, depending simply upon the extracting solution used. The most commonly used methods are: (a) sum of cations displaced by a neutral salt solution, such as 1 $N$ KCl (pH can be <5 in acid soils); (b) $NH_4^+$ adsorbed from neutral, 1 $N$ $NH_4OAc$; (c) sum of cations displaced plus acidity titrated by a 0.05 $N$ $BaCl_2$-0.55 $N$ triethanolamine solution at pH 8.0; (d) $Na^+$ adsorbed from 1 $N$ NaOAc at pH 8.2.

### 8.2.4.3 Lime Rate, CEC, and Base Saturation

The amount of lime required to raise soil pH a specific amount is the soil's "buffer capacity" and is directly proportional to the CEC. However, it is inversely proportional to base saturation and is highest at low soil pH. An example of buffer capacity as a function of soil pH is given in Fig. 8.2-4. At low base saturation (and low pH), soils are highly buffered because of the hydrolytic reaction of $Al^{3+}$, after it has exchanged for $Ca^{2+}$ or $Mg^{2+}$, according to the reaction

$$Al^{3+} + 3H_2O = Al(OH)_3 + 3H^+ \quad \dots\dots\dots\dots\dots\dots\dots\dots\dots \quad [8.2\text{-}6]$$

The buffer capacity decreases sharply when all exchangeable $Al^{3+}$ has been neutralized, and it becomes a function primarily of pH-dependent exchange sites between pH $\sim$ 5.0 and pH $\sim$ 6.0. Buffer capacity again increases as soil pH increases above pH $\sim$ 6.0 because of the increasing role of $HCO_3^-$. Thus, the amount of lime required to change soil pH by one unit is least in the pH range of about 5 to 6, where exchangeable $Al^{3+}$ is nil and the $HCO_3^-$ system is insignificant.

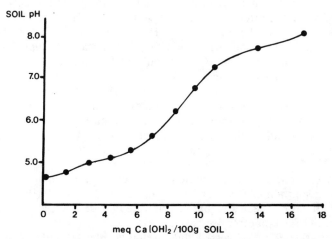

FIG. 8.2-4 Titration curve of an acid Boswell clay loam (*Vertic Paleudalfs*) (unpublished data of N. V. Hue and F. Adams, Auburn University).

#### 8.2.4.4 Laboratory. Methods

Lime requirement methods need not be highly accurate for their successful application because of the insensitiveness of most plants to a wide range in soil pH. It is only necessary that a soil be limed beyond a minimum, "critical" pH value, but short of wastefulness and a pH value that might induce a nutrient deficiency. Lime is most efficient in raising soil pH up to pH $\sim$ 6.0, but it becomes progressively less effective as pH increases above this value. Consequently, large amounts of lime are required to raise soil pH from 6.0 to 7.0 or more, a practice that should be discouraged generally.

Laboratory precision in measuring lime requirements is often wasted because of problems related to pH adjustment on a field basis. Major field-application problems are (a) natural soil variation, (b) calibration of lime spreader, (c) lime-spreading pattern, (d) lime quality, and (e) degree of mixing between lime and soil. Because of these field-related problems, there is little advantage in using a highly accurate laboratory method for determining lime requirements for advisory purposes.

Probably, the most accurate method is the $Ca(OH)_2$ titration method (Abruna and Vicente, 1955), as illustrated by data in Fig. 8.2-4. However, this method is not suitable for soil testing laboratories that handle large numbers of samples and quick results are needed. It is more suited to the researcher, where accuracy is more important than time or volume of samples.

Soil testing laboratories have found buffered solutions to be the quickest and easiest methods for estimating lime requirements. Because the chemistry behind buffered-solution methods is quite complex (McLean, 1978), the methods are based on empirical relationships.

The SMP method (Shoemaker et al., 1961) is based on the pH response of selected soils to $CaCO_3$ rates and the amount of soil $H^+$ titrated by a buffered solution at pH 7.50. The solution is a mixture of p-nitrophenol, triethanolamine, $K_2CrO_4$, $CaCl_2$, and $Ca(OAc)_2$. Its application is limited to soils with similar acidic properties and CECs.

The Adams-Evans method (Adams and Evans, 1962) is based on the empirical "soil pH"-"base saturation" relationship of similar soils and the amount of soil $H^+$ titrated by a buffered solution at pH 8.00. The buffered solution is composed of p-nitrophenol, $H_3BO_3$, and KCl. The method was designed primarily for soils with low CEC, but is readily adapted to other soils where the "soil pH"-"base saturation" relationship has been established over the necessary pH range.

The "double buffer" method (Yuan, 1974, 1976; McLean et al., 1978) is based on the difference in soil $H^+$ measured by two buffered solutions at different pHs. The two measurements estimate the buffer capacity, hence the lime required to adjust soil pH to a particular value. Yuan's buffered solution is a mixture of tris(hydroxymethyl)-aminomethane, imidazol, pyridine, $K_2CrO_4$, and $CaCl_2$.

An older, but less used method, is based on the amount of soil $H^+$ titrated by a solution of $BaCl_2$ and triethanolamine at pH 8.0 (Peech, 1965). This method has not been adequately calibrated for field-application use and is more suited for laboratory-research use.

Kamprath (1970) suggested that exchangeable Al could serve as the single criterion for determining lime requirements for the more weathered soils. The method has merit, but it does not appear to be well-suited for use in soil testing laboratories.

All lime-requirement methods assume that lime is ⸗ specified soil volume of average bulk density. They furth about 50 to 70 percent of the field-applied lime will ac soil. Thus, the recommended rate of lime will assume ⸗ will be mixed to a certain soil depth, e.g., 20 cm, and that abou⸗ ⸗⸗ the lime will not affect soil pH.

## 8.3 ALUMINUM IN ACID SOILS

The role of Al in causing infertility of acid soils was recognized before 1920 (Miyake, 1916) and received considerable attention until the early 1930s. After 1930, interest in soil Al almost disappeared from the literature, and writers began to assume that exchangeable acidity was $H^+$ rather than $Al^{3+}$. Interest in exchangeable Al was rekindled in the late 1940s, and many of the early concepts of soil acidity and exchangeable Al were soon republished (Jenny, 1961). Soluble Al is now considered to be the most universal factor of acid-soil infertility.

### 8.3.1 Aluminum Phytotoxicity

The pH-dependent solubility of Al-containing soil minerals is such that phytotoxic levels of soil-solution Al can be expected in most mineral soils when soil pH is <5.0 to 5.5. Solution Al need be no more than a fraction of a ppm for phytotoxic responses to be exhibited by sensitive species.

Symptoms of Al toxicity are remarkably similar for most plants. They are most pronounced on the root system, where all roots tend to be shortened and swollen. The entire root system has a characteristically stubby appearance because lateral-root growth is inhibited to a greater extent than primary-root growth (Foy, 1974). With increasing severity of Al toxicity, roots become sparser, shorter, and more swollen, and fine branching vanishes.

Moderate levels of Al toxicity show no characteristic foliage symptoms and tend to go undetected. Under severe Al stress, however, plant tops are stunted and are often indistinguishable from P-deficiency symptoms. There is considerable experimental evidence that Al interferes with the normal metabolism of P by plants (Clarkson, 1969). Although Al phytotoxicity is not readily recognizable in the field by foliage symptoms, it can be readily diagnosed by the appearance of damaged root systems.

Plants differ markedly in their tolerance levels to solution Al (Foy, 1974, 1976). Undoubtedly, selection pressures under both natural and agricultural ecosystems have produced an array of plants that are quite different in their responses to solution Al. Even cultivars of the same species can differ widely in their tolerance to Al (Devine, 1976; Reid, 1976). In general, plants that grow best in soil at pH 6 or more are least tolerant of Al while those that grow equally well at about pH 5 or less are highly tolerant of Al.

In determining the Al concentration in solution required for phytotoxic symptoms, it is necessary to know the Al concentration actually present in solution. This normally requires that the intended Al concentration not exceed the solubility of amorphous $Al(OH)_3$ or $AlPO_4 \cdot 2H_2O$; Al concentrations are not readily definable if solutions are supersaturated with hydroxy Al polymers. Such supersaturated solutions can be prepared in laboratories (Bartlett and Riego, 1972a) but are not likely to exist for long in most soluns. Since soil solutions are expected to obey the solubility-product princi-ple, culture-solution techniques are invalid unless they recognize the same principle.

Threshold phytotoxic levels of Al vary from trace amounts to more than 1 ppm, depending upon the sensitiveness of the plant (Andrew et al., 1973). The presence of the cations Mg (Dios and Broyer, 1962) and Ca (Lund, 1970) may modify the severity of Al phytotoxicity. Merely increasing electrolyte concentration and ionic strength will reduce both Al phytotoxicity and solution-Al activity. Adding a fertilizer salt (except P) to an acid soil, however, will increase both solution Al and Al phytotoxicity. This is because the cation-exchange reactions between exchangeable $Al^{3+}$ and the fertilizer cations increase solution Al to a greater extent than the fertilizer ions tend to detoxify Al. In spite of the quantitative uncertainties, it seems quite likely that Al phytotoxicity is more a function of solution Al activity than of Al con-centration (Adams and Lund, 1966). The ameliorating effect of adding citrate to a phytotoxic Al solution can be explained by the fact that Al activity is greatly reduced by chelation (Bartlett and Reigo, 1972b).

## 8.3.2 Solubility of Al-Containing Soil Minerals

Soil-solution Al activity is determined by the solubility of the particular Al compounds in quasi-equilibrium with the solution. The most likely Al compounds are hydroxides (or hydrated oxides), silicates, phosphates, and hydroxy-sulfates. Thus, soil-solution Al activity will be determined by the solution activities of $OH^-$, $H_4SiO_4$, $H_2PO_4^-$, and $SO_4^{2-}$. Only under fertilized ecosystems will $H_2PO_4^-$ and $SO_4^{2-}$ normally be expected to significantly in-fluence solution Al. A major exception would be the high sulfate, acid soils formed from reclaimed lands.

Total Al concentration in soil solution may exceed Al activity by several fold because of the powerful effect that ionic strength has on the activity coef-ficient of $Al^{3+}$ and because of the propensity of $Al^{3+}$ to form ion-pairs with $OH^-$, $H_2PO_4^-$, and $SO_4^{2-}$. Solution Al consists of several ionic species, such as $AlOH^{2+}$, $Al(OH)_2^+$, $Al(OH)_3^0$, $AlSO_4^+$, $AlH_2PO_4^{2+}$. It is not known if all species would be equally phytotoxic because no work has been reported in this area.

### 8.3.2.1 Hydrated Al Oxides

Considerable soil research has been based on the premise that soil-solution Al in acid soil is controlled by the solubility of $Al(OH)_3$ (Richburg and Adams, 1970; Marion et al., 1976), which implies that $Al^{3+}$ activity is a function only of pH. The dissolution of $Al(OH)_3$ can be written as

$$Al(OH)_3 \rightleftharpoons Al^{3+} + 3OH^- \quad \dots \dots \dots \dots \dots \dots \dots \dots \dots [8.3-1]$$

and its solubility product as

$$K = (Al^{3+})(OH^-)^3 \quad \dots \dots \dots \dots \dots \dots \dots \dots \dots \dots \dots [8.3-2]$$

The solubility product of $Al(OH)_3$ varies from a pK of 32.3 for the amor-phous hydroxide to 34.0 for crystalline gibbsite, while soil-solution and soil-extract $pAl(OH)_3$ values have generally ranged between 32.0 and 35.0, with pK being inversely proportional to pH. However, it is not unlikely that much of the discrepancy in apparent $pAl(OH)_3$ values for soils has been the result of faulty distribution of solution Al into its various ionic components (Marion et al., 1976).

### 8.3.2.2 Aluminosilicates

In addition to pH, hydrated solution $SiO_2$ can be a major controller of $Al^{3+}$ activity. Quartz is the most stable and least soluble of the $SiO_2$ forms, but its dissolution and precipitation rates are too slow to be of consequence in affecting soil-solution Al (Kittrick, 1969). Most soil solutions will have silica levels in excess of that provided by the solubility of quartz, which is 0.18 m$M$. Amorphous silica has a solubility of 1.9 to 2.5 m$M$ and probably controls soil-solution $H_4SiO_4$ concentration in most soils.

If dilute, acid solutions of Al and silica are allowed to react at 25 °C, an amorphous precipitate having the composition of halloysite forms. The solubility of the precipitate can be written as

$$\tfrac{1}{2}Al_2 Si_2 O_5 (OH)_4 + 2\tfrac{1}{2}H_2 O \rightleftharpoons Al^{3+} + H_4 SiO_4 \,(aq) + 3OH^- \quad \dots \dots \quad [8.3\text{-}3]$$

and its solubility product as

$$K = (Al^{3+}) (OH^-)^3 (H_4 SiO_4) \quad \dots \dots \dots \dots \dots \dots \dots \dots \dots \dots \quad [8.3\text{-}4]$$

where pK is about 36.4 (Hem et al., 1973). If $pH_4SiO_4$ is near 3.0 (which is reasonable for many soil solutions), then $pAl(OH)_3$ would be about 33.4, a value similar to those reported for many soils. Crystalline halloysite is slightly less soluble than the amorphous material, having a pK of 36.9 (Kittrick, 1969). If $pH_4SiO_4$ is 3.0, then its $pAl(OH)_3$ would be 33.9.

Kaolinitic soils release both Al and silica upon acidification acording to the reaction

$$Al_2 Si_2 O_5 (OH)_4 + 6H^+ \rightleftharpoons 2Al^{3+} + 2H_4 SiO_4 + H_2 O \quad \dots \dots \dots \dots \quad [8.3\text{-}5]$$

Substituting $K_w/(OH^-)$ for $(H^+)$ gives a solubility product of

$$K = (Al^{3+}) (OH^-)^3 (H_4 SiO_4) \quad \dots \dots \dots \dots \dots \dots \dots \dots \dots \dots \quad [8.3\text{-}6]$$

with a pK of 38.75 (Kittrick, 1969). With $pH_4SiO_4$ at about 3, than $pAl(OH)_3$ is 35.7 at equilibrium. Montmorillonite behaves in a similar fashion to kaolinite but yields a lower $Al(OH)_3$ activity in solution at equilibrium

The relative solubilities of Al hydroxides and aluminosilicates show that soil-solution $Al^{3+}$ activity, at a fixed pH, will vary according to the solubility of the solution-controlling mineral, as follows: $Al(OH)_3$ (amorph) > $Al_2Si_2O_5(OH)_4$ (amorph) > halloysite > gibbsite > kaolinite > montmorillonite. Consequently, the manifestation of Al phytotoxicity can be expected to vary with soil pH, depending upon which mineral is controlling solution Al.

### 8.3.2.3 Aluminum Phosphate

Under nonagricultural ecosystems, soil solutions generally contain too little P to affect solution Al. However, P fertilizer at high rates can be an effective agent for lowering soil-solution Al to non-toxic levels by forming such insoluble precipitates as $Al(OH)_2H_2PO_4$. It has been recognized since the

1920's that high rates of superphosphate would alleviate symptoms of Al phytotoxicity. The amount of P required to detoxify Al temporarily depends upon the level of exchangeable Al. In a study of 60 acid subsoils, Coleman et al. (1960) found a high correlation ($r = 0.84$) between the exchangeable Al level and the amount of P converted to a variscite-like mineral. Because the solubility of Al phosphate follows the equation

$$Al(OH)_2 H_2 PO_4 \rightleftharpoons Al^{3+} + 2OH^- + H_2 PO_4^-, \dots\dots\dots\dots\dots \quad [8.3\text{-}7]$$

increasing amounts of $H_2PO_4^-$ will lower solution Al to satisfy the solubility product of amorphous $Al(OH)_2H_2PO_4$. Veith (1978) reported a pK of 28.06 for this amorphous compound at 50 °C.

#### 8.3.2.4 Aluminum-Hydroxy-Sulfates

Sulfate reacts with Al in dilute, acid solutions to form insoluble compounds having the approximate chemical composition of basaluminite, $Al_4(OH)_{10}SO_4 \cdot 5H_2O$, or alunite, $KAl_3(OH)_6(SO_4)_2$ (Adams and Hajek, 1978; Adams and Rawajfih, 1977). Soil conditions favorable for the formation of these compounds are found when soils are made strongly acid by the oxidation of $(NH_4)_2SO_4$ or by the oxidation of S and sulfides, as might be found in acid-sulfate soils and in reclaimed coal-mine spoils.

The dissolution of basaluminite follows the equation,

$$Al_4(OH)_{10} SO_4 \cdot 5H_2 O \rightleftharpoons 4Al^{3+} + 10OH^- + SO_4^{2-} + 5H_2 O, \quad \dots\dots \quad [8.3\text{-}8]$$

which shows that increasing solution $SO_4^{2-}$ causes a reduction in $Al^{3+}$ activity because of the solubility-product principle. In the presence of high $SO_4^{2-}$ concentrations, soil-solution $Al^{3+}$ will be lower than that allowed by the solubility of $Al(OH)_3$ or the more soluble aluminosilicates.

#### 8.3.3 Soil pH, Exchangeable Al, and Plant Growth

Although $Al^{3+}$ phytotoxicity is a function of soil-solution $Al^{3+}$ activity, it is generally more convenient to express phytotoxicity as a function of exchangeable Al. Exchangeable Al is easier to determine than solution $Al^{3+}$ activity, and there is a positive correlation between the two for a particular soil. Unfortunately, the correlation is disrupted by the addition of soluble fertilizer salts. Because a salt addition increases solution $Al^{3+}$ and lowers pH without greatly affecting exchangeable Al, there is a high correlation between pH and the ratio of $Al^{3+}$/salt in solution for a particular soil (Brenes and Pearson, 1973).

Absolute levels of exchangeable Al are poorly correlated with solution $Al^{3+}$ and the severity of Al phytotoxicity, probably because of the wide variations in CEC among soils as well as differences in the nature of cation-exchange sites (Table 8.3-1). A better correlation is found by expressing exchangeable Al in terms of Al saturation, i.e., the fraction of CEC that is satisfied by adsorbed Al. However, this relationship, too, is confounded by the soil-solution electrolyte concentration.

Soil solutions of low ionic strength have low levels of solution $Al^{3+}$ at all Al-saturation values. With increasing ionic strengths, generally with soluble fertilizers, solution Al increases proportionally because of cation-exchange reactions. Thus, the concentration of soil-solution Al is changed markedly by the addition of fertilizers and by the leaching action of percolating wat(

The direct relationship between phytotoxicity and Al saturation is rea( ly demonstrable for a particular soil at a constant fertilizer rate. However, this relationship fails to yield the ideal "critical" Al-saturation value because of differences in genetically controlled tolerances to Al as well as to inherent differences in soil chemical properties. For example, in an effort to minimize the genetic factor, a very uniform selfed cotton line, 'Empire', was grown in several different soils under the same environmental conditions. Even with an apparently constant genetic factor for Al phytotoxicity, "critical" Al saturation for maximum rate of root growth in nine Alabama soils was found to range from about 5 to 25 percent (Table 8.3-1).

Work in North Carolina (Nye et al., 1961; Evans and Kamprath, 1970) suggests that "critical" Al saturation is about 60 percent. The data in Table 8.3-1, along with other data, however, show that Al phytotoxicity occurs at much lower values than that (Adams and Pearson, 1967; Martini et al., 1977; Pearson, 1975). The reason for different "critical" Al-saturation values is obscure, but it may be related to the method of extracting exchangeable Al from different soil materials or it may be due to the presence of an unrecognized growth factor that functions in conjunction with Al. Whatever the reasons, a universal critical value seems unlikely with our present state of knowledge about soil Al.

The negative correlation between soil pH and Al saturation is the collateral of the positive correlation between soil pH and base saturation. Generally, Al saturation increases sharply as soil pH decreases below 5.0 to 5.5; above pH 5.0 to 5.5, Al saturation rapidly vanishes. Because of the relationship between soil pH and Al saturation, soil pH is often a very useful predictor of likely phytotoxic levels of Al.

## 8.4 MANGANESE IN ACID SOILS

Excess soluble Mn has long been suspected as a major factor of acid-soil infertility (Funchess, 1919), and foliage symptoms of Mn phytotoxicity are easily recognized. However, predicting phytotoxic response based on soil Mn levels has not been particularly fruitful because of difficulties in determining soil-solution $Mn^{2+}$ levels of the rhizosphere while plants are growing and absorbing nutrients.

### 8.4.1 Manganese Phytotoxicity

Phytotoxic symptoms of Mn are usually characteristic for a particular species but vary widely among species. The most commonly reported visual symptoms include the following:

1  Marginal leaf chlorosis or necrosis, appearing first on older leaves; example: alfalfa (*Medicago sativa* L.), lespedeza (*Lespedeza striata* Thumb.), peanut (*Arachis hypogaea* L.), sweet clover (*Melilotus indica* L.) (Morris and Pierre, 1949; Ouellette and Dessureau, 1958).

2  Interveinal brown spots; example: barley (*Hordeum vulgare* L.) (Williams and Vlamis, 1957).

3  Stunted, chlorotic leaves with necrotic spots; example: tobacco (*Nicotiana tabacum* L.) (Bortner, 1935).

4  Stunted, cupped or crinkled leaves; example: cotton (*Gossypium hirsutum* L.) (Adams and Wear, 1957).

TABLE 8.3-1. RELATIONSHIP BETWEEN SOIL pH,
EXCHANGEABLE Al, Al SATURATION, AND GROWTH RATES
OF PRIMARY COTTON ROOTS (*GOSSYPIUM HIRSUTUM* L.,
CUL. 'EMPIRE') (Adams and Lund, 1966; Richburg and Adams,
1970; unpublished data of F Adams)

| Soil type | Soil pH | Exch. Al, meq/100g | Al sat.*, percent | Relative root growth†, percent |
|---|---|---|---|---|
| Benndale s.l | 4.9 | 0.49 | 51 | 77 |
| (*Typic paleudults*) | 5.0 | 0.39 | 46 | 83 |
| | 5.4 | 0.25 | 25 | 88 |
| | 6.0 | 0.01 | 0 | 100 |
| Dothan s.l. (1) | 4.8 | 0.71 | 38 | 36 |
| (*Plinthic paleudults*) | 5.3 | 0.29 | 16 | 73 |
| | 5.4 | 0.13 | 6 | 84 |
| | 5.6 | 0.06 | 3 | 100 |
| Dothan s.l. (2) | 5.0 | 0.62 | 52 | 34 |
| (*Plinthic paleudults*) | 5.5 | 0.19 | 16 | 65 |
| | 6.4 | 0.00 | 0 | 100 |
| Lucedale s.l. (1) | 4.7 | 1.08 | 61 | 60 |
| (*Rhodic paleudults*) | 4.9 | 0.80 | 46 | 77 |
| | 5.1 | 0.49 | 29 | 90 |
| | 5.2 | 0.44 | 27 | 95 |
| | 5.7 | 0.09 | 4 | 100 |
| Lucedale s.l. (2) | 4.9 | 1.01 | 43 | 26 |
| (*Rhodic paleudults*) | 5.4 | 0.12 | 5 | 51 |
| | 6.0 | 0.00 | 0 | 100 |
| Lucedale s.l. (3) | 4.2 | 1.68 | 56 | 17 |
| (*Rhodic paleudults*) | 4.4 | 1.70 | 53 | 43 |
| | 5.1 | 0.33 | 12 | 56 |
| | 5.3 | 0.19 | 7 | 88 |
| | 5.7 | 0.01 | 0 | 100 |
| Lucedale s.l. (4) | 4.3 | 2.38 | 75 | 17 |
| (*Rhodic paleudults*) | 4.6 | 2.35 | 61 | 23 |
| | 5.4 | 0.10 | 3 | 100 |
| Dickson si.l. | 4.4 | 2.73 | 53 | 41 |
| (*Glossic fragiudults*) | 4.7 | 1.99 | 38 | 72 |
| | 5.0 | 1.01 | 19 | 100 |
| Bladen c.l. | 4.6 | 7.33 | 73 | 27 |
| (*Typic albaquults*) | 4.8 | 5.00 | 52 | 57 |
| | 4.9 | 2.55 | 26 | 88 |
| | 5.3 | 0.79 | 8 | 100 |

*CEC was sum of exchangeable Al, Ca, Mg, Mn, K.
†Average length of primary root 48 h after entering the acid soil
below a well-limed and fertilized surface soil where seeds were
planted.

5 Black specks on lower stems and petioles; premature leaf crop; example: potato (*Solanum tuberosum* L.) (Lee and MacDonld, 1977).
6 Interveinal chlorosis because of induced Fe deficiency (Hewitt, 1963).

Root systems do not appear to be directly affected by toxic levels of Mn, although blackened root tips have been reported where plants were grown in solutions high in Mn (Hiatt and Ragland, 1963; Morris and Pierre, 1949). However, it is not improbable that the blackened root tips were caused by Mn oxide accumulation. Root systems appear to remain healthy and active while Mn accumulations in the aerial plant parts cause severe growth distortions, chlorosis, and necrosis.

There is no universal "critical" Mn concentration in soil solution required for phytotoxic symptoms because of differences in genetically controlled tolerances and because the ionic strength of the solution affects the severity of the symptoms. Lohnis (1951) reported 10 ppm Mn to be toxic to beans (*Phaseolus vulgaris* L.) in his most dilute solutions but not in his most concentrated. Similarly, 0.5 ppm Mn was toxic to barley in a 1/5 Hoagland nutrient solution but nontoxic in full-strength Hoagland (Williams and Vlamis, 1957). In trying to define a "critical" level of Mn for toxicity, most solution experiments have been poorly designed for that purpose because they failed to maintain constant Mn concentrations and constant ionic strengths.

Available data show that incipient Mn phytotoxicity can occur with as high as 10 ppm for the more tolerant species. Very little work has been done with soil solutions *per se*, but incipient phytotoxic concentrations can be expected to be about the same as in culture solutions.

The typical effect of Mn phytotoxicity is an exponential decrease in growth with increasing solution Mn concentration above the "critical" value. The effect of excess Mn on growth is reflected in leaf content of Mn, where phytotoxic levels vary widely. For example, the "critical" Mn content at flower initiation for the entire aerial part of plants was reported to be 1000 ppm for beans (*Phaseolus vulgaris,* L.), 550 ppm for peas (*Pisum sativum* L.), and 200 ppm for barley (*Hordeum vulgare,* L.) (White, 1970). Ohki (1976) reported only 160 ppm Mn in the third leaf from the terminal was required for a yield reduction in 'Bragg' soybeans.

The best potential index for diagnosing Mn phytotoxicity probably lies in leaf analysis. However, no suitable index is yet available because of inadequate data. Obtaining the required data for a single species is greatly handicapped because Mn contents of leaves vary with the particular leaf sampled, plant age, and even cultivar. In a comparative study of 12 legume species, Andrew and Hegarty (1969) found critical Mn contents of aerial plant parts to range from 380 ppm in alfalfa to 1600 ppm in *Centrosema pubescens*. Even higher threshold values have been reported for other species, e.g., 3000 ppm in tobacco (Hiatt and Ragland, 1963) and cotton (Adams and Wear, 1957).

### 8.4.2 Soil pH, Redox Potential, and Solution Mn

Manganese occurs in several positive valency states, but only the two, three, and four valences are of consequence in soils. Trivalent and tetravalent Mn occur as insoluble oxides. Soil-solution and exchangeable Mn are present as $Mn^{2+}$ ions. Except for $OH^-$ and $CO_3^{2-}$, the salts of $Mn^{2+}$ that normally occur in soils are highly water soluble.

The nature of Mn oxides in soils is somewhat uncertain, but $MnO_2$ is the stable form. Although pyrolusite is the thermodynamically stable phase of $MnO_2$, its formation may be prevented in soils by the presence of foreign ions (McKenzie, 1972). The reactive $MnO_2$ in soils is probably present as hydrated, amorphous compounds (Ross et al., 1976). The dissolution of $MnO_2$ is dependent upon the reduction of $Mn^{4+}$ to $Mn^{2+}$ in the reaction

$$MnO_2 + 4H^+ + 2e^- \rightleftharpoons Mn^{2+} + 2H_2O \dots\dots\dots\dots\dots [8.4-1]$$

When $Mn^{2+}$ is precipitated as $Mn(OH)_2$ by an alkali, enough oxygen is taken up from the air to convert the $Mn(OH)_2$ completely to $MnO_2$. Although this reaction occurs within a few days in solution, it takes considerably longer in a soil. In fact, the $Mn(OH)_2$ may first be oxidized to $MnOOH$, a brown material which may subsequently undergo dismutation (Dion and Mann, 1946), probably by the reaction

$$eq \; 2MnOOH = Mn^{2+} + MnO_2 + 2OH^- \; \dots\dots\dots\dots\dots\dots\dots\dots \; [8.4\text{-}2]$$

The equilibrium between solid-phase Mn oxides and solution $Mn^{2+}$ is subject to rapid shifts in the soil. As shown by equation [8.4-1], the shift toward $Mn^{2+}$ is favored by lower pH and lower redox potential; it is also favored by drying and by heating the soil (Fujimoto and Sherman, 1945). Not all $MnO_2$ in soil is readily reducible to $Mn^{2+}$. In order to identify the reactive portion of $MnO_2$ and $MnOOH$, a reducing agent, such as hydroquinone, is reacted with the soil to yield what is termed "easily reducible Mn" (Adams, 1965). It is the easily reducible Mn that determines the potential for Mn phytotoxicity in acid soils. Reducible Mn contents vary in soils from almost nil to several 1000 ppm (Adams and Pearson, 1967). Although difinitive soil data are unavailable, it appears that minimum toxic level of easily reducible Mn is about 50 to 100 ppm.

Since $Mn^{2+}$ is the available form of Mn, its phytotoxicity is a function of soil-solution $Mn^{2+}$ concentration. Soil-solution Mn is known as "water-soluble Mn" and is the ideal component for defining the threshold soil-Mn level required for incipient phytotoxicity. Unfortunately, the water-soluble fraction is difficult to capture in situ because of the propensity of Mn to change its valency state during sample handling (Table 8.4-1). Efforts to correlate phytotoxicity with exchangeable Mn suffer from the same handicaps, and exchangeable Mn is not a reliable index for predicting phytotoxicity (Fergus, 1953; White, 1970). It cannot be considered superior to the water-soluble fraction.

Phytotoxicity is best predicted by combining soil pH with easily reducible Mn (Fergus, 1953). If a soil contains at least 50-100 ppm of reducible Mn, phytotoxicity is a distinct probability when soil pH is <5.5 (Adams and Pearson, 1967). The lower the soil pH, the more likely phytotoxicity will occur.

Equation [8.4-1] shows that the dissolution of $MnO_2$ involves electron transfers as well as $H^+$ ions and that the equilibrium shifts toward $Mn^{2+}$ with decreasing redox potentials. Lower soil redox potentials are favored by water-logged conditions, particularly when accompanied by the rapid decomposition of organic matter (Redman and Patrick, 1965; Cheng and Ouellette, 1971). Consequently, Mn phytotoxicity occurs at higher pHs under poorly aerated conditions than under well aerated conditions.

Normally, Mn and Al phytotoxicities can be expected to occur concurrently in acid soils because of the similarity in the pH-dependent solubility of Mn oxides and Al-containing soil minerals. Since Al phytotoxicity symptoms are not generally evident on aerial plant parts, the striking aerial symptoms of Mn phytotoxicity may give Mn more credit than it deserves for poor plant growth (Jones and Nelson, 1978). In fact, symptoms of Mn phytotoxicity may have to be quite severe before plant growth is significantly inhibited (Hiatt and Ragland, 1963).

TABLE 8.4-1. EFFECT OF DRYING AND HEATING ON THE
LEVEL OF EXCHANGEABLE Mn IN WELL-AERATED
HAWAIIAN SOILS (Fugimoto and Sherman, 1945)

| Sample no. | Soil pH | Field cond. | Exchangeable Mn, meq/100g | | |
|---|---|---|---|---|---|
| | | | Air-dried | Oven-dried | Autoclaved |
| 44-166 | 4.2 | 0.012 | 0.016 | 2.26 | 13.65 |
| 44-162 | 4.4 | 0.076 | 0.086 | 1.11 | 4.46 |
| 44-161 | 4.7 | 0.023 | 0.034 | 1.05 | 6.63 |
| 44-163 | 4.8 | 0.093 | 0.098 | 1.69 | 7.71 |
| 44-164 | 5.1 | 0.106 | 0.089 | 0.85 | 6.16 |
| 44-178 | 5.3 | 0.018 | 0.042 | 0.59 | 0.58 |
| 44-177 | 5.6 | 0.001 | 0.007 | 0.31 | 1.71 |
| 44-175 | 5.7 | 0.005 | 0.011 | 0.22 | 0.26 |
| 44-167 | 5.9 | 0.006 | 0.007 | 0.91 | 4.89 |

## 8.5 CALCIUM IN ACID SOILS

Calcium is the dominant exchangeable cation in normally productive soils. With the development of increasing soil acidity, exchangeable Ca is increasingly replaced by exchangeable Al and sometimes Mn. In soils with low CEC, however, low Ca levels may cause Ca deficiency problems with certain crops. For example, the major effect of liming loamy Ultisols when growing peanuts in the southern USA is to supply Ca for the pod-filling process (Fig. 8.5-1).

The sensitivity of peanuts to low Ca levels is attributed to its unusual fruiting habit. However, Andrew and Norris (1961) compared the effect of lime on normal-fruiting legumes in a sandy soil at pH 5.5 and found typical Ca-deficiency symptoms on some species. The tropical Oxisols can be particularly low in exchangeable Ca without a concurrent phytotoxic level of soluble Al (Hortenstine and Blue, 1968; Martini et al., 1974; Spain et al., 1975).

Liming experiments have seldom been concerned with the Ca requirement for root growth, although the need for a minimum Ca level has been clearly demonstrated (Howard and Adams, 1965). The likelihood is strong for Ca deficiency to occur in highly leached, acid, sandy horizons and in highly weathered Oxisols. Subsurface horizons in Oxisols are particularly prone toward Ca deficiency because base retention is almost nil at low pH. Calcium deficiency for root growth is manifested typically by the death of the meristematic root tissue, and the inability of plants to translocate Ca within the phloem tissue means that roots can not grow into a soil zone that is Ca deficient.

Inadequate surface-soil Ca is dramatically evident on plant growth, but inadequate subsoil Ca may go undiagnosed. If Ca deficiency occurs only in the subsoil, it will be manifested by more drought-sensitive plants because the stored subsoil moisture will have become positionally unavailable to the crop. Roots simply die as they attempt to extend into the Ca-deficient soil. Whereas surface-soil Ca deficiency is readily correctable by liming, subsoil Ca deficiency is not.

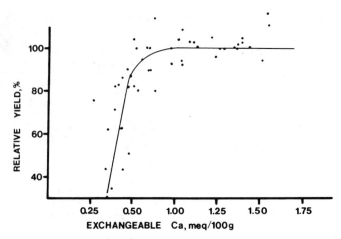

FIG. 8.5-1 Effect of exchangeable Ca upon yield of runner peanuts (*Arachis hypogea* L.) on coarse-textured soils of southeastern Alabama (Hartzog and Adams, 1973).

## 8.6 LIMING PRACTICES

Regardless of which factor is responsible for the infertility of a particular acid soil, it is always correctable by a good liming program. Liming has been practiced since antiquity, but it continues to plague farmers with difficulties. Although laboratory lime-requirement methods may be highly exact, translating that information into application techniques for equally exact field applications has eluded the average lime applicator. All needed knowledge is available, but meshing the various facets of application into a practical program for farmers still falls short of the desired goal. The effectiveness of applied agricultural lime is dependent upon the quality of the material, the amount applied, the soil pH, the uniformity of the spread, and the extent of soil-lime mixing.

### 8.6.1 Quality of Liming Materials

The quality of an agricultural liming material is judged by its effectiveness in neutralizing soil acidity. Its potential effectiveness, in turn, is determined by its chemical composition and by its particle-size distribution. Since agricultural liming materials are defined as those products whose Ca and Mg compounds are capable of neutralizing soil acidity, liming materials are restricted to the Ca and Mg salts of carbonate, hydroxide, and silicate.

The most common agricultural liming materials are ground limestones, marl, chalk, slag from iron and steel making, flue dust from cement plants, and refuse from sugar-beet factories, paper mills, and Ca-carbide plants. The bulk of agricultural lime comes from ground limestone. .

The chemical potential for neutralizing soil acidity is defined in terms of $CaCO_3$ equivalence, i.e., grams of $CaCO_3$ required to equal the reactiveness of 100 g of material. For example, pure calcite has a $CaCO_3$ equivalent of 100, and pure dolomite has a $CaCO_3$ equivalence of 109.

The effectiveness of agricultural lime in soil is also highly dependent upon the size distribution of lime particles. Since lime particles are barely soluble in water but readily soluble in acid, the extent of dissolution is dependent upon the amount of contact between lime-particle surface and the acid soil solution. Following saturation of the ambient solution with dissolving lime, further dissolution is dependent upon diffusion of $Ca^{2+}$ away from the saturated solution and the diffusion of $H^+$ into the saturated zone. Such diffusion processes are extremely slow and extend only over very short distances. For example, DeTurk (1938) found the maximum distance of Ca diffusion from a large piece of limestone in an acid soil was only 6 mm after 260 days.

Although numerous chemical methods have been used to measure the reactiveness of limestone, reliance is still placed almost solely on particle-size distribution of the limestone as a guide to expected neutralization of acid soils (Schollenberger and Whittaker, 1962). In an effort to explain particle-size effects, a useful hypothesis has been that the diameter of all lime particles is reduced at the same rate (Schollenberger and Salter, 1943). The hypothesis holds that the proportion of lime remaining (R) after particles of "d" diameter have lost a shell of "a" thickness is given by the equation

$$R = [(d-a)/d]^3 \dots\dots\dots\dots\dots\dots\dots\dots\dots\dots\dots\dots\dots\dots \quad [8.6-1]$$

Application of the diameter-reduction hypothesis requires that the soil remains undisturbed and that "R" for each particle-size separate be summed. Since the rate of diameter reduction is equal for all particles, doubling the diameter of a particle doubles the time required for complete dissolution.

The literature since about 1920 is replete with greenhouse and field experiments evaluating the quality of lime (Barber, 1967). All show that reaction rate is inversely proportional to particle size, but that particle size has a practical lower limit that should govern lime use in the field. Results of most experiments correspond to those of Meyer and Volk (1952) in which 50- and 60-mesh material was just as effective as much finer material (Fig. 8.6-1). Their data also illustrate the inertness of coarse particles, i.e., those failing to pass a 10-mesh screen have practically no effect on soil pH.

The rate at which lime reacts in the soil is also dependent upon the extent of soil-lime mixing. Ideally, each lime particle should be surrounded by soil particles. In practice, of course, this cannot be realized because of adhesive forces between small lime particles and because of physical limitations in stirring of the soil mass. However, sufficient manipulation of soil is possible to mix lime fairly thoroughly with the top 15 to 20 cm of soil. With thorough mixing of lime with the topsoil, agricultural-grade limestone will dissolve sufficiently rapidly that maximum soil pH should be achieved within 1 to 2 yr after application.

Most state laws regulating the sale of agricultural lime require that two screen sizes be used to define minimum particle-size specifications, e.g., 90 percent passes a 10-mesh screen and 50 percent passes a 60-mesh screen. If the percentages passing two screen sizes are known, percentages passing other screen sizes can be closely approximated (Schollenberger and Salter, 1943). An example of particle-size distribution in samples of five ground limestones is shown in Fig. 8.6-2. This principle is the basis for lime laws and regulations requiring only two screen-size specifications.

### 8.6.2 Neutralization Reactions

The overall chemical reaction between lime and soil acidity is the net ef-

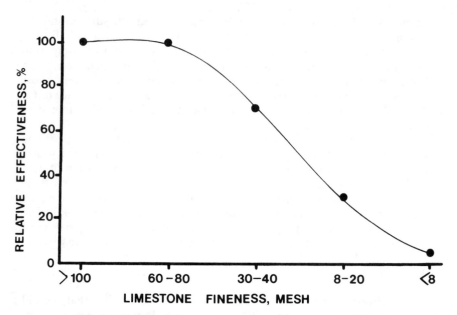

FIG. 8.6-1 Effect of limestone particle-size upon the reactivity of ground dolomitic limestone in an acid soil (data from Meyer and Volk, 1952).

fect of two separate reactions. One reaction is the release of $Al^{3+}$ and $H^+$ ions to the soil solution from exchange sites; the other is lime dissolution and the hydrolysis of $CO_3^{2-}$ ions. Neutral-salt exchangeable acidity is essentially $Al^{3+}$ while $H^+$ is released from surface ligands as soil pH is increased.

Practically all soil acidity is located at exchange sites. Although a soil may be quite acid, only an insignificant fraction of total acidity is present in the soil solution. For example, a soil solution at pH 5.0 contains approximately 10 micromoles of $H^+$ per liter. If the soil has a bulk density of 1.3 g per $cm^3$ and a moisture content of 20 percent, a hectare to a depth of 30 cm will contain 793 000 liters of soil solution and only 7.93 equivalents of solution $H^+$. If the soil also contains 5 meq of exchangeable acidity per 100 g soil, total exchangeable acidity in this soil volume would be about 198,000 equivalents or 25,000 times more than that in the soil solution.

When exchangeable $Al^{3+}$ is displaced by $Ca^{2+}$ from the lime, it undergoes stepwise hydrolysis until neutralized as $Al(OH)_3$. The overall reaction is expressed by the equation

$$2Al\text{-soil} + 3Ca^{2+} + 6H_2O \rightleftharpoons 3Ca\text{-soil} + 2Al(OH)_3 + 6H^+ \quad \dots\dots [8.6\text{-}2]$$

where $H^+$ in soil solution comes from $Al^{3+}$ hydrolysis and will react with lime.

Undissociated $H^+$ in soils is the pH-dependent component of exchangeable acidity. The mineral fraction of soils may contain positive-charged Al and Fe oxides which do not contain their full complement of $OH^-$. The reaction of their surface with an alkaline solution can be illustrated by the expression

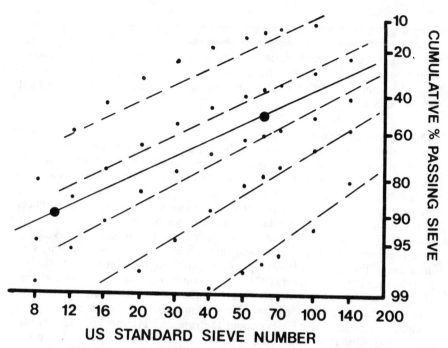

US STANDARD SIEVE NUMBER

FIG. 8.6-2 The cumulative percentages of five different samples of ground limestone (dashed lines) passing a series of U.S. standard sieve sizes (Schollenberger and Salter, 1943). The solid line represents the hypothetical particle-size distribution of ground limestone where 90 percent passes a No. 10 sieve and 50 percent passes a No. 60 sieve.

$$\begin{bmatrix} \diagdown \underset{\diagup}{\overset{\diagup}{\underset{|}{Al}}} \diagdown \overset{OH_2}{\underset{OH_2}{}} \end{bmatrix}^{+} + OH^{-} \;\rightleftharpoons\; \begin{bmatrix} \diagdown \underset{\diagup}{\overset{\diagup}{\underset{|}{Al}}} \diagdown \overset{OH}{\underset{OH_2}{}} \end{bmatrix}^{0} + H_2O \;\ldots\ldots\; [8.6\text{-}3]$$

The organic fraction of soils contains many acid functional groups in which the degree of $H^+$ dissociation increases with increasing pH. The carboxyl group probably dominates in acid soils and its release of $H^+$ to soil solution is shown by the equation

$$\underset{R\text{-}\overset{\overset{O}{\|}}{C}\text{-}OH}{} \;\rightleftharpoons\; \underset{R\text{-}\overset{\overset{O}{\|}}{C}\text{-}O^{-}}{} + H^{+} \;\ldots\ldots\ldots\ldots\ldots\ldots\ldots\; [8.6\text{-}4]$$

As soil pH is increased by liming, proportionately more of the $H^+$ becomes dissociated from the carboxyl groups.

Dissolved lime also undergoes hydrolytic reactions in soil solutions. The $CO_3^{2-}$ from $CaCO_3$ reacts stepwise with water to give the overall reaction

$$CO_3^{2-} + 2H_2O \rightleftharpoons H_2CO_3 + 2OH^- \;\ldots\ldots\ldots\ldots\ldots\ldots\; [8.6\text{-}5]$$

Similarly, $SiO_3^{2-}$ from $CaSiO_3$ slags react with $H_2O$ to yield $H_4SiO_4$ and $OH^-$ ions. Combining equations [8.6-2] and [8.6-5] yields the overall neutralization reaction

$$2Al\text{-soil} + 3CaCO_3 + 3H_2O \rightleftharpoons 3Ca\text{-soil} + 2Al(OH)_3 + 3CO_2 \quad \ldots \ldots \quad [8.6\text{-}6]$$

With thorough and intimate mixing of lime and acid soil, the neutralization reaction of equation [8.6-6] is quite efficient in raising soil pH to $\sim$ 6.0. Lime becomes progressively less effective in dissolving and raising soil pH beyond this value.

### 8.6.3 Liming Cultivated Crops

Liming of cultivated crops may be done anytime between harvesting of one crop and planting of the next. The lime may be broadcast on undisturbed soil or on turned soil, depending on such practical considerations as lime availability and spreading capabilities. The key to effective liming is a uniform spread of quality material with a reasonable control of the application rate. Equipment is available that provides uniform and accurate application, but too often it is not used at all or is used improperly. In a study of lime vending in the Tennessee Valley area, it was found that most lime vendors used discarded fertilizer spreaders with no functioning rate-control mechanisms (W. S. Stewart, 1977. An economic analysis of agricultural lime vending—a case study. Ph.D. Thesis. University of Mississippi, Oxford). The spreader operators were found to be unaware of the basic principles of liming, and most were careless about the rate and evenness of the spread.

The combination of poor equipment and untrained operators has probably resulted in a great deal of poorly spread lime; some areas in the field are skipped while other areas receive more than needed. Much improvement is needed in the area of lime spreading on farmers' fields.

Lime should be added at a rate calculated to raise pH of the entire plowlayer to the desired level. It should be thoroughly mixed with the surface soil, where roots are most active and where most nutrient absorption occurs. Mixing of lime with soil is done by plowing, disking, harrowing, and cultivation. None of these operations will achieve complete and uniform mixing with the plow layer. Plowing buries most of the lime at the plow sole; disking gives satisfactory mixing in the top 6 to 10 cm; a rotary tiller gives good mixing to an uncertain depth; split-lime applications before and after turning of soil, followed by disking, gives fair vertical distribution but poor horizontal distribution (Hulburt and Menzel, 1953). A farmer must choose the method of incorporation that matches his equipment and the soil's need for lime.

It is generally recommended that lime be applied 2 to 3 months prior to planting to allow time for the lime to react. This is based partially on the dissolution rate of lime but mostly on experience from earlier days when implements for thorough mixing with soil were unavailable. If the recommended rate of agricultural-grade limestone is properly mixed with the soil, planting may follow without delay because enough fines will be present to raise soil pH immediately above phytotoxic levels of Al and Mn and to correct a Ca deficiency. Consequently, time of application should not be based on planting dates but upon lime availability.

If the plow layer has become so acid that root growth is severely inhibited, lime should be applied in split applications to encourage the deepest root system possible. One-half should be surface applied, disked or mixed by other means, then turned by plow. The other half should then be surface applied and worked into the soil.

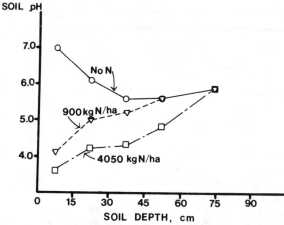

FIG. 8.6-3 Effect of $(NH_4)_2SO_4$ on soil-profile pH after a 3-yr experiment with napiergrass (*Pennisetum purpureum* Schumach.) on a Toa clay loam in Puerto Rico (Adams and Pearson, 1967).

A significant portion of the cost of agricultural lime is spent on transportation. Consequently, distant hauling is prohibitively expensive, and prices are usually controlled by the availability of local lime sources. Minor differences in lime quality are usually of no economic consequence. For example, calcitic limestones tend to dissolve more rapidly than dolomitic limestones, but either dissolves rapidly enough to satisfy crop needs. Dolomitic, of course, can have a real advantage on acid soils that are deficient in Mg. The value of plant nutrients, other than Ca and Mg, present in some industrial by-product limes may be significant. Flue dust from cement plants may contain appreciable K while other materials may contain P, Zn, or other micronutrients.

### 8.6.4 Liming Sod and No-Till Crops

Liming for the establishment of sod crops does not differ from that for cultivated crops. After sod establishment, however, liming must be accomplished without disturbing the sod, which means no mechanical mixing of lime with soil. Lime is simply spread atop the sod and left to filter down to the soil's surface. Liming grass sods where high rates of ammonium fertilizers are used is particularly important to hay growers.

Where $NH_4$-containing fertilizers are used in the absence of an adequate liming program, soil profiles can become quite acid to considerable depths (Fig. 8.6-3) (Albruna et al., 1958; Abruna et al., 1964; Adams et al., 1967b, c; Adams and Pearson, 1969). Although the acidity created by nitrification is located near the soil's surface, leaching will cause it to move downward in the profile as exchangeable Al becomes significant in the uppermost layers. If excess acidity is allowed to accumulate underneath an undisturbed sod, it is not readily amenable to correction by conventional liming practices.

The effect of the intense acidity that can develop from $NH_4$ fertilizers is shown by no-till experiments in Virginia (Moschler et al., 1973) and in Kentucky (Blevins et al., 1978). Their data showed that the acidity from nitrification accumulated in the upper 5 cm or so of soil and that incorporating lime to plow depth at the beginning of the experiment was not always satisfactory. Because root density is greatest in the upper few cm of soil, the accumulation of acidity in that zone was quite detrimental to corn yields. Blevins et al. (1978) recommended that lime be surface applied with shallow or no incorporation where no-till corn is to be continued for more than 1 or 2 yr.

If acidity beneath undisturbed sods is to be corrected, the following basic principles must be considered: (a) the relative insolubility of limestone; (b) the effect of soil-solution anion concentration on cation mobility; (c) the preferential absorption by plant roots of $NO_3$ ions relative to $Ca^{2+}$ and $Mg^{2+}$ ions.

In an experiment with bermudagrass sods, Adams and Pearson (1969) found that acidity from low rates of $NH_4NO_3$ was effectively neutralized by the concurrent application of normal lime rates. Apparently, all N was used by the crop, and none was available for leaching. At N rates which exceeded the crop's ability to absorb N, however, lime became less efficient at counteracting the accumulation of profile acidity. Surface-applied lime was particularly ineffective in countering the acidity from high rates of $(NH_4)_2SO_4$. Whereas plants preferentially absorb $NO_3^-$ over $Ca^{2+}$ and $Mg^{2+}$, they do not preferentially take up $SO_4^{2-}$. Consequently, the acidity associated with the $SO_4^{2-}$ component of $(NH_4)_2SO_4$ remained largely intact in the soil. The authors concluded from the study that effectiveness of lime on a grass sod was dependent upon (a) soil type, (b) lime rate, (c) N source, and (d) N rate. Lime was more effective on coarse-textured soil, possibly because of some physical movement of lime particles. Lime was less effective at high N rates, probably because more nitrification occurred below the limed layer. Lime was more effective with $NH_4NO_3$ than with $(NH_4)_2SO_4$, probably because the residual acidity associated with the $SO_4^{2-}$ ion was not neutralized by preferential ion uptake, as it was with $NO_3^-$. The difficulty of maintaining soil pH beneath a highly fertilized grass sod is illustrated by the data in Fig. 8.6-4.

Surface-applied lime does not quickly correct acidity that has accumulated beneath an undisturbed sod. Normal lime rates are quite ineffective during the first few years after application. However, Adams and Pearson (1969) found that adding lime in considerable excess did have a slight ameliorating effect on profile acidity after 2 or 3 yr.

Although some downward migration of fine lime particles may occur in coarse-textured soils, major movement must occur as dissolved $Ca(HCO_3)_2$ in the percolating waters. Movement as $Ca(HCO_3)_2$ can be appreciable when soil pH is 6 or more; at lower pHs, $HCO_3^-$ concentration becomes increasingly insignificant. With the use of N fertilizers, most downward movement of $Ca^{2+}$ is probably as $Ca(NO_3)_2$. If the leached $NO_3^-$ is subsequently absorbed by roots, there will be a corresponding increase in soil pH. Consequently, surface-applied lime will slowly neutralize soil acidity beneath an undisturbed sod if proper management is practiced.

### 8.6.5. Subsoil Acidity

The physical inaccessibility of subsoil to direct liming treatment makes prevention of subsoil acidity a desirable part of any liming program. Preventing the development of subsoil acidity is achieved by maintaining a favorable surface-soil pH. This is illustrated by the data in Table 8.6-1 for a long-term experiment (32 yr) in which surface-soil pH was maintained above 6 by some

TABLE 8.6-1. EFFECT OF LONG-TERM N AND LIME RATES ON SOIL PROFILE ACIDITY IN TWO 2-YR COTTON-CORN ROTATION EXPERIMENTS IN ALABAMA, 1930-1962 (Unpublished data of F.Adams)

| N and lime treatment[†] | Soil pH at different depths after 32 years[*] | | | | | | | |
| | Dothan loamy sand[‡] | | | | Lucedale sandy loam[‡] | | | |
| | 0-15 cm | 15-30 cm | 30-45 cm | 45-60 cm | 0-15 cm | 15-30 cm | 30-45 cm | 45-60 cm |
| $(NH_4)_2SO_4$ | 4.8 | 4.8 | 4.3 | 4.3 | 4.8 | 4.8 | 4.9 | 4.9 |
| $(NH_4)_2SO_4$ + Lime | 6.2 | 5.4 | 4.7 | 4.6 | 6.0 | 5.5 | 5.3 | 5.1 |
| $NaNO_3$ | 5.7 | 5.5 | 5.1 | 5.0 | 5.8 | 5.5 | 5.3 | 5.1 |
| $NaNO_3$ + Lime | 6.4 | 6.4 | 6.3 | 6.3 | 6.8 | 6.6 | 5.8 | 5.2 |

[*]Initial surface-soil pH was 6.0 in 1930 at both locations.
[†]N rates were 40 kg/ha during 1930-1945 and 53 kg/ha during 1946-1962. Lime source was basic slag applied at 500 kg/ha annually during 1930-1945 and 800 kg/ha annually during 1946-1962.
[‡]Dothan loamy sand is a fine-loamy, siliceous, thermic *Plinthic paleudults*
  Lucedale sandy loam is fine-loamy, siliceous, thermic *Rhodic paleudults*

treatments and allowed to become strongly acid in others. Both surface-soil and subsoil pH remained mostly unaffected even with no lime applied when N was supplied as $NaNO_3$. In contrast, $(NH_4)_2SO_4$ acidified the profile to considerable depths. When a liming material was used in conjunction with $NaNO_3$, subsoil pH as well as surface-soil pH was increased substantially over the years.

Utilizing surface-applied lime to raise subsoil pH usually requires several years to be effective. Data in Fig. 8.6-4 show how it can be accomplished in a shorter time by unusually high rates of both lime and

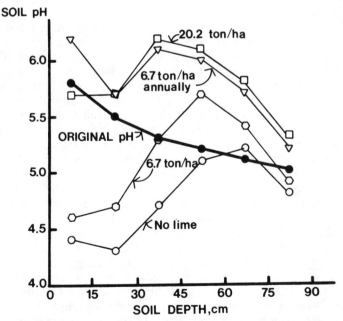

FIG. 8.6-4 Effect of rates of calcitic limestone on soil profile acidity after 4 yr under a Coastal bermudagrass (*Cynodon dactylon* (**L.**) **Pers.**) sod fertilized with $NH_4NO_3$ at annual N rate of 900 kg/ha (Adams et al., 1967c). Single lime applications were made at the beginning of the experiment except for the treatment labeled "6.7 ton/ha annually".

TABLE 8.6-2. EFFECT OF Ca(NO$_3$)$_2$ RATES ON SOIL
PROFILE pH OF A LUCEDALE SANDY LOAM*
BENEATH A 'COASTAL' BERMUDAGRASS
(Cynodon dactylon (L.) Pers.) SOD AFTER 4 YR
(Adams and Pearson, 1969)

| Soil depth, cm | Initial soil pH | Final soil pH at different annual N rates | | |
|---|---|---|---|---|
| | | 450 kg/ha | 900 kg/ha | 1350 kg/ha |
| 0-15 | 5.7 | 6.1 | 6.1 | 5.9 |
| 15-30 | 5.4 | 6.3 | 6.1 | 6.1 |
| 30-45 | 5.2 | 6.1 | 6.1 | 5.9 |
| 45-60 | 5.2 | 5.7 | 5.8 | 5.7 |
| 60-75 | 5.2 | 5.5 | 5.5 | 5.5 |

*Lucedale sandy loam is fine-loamy siliceous, thermic, Rhodic paleudults

NH$_4$NO$_3$. If subsoil pH needs to be raised within 1 or 2 yr, this can be accomplished by the use of high rates of NaNO$_3$ or Ca(NO$_3$)$_2$. The Ca$^{2+}$ and NO$_3^-$ ions move downward into the subsoil together, after which roots absorb considerably more NO$_3^-$ than Ca$^{2+}$. This imbalance of anion-cation uptake causes a corresponding rise in ambient pH. The efficiency of this approach is shown by the data in Table 8.6-2. The N rate must be high enough to allow for considerable migration of NO$_3^-$ to the subsoil, but adding beyond the capability of the crop's ability to absorb NO$_3^-$ ions has no further effect on pH. Similar results for high rates of NaNO$_3$ on two other soil types have also been reported (Adams and Pearson, 1969).

Plant roots concentrate in the surface soil where oxygen and mineral nutrients are most available. Subsoils need not serve as a major supplier of nutrients, although they sometimes do supply significant amounts of certain nutrients. Since roots actively compete with aerial plant parts for photosynthates (Troughton, 1957), it seems likely that root growth in subsoils should not be as vigorous or extensive as in surface soils. Instead, subsoils are needed primarily to serve as major reservoirs of stored soil moisture for plant use between rains. In nonirrigated lands, roots must have ready access to this moisture reservoir if maximum production is to be realized.

A very effective barrier to root growth into subsoils is low pH because of extreme Al phytotoxicity or Ca deficiency. The effectiveness of Al phytotoxicity in denying subsoil moisture to a cotton crop is shown by the data in Fig. 8.6-5, where root growth into the subsoil was inversely proportional to soil-solution Al concentration (Adams et al., 1967a).

Acid subsoils may restrict root growth because of Ca deficiency as well as by Al phytotoxicity. Calcium is the one nutrient that must be present in adequate amounts in the ambient solution for each growing root. Calcium deficiency in subsoils is most likely to occur in acid, coarse-textured, highly leached, subsoil horizons and in some highly weathered Oxisols.

Some correction of subsoil acidity can be made through direct incorporation of lime into the subsoil. However, this procedure is fraught with hurdles that appear to be insurmountable. A major problem is mixing the lime with enough soil to affect root growth significantly. This is probably the reason that there has been such poor plant response to attempted deep placement of lime in field experiments (Barber, 1967).

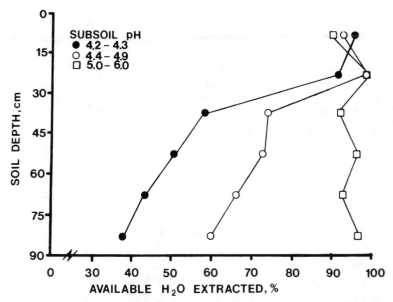

FIG. 8.6-5 Effect of subsoil pH on amount of available water extracted by cotton (*Gossypium hirsutum* L.) from a Greenville fine sandy loam (*Rhodic Paleudults*) profile between July 17 and August 2 (Adams et al., 1967a).

Complete mixing of lime with subsoil is simply impractical. In an effort to determine just how little mixing might be permissible in the subsoil, Pearson et al. (1973) found that root growth of cotton plants in a growth-chamber experiment was satisfactory where limed zones were not separated by more than 7.5 cm of highly acid soil. In a similar-type study, Kauffman and Gardner (1978) obtained maximum growth of wheat (*Triticum aestivum* L.) when lime was thoroughly mixed with only 30 percent of the soil volume.

Translating results of such glass-box and rhizotron experiments to field conditions is a major challenge to machinery engineers. Dumas and Doss (1976) attempted to incorporate lime to a depth of 45 cm with the aid of moldboard plows and rotary tillers but achieved barely more than 30 cm. It appears that the most plausible means of preventing or correcting subsoil acidity lies in the judicious choice of surface-soil amendments. Trying to incorporate lime directly into the subsoil is simply fraught with too many handicaps to be considered a viable option to surface-soil treatments.

### References

1   Abruna, F., J. Vicente-Chandler, and R. W. Pearson. 1964. Effects of liming on yields and composition of heavily fertilized grasses and on soil properties under humid tropical conditions. Soil Sci. Soc. Am. Proc. 28:657-661.

2   Abruna, F., R. W. Pearson, and C. B. Elkins. 1958. Quantitative evaluation of soil reaction and base status changes resulting from field application of residually acid-forming nitrogen fertilizers. Soil Sci. Soc. Am. Proc. 22:539-542.

3   Abruna, F., and J. Vicente. 1955. Refinement of a quantitative method for determining the lime requirements of soils. Univ. Puerto Rico J. Agric. 39:41-45.

4   Adams, Fred. 1965. Manganese. In: C. A. Black (ed.) Methods of soil analysis. Agronomy 9:1011-1018.

5   Adams, Fred. 1978. Liming and fertilization of Ultisols and Oxisols. p. 377-394. In:

C. S. Andrew and E. J. Kamprath (ed.) Mineral nutrition of legumes in tropical and subtropical soils. CSIRO. Melbourne, Australia.

6  Adams, Fred, and C. E. Evans. 1962. A rapid method for measuring lime requirement of red-yellow Podzolic soils. Soil Sci. Soc. Am. Proc. 26:355-357.

7  Adams, Fred, and B. F. Hajek. 1978. Effects of solution sulfate, hydroxide, and potassium concentrations on the crystallization of alunite, basaluminite, and gibbsite from dilute aluminum solutions. Soil Sci. 126:169-173.

8  Adams, Fred, and Z. F. Lund. 1966. Effect of chemical activity of soil solution aluminum on cotton root penetration of acid subsoils. Soil Sci. 101:193-198.

9  Adams, Fred, and R. W. Pearson. 1967. Crop response to lime in the southern United States and Puerto Rico. In: R. W. Pearson and F. Adams (ed.) Soil acidity and liming. Agronomy 12:161-206.

10  Adams, Fred, and R. W. Pearson. 1969. Neutralizing soil acidity under bermudagrass sod. Soil Sci. Soc. Am. Proc. 33:737-742.

11  Adams, Fred, R. W. Pearson, and B. D. Doss. 1967a. Relative effects of acid subsoils on cotton yields in field experiments and on cotton roots in growth-chamber experiments. Agron. J. 59:453-456.

12  Adams, Fred, and Z. Rawajfih. 1977. Basaluminite and alunite: A possible cause of sulfate retention by acid soils. Soil Sci. Soc. Am. J. 41:686-692.

13  Adams, Fred, and J. I. Wear. 1957. Manganese toxicity and soil acidity in relation to crinkle leaf of cotton. Soil Sci. Soc. Am. Proc. 21:305-308.

14  Adams, W. E., R. W. Pearson, W. A. Jackson, and R. A. McCreery. 1967b. Influence of limestone and nitrogen on soil pH and 'Coastal' bermudagrass yield. Agron. J. 59:450-453.

15  Adams, W. E., A. W. White, Jr., and R. N. Dawson. 1967c. Influence of lime sources and rates on 'Coastal' bermudagrass production, soil profile reaction, exchangeable Ca and Mg. Agron. J. 59:147-149.

16  Andrew, C. S., and M. P. Hegarty. 1969. Comparative responses to manganese excess of eight tropical and four temperate pasture legume species. Aust. J. Agric. Res. 20:687-696.

17  Andrew, C. S., A. D. Johnson, and R. L. Sandland. 1973. Effect of aluminum on the growth and chemical composition of some tropical and temperate pasture legumes. Aust. J. Agric. Res. 24:325-340.

18  Andrew, C. S., and D. O. Norris. 1961. Comparative responses to calcium of five tropical and four temperate pasture legume species. Aust. J. Agric. Res. 12:40-55.

19  Barber, S. A. 1967. Liming materials and practices. In: R. W. Pearson and F. Adams (ed.) Soil acidity and liming. Agronomy 12:125-160.

20  Bartlett, R. J., and D. C. Riego. 1972a. Toxicity of hydroxy aluminum in relation to pH and phosphorus. Soil Sci. 114:194-200.

21  Bartlett, R. J., and D. C. Riego. 1972b. Effect of chelation on the toxicity of aluminum. Plant Soil 37:419-423.

22  Blevens, R. L., L. W. Murdock, and G. W. Thomas. 1978. Effect of lime application on no-tillage and conventionally tilled corn. Agron. J. 70:322-326.

23  Bortner, C. E. 1935. Toxicity of manganese to Turkish tobacco in acid Kentucky soils. Soil Sci. 39:15-33.

24  Brenes, Eduardo, and R. W. Pearson. 1973. Root responses of three gramineae species to soil acidity in an Oxisol and an Ultisol. Soil Sci. 116:295-302.

25  Cheng, B. T., and G. J. Ouellette. 1971. Effects of organic amendments on manganese toxicity in potatoes as measured by sand and soil culture studies. Plant Soil. 34:165-181.

26  Clarkson, D. T. 1969. Metabolic aspects of aluminum toxicity and some possible mechanism for resistance. p. 321-397. In: I. H. Rorison et al. (ed.) Ecological aspects of the mineral nutrition of plants. Blackwell Scientific Publ., Oxford.

27  Coleman, N. T., and G. W. Thomas. 1967. The basic chemistry of soil acidity. In: R. W. Pearson and F. Adams (ed.). Soil acidity and liming. Agronomy 12:1-42.

28  Coleman, N. T., J. T. Thorup, and W. A. Jackson. 1960. Phosphate-sorption reactions that involve exchangeable Al. Soil Sci. 90:1-7.

29  Davies, B. E. 1971. A statistical comparison of pH values of some English soils after measurement in both water and 0.01$M$ calcium chloride. Soil Sci. Soc. Am. Proc. 35:551-552.

30  DeTurk, E. E. 1938. Properties of Illinois soils which are related to their need for limestone and factors controlling effectiveness of limestone. Illinois Geol. Surv. Circ. 23:191-203.

31  Devine, T. E. 1976. Aluminum and manganese toxicities in legumes. p. 65-72. In: M. J. Wright (ed.) Plant adaptation to mineral stress in problem soils. Cornell Univ. Agric. Exp. Stn., Ithaca, New York.

32  Dion, H. G., and P. J. G. Mann. 1946. Three-valent manganese in soils. J. Agric. Sci. 36:239-245.

33  Dios, R., and T. C. Broyer. 1962. The effect of high levels of manganese on aluminum absorption and growth of corn plants in nutrient medium. An. Edafol. Agrobiol. 21:13-30.

34  Dumas, W. T., and B. D. Doss. 1976. Correction of subsoil acidity in cotton production on Coastal Plain soils. Highlights Agric. Res. 23(No. 4). Auburn Univ. (AL) Agric. Exp. Stn.

35  Evans, C. E., and E. J. Kamprath. 1970. Lime response as related to percent Al saturation, solution Al and organic matter content. Soil Sci. Soc. Am. Proc. 34:893-896.

36  Fergus, I. F. 1953. Manganese toxicity in an acid soil. Queensland J. Agric. Sci. 10:15-27.

37  Foy, C. D. 1974. Effects of aluminum on plant growth. p. 601-642. In: E. W. Carson (ed.) The plant root and its environment. Univ. Press Virginia, Charlottesville.

38  Foy, C. D. 1976. General principles involved in screening plants for aluminum and manganese tolerance. p. 255-267. In: M. J. Wright (ed.) Plant adaptation to mineral stress in problem soils. Cornell Univ. Agric. Exp. Stn., Ithaca, NY.

39  Fujimoto, C. K., and G. D. Sherman. 1945. The effect of drying, heating, and wetting on the level of exchangeable manganese in Hawaiian soils. Soil Sci. Soc. Am. Proc. 10:107-112.

40  Funchess, M. J. 1919. Acid soils and the toxicity of manganese. Soil Soc. 8:69.

41  Hartzog, D., and F. Adams. 1973. Fertilizer, gypsum, and lime experiments with peanuts in Alabama. Auburn Univ. (AL) Agric. Exp. Stn. Bull. 448.

42  Hem, J. D., C. E. Roberson, C. J. Lind, and W. L. Polzer. 1973. Chemical interactions of aluminum with aqueous silica at 25 °C. U.S. Geol. Sur. Water-Supply Paper 1827-E.

43  Hester, J. B., and F. A. Shelton. 1933. Seasonal variation of pH in field soils—a factor in making lime recommendations. J. Am. Soc. Agron. 25:299-300.

44  Hewitt, E. J. 1963. The essential nutrient elements: requirements and interactions in plants, p. 137-329. In: F. C. Steward (ed.) Plant physiology: a treatise. III. Inorganic nutrition of plants. Academic Press, NY.

45  Hiatt, A. J., and J. L. Ragland. 1963. Manganese toxicity of burley tobacco. Agron. J. 55:47-49.

46  Hiltbold, A. E., and Fred Adams. 1960. Effect of nitrogen volatilization on soil acidity changes due to applied nitrogen. Soil Sci. Soc. Am. Proc. 24:45-47.

47  Hortenstine, C. C., and W. G. Blue. 1968. Growth responses in three plant species to lime and phosphorus applied to Puleton loamy fine sand. Soil Crop Sci. Soc. Fla. Proc. 28:23-28.

48  Howard, D. D., and Fred Adams. 1965. Calcium requirement for penetration of subsoils by primary cotton roots. Soil Sci. Soc. Am. Proc. 29:558-562.

49  Hulburt, W. C., and R. G. Menzel. 1953. Soil mixing characteristics of tillage implements. AGRICULTURAL ENGINEERING 34(10):702-708.

50  Jackson, W. A. 1967. Physiological effects of soil acidity. In: R. W. Pearson and F. Adams (ed.) Soil acidity and liming. Agronomy 12:43-124.

51  Jenny, H. 1961. Reflections on the soil acidity merry-go-round. Soil Sci. Soc. Am. Proc. 25:428-432.

52  Jones, W. F., and L. E. Nelson. 1978. Response of field-grown soybeans to lime. Commun. Soil Sci. Plant Anal. 9:607-614.

53  Kamprath, E. J. 1970. Exchangeable aluminum as a criterion for liming leached mineral soils. Soil Sci. Soc. Am. Proc. 34:252-254.

54  Kauffman, M. D., and E. H. Gardner. 1978. Segmental liming of soil and its effect on the growth of wheat. Agron. J. 70:331-336.

55  Kittrick, J. A. 1969. Soil minerals in the $Al_2O_3$-$SiO_2$-$H_2O$ system and a theory of their formation. Clays Clay Miner. 17:157-167.

56  Lee, C. R., and M. L. MacDonald. 1977. Influence of soil amendments on potato growth, mineral nutrition, and tuber yield and quality on very strongly acid soils. Soil Sci. Soc. Am. J. 41:573-577.

57  Lohnis, M. 1951. Manganese toxicity in field and market garden crops. Plant Soil 3:193-221.

58  Lund, Z. F. 1970. The effect of calcium and its relation to some other cations on soybean root growth. Soil Sci. Soc. Am. Proc. 34:456-459.

59  Marion, G. M., D. M. Hendricks, G. R. Dutt, and W. H. Fuller. 1976. Aluminum and silica solubility in soils. Soil Sci. 121:76-85.

60  Martini, J. A., R. A. Kockhann, O. J. Siqueira, and C. M. Borkert. 1974. Response of soybeans to liming as related to soil acidity, Al and Mn toxicities, and P in some oxisols of Brazil. Soil Sci. Soc. Am. Proc. 38:616-620.

61  Martini, J. A., R. A. Kockhann, E. P. Gomes, and F. Langer. 1977. Response of wheat cultivars to liming in some high Al Oxisols of Rio Grande do Sul, Brazil. Agron. J. 69:612-616.

62  McKenzie, R. M. 1972. The manganese oxides in soils—a review. Z. Pflanz. Bodenkd. 131:221-242.

63  McLean, E. O. 1978. Principles underlying the practice of determining lime requirements of acid soils by use of buffer methods. Commun. Soil Sci. Plant Anal. 9:699-715.

64  McLean, E. O., D. J. Eckert, G. Y. Reddy, and J. F. Trierweiler. 1978. An improved SMP soil lime requirement method incorporating double-buffer and quick-test features. Soil Sci. Soc. Am. J. 42:311-316.

65  Meyer, T. A., and G. W. Volk. 1952. Effect of particle size of limestone on soil reaction, exchangeable cations, and plant growth. Soil Sci. 73:37-52.

66  Miyake, K. 1916. The toxic action of soluble Al salts upon the growth of the rice plant. J. Biol. Chem. 25:23-28.

67  Morris, H. D., and W. H. Pierre. 1949. Minimum concentration of manganese necessary for injury to various legumes in culture solution. Agron. J. 41:107-113.

68  Moschler, W. W., D. C. Martens, C. I. Rich, and G. M. Shear. 1973. Comparative lime effects on continuous no-tillage and conventionally tilled corn. Agron. J. 65:781-783.

69  Moschler, W. W., R. K. Stivers, D. L. Hallock, R. D. Sears, H. M. Camper, M. J. Rogers, G. D. Jones, M. T. Carter, and F. S. McClaugherty. 1962. Lime effects on soil reaction and base content of eleven soil types in Virginia. Virginia Agric. Exp. Stn. Tech. Bull. 159.

70  Munns, D. N. 1978. Soil acidity and nodulation. p. 247-264. In: C. S. Andrew and E. J. Kamprath (ed.) Mineral nutrition of legumes in tropical and subtropical soils. CSIRO. Melbourne, Australia.

71  Nyatsanga, Titus, and W. H. Pierre. 1973. Effect of nitrogen fixation by legumes on soil acidity. Agron. J. 65:936-940.

72  Nye, P., D. Craig, N. T. Coleman, and J. L. Ragland. 1961. Ion exchange equilibria involving aluminum. Soil Sci. Soc. Am. Proc. 25:14-17.

73  Ohki, K. 1976. Manganese deficiency and toxicity levels for 'Bragg' soybeans. Agron. J. 68:861-864.

74  Ouellette, G. J., and L. Dessureaux. 1958. Chemical composition of alfalfa as related to degree of tolerance to manganese and aluminum. Can. J. Plant Sci. 38:206-214.

75  Pearson, R. W. 1975. Soil acidity and liming in the humid tropics. Cornell Univ. (NY) Int. Agric. Bull. 30.

76  Pearson, R. W., and F. Adams. 1967. Soil acidity and liming. Am. Soc. Agron., Madison, WI.

77  Pearson, R. W., J. Childs, and Z. F. Lund. 1973. Uniformity of limestone mixing in acid subsoil as a factor in cotton root penetration. Soil Sci. Soc. Am. Proc. 37:727-732.

78  Peech, M. 1965. Lime requirement. In: C. A. Black (ed.) Methods of soil analysis. Agronomy 9:927-932.

79  Pierre, W. H. 1928. Nitrogenous fertilizers and soil acidity: I. Effect of various nitrogenous fertilizers on soil reaction. J. Am. Soc. Agron. 20:2-16.

80  Pierre, W. H., and W. L. Banwart. 1973. Excess-base and excess-base/N ratio of various crop species and parts of plants. Agron. J. 65:91-96.

81  Pierre, W. H., J. Meisinger, and J. R. Birchett. 1970. Cation-anion balance in crops as a factor in determining the effect of nitrogen fertilizers on soil acidity. Agron. J. 62:106-111.

82  Puri, A. N., and A. G. Asghar. 1938. Influence of salts and soil-water ratio on pH value of soils. Soil Sci. 46:249-257.

83  Redman, F. H., and W. H. Patrick, Jr. 1965. Effect of submergence on several biological and chemical soil properties, Louisiana Agric. Exp. Stn. Bull. 592.

84  Reid, D. A. 1976. Genetic potentials for solving problems of soil mineral stress: Aluminum and manganese toxicities in the cereal grains. p. 55-64. In: M. J. Wright (ed.) Plant adaptation to mineral stress in problem soils. Cornell Univ. Agric. Exp. Stn., Ithaca, NY.

85  Richburg, J. S., and F. Adams. 1970. Solubility and hydrolysis of aluminum in soil solutions and saturated-paste extracts. Soil Sci. Soc. Am. Proc. 34:728-734.

86  Rogers, H. T., F. Adams, and D. L. Thurlow. 1973. Lime needs of soybeans on Alabama soils. Auburn Univ. (AL) Agric. Exp. Stn. Bull. 452.

87   Ross, S. J., D. P. Franzmeier, and C. B. Roth. 1976. Mineralogy and chemistry of manganese oxides in some Indiana soils. Soil Sci. Soc. Am. J. 40:137-143.

88   Schofield, R. K., and A. W. Taylor. 1955. The measurement of soil pH. Soil Sci. Soc. Am. Proc. 19:164-167.

89   Schollenberger, C. J., and R. M. Salter. 1943. A chart for evaluating agricultural limestone. J. Am. Soc. Agron. 35:955-966.

90   Schollenberger, C. J., and C. W. Whittaker. 1962. A comparison of methods for evaluating activities of agricultural limestones. Soil Sci. 93:161-171.

91   Shoemaker, H. E., E. O. McLean, and P. F. Pratt. 1961. Buffer methods for determining lime requirement of soils with appreciable amounts of extractable aluminum. Soil Sci. Soc. Am. Proc. 25:274-277.

92   Spain, J. M., C. A. Francis, R. H. Howeler, and F. Calvo. 1975. Differential species and varietal tolerance to soil acidity in tropical crops and pastures. p. 308-329. In: E. Bornemisza and A. Alvarado (ed.) Soil management in Tropical America. North Carolina State Univ., Raleigh.

93   Spurway, C. H. 1941. Soil reaction (pH) preferences of plants. Michigan Agric. Exp. Stn. Spec. Bull. 306.

94   Troughton, A. 1957. The underground organs of herbage plants. Commonwealth Agric. Bureaux Bull. 44. Bucks, England.

95   Veith, J. A. 1978. Formation of X-ray amorphous aluminum o-phosphates from precipitation and secondary precipitation. Z. Pflanz. Bodenkd. 141:29-42.

96   Volk, G. M., and C. E. Bell. 1944. Soil reaction (pH). Some critical factors in its determination, control and significance. Florida Agric. Exp. Stn. Tech. Bull. 400.

97   White, R. P. 1970. Effects of lime upon soil and plant manganese levels in an acid soil. Soil Sci. Soc. Am. Proc. 34:625-629.

98   Williams, D. E., and J. Vlamis. 1957. Manganese and boron toxicities in standard culture solutions. Soil Sci. Soc. Am. Proc. 21:205-209.

99   Woodruff, C. M. 1967. Crop response to lime in the midwestern United States. In: R. W. Pearson and F. Adams (ed.) Soil acidity and liming. Agronomy 12:207-231. Am. Soc. Agron., Madison, WI.

100   Wright, M. J. 1976. Plant adaptation to mineral stress in problem soils. Cornell Univ. Agric. Exp. Stn., Ithaca, NY.

101   Yuan, T. L. 1974. A double buffer method for the determination of lime requirement of acid soils. Soil Sci. Soc. Am. Proc. 38:437-440.

102   Yuan, T. L. 1976. Anomaly and modification of pH-acidity relationship in the double buffer method for lime requirement determination. Soil Sci. Soc. Am. J. 40:800-802.

# chapter 9 ▬▬▬▬▬▬▬▬▬

**ALLEVIATING SALINITY STRESS**

**9**

# 9

## ALLEVIATING SALINITY STRESS

by    Glenn J. Hoffman, U.S. Salinity Laboratory,
      USDA, SEA/AR, Riverside, CA

### 9.1 INTRODUCTION

The rootzone of all soils contains a mixture of soluble salts that reduce crop productivity in excessive concentrations. Yield reductions may result from osmotic stress caused by the total soluble salt concentration; from toxicities or nutrient imbalances created when specific solutes become excessive; or from a reduction of water penetration through the rootzone caused by excess sodium inducing a deterioration of soil structure. The occurrence of a salinity stress depends on the salt content of the applied water, the soil, and the groundwater; on water table elevations, crop salt tolerance and water use, geologic and climate conditions; and on management.

Salts are leached whenever water applications exceed evapotranspiration, provided soil infiltration and drainage rates are adequate. In some regions, rainfall periodically flushes salts; in others, provisions must be made for adequate leaching. The key to alleviating salinity stress is a net downward movement of soil water through the rootzone. Even in well-managed, high-yielding fields, however, the soil water may be several times more saline than the irrigation water because evapotranspiration selectively removes water, concentrating the salts in the remaining soil water. With insufficient leaching, this ratio can easily increase to tenfold or more, with resultant damage to crops.

The original source of salt in soils is the weathering of rocks and minerals in the earth's crust. During weathering, bonds between elements are broken and new combinations are formed, including soluble salts. Throughout geologic time, soluble salts have been exchanged between land and sea through the water cycle. The most significant transfer mechanism for salts from oceans to the continents is by the accumulation of salts in marine sediments, followed by periods of uplift, and exposure of the sediments at the earth's surface. The Mancos shales of Colorado, Utah, and Wyoming, the Permian Red Beds of Kansas and Oklahoma, the Green River formation of Utah, and the marine sediments along the west side of the San Joaquin Valley in California are prominent examples of exposed marine deposits that contain large quantities of salt. The ocean, of course, is also a direct source of salts from sea water intrusion. Completing the water cycle, soluble salts are transported from the continents to the oceans by water moving over and under the landscape to water courses that terminate in the sea.

Salinity is estimated to be a potential threat to about half of the 20 million ha of irrigated land in the western United States, with crop production being limited by salinity on about one-fourth of this land (Wadleigh, 1968). The total area of salt-affected soils throughout the world is not known precisely, but it is large. A recent survey indicated that of the 756 million ha cultivated and the 91 million irrigated in 24 countries, over 50 million ha is affected by salinity (Shalhevet and Kamburov, 1976).

Although inevitable in arid regions, salinity hazards are not confined to irrigated agriculture. Rainfed agriculture in semi-arid regions can encounter severe salinity problems where rainfall is approximately equal to evapotranspiration and soluble salts are present from marine deposits or from soil weathering. Such problems may be due to over-irrigation in nearby areas, to a high water table resulting from restricted natural drainage, or to upward flow from great depths associated with artesian pressures in underlying aquifers. Saline seeps, common in North America and Australia, result from excess water percolating below the rootzone because of changes in cropping patterns or replacement of natural vegetation. This percolating soil water becomes saline, is intercepted by impermeable horizontal layers, and conducted laterally to hillside surfaces or landscape depressions (Doering and Sandoval, 1976). In the Red River Valley of North Dakota, a salinity problem is caused by upward migration of dissolved salts with groundwater under artesian pressure (Benz et al., 1976). Salt problems have also developed in the Columbia River Basin, even though the salt content of the irrigation water is low, because inadequate drainage has resulted in a shallow, saline water table. In more humid areas, salinity problems are encountered near oceans, as in the delta of the Senegal River in Africa. The solution to such problems differs greatly in detail, depending on specific circumstances. If a generalization is possible, it is that the management objective should be to establish and maintain a downward flux of water to remove soluble salts.

This chapter outlines the principles controlling salt movement in the rootzone, summarizes the influence of soil salinity on crop production, describes principles and techniques for diagnosing a salinity hazard, introduces management techniques to cope with salinity in the rootzone, and outlines major areas where our knowledge of techniques for minimizing salinity stress is still lacking.

## 9.2 DYNAMICS OF SALT IN THE ROOTZONE

Dissolved salts move with the water in which they are contained. In the rootzone, dissolved salts move with the soil water as it is pulled by the force of gravity, by adsorptive forces arising from soil particle surfaces, or along energy gradients to plant roots or the soil surface. The dissolved salts, however, are left behind at evaporation sites and the major fraction of dissolved salts is also excluded from transpiration by plant roots. As a consequence, the distribution of salts and water are seldom uniform in the rootzone and depend not only upon the quantity of salt and water, but also upon a dynamic balance among the forces that govern their rate and direction of movement.

### 9.2.1 Steady State

The major inflows and outflows from a unit surface of a soil profile in a

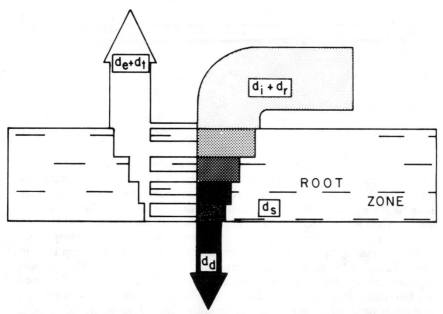

**FIG. 9.2-1 Major inflows and outflows from a unit surface area of a soil profile in a given time. The width of the arrows indicates amount of flow and their degree of shading indicates salt content.**

given time is illustrated in Fig. 9.2-1. The width of the arrows indicates amount of flow and their degree of shading indicates salt content. Volumes of water are expressed as equivalent depths per unit surface area. At the soil surface, inflows are from irrigation ($d_i$) and rainfall ($d_r$). Outflows are from evaporation ($d_e$), transpiration ($d_t$), and drainage ($d_d$). Because the process of evaporation and transpiration remove essentially pure water, the amount of salt leaving the profile through the upper surface is negligible. In the example illustrated, $d_i + d_r$ exceeds $d_e + d_t$ so water is available for soil storage ($d_s$) or drainage. For a given rootzone depth, the difference between inflows and outflows must equal storage. That is

$$d_s = d_i + d_r - d_e - d_t - d_d. \quad\dots\dots\dots\dots\dots\dots\dots\dots\dots\dots\dots\dots\dots\dots\quad [9.2\text{-}1]$$

The change in mass of salt stored per unit area within a given rootzone ($m_s$) is

$$m_s = c_i d_i - c_d d_d \quad\dots\dots\dots\dots\dots\dots\dots\dots\dots\dots\dots\dots\dots\dots\dots\dots\quad [9.2\text{-}2]$$

where the $c$'s represent the mean salt concentration during the time period. This equation assumes negligible salt concentration in rainfall, evaporation, and transpiration. For sufficiently long time periods, $m_s$ is normally small compared to the other terms and, therefore, can be neglected. The resulting steady-state equation is

$$c_i d_i = c_d d_d \quad\dots\dots\dots\dots\dots\dots\dots\dots\dots\dots\dots\dots\dots\dots\dots\dots\dots\dots\quad [9.2\text{-}3]$$

The reader is cautioned that this equation, as well as other steady-state equations that follow, applies only to salt constituents that remain dissolved.

The leaching fraction (LF), defined as the steady-state ratio of water leaving the profile as drainage to that entering as irrigation or rain water, can be expressed as

$$LF = d_d/(d_i+d_r). \quad \dots\dots\dots\dots\dots\dots\dots\dots\dots\dots\dots [9.2-4]$$

By substitution from the equation above,

$$c_d - c_i d_i/[LF(d_i + d_r)] \dots\dots\dots\dots\dots\dots\dots\dots\dots\dots [9.2-5]$$

This equation gives the mean concentration of the water leaving the root-zone at steady state as a function of leaching fraction and the depths and concentrations of soluble salts for irrigation and rain. It must be modified to account for chemical precipitation and dissolution where slightly soluble solutes, such as gypsum and lime, are present (Rhoades et al., 1974).

The relationship between leaching fraction and rootzone salinity is important in managing soil salinity. The minimum value of LF a crop can endure without yield reduction is the leaching requirement ($L_r$). The leaching requirement depends on the salt content of the applied water as well as crop salt tolerance. Because several factors can influence the salt tolerance of a crop, $L_r$ is not a single value for a given crop. Knowledge of the values of $L_r$ for various crops is one area of salinity management where data are completely lacking.

### 9.2.2 Non-Steady-State Adjustments

If ($d_i + d_r$) is less than ($d_e + d_t$), the soil water deficit is first met by extraction from soil storage ($d_s$). As $d_s$ is depleted, the soil dries and the crop becomes water-stressed which reduces evaporation ($d_e$) and transpiration ($d_t$). These processes initially bring the water loss from the rootzone in balance with the water supply at zero drainage. However, without drainage, salt stored in the rootzone concentrates in the remaining volume of stored water. This increase causes plant osmotic stress, further reducing transpiration. This further reduction continues until the plant dies or until transpiration is reduced to the point that excess water is again present in the profile and drainage begins, carrying salt out of the rootzone. The steady leaching fraction that results depends on the capability of the crop to extract water from a saline rootzone. This represents the absolute minimum leaching fraction of the crop. This leaching fraction, however, is far less than the leaching requirement because yield reductions are significant.

In the presence of a water table, deficiencies in ($d_i + d_r$) may be offset by upward flow from the groundwater. In this case $d_d$ is negative. This situation, however, does not provide salt export from the profile and it cannot endure indefinitely. Ultimately soil salinity will reduce crop water consumption to the point that upward flow ceases and some leaching occurs or the crop dies.

Temporary use of water from soil storage or from a shallow water table can be useful as a strategy for managing water, but over the long term where salinity is a hazard crops can be grown only with a net downward flow of water through the rootzone. Rarely will conditions controlling inflow and outflow of water from the rootzone prevail long enough for true steady state to occur. As a consequence, the quantity and salt content of stored water in the rootzone fluctuates continually. The goal of water management is to keep this fluctuation within limits that neither allows excess drainage nor reduces crop growth by stress.

### 9.2.3 Salt Distribution in the Rootzone

After an irrigation, salts contained in the rootzone are carried deeper into the soil profile. If a soil is irrigated with saline water and not enough is applied each time to leach the salts below the rootzone, salt accumulates near the depth to which the soil is wet during each irrigation. If a saline soil is leached with salt-free water, the salt is carried down with the water and much of it carried below the rootzone. A considerable amount of mixing of the incoming water with that already in the soil occurs, and although there is a higher concentration of salt near the wetting front, some salt remains distributed in the rootzone. This occurs because of the difference in the flow velocity of water between the larger soil pores or channels and the fine pores, which allows some of the penetrating water to bypass some salt and facilitates mixing and lateral diffusion of salt.

Salt distribution within the rootzone is also greatly influenced by the

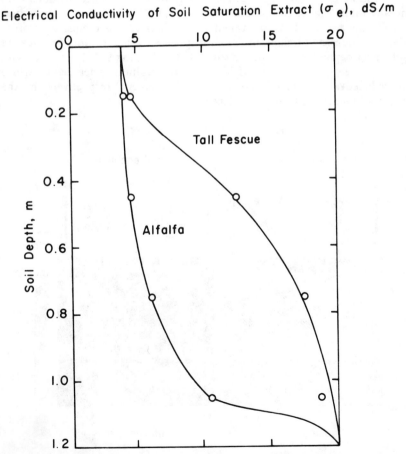

FIG. 9.2-2 Steady-state salinity profiles for tall fescue and alfalfa. (Adapted from Bower et al., 1969, 1970).

crop's pattern of water extraction. This is illustrated in Fig. 9.2-2 for deep-rooted alfalfa compared with shallow-rooted tall fescue. Although both soil (Pachappa sandy loam) profiles were treated identically, the electrical conductivity of the irrigation water ($\sigma_i$) was 4 dS/m and the leaching fraction (LF) was 0.2, the grass extracted most of its water above the 0.5-m depth, and almost none below 1.0 m. Thus, because plant roots extract relatively pure water, the salt concentrated where water was extracted. Alfalfa extracted water from much deeper, and significant quantities of salt accumulated below the 1.0-m depth.

Fig. 9.2-3 illustrates how leaching can influence soil salinity distribution. Here, water was applied daily by drip irrigation along every other plant row. The soil salinity distributions, as measured by soil-water chloride concentration, are given for 17- and 2- percent leaching. The chloride concentration of the irrigation water was 7.5 mol/m³ and $\sigma_i$ was 2.2 dS/m. Thus, assuming no significant amount of chloride uptake by the crop, a soil-water chloride concentration of 15 mol/m³ would be associated with locations in the rootzone where half of the irrigation water had been evapotranspired. A soil-water chloride concentration of 75 mol/m³ would indicate where 90 percent of the water had been evapotranspired. While salt concentrations were generally much higher for 2- than for 17-percent leaching, the salinity patterns for the two leaching treatments were similar, and typical of profiles under line sources of irrigation. Yield responses to such salinity profiles are highly crop dependent, as will be discussed in Section 9.3. Beginning with the second year of treatment, yields of grain sorghum, wheat, and lettuce were reduced approximately 10, 30, and 70 percent, respectively, by the reduction in leaching from 17 to 2 percent.

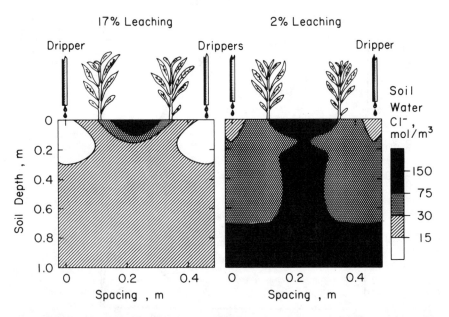

FIG. 9.2-3 Steady-state soil-water chloride profiles for 17- and 2-percent leaching under trickle irrigation. (Hoffman et al., 1979).

### 9.2.4 Sodium-Induced Reduction of Soil Water Penetration

One of the major factors affecting the stability of soil structure, and hence water penetration, is the solute composition of the soil water. Of the cations commonly found in natural waters, sodium is most apt to reduce water penetration when present in excess on the cation-exchange sites of soil particles. When calcium and/or magnesium are the predominant cations adsorbed on the soil exchange complex, the soil tends to have a granular structure that is easily tilled and readily permeable. When the exchangeable-sodium-percentage (ESP) exceeds 10 percent, soil mineral particles tend to disperse and hydraulic conductivity decreases. ESP is defined as

$$ESP = 100\ X_{Na}/X_c, \dotfill [9.2\text{-}6]$$

where $X_c$ is the cation-exchange-capacity of the soil and $X_{Na}$ is the amount of exchangeable sodium. Because drainage is essential for salinity management and reclamation, good soil structure must be maintained.

Permeability becomes a problem when the rate of soil water penetration is reduced to the point that the crop is not adequately supplied with water. Sodium may also add to cropping difficulties through crusting of seed beds, temporary saturation of surface soil, and possible disease, weed, oxygen, nutritional, and salinity problems. Water of very low salt content can aggravate a sodium problem because it allows maximal swelling and dispersion of soil minerals and organic matter, and also has a tremendous capacity to dissolve and remove calcium. The sodium-adsorption-ratio (SAR) of soil saturation extracts is generally a good indicator of how much exchangeable sodium is present and is defined as

$$SAR = Na/\sqrt{(Ca + Mg)}, \dotfill [9.2\text{-}7]$$

where all ion concentrations are in $mol/m^3$. In some cases, SAR of the irrigation water has not proved a satisfactory guide to potential soil permeability problems because of the influence of carbonates and bicarbonates on the precipitation of calcium and magnesium. Thus the relationship for adjusted SAR,

$$adj\ SAR = \frac{Na}{\sqrt{(Ca + Mg)}}\ [1 + (8.4 - pH_c)] \dotfill [9.2\text{-}8]$$

was developed (Rhoades, 1972) to estimate the potential permeability hazard as a result of irrigation waters, taking into account the relative concentration of sodium, calcium, magnesium, carbonate, and bicarbonate in the irrigation water. All ion concentrations for the relationship are in $mol/m^3$. Values for $pH_c$, a theoretical pH for the water if in contact with solid-phase lime at atmospheric levels of carbon dioxide, can be obtained from Ayers and Westcot (1976).

Most crops show small or only moderate losses in yield when favorable soil structure is maintained in the presence of high levels (15 to 30 percent) of exchangeable sodium (Bernstein and Pearson, 1956; Pearson and Bernstein, 1958). In general, higher SAR levels can be tolerated in coarse-textured than in fine-textured soils. Data on crop response and changes in hydraulic conductivity with changes in SAR are limited and additional research is needed. The crusting problem caused by using snow melt for irrigation is of particular concern in several irrigation areas and warrants research.

FIG. 9.3-1 Influence of salinity on crop growth or yield. Decline in onion bulb size due to salinity levels of 1, 6, 11, and 17 dS/m from left to right (upper left). Growth reduction of cotton for salinity levels of 1, 14, 28, and 42 dS/m from left to right (upper right). Reduction in growth of tomato and oats in lower photographs from 17-percent leaching on the left to 2-percent leaching on the right with irrigation water having a salinity level of 2.2 dS/m.

## 9.3 PLANT RESPONSE TO SALT-AFFECTED SOILS

The predominant influence of salinity stress on plants is growth suppression (Fig. 9.3-1). Typically, growth decreases linearly as salinity increases beyond a threshold level, and the effect is similar whether salinity is increased by raising the concentration of nutrients or by adding non-nutrient salts, such as sodium chloride, sodium sulfate, or calcium chloride, which are common in saline soils (Hayward and Long, 1941). Growth suppression typically is a nonspecific salt effect, depending more on osmotic stress created by the total concentration of soluble salts than on the level of specific solutes. Excessive concentrations of some individual solutes, as described in Section 9.3.4, can also be detrimental to some crop species in addition to their osmotic effect.

### 9.3.1 Plant Growth Characteristics

Soil salinity is typically so variable that areas of a growing crop can show distinctively different symptoms within the same field. Bare spots, poor spotty stands, and severely stunted plants are all signs of a serious salinity problem (Fig. 9.3-2). However, moderate salinity levels often restrict plant growth without any overt injury symptoms. Salt-affected plants usually appear normal although they are stunted in stature. Salt-affected leaves may be smaller and may have a darker blue-green color than normal leaves. Increased succulence (defined as water content per unit leaf area) is also symptomatic, particularly as a response to chloride salinity. Chlorosis (the yellowing or blanching of green plant parts) is not a typical characteristic of salt-affected plants.

**FIG. 9.3-2 Salt-affected fields in Colorado (upper left), Texas (lower left), and California (upper and lower right).**

Wilting is not a regular characteristic of salt-affected plants because it typically occurs when water availability decreases rather abruptly, as in a drying soil. Under saline conditions moderately low water potentials are always present and water potential changes are usually gradual. Plants are, therefore, hardened by the continual stress and are less apt to exhibit abrupt changes in turgor.

Acute injury symptoms, such as marginal or tip burn of leaves, occur as a rule only in woody plants in which these symptoms indicate toxic accumulations of chloride or boron. Most non-woody plants do not exhibit leaf injury symptoms even though some accumulate as much chloride or boron as woody species that show injury. Sprinkling with saline water, however, induces leaf burn symptoms in both woody and non-woody species. In rare instances, excess sodium may cause calcium deficiency symptoms.

Salinity reduces root growth as well as shoot growth but generally to a lesser degree. Root maturation occurs closer to the root tip in saline soil because root growth is retarded. Roots in saline media generally appear normal, but at high salinity levels they sometimes develop bulbous regions just behind the tip.

Besides generally stunting plant growth, salinity causes many specific anatomical changes that are often related to the ionic composition of the root media (Maas and Nieman, 1978). Chloride salinity may cause larger epidermal cells, fewer stomata per unit leaf area, and a poorly developed xylem system. In contrast, sulfate salinity produces smaller cells and an increased number of stomata per unit leaf area. Salinity retards all plant growth processes that have been measured — cell enlargement and cell division; the production of proteins and nucleic acids; and the increase in plant mass (Maas and Nieman, 1978). On the subcellular level, salinity affects membrane structure and integrity, causes swelling and structural distortion in chloroplasts, and causes swelling of mitochondria and the golgi system. Although many salt-induced structural changes are certainly detrimental to the plant, some undoubtedly are beneficial adaptations that contribute to survival.

In spite of the vast knowledge gained on the effects of salinity on plant growth, much still is unknown. For example, only recently has insight been gained into the influence of salinity on the bioenergetics of plant growth (Nieman, 1980).

### 9.3.2 Crop Salt Tolerance

The relative salt tolerances of selected agricultural crops are given in Table 9.3-1, as taken from Maas and Hoffman (1977). This alphabetical crop list provides two essential parameters sufficient to evaluate salt tolerance: (a) the threshold salinity level [the maximum allowable salinity that does not reduce yield measurably below that of a nonsaline control treatment] and (b) the percent yield decrease per unit of salinity increase beyond the threshold. All salinity values are reported as $\sigma_e$ [the electrical conductivity of soil saturation extracts reported in units of dS/m and corrected for temperature to 25 °C] and rounded to two significant digits. A qualitative salt-tolerance rating is also given for quick, relative comparisons among crops. These ratings are defined by the boundaries shown in Fig. 9.3-3. Unfortunately, Table 9.3-1 is far from complete; many crops are not included or only qualitative ratings are given because data are lacking.

The data presented in Table 9.3-1 were normally obtained by artificially salinizing field plots and using cultural practices that closely simulated field conditions. Thus, they indicate the salt tolerance to be expected under

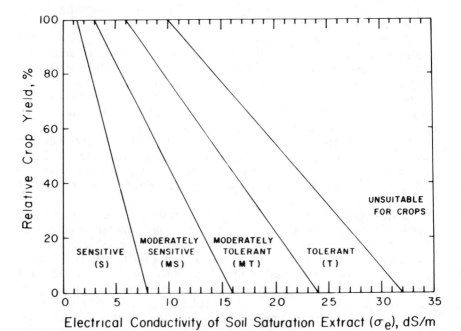

FIG. 9.3-3 Divisions for qualitative salt-tolerance ratings of agricultural crops. Symbols given compare with those in Table 9.3-1. (Maas and Hoffman, 1977).

TABLE 9.3-1. SALT TOLERANCE OF AGRICULTURAL CROPS AS A
FUNCTION OF SOIL SATURATION EXTRACT SALINITY ($\sigma_e$)
WHERE RELATIVE YIELD (Y) IN PERCENT = $100 - B(\sigma_e - A)$
(From Maas and Hoffman, 1977)

| Crop | Salinity* at initial yield decline (threshold) (A), dS/m | Percent yield decrease per unit increase in salinity beyond threshold (B), %/(dS/m) | Qualitative salt tolerance rating† |
|------|------|------|------|
| Alfalfa | | | |
|   *Medicago sativa* | 2.0 | 7.3 | MS |
| Almond | | | |
|   *Prunus dulcis* | 1.5 | 19 | S |
| Apple | | | |
|   *Malus sylvestris* | — | — | S |
| Apricot‡ | | | |
|   *Prunus armeniaca* | 1.6 | 24 | S |
| Avocado‡ | | | |
|   *Persea americana* | — | — | S |
| Barley (forage)§ | | | |
|   *Hordeum vulgare* | 6.0 | 7.1 | MT |
| Barley (grain)§ | | | |
|   *Hordeum vulgare* | 8.0 | 5 | T |
| Bean | | | |
|   *Phaseolus vulgaris* | 1.0 | 19 | S |
| Beet, garden‖ | | | |
|   *Beta vulgaris* | 4.0 | 9 | MT |
| Bentgrass | | | |
|   *Agrostis palustris* | — | — | MS |
| Bermudagrass# | | | |
|   *Cynodon dactylon* | 6.9 | 6.4 | T |
| Blackberry | | | |
|   *Rubus* spp. | 1.5 | 22 | S |
| Boysenberry | | | |
|   *Rubus ursinus* | 1.5 | 22 | S |
| Broadbean | | | |
|   *Vicia faba* | 1.6 | 9.6 | MS |
| Broccoli | | | |
|   *Brassica oleracea botrytis* | 2.8 | 9.2 | MS |
| Bromegrass | | | |
|   *Bromus inermis* | — | — | MT |
| Cabbage | | | |
|   *Brassica oleracea capitata* | 1.8 | 9.7 | MS |
| Canarygrass, reed | | | |
|   *Phalaris arundinacea* | — | — | MT |
| Carrot | | | |
|   *Daucus carota* | 1.0 | 14 | S |
| Clover, alsike, ladino, red, strawberry | | | |
|   *Trifolium* spp. | 1.5 | 12 | MS |
| Clover, berseem | | | |
|   *T. alexandrinum* | 1.5 | 5.7 | MS |
| Corn (forage) | | | |
|   *Zea mays* | 1.8 | 7.4 | MS |
| Corn (grain) | | | |
|   *Zea mays* | 1.7 | 12 | MS |
| Corn, sweet | | | |
|   *Zea mays* | 1.7 | 12 | MS |
| Cotton | | | |
|   *Gossypium hirsutum* | 7.7 | 5.2 | T |
| Cowpea | | | |
|   *Vigna unguiculata* | 1.3 | 14 | MS |

Table 9.3-1 cont'd

| | | | |
|---|---|---|---|
| Cucumber | | | |
| *Cucumis sativus* | 2.5 | 13 | MS |
| Date palm | | | |
| *Phoenix dactylifera* | 4.0 | 3.6 | T |
| Fescue, tall | | | |
| *Festuca elatior* | 3.9 | 5.3 | MT |
| Flax | | | |
| *Linum usitatissimum* | 1.7 | 12 | MS |
| Grape ‡ | | | |
| *Vitis* spp. | 1.5 | 9.6 | MS |
| Grapefruit‡ | | | |
| *Citrus x paradisi* | 1.8 | 16 | S |
| Hardinggrass | | | |
| *Phalaris tuberosa* | 4.6 | 7.6 | MT |
| Lemon‡ | | | |
| *Citrus limon* | — | — | S |
| Lettuce | | | |
| *Lactuca sativa* | 1.3 | 13 | MS |
| Lovegrass** | | | |
| *Eragrostis* spp. | 2.0 | 8.4 | MS |
| Meadow Foxtail | | | |
| *Alopecurus pratensis* | 1.5 | 9.6 | MS |
| Millet, Foxtail | | | |
| *Setaria italica* | — | — | MS |
| Okra | | | |
| *Abelmoschus esculentus* | — | — | S |
| Olive | | | |
| *Olea europaea* | — | — | MT |
| Onion | | | |
| *Allium cepa* | 1.2 | 16 | S |
| Orange | | | |
| *Citrus sinensis* | 1.7 | 16 | S |
| Orchardgrass | | | |
| *Dactylis glomerata* | 1.5 | 6.2 | MS |
| Peach | | | |
| *Prunus persica* | 1.7 | 21 | S |
| Peanut | | | |
| *Arachis hypogaea* | 3.2 | 29 | MS |
| Pepper | | | |
| *Capsicum annuum* | 1.5 | 14 | MS |
| Plum‡ | | | |
| *Prunus domestica* | 1.5 | 18 | S |
| Potato | | | |
| *Solanum tuberosum* | 1.7 | 12 | MS |
| Radish | | | |
| *Raphanus sativus* | 1.2 | 13 | MS |
| Raspberry | | | |
| *Rubus idacus* | — | — | S |
| Rhodesgrass | | | |
| *Chloris gayana* | — | — | MS |
| Rice, paddy § | | | |
| *Oryza sativa* | 3.0 | 12 | MS |
| Ryegrass, perennial | | | |
| *Lolium perenne* | 5.6 | 7.6 | MT |
| Safflower‡ | | | |
| *Carthamus tinctorius* | — | — | MT |
| Sesbania § | | | |
| *Sesbania exaltata* | 2.3 | 7 | MS |
| Sorghum | | | |
| *Sorghum bicolor* | — | — | MS |
| Soybean | | | |
| *Glycine max* | 5.0 | 20 | MT |
| Spinach | | | |
| *Spinacia oleracea* | 2.0 | 7.6 | MS |
| Strawberry | | | |
| *Fragaria* spp. | 1.0 | 33 | S |
| Sudangrass | | | |
| *Sorghum sudanense* | 2.8 | 4.3 | MT |
| Sugarbeet‖ | | | |
| *Beta vulgaris* | 7.0 | 5.9 | T |

Table 9.3-1 cont'd

| | | | |
|---|---|---|---|
| Sugarcane | | | |
| Saccharum officinarum | 1.7 | 5.9 | MS |
| Sweet potato | | | |
| Ipomoea batatas | 1.5 | 11 | MS |
| Timothy | | | |
| Phleum pratense | — | — | MS |
| Tomato | | | |
| Lycopersicon, L. | | | |
| Lycopersicum | 2.5 | 9.9 | MS |
| Trefoil, Big | | | |
| Lotus uliginosus | 2.3 | 19 | MS |
| Trefoil, Birdsfoot narrowleaf‡‡ | | | |
| L. corniculatus | | | |
| tenuifolium | 5.0 | 10 | MT |
| Vetch | | | |
| Vicia sativa | 3.0 | 11 | MS |
| Wheat §,†† | | | |
| Triticum aestivum | 6.0 | 7.1 | MT |
| Wheatgrass, crested | | | |
| Agropyron desertorum | 3.5 | 4 | MT |
| Wheatgrass, fairway | | | |
| A. cristatum | 7.5 | 6.9 | T |
| Wheatgrass, slender | | | |
| A. trachycaulum | — | — | MT |
| Wheatgrass, tall | | | |
| A. elongatum | 7.5 | 4.2 | T |
| Wildrye, Altai | | | |
| Elymus angustus | — | — | T |
| Wildrye, beardless | | | |
| E. triticoides | 2.7 | 6 | MT |
| Wildrye, Russian | | | |
| E. junceus | — | — | T |

*Salinity expressed as $\sigma_e$ (dS/m = decisiemens per meter = 1 millimho/cm, referenced to 25 °C).

†Ratings are defined by the boundaries in Fig. 9.3-3.

‡Tolerance is based on growth rather than yield.

§Less tolerant during emergence and seedling stages. $\sigma_e$ should not exceed 4 to 5 dS/m at these times.

‖Sensitive during germination. $\sigma_e$ should not exceed 3 dS/m at this time for garden beet and sugarbeet.

#Average of several varieties. Suwannee and Coastal are about 20 percent more tolerant, and Common and Greenfield are about 20 percent less tolerant than the average.

**Average for Boer, Wilman, Sand, and Weeping varieties. Lehmann seems about 50 percent more tolerant than the other varieties tested.

††The salt tolerance of some semidwarf varieties may be higher.

‡‡Broadleaf birdsfoot trefoil seems less tolerant than narrowleaf trefoil.

representative cultural conditions for that particular crop. Typically, salinity treatments were imposed at the seedling stage, so the data do not represent salt tolerance for the germination and early seedling growth stages. Soil salinity was maintained relatively uniform by irrigating frequently at a leaching fraction of about 0.5 with waters of a given salt concentration. Crop yields were correlated with $\sigma_e$ of soil samples from the major part of the crop rootzone.

In general, decreases in yield with increasing salinity depart from linearity only in the lower part of the relative yield curves where yields are commercially unacceptable. With some crops, e.g., bean, onion, clover, and pepper, yield approaches zero asymptotically; with a few others, yields decrease linearly with increasing salinity to a point beyond which yields drop sharply to zero as the plants die. Salinity at zero yield is of interest because it may be used to estimate the leaching requirement (van Schilfgaarde et al., 1974).

Relative yield (Y) in percent at any given soil salinity, expressed as the electrical conductivity of the soil saturation extract ($\sigma_e$), can be calculated by the equation

$$Y = 100 - B \, (\sigma_e - A), \quad \dotfill \quad [9.3\text{-}1]$$

where A is the salinity threshold value and B is the yield decrease per unit salinity increase, as given in Table 9.3-1. For example, alfalfa yields decrease approximately 7.3 percent per dS/m when the soil salinity exceeds 2.0 dS/m. Therefore, at a soil salinity of 5.4 dS/m,

$$Y = 100 - 7.3 \, (5.4 - 2.0) = 75\%. \quad \dotfill \quad [9.3\text{-}2]$$

Division boundaries for the salt tolerance ratings defined in Fig. 9.3-3 were chosen to approximate the family of linear curves that represent most of the crops reported. Four divisions were labeled to correspond with previously published terminology ranging from sensitive to tolerant. With few exceptions, the linear salt tolerance curves for each crop remained within a single division. Where the linear salt tolerance curve for a crop happened to cross division boundaries, the crop was rated based on its relative tolerance at the lower salinity levels.

### 9.3.3 Factors Influencing Salt Tolerance

**9.3.3.1 Growth Stage.** Many crops seem to tolerate salinity equally well during seed germination and later growth stage (Bernstein and Hayward, 1958). Germination failures that occur on saline soils are frequently not due to crops being especially sensitive during germination, but to exceptionally high concentrations of salt where the seeds are planted. These high salt concentrations at the soil surface are a consequence of upward water movement and evaporation in the absence of water applications.

The salt tolerance of some crops, however, does change with growth stage (Maas and Hoffman, 1977). For example, barley, wheat, and corn are more sensitive to salinity during early seedling growth than during germination or later growth stages, whereas sugarbeet and safflower are relatively sensitive during germination. The tolerance of soybean may either increase or decrease from germination to maturity, depending on the variety.

**9.3.3.2 Varieties and Rootstocks.** Varietal differences, while not common, must be considered in evaluating crop salt tolerance. Although most known varietal differences are among grass species (i.e., bermudagrass, bromegrass, creeping bentgrass, barley, rice, and wheat), some variation has been noted among legumes (birdsfoot trefoil, soybean, and berseem clover) (Maas and Hoffman, 1977). Increased varietal differences are anticipated in the future as new varieties are developed based upon their salt tolerances.

Rootstocks are also an important factor affecting the salt tolerance of tree and vine crops. Varieties and rootstocks of avocado, grapefruit, and orange that differ in their ability to absorb and transport sodium and chloride have different salinity tolerances (Cooper, 1951, 1961). Similar effects of rootstocks on salt accumulation and tolerance have been reported for stone-fruit trees and grapes.

**9.3.3.3 Nutrition.** Apparent salt tolerance may also vary with soil fertility. Crops grown on infertile soils may have abnormally high apparent salt tolerances when compared with crops grown on fertile soils, because yields on nonsaline soil are severely limited by inadequate fertility. Obviously, proper fertilization would increase absolute crop yields even though apparent salt tolerance (i.e., relative yield) would be decreased.

Unless salinity causes specific nutritional imbalances, fertilization in excess of nutritional need generally has little effect on, and can even reduce, relative salt tolerance. Apparent decreases in salt tolerance accompanying excessive nitrogen fertilization have been reported for corn, cotton, rice, wheat, and spinach (Maas and Hoffman, 1977). Bernstein et al. (1974) concluded from sand culture studies that high nitrogen or potassium levels did not improve the salt tolerance of wheat, barley, corn, or six vegetable crops (garden beet, broccoli, cabbage, carrot, lettuce, and onion).

**9.3.3.4 Irrigation Management.** Irrigation practices frequently influence salinity damage. Assuming that plants respond primarily to the soil water salinity in that part of the rootzone with the highest total water potential, then time-integrated salinity measured in the zone of maximum water uptake should correlate best with crop response. When irrigated frequently, this zone corresponds primarily to the upper part of the rootzone where soil salinity is influenced primarily by the salinity of the irrigation water (Bernstein and Francois, 1973b). With infrequent irrigation, the zone of maximum water uptake becomes larger as the plant extracts water from increasingly saline solutions at greater depths. The decreasing soil matric potential during a typical irrigation cycle is an additional inhibitory factor to plant growth, the effect of which seems additive to that of decreased soil water osmotic potential (Wadleigh and Ayers, 1945).

**9.3.3.5 Aerial Environment.** Environmental factors, such as temperature, atmospheric humidity, and air pollution, markedly influence crop salt tolerance. Many crops seem less salt tolerant when grown in hot, dry environments than in cool, humid ones (Magistad et al., 1943). High atmospheric humidity alone tends to increase the salt tolerance of some crops, with high humidity generally benefitting salt-sensitive more than salt-tolerant crops (Hoffman and Rawlins, 1971; Hoffman and Jobes, 1978).

A strong interaction between salinity and ozone, a major air pollutant, has been found for pinto bean, garden beet, and alfalfa. At ozone concentrations often prevalent in agricultural settings near metropolitan areas, alfalfa yields may be increased by maintaining moderate, but not detrimental, salinity levels (Hoffman et al., 1975). Salinity also reduced ozone damage to pinto bean and garden beet, but effects were beneficial only at salinity and ozone levels too high for economical crop production. These results, however, indicate that salinity-ozone interactions may be commercially important for leafy vegetable and forage crops.

### 9.3.4 Specific Solute Effects
**9.3.4.1 Toxicity.** Unlike most annual crops, trees and other woody perennials may be specifically sensitive to chloride, which is taken up with soil water, moves with the plant transpiration stream, and accumulates in the leaves. Crop, varietal, and rootstock differences in tolerance to chloride de-

pend largely upon the rate of transport from the soil to the leaves. In general, the slower the chloride absorption, the more tolerant the plant to this solute. Leaf injury symptoms appear in chloride-sensitive crops when leaves accumulate about 0.3- to 0.5-percent chloride on a dry-weight basis. Symptoms develop as leaf burn or drying of leaf tissue, typically occurring first at the extreme tip of older leaves and progressing back along the leaf edges. Excessive leaf burn is often accompanied by defoliation. Chemical analysis of soil or leaves can be used to confirm probable chloride toxicity. The maximum permissible concentrations of chloride in the soil saturation extract or in plant leaves for several sensitive crops are given in Table 9.3-2.

Irrigation methods that wet plant leaves, such as overhead sprinkling, can cause specific solute toxicity problems at solute concentrations lower than those that cause problems with surface irrigation methods. This problem is acute during periods of high temperature and low humidity. Intermittent sprinkling so that the leaves are alternately wet and dry further aggravates the problem because of the rapid increase in the salt concentration

TABLE 9.3-2. CHLORIDE TOLERANCES OF FRUIT CROP
ROOTSTOCKS AND VARIETIES IF LEAF INJURY IS
TO BE AVOIDED

| | | Maximum permissible chloride concentration | |
| | | Soil saturation extract*, mol/m$^3$ | Plant leaf analysis,† %(kg/kg) |
| Crop | Rootstock or variety | | |
|---|---|---|---|
| | Rootstocks | | |
| Citrus | | | |
| Citrus spp. | Rangpur lime, mandarin | 25 | 0.7 |
| | Rough lemon, tangelo sour orange | 15 | |
| | Sweet orange, citrange | 10 | |
| Stone fruit | Marianna | 25 | 0.3 |
| Prunus spp. | Lovell, Shalil | 10 | |
| | Yunnan | 7 | |
| Avocado | West Indian | | |
| Persea americana | Mexican | 8 | 0.25 to 0.50 |
| Grape | Salt creek, 1613-3 | 40 | 0.5 |
| Vitis spp. | Dog ridge | 30 | |
| | Varieties | | |
| Grape | Thompson seedless, Perlette | 25 | 0.5 |
| Vitis spp. | Cardinal, Black rose | 10 | |
| Olive | | | |
| Olea europaea | | | 0.5 |
| Berries‡ | | | |
| Rubus spp. | Olallie blackberry | 10 | |
| | Indian summer raspberry | 5 | |
| Strawberry | Lassen | 8 | |
| Fragaria spp. | Shasta | 5 | |

*From Bernstein (1965).
†From Reisenauer (1976).
‡Data available for a single variety of each crop only.

of the drops that remain on the leaves after each wetting. Sprinkler-irrigated citrus in several California valleys has been damaged with chloride and sodium concentrations as low as 3 mol/m³. These same concentrations caused no toxic effects with furrow or flood irrigation (Harding et al., 1958). Slight damage has been reported for more tolerant crops, such as alfalfa sprinkled under extremely high evaporative conditions with water containing chloride and sodium concentrations in excess of 6 mol/m³. In contrast, waters with sodium and chloride levels of 24 and 37 mol/m³, respectively, have caused little or no damage when evaporation rates were low (Nielson and Cannon, 1975). Keeping the foliage wet continuously during an irrigation, sprinkling at night, and using below-canopy sprinklers may reduce or eliminate foliar damage.

Boron, although an essential minor element, is phytotoxic if present in excess. Most boron problems arise from high concentrations in well waters or springs located near geothermal areas or geological faults. Few surface waters contain enough boron to cause toxicity. Sensitivity to boron is not limited to woody perennials, but affects a wide variety of crops. Boron toxicity symptoms typically appear at the tip and along the edges of older leaves as yellowing, spotting, or drying of leaf tissue. The damage gradually progresses interveinally toward midleaf. A gummosis or exudate on limbs or trunks is sometimes noticeable on boron-affected trees, such as almond. Many sensitive crops show toxicity symptoms when boron concentrations in leaf blades exceed 250 mg/kg, but not all sensitive crops accumulate boron in their leaves. Stone fruits (peach, plum, almond) and pome fruits (pear, apple, and others) may not accumulate enough boron in leaf tissue that leaf analysis is a reliable toxicity indicator.

A wide range of crops have been tested for boron tolerance in sand cultures. The results of these tests are summarized in Table 9.3-3. The crops have been grouped according to their relative tolerance to boron in the irrigation water. These data are based on the boron level at which toxicity symptoms were observed, and do not necessarily indicate corresponding reductions in yield. Establishment of the relationships between boron concentration and crop yield must await further research.

**9.3.4.2 Nutritional Imbalance.** The concentrations of some solutes in saline soils may be several orders of magnitude greater than concentrations of some essential nutrients. With differences of this magnitude, it is surprising that plant nutritional disturbances are not more common in salt-affected soils. However, with the exception of the disturbance of orthophosphate utilization (Nieman and Clark, 1976) salinity has not been found to influence the utilization of essential nutrients by plants (Bernstein et al., 1974).

In some instances, however, if the proportion of calcium to sodium becomes either extremely high or low, nutritional imbalances can occur that reduce crop yield below that expected from osmotic effects alone (Bernstein, 1964). Bean plants, which accumulate calcium, cannot tolerate excess calcium. In contrast, calcium deficiency under some saline conditions results in blossom-end rot of tomatoes, internal browning of lettuce, and reduced corn growth. Because soil salinity in the field normally involves a mixture of salts, the effects of specific solutes on crop nutrition tend to be minimized so that the osmotic effect usually predominates.

**TABLE 9.3-3. RELATIVE TOLERANCE OF CROPS TO BORON***
**(ADAPTED FROM WILCOX, 1960)**

| Tolerant† 4.0 mg/L of boron | Semitolerant 2.0 mg/L of boron | Sensitive 1.0 mg/L of boron |
|---|---|---|
| Asparagus *Asparagus officinalis* | Sunflower, native *Helianthus annus* | Pecan *Carya illinoinensis* |
| Date palm *Phoenix dactylifera* | Potato *Solanum tuberosum* | Walnut, black and Persian or English *Juglans* spp. |
| Sugarbeet *Beta vulgaris* | Cotton, Acala and Pima *Gossypium* spp. | Jerusalem artichoke *Helianthus tuberosus* |
| Garden beet *Beta vulgaris* | Tomato *Lycopersicon, L. Lycopersicum* | Navy bean *Phaseolus vulgaris* |
| Alfalfa *Medicago sativa* | Radish *Raphanus sativus* | Plum *Prunus domestica* |
| Broadbean *Vicia faba* | Field pea *Pisum sativum* | Pear *Pyrus communis* |
| Onion *Allium cepa* | Olive *Olea europaea* | Apple *Malus sylvestris* |
| Turnip *Brassica rapa* | Barley *Hordeum vulgare* | Grape *Vitis* spp. |
| Cabbage *Brassica oleracea capitata* | Wheat *Triticum aestivum* | Kadota fig *Ficus carica* |
| Lettuce *Lactuca sativa* | Corn *Zea mays* | Persimmon *Diospyros virginiana* |
| Carrot *Daucus carota* | Sorghum *Sorghum bicolor* | Cherry *Prunus* spp. |
| | Oat *Avena sativa* | Peach *Prunus persica* |
| | Pumpkin *Cucurbita* spp. | Apricot *Prunus armeniaca* |
| | Beli pepper *Capsicum annuum* | Thornless blackberry *Rubus* spp. |
| | Sweet potato *Ipomoea batatas* | Orange *Citrus sinensis* |
| | Lima bean *Phaseolus lunatus* | Avocado *Persea americana* |
| | | Grapefruit *Citrus X paradisi* |
| | | Lemon *Citrus limon* |
| 2.0 mg/L of boron | 1.0 mg/L of boron | 0.3 mg/L of boron |

*Relative tolerance is based on the boron concentration in irrigation water at which boron toxicity symptoms were observed when plants were grown in sand culture. It does not necessarily indicate a reduction in crop yield.
†Tolerance decreases in descending order in each column between the stated limits.

## 9.4 DIAGNOSING SALINITY HAZARDS

Proper management of salt-affected soils and waters depends upon knowledge of the nature and severity of the salinity problem. Visual observations of soils or crops (Fig. 9.3-2), are rarely sufficient for diagnosis because crop yields can be reduced significantly without any visible symptoms of excess salt. For example, gypsum, a salt that is harmless to crops when not dissolved in soil water, may form a white crust on the soil surface that can lead to a false visual diagnosis of a salinity problem. Reliable diagnosis requires appropriate field and laboratory tests on representative samples of the soil or water in question.

The types of salinity problems commonly confronted are osmotic stress, reduction in soil water penetration, and toxicity or imbalance of specific solutes. Salinity problems are normally diagnosed by laboratory determinations of the collective and individual concentrations of dissolved salts contained in samples of soil, water, or plant tissue. For economic reasons, approximate methods and tests are often employed first to characterize the problem; detailed appraisals are made only as needed. Osmotic stress should be evaluated first because it frequently reduces yield. If osmotic stress is not the problem, then tests for the other types of salinity problems are conducted. Where salinity is severe, visual observations of the crops (Section 9.3) can help to determine what type is likely to be a problem. Occasionally, more than one type occurs simultaneously.

### 9.4.1 Diagnosing Soil Salinity

The first diagnostic step is the measurement of the electrical conductivity of a saturation extract ($\sigma_e$) from a soil sample in the laboratory or by four-electrode units (described in Section 9.4.2) directly in the field. If $\sigma_e$ is within the range that can be tolerated by the crop (Table 9.1-3), osmotic stress is not a problem and the diagnosis should be continued for the other types of salinity problems. If the osmotic stress is excessive and changing to a more tolerant crop is not prudent, then the soil must be leached.

Before leaching, the sodium-adsorption-ratio (SAR), defined in Section 9.2.4, must be determined. This requires laboratory determinations of the concentrations of sodium, calcium, and magnesium in the soil profile. If the SAR is less than 10 and the $\sigma$ of the leaching water is greater than 0.3 dS/m, there is little chance of creating a sodic soil by leaching. In this case leaching can proceed as described in Section 9.5.4.1. If the SAR exceeds 10, the soil may become sodic if leached and water penetration through the soil using the leaching water must be determined. If the daily rate of water penetration exceeds that required to meet the evapotranspiration rate of the crops for an irrigation period, the soil may be reclaimed simply by leaching. If water penetration is less, the soil may still be reclaimed by amending the soil or the leaching water, provided the soil is not excessively high in clay or of poor aggregation. The procedure in Section 9.5.4.3 may be followed to estimate the amount and type of amendment required.

Toxicity of chloride or boron can be assessed by comparing the amount found in laboratory determinations with permissible crop values (Tables 9.3-2 and 9.3-3). Crop appearance, although not definitive, is helpful in establishing the likelihood of toxic or nutrient imbalance problems. Nutrient imbalances normally occur only in sodic soils because excess sodium is concomitant with a reduction in calcium and magnesium. Calcium deficiency can occur at concentrations below 1 mol/m$^3$.

### 9.4.2 Monitoring Soil Salinity

Soil salinity must be measured for diagnosis and management of salt-affected soils. The amount of soluble salts in a soil can be determined or estimated from measurements made on water extracts from soil samples, on direct samples of soil water, and in soil directly with salinity sensors or four-electrode units. Ideally, knowledge of the concentration of each individual solute in the soil water over the entire range of field water contents is desirable. At present, however, field methods are capable of measuring only the total solute concentration. Nevertheless these measurements are extremely valuable for surveying soil, for monitoring irrigation and drainage needs, and for estimating reclamation requirements.

If knowledge of the individual solute concentrations is required, soil water samples must be collected for laboratory analysis. At present, simple methods of obtaining soil water samples at field water contents are not available. Thus, soil solution extracts are usually obtained at water contents higher than those in the field at the time of sampling. The concentration of each solute, of course, is influenced by the soil/water ratio and hence this ratio must be standard for universal application. By convention, soil salinity parameters are measured on water extracts of saturated soil pastes (U.S. Salinity Laboratory Staff, 1954). This ratio was selected because it is the lowest reproducible ratio from which a sufficient volume of solution can be extracted by pressure or vacuum for laboratory analysis and because it is normally related in a predictable manner to field soil water contents. For these reasons, crop salt tolerance is typically expressed on the basis of the electrical conductivity of a soil saturation extract (Section 9.3.2). Once soil saturation extract samples are obtained, laboratory determinations can be made for the same solutes as for irrigation waters (Section 9.4.3).

If the soil is wet, soil water samples can be collected directly from the soil by vacuum extraction through porous ceramic cups. Although the available range of soil water may extend to $-150$ m of soil matric potential for some crops, soil water movement is very slow at matric potentials greater than $-3$ m. Thus, this sampling method is useful for extracting soil water samples only if the soil matric potential is less than $-3$ m.

For many diagnostic and management purposes, a measure of the total salt concentration of the soil water is sufficient. For such cases, devices that can be installed in the field are advantageous. Two types are presently available, the porous-matrix salinity sensor and four-electrode units. The salinity sensor, illustrated in Fig. 9.4-1, measures the electrical conductivity ($\sigma$) of soil water which has been imbibed into a 1-mm thick ceramic disc between two platinum screen electrodes (Richards, 1966). A thermistor is incorporated into the sensor so that $\sigma$ may be referred to a standard temperature (25 °C). Salinity sensors are well suited for continuous monitoring if significant changes in salinity occur over a period of several days or longer. The accuracy of commercial units is about $\pm$ 0.5 dS/m (Oster and Willardson, 1971). Desaturation of the ceramic at soil matric potentials below 20 m causes inaccurate readings in dry soils.

Four-electrode units measure the electrical conductivity of the bulk soil by measuring the resistance to current flow between a pair of electrodes inserted into the soil while an electrical current is passed through the soil between a second pair of electrodes (Rhoades and Ingvalson, 1971). Soil minerals are insulators, so electrical conduction is primarily through the soil water that contains dissolved salts. The electrical conductivity of the bulk soil depends upon the salt content of the soil water, the volumetric water content, the soil texture, and the surface conductance of soil particles. Thus, soil salinity can be determined from measurements taken at a reference soil water content in a given soil type. A distinct advantage of the four-electrode unit is that the spacing among the electrodes can be altered to change the volume of soil measured. Thus, an average $\sigma$ for an entire rootzone or a fraction thereof can be determined.

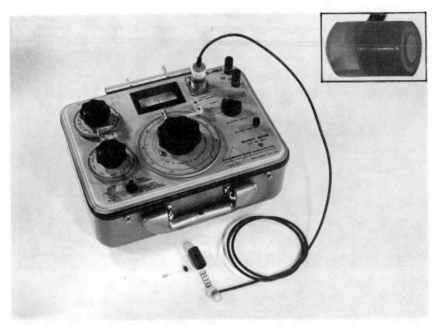

FIG. 9.4-1 Porous-matrix salinity sensor (enlarged view of sensor is shown in insert) with meter.

For large scale measurements the electrodes are inserted into the soil surface in a straight line (Fig. 9.4-2). The depth of current penetration and thus the depth of measurement is approximately equal to one-third of the distance between the outer electrodes. Thus, the average soil salinity can be measured to different depths by varying the electrode spacing. One major advantage of this method is the much larger volume of soil measured compared with the other techniques described.

For information on the salinity distribution within a specified fraction of the rootzone or for permanently installed units, the four-electrode salinity probe (Rhoades and van Schilfgaarde, 1976) and burial type probe (Rhoades, 1979) are recommended, respectively. These probes, shown in Fig. 9.4-2, consist of four annular rings molded into a plastic or rubber cylinder that is tapered slightly so it can be inserted into a hole cored into the soil while retaining soil contact. In the portable version, the probe is attached to a shaft through which the electrical leads are passed to the meter. In the burial unit the leads are simply led to the soil surface. The volume of soil measured can be varied by changing the spacing among the annual rings but the standard commercial unit measures within a soil volume of about 2 L.

## 9.5 MANAGEMENT OF SALT-AFFECTED SOILS

The principles relating to soil properties and plant responses apply equally to salt-affected irrigated cropland and dryland agriculture. The primary difference is that in nonirrigated regions precipitation normally exceeds evapotranspiration so that, if drainage is adequate and the water table remains low enough, excess soluble salts are leached from the rootzone. Where salinity is a hazard and precipitation is not always adequate, measures must be taken to preserve and accumulate rainfall to ensure

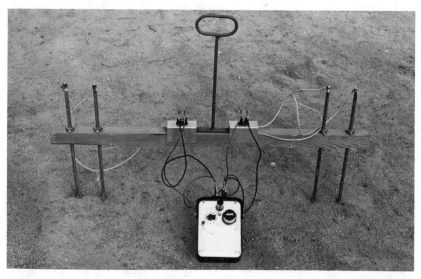

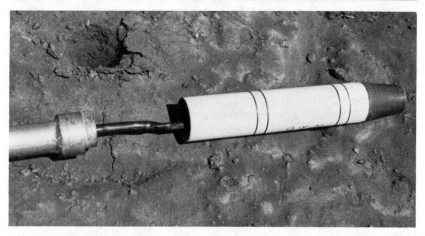

FIG. 9.4-2 Four-electrode units (top) for large scale measurements from the soil surface, (center) for portable measurements with a soil auger, and (bottom) for direct burial.

leaching. In irrigated regions, salinity control is influenced by the quality and quantity of the applied water, the irrigation system and its management, drainage conditions, and agronomic techniques. These factors are often interrelated so that the solution to a salinity problem may not be obvious without proper diagnosis. The key to salinity control is a properly integrated total water management system.

### 9.5.1 Irrigation Management

The primary objective of any irrigation method is to supply water to the soil so that it will be readily available at all times for crop growth, but soil salinity is definitely an influencing factor. The soil salinity profile that develops as water is transpired or evaporated, depends in part on the water distribution pattern inherent with the irrigation method. Distinctly different salinity profiles develop for various irrigation methods, and significantly different profiles can develop for each method within a given field because of differences in soil properties or in the management of the system. Each irrigation method has specific advantages and disadvantages for salinity management; they are only abstracted here.

The major types of flood irrigation are wild flooding, borders, and basins. Wild flooding is not practiced extensively, but is used for pasture or other low value crops where the cost of land preparation for other systems is not warranted. It is normally a very inefficient system that should not be used where salinity is a problem. Border methods utilize levees for control of the water. The water is not impounded, except perhaps at the lower end of the strip, but is flooded over the surface and downslope. This method is suitable for use with alfalfa and grains and in orchards. Excessive water penetration near the levees and at the ends of the strips is common and water penetration tends to be inadequate midway down the strip, which may result in salt accumulations. The basin method of flooding is common where water can be impounded in level basins. This method provides better control of the depths of water applied and greater uniformity in application than other flooding methods. Irrigating orchards by filling small basins beneath each tree from bubblers that are fed from large-diameter, thin-wall plastic pipe laterals is currently the most efficient method of applying precisely the desired amount of water to each tree (Rawlins, 1977).

Sprinkling is an efficient irrigation method that is applicable to most crops. The depth of water applied can be controlled, and when the system is properly designed, water distribution is adequate. Wind, however, causes serious distortion of the water distribution pattern, and in some cases solutes left on the leaves cause foliar damage.

Surface flooding and sprinkler irrigation systems that wet the entire soil surface create a profile that steadily increases in salinity with soil depth to the bottom of the rootzone, if leaching is not excessive, application is uniform, and no shallow, saline water table is present. If irrigation is infrequent, the salt concentration in the soil solution increases with time between irrigation, particularly near the soil surface.

Furrow irrigation is well suited to row crops and can be used where the topography is too uneven or steep for other gravity methods. With this method salts tend to accumulate in the ridges, because leaching occurs only through the furrows. If the surface soil is mixed occasionally and the irrigation water is not too saline, the increase in salt over a period of time may not be serious. If excess salt does accumulate, special agronomic techniques described in Section 9.5.3 can be helpful or leaching with another irrigation system may be required. In the furrow and border methods, the length of run, size of stream, slope of the land, and time of application are factors that govern the depth and uniformity of application. Proper balance among these factors is therefore directly related to leaching and salinity control.

Trickle or drip irrigation systems provide water through porous tubes or emitters spaced according to plant density. Their advantage is the maintenance of high soil water contents in the rootzone by frequent but small water applications. Plant roots tend to proliferate in the leached zone of high soil water content beneath the water sources. This allows water of relatively high salt content to be used successfully in many cases. Trickle irrigation is appropriate for orchards and for some widely spaced or valuable row crops, but it is usually too expensive for more densely planted crops. Possible emitter clogging, the increased flow required to germinate seeds, and the accumulation of salt at the soil surface between emitters are potential management concerns.

The salinity profile under line sources, such as furrows and either porous or multi-emitter drip irrigation systems, has clearly recognizable lateral and downward components. The typical cross-sectional profile has an isolated pocket of accumulated salts at the soil surface midway between the line sources and a second, deep zone of accumulation, depending on the degree of efficiency of leaching. Directly beneath the line source is a leached zone whose size depends on the irrigation rate and frequency, and the crop's water extraction pattern (Fig. 9.2-3).

Whereas the salt distribution from line sources increases laterally and downward, the distribution from point irrigation sources, such as microbasins and drip systems with widely spaced emitters, increases radially in all directions below the soil surface. As the rate of water application increases, the shape of the salinity distribution changes. Based on the mathematical model of Bresler (1975), the salinity distribution in uniform, isotropic sand changes from elliptical (with the maximum movement vertical) to more circular as the rate of water application increases. In isotropic and layered soils, the horizontal rate of water movement increases relative to the vertical, resulting in relatively shallow salt accumulations. For tree crops irrigated with relatively few drip emitters per tree, the wetting patterns may overlap, thereby reducing the level of salt accumulation midway between emitters under a tree.

Subsurface irrigation is the least common of the various methods of irrigation. The continuous upward water movement from a subsurface system results in rapid salt accumulation near the soil surface as water is lost by evapotranspiration. The degree of salt accumulation at shallow depths is a function of the depth of the subsurface system and the application rate; the more shallow the system and the larger the application rate, the smaller the amount of salts that accumulate above the system. Subsurface systems provide no means of leaching these shallow accumulations. Unless the soil is leached periodically by rainfall or surface irrigations, salt accumulations are a certainty.

The influence of an irrigation system on the soil salinity profile is sometimes difficult to observe in the field because of the subtle, but often significant, differences in cultural practices associated with each system. For example, the addition of fertilizer with the water in one system but not another can cause unassessable fertilizer effects on plant growth and water use. Differences in water application efficiency, even though equal amounts of water are applied, can cause significant differences in salt distribution. The experiment of Bernstein and Francois (1973a) as summarized in Fig. 9.5-1, shows that pepper plants drip irrigated with good quality water ($\sigma_i$ = 0.6 dS/m), outyielded plants irrigated with the same amount of water by sprinkler and furrow by about 50 percent. With saline water ($\sigma_i$ = 3.8

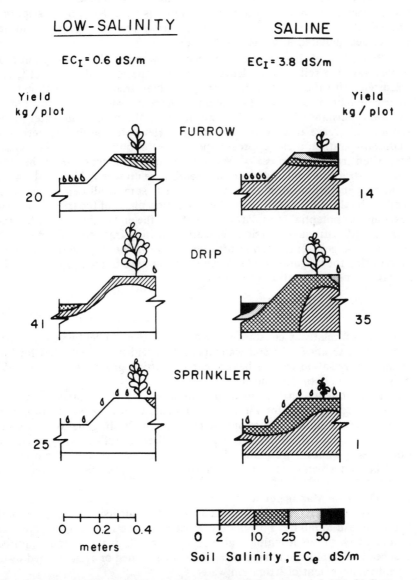

**FIG. 9.5-1 Influence of the irrigation system on the soil salinity pattern and yield of bell pepper at two levels of irrigation water quality. (Bernstein and Francois, 1973a).**

dS/m), the yield differences can be explained by more frequent irrigation with the drip system and foliar salt damage from sprinkling.

The salt distribution profiles shown in Fig. 9.5-1 are typical of those just described. The lateral distributions of salts under sprinkler irrigation was relatively uniform, with evaporation accounting for the salt accumulation near the soil surface. The salinity distributions for both line source systems (furrow and drip) were similar, with salinity levels relatively low beneath the water sources and relatively high midway between the sources. Observed differences in salt accumulation below the side slopes of the planting bed were no doubt caused by additional leaching during furrow irrigation. Of course, the more saline irrigation treatment accounted for the higher salt concentration in each case. Because the salt distributions were determined after only one irrigation season for soil that had previously been well leached, salt had not yet accumulated at greater soil depths.

Regardless of the period of time between water applications, plant roots concentrate the soil solution leaving the rootzone to a salinity level that is consistent with the leaching fraction. With high-frequency irrigation, both matric potential and osmotic potential decrease with depth but remain relatively constant with time. This means that, although plant water potential must decrease at least to the drainage water potential during periods of high transpiration, it can approach the high water potential near the soil surface when transpiration is low. With low-frequency irrigation, on the other hand, matric potential and osmotic potential are uniformly high to the depth of water penetration after an irrigation. Their sum at all depths, however, tends to be the same at the end of the extraction phase of the irrigation cycle, reaching the potential of the drainage water at the end of this phase. In other words, for low-frequency irrigation with a leaching fraction (LF) of 0.1, the soil water ranges from the concentration of salts in the irrigation water ($c_i$) to 10 $c_i$ during the irrigation cycle. In contrast, for high-frequency irrigation the average salt concentration given by

$$c = [\ln(1/LF)/(1-LF)]\,c_i \dots \dots \dots \dots \dots \dots \dots \dots \dots \dots \dots \quad [9.5\text{-}1]$$

remains at a relatively constant value of 2.5 $c_i$. Because leaf water potential can never rise above the soil water potential, leaf water potential for low-frequency irrigation can be below that permitting growth for an extended period of time both day and night.

This reduction in the mean soil salinity with high-frequency irrigation accounts in large part for the increased yields with trickle irrigation compared to less frequent irrigations by other methods. In general, increasing the frequency of irrigation helps maintain full crop production as the irrigation water becomes more saline. Of course, all crops have a threshold salinity value beyond which yields are reduced regardless of the irrigation frequency.

### 9.5.2 Drainage Management

Without drainage, salination is a foregone conclusion and irrigation is bound to lead to failure. Fortunately, most soils have some degree of natural drainage to facilitate leaching, but supplementary drainage is often required. The need for supplementary drainage may be lessened or even avoided by efficient management of irrigation water.

For salinity control, the leaching requirement establishes the minimum drainage requirement. Additional drainage, however, may be necessary for excess precipitation, for seepage from adjacent irrigated areas or from water conveyance structures, or for nonuniformity of irrigation applications and similar sources of water that are not directly related to leaching. During reclamation of saline land, the drainage requirement is high, but the design of a drainage system is generally based on the drainage requirements following reclamation.

The drainage requirement for salt removal under steady-state conditions is given by

$$d_d = L_r \, d_{et}/(1-L_r) \quad\dots\dots\dots\dots\dots\dots\dots\dots\dots\dots \quad [9.5\text{-}2]$$

where $d_d$ is the equivalent depth of drainage per unit land area, $L_r$ is the leaching requirement, and $d_{et}$ is the equivalent depth of evapotranspiration per unit land area. In terms of the electrical conductivities of the irrigation ($\sigma_i$) and drainage ($\sigma_d$) waters,

$$d_d = \sigma_i \, d_{et}/(\sigma_d - \sigma_i). \quad\dots\dots\dots\dots\dots\dots\dots\dots\dots\dots \quad [9.5\text{-}3]$$

Crop salt tolerance dictates the selection of the leaching requirement in one equation and the permissible value of $\sigma_d$ in the other. To illustrate the effect of irrigation water quality on the drainage requirement, consider a moderately tolerant crop ($\sigma_d = 12$ dS/m). For irrigation water having an $\sigma_i$ of 0.6, 2.4, or 4.0 dS/m, the amounts of drainage that must pass through the rootzone to prevent a reduction in crop yield are 5, 25, or 50 percent of the evapotranspiration, respectively.

The value of $d_d$ refers only to the minimum amount of water that must leave the rootzone at any point of a field. It is not the total amount of water that must be removed. Furthermore, it is clear that the actual amount will vary significantly with the crop considered. Where a range of crops is grown in rotation, or where the cropping pattern changes over time, the effect of these changes on the drainage requirement must be considered. In planning a drainage system, it is desirable to base the design on the most demanding crop to be grown.

Although the rootzone is the region of primary concern in salinity control, the region that makes up the drainage system is much larger. To define it, one must establish its geometric boundaries, identify layers that have different hydraulic transmission and water retention properties, and locate any sources of water, such as springs, artesian aquifers, or seepage into the area of concern. In addition to delineating the physical boundaries, the boundary conditions also establish either the flow conditions or the hydraulic potential distribution along the boundaries. An important boundary is the water table, defined as the locus of points at which the pressure potential is zero.

Supplementary drainage systems may consist of intercepting or relief-type drains, depending upon their location and function. Intercepting drains collect and divert water before it reaches the land under construction; relief drains are placed directly in the problem area. Pumped wells, horizontal conduits, or open drains may serve either of these purposes. Relief-type drains are used in broad valleys where the land has little slope, whereas intercepting drains more often are used in areas where topography is irregular.

In areas of rolling or irregular topography, water that percolates downward through the surface soil often flows laterally through materials in the direction of the land slope. In these areas, seeps may be caused by a decrease in grade, a decrease in soil permeability, a thinning out of permeable underlying layers, the occurrence of dikes or water barriers, or the outcropping of relatively impermeable layers or hardpans. If the seepage water cannot be eliminated at its source, and if a suitable outlet exists, drains immediately above the seep to intercept such flows is usually the most effective solution.

In nonuniform soils, drainage systems may best be designed by considering the nature and extent of subsoil layers and by locating the drains with respect to these subsoil materials. Generally, drains should be oriented perpendicular to the direction of groundwater flow, and, where possible, should connect with sand and gravel layers or deposits. In alluvial soils, both permeable and impermeable deposits may be oriented so that a few well-placed drains can control groundwater over a much larger area than the same length of drains installed with uniform spacing in accordance with some regular pattern.

In areas where artesian conditions occur, drainage by pipe and open drains is generally impractical. Although the quantity of upward flow from an artesian source may be small, it usually exerts an important controlling effect on the height of the water table between drains. Reduction of the water pressure in these aquifers by pumping or other means should be a first consideration.

Relief-type drains are commonly placed at much greater depth and farther apart in irrigated than in humid areas. The optimum depth of the water table in irrigated areas ensures that the net flow of soil water in the rootzone will be downward. Where salts are present, a water-table level within less than 2 m of the ground surface generally requires special management techniques to overcome this potential salinity hazard. The greater the drain depth, the wider the spacing; however, below normal trenching depths, the expense of deeper drains becomes prohibitive. The most satisfactory depth is between 2 and 3 m (Willardson and Donnan, 1978).

Several drain-spacing equations have been developed to properly space relief-type drains, but designers rely heavily on local experiences. Drain spacing may vary from 30 to 200 m in irrigated areas; the proper spacing depends, to a large extent, on the nature of the subsoil materials. The solution of drainage flow problems has led to equations based on steady-state conditions and others that predict the transient conditions of a moving water table. Generally, these equations describe the height of the water table without referring explicitly to the soil water content. In recent years, a number of equations have appeared that take explicit account of the soil water content, but these require computer programs and cannot be presented in simple, closed form. For a detailed review of drainage theory and drain spacing equations, the reader may refer to several sources (van Schilfgaarde, 1974).

Porous material is commonly placed as an envelope around subsurface drains, especially in arid regions with young soils. An envelope increases the effective diameter of the drain and thus its effectiveness; it keeps fine soil particles from washing into the drain; and it helps to stabilize the soil during drain construction, thus providing suitable bedding for drain placement. Many fine-textured soils have cohesive properties that appear to make envelopes unnecessary. However, alluviums found in arid and semiarid regions contain strata of very fine sand or silt that have little or no cohesive properties and that require an envelope to prevent soil particles from entering the drain. In the past, topsoil was often used for the envelope, but more recently many organic and inorganic materials have been used successfully (Willardson, 1974). For all envelope materials it is recommended that the material be placed completely around the drain. Normally, the bulk of the water movement occurs upward into the drain; thus, it is imperative to have envelope material below the drain.

### 9.5.3 Agronomic Management

**9.5.3.1 Crop Selection.** Many crop failures on salt-affected soils result from growing crops that have low salt tolerance. There is about a tenfold range in salt tolerance among agricultural crops (Table 9.3-1). In areas where only saline irrigation water is available, where shallow, saline water tables prevail, or where soil permeability is low, achieving nonsaline conditions may not be economically feasible. In such areas, crops that produce satisfactory yields under saline conditions must be selected.

Owing to the variations among cropping seasons and irrigation management, soils may tend to salinize during one crop and to desalinize during a following crop. Thus, by selecting an appropriate crop rotation, productivity can sometimes be maintained where, in the absence of a rotation, the soil would become unproductive. To illustrate, the irrigation of alfalfa in the Imperial Valley of California is often just sufficient to meet the crop's evapotranspiration because of low soil permeability and high water use by the crop, but rotating alfalfa with winter vegetable crops permits leaching during this period of low evapotranspiration. For soils in the San Joaquin Valley of California with low permeability, rice is often included in the crop rotation to facilitate leaching.

**9.5.3.2 Seed Placement.** Obtaining a satisfactory stand of furrow-irrigated crops on saline soils or when using saline water is often a problem. Growers sometimes compensate for poor germination by planting two or three times as much seed as would normally be required. In other cases, planting procedures are adjusted to ensure that the soil area around the germinating seeds is low in salinity. This can be done by selecting suitable planting practices, bed shapes, or irrigation management.

In furrow-irrigated soils, dissolved salts move up with the applied water into the raised beds from each wet furrow where, upon evapotranspiration, salt accumulates. Planting seeds in the center of a single-row, raised bed places the seeds exactly in the area where salts are expected to concentrate most (Fig. 9.5-2). A double-row, raised planting bed places the seeds near the shoulder of the bed and away from the area of greatest salt accumulation. By this method, higher soil salinities can be tolerated at germination than with the single-row plantings because the water moves the salts through and away from the seed area toward the center of the ridge.

There are other alternatives as well. Alternate-furrow irrigation may help in some cases. If beds are wetted from both sides, the salt accumulates in the top and center of the bed. If alternate furrows are irrigated, however, salts often can be moved beyond the single-seed row to the nonirrigated side of the planting bed. Salts may still accumulate, but less accumulates at the center of the bed. The longer the water is held in the furrow, the less salt accumulates at the mid-bed seed area. Off-center, single-row planting on the

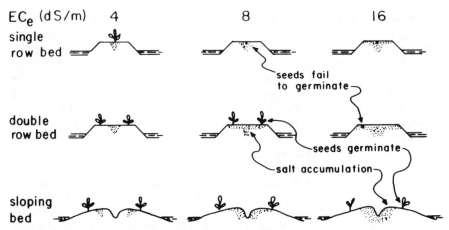

**FIG. 9.5-2 Pattern of salt build-up as a function of seed placement, bed-shape, and irrigation water quality (taken from Bernstein, Fireman, and Reeve, 1955).**

shoulder of the bed, close to the water furrow, has also been used as an aid to germination under saline conditions. Double-row planting under alternate-furrow irrigation is not recommended because salt accumulates on the edge of the bed away from the irrigation furrow.

With either single- or double-row plantings, increasing the depth of water in the furrow can also improve germination in salt-affected soils. Salinity can be controlled even better by use of sloping beds, with the seeds planted on the sloping side just above the water line. Irrigation is continued until the wetting front had moved well past the seed row. During the first cultivation following planting, the sloped bed can be converted to a conventional raised bed if desired.

**9.5.3.3 Soil Profile Modification.** In situations where soils have layers that impede or inhibit root and water penetration, water management and salinity control can be greatly simplified if these layers are fractured, destroyed, or at least rendered more permeable. Deep tillage has been used successfully on many of these soils. In general, the improvement of internal drainage after subsoiling or chiseling has been short-lived. Deep plowing, however, often improves the soil permanently. This drastic treatment often brings saline subsoils up into the rootzone, so that an annual crop such as barley must be grown the first year under heavy leaching and then the soil must be regarded.

In some instances fields are not graded accurately enough to permit satisfactory water distribution by surface irrigation methods. High spots in the field reduce water-intake and may lead to salinity problems. Precise grading, as is possible with laser equipment, may be required to ensure uniform water distribution with surface irrigation methods. As an alternative, sprinkler or drip irrigation can be used without precise grading.

Whenever possible, organic residues should be incorporated into the surface soil to improve water penetration. Crop residues and particularly winter cover crops or green mulches are beneficial. Animal manure from feedlots is of dubious benefit because of its high salt content. Rarely are the benefits of organic residues sufficient to make their purchase economical.

**9.5.3.4 Preserving Rainfall.** Salt-affected soils in nonirrigated areas of semiarid and subhumid regions may persist because geographical or economical considerations prohibit leaching with applied water or the use of chemical amendments. In these situations, mulching or leaving the soil fallow for a time to preserve rainfall may be an effective method of reducing soil salinity. Mulching reduces runoff, decreases soil surface sealing, increases infiltration, and reduces evaporation. Almost any organic material, sand, or rough cultivated soil itself can be an effective mulch. Organic materials that have been used successfully as mulches include manure, cereal straw, cut grass, cotton burs, and other types of crop residue.

Where soil salinity is too high for commercial crop production because of insufficient rainfall or because of a shallow, saline water table, it can be reduced during a fallow period if rainfall is preserved. Without a crop, much of the rainfall penetrates and leaches soluble salts. To be most effective, the soil should be cultivated several times during this fallow period to control weeds and maintain a mulch. During a four-year study in northeastern North Dakota on silt loamy and silty clay loam soil influenced by shallow groundwater (Sandoval and Benz, 1966), over half of the soluble salts in the 0.15- to 0.60-m depth were removed under fallow conditions while salinity levels fluctuated and tended to increase under barley or grass. Likewise, in a study in the Lower Rio Grande Valley of Texas reclamation was more successful when the soil was left bare rather than cropped to Coastal Bermudagrass (Heilman et al., 1968). Applying a 15-mm thick sand mulch or 45 Mg/ha of cotton burs as a mulch, however, was better than leaving the clay loam soil bare.

As another example, sugar cane could not be grown commercially on a saline-sodic soil in the semiarid, tropic area of the Barbados Island in the West Indies (Eavis and Cumberbatch, 1977) until a grass mulch (applied at the rate of 10 Mg/ha) and adequate fertilization were applied. Mulching increased the average volumetric water content from about 20 percent on bare land to nearly 25 percent.

## 9.5.4 Reclamation of Salt-Affected Soils

Reclamation is discussed separately from other management techniques to emphasize the differences between management procedures required to control salinity on a continuous basis and infrequently required reclamation procedures to restore productivity. Leaching is the only proven method of reclaiming salt-affected soils. To reclaim sodic soils, the excess sodium ions that are adsorbed on soil exchange sites must first be replaced with a divalent cation, such as calcium, through a chemical reaction, which improves water penetration; only then can leaching be effective. Soils high in boron also present special reclamation difficulties.

Reclamation requires adequate drainage. Natural internal drainage may be adequate if the soil profile is permeable and provides sufficient storage capacity or if permeable layers are present below the rootzone to provide drainage to a suitable outlet. Where such natural drainage is lacking, an artificial system must be constructed. Since shallow, saline water tables often are the cause of salt-affected soils, providing drainage prior to reclamation is good insurance that the problem will not recur.

**9.5.4.1 Removal of Soluble Salts.** The amount of water that must be applied to remove soluble salts from the rootzone depends primarily on the initial soil salinity level and the technique of applying water. Water suitable for irrigation is normally adequate for reclamation. Typically, about 70 percent of the soluble salts initially present in a saline soil profile that has no shallow water table will be removed by leaching with a depth of water equivalent to the depth of soil to be reclaimed if water is ponded continuously on the soil surface. The relationship between the fraction of salt remaining in the profile, $c/c_0$, where $c_0$ is the initial salt concentration and $c$ is the salt concentration during reclamation, and the amount of water applied per unit depth of soil, $d_w/d_s$, by continuous ponding (Hoffman, 1980) can be approximated adequately by

$$(c/c_0) \cdot (d_w/d_s) = 0.3. \quad\dotfill \quad [9.5\text{-}4]$$

This relationship is illustrated in Fig. 9.5-3. This relationship is for fine-textured soils. The value of 0.3 can be reduced to 0.1 for coarse-textured

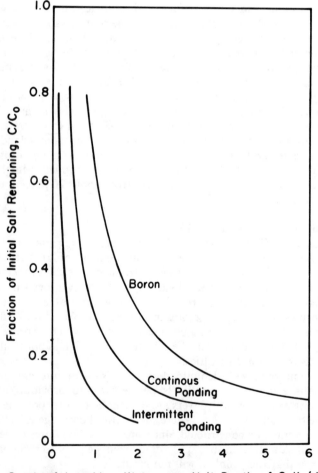

**Depth of Leaching Water per Unit Depth of Soil ($d_w/d_s$)**

FIG. 9.5-3 Depth of water per unit depth of soil required to leach a saline soil by continuous or intermittent ponding or to leach a soil inherently high in boron.

soils. The equation can be refined by taking the salt concentration of the applied water $(c_i)$ into account by substituting $(c-c_i)/(c_0-c_i)$ for $c/c_0$. This refinement improves the assessment of $d_w$ as $c_i$ increases or as complete reclamation $(c \rightarrow c_i)$ is approached.

The amount of water required for leaching soluble salts can be reduced by intermittent applications of ponded water or by sprinkling. The differences in leaching efficiency among the leaching methods are caused by differences in hydrodynamic dispersion, molecular diffusion, and negative adsorption between saturated and unsaturated flow. The amount of solution withheld is considerable for saturated soils and decreases with decreasing soil water content. Consequently, the drier the soil, the larger the percentage of water flowing through fine pores and the more efficiently the leaching water displaces the saline soil solution.

The relationship between $c/c_0$ and $d_w/d_s$ for intermittent ponding (Hoffman, 1980), illustrated in Fig. 9.5-3, can be approximated by

$$(c/c_0) \cdot (d_w/d_s) = 0.1. \dots\dots\dots\dots\dots\dots\dots\dots\dots\dots\dots\dots\dots [9.5\text{-}5]$$

To remove about 70 percent of the soluble salts initially present by intermittent ponding, a depth of water equal to about one-third of the depth of soil to be reclaimed is required—one-third of that required if continuous ponding is used.

Leaching efficiency by sprinkling is similar to that for intermittent ponding and, in some cases, the efficiency may be improved, particularly where low application rates are maintained or intermittent sprinkling is practiced. The sprinkling method has the added advantage over ponding that precise land leveling is not required. One possible disadvantage of intermittent ponding or sprinkling is that both methods may take longer than continuous ponding. In one comparison, leaching a clay loam soil took twice as long with intermittent than with continuous ponding (Miller et al., 1965). In another comparison, however, the same degree of leaching was achieved with intermittent ponding or sprinkling as with continuous ponding in the same time on a silty clay soil (Oster et al., 1972).

**9.5.4.2 Boron Removal.** Excess boron is generally more difficult to leach than other salts because it can be tightly adsorbed on soil particles. Soils inherently high in boron seem to withhold boron with more tenacity than soils where boron has been added in the irrigation water. Thus, the origin of the boron determines the amount of water required for reclamation. Soils inherently high in boron require more leaching for reclamation and require periodic leaching to remove additional boron released from the soil. Of course, the problem of irrigation water high in boron can only be resolved by changing the water supply. Once the supply is changed, the boron accumulated from the irrigation water can be readily leached and the problem should not recur.

As in leaching soluble salts, the relationship between $c/c_0$ and $d_w/d_s$ in leaching soils inherently high in boron (Hoffman, 1980), illustrated in Fig. 9.5-3, can be approximated by

$$(c/c_0) \cdot (d_w/d_s) = 0.6. \dots\dots\dots\dots\dots\dots\dots\dots\dots\dots\dots\dots\dots [9.5\text{-}6]$$

Thus, for soils inherently high in boron, the amount of water required to remove a given fraction of boron is about twice that required to remove soluble salts by continuous ponding. Boron-leaching efficiency does not appear to be significantly influenced by the method of water application. Bingham et al. (1972) reported the same leaching efficiency at two sprinkler application rates as was achieved by Reeve et al. (1955) using flooding.

An extreme example of the tenacity by which boron may be held by soil was reported by Penman (1966) in southeastern Australia. There, in Malbe soils, boron concentrations as high as 3 g/m³ were still obtained from a subsurface drainage system after 30 yr of 20- to 30-percent leaching. In contrast, the boron concentration of a soil that had been degraded by boron in the irrigation water was reduced by one-half with less than half as much water as required for leaching soils inherently high in boron (Meyer and Ayers, 1968). Since the water supply was changed, the boron problem has not recurred. Likewise, laboratory soil-column studies have demonstrated that soils presaturated with boron before elution can be leached more quickly than soils inherently high in boron (Rhoades et al., 1970).

### 9.5.4.3 Reclamation of Sodic Soils.

The reclamation of sodic soil usually requires that soil-water penetration be improved by exchanging excess sodium with calcium so that leaching can proceed. The typical source of calcium is an amendment that either contains soluble calcium or dissolves calcium upon reaction in the soil. Occasionally, calcium present in the subsoil can be mixed with a shallow, sodic layer by deep tillage, eliminating or reducing the need for amendments. Economics normally dictates that an inexpensive amendment be used in conjunction with leaching and possibly deep tillage to reclaim sodic soils. When rapid reclamation is desired, the high-salt-water dilution method is an effective procedure (Reeve and Doering, 1966).

Chemical amendments that provide soluble calcium include gypsum, limestone, calcium chloride, and calcium nitrate. Common examples of amendments that produce calcium in calcareous soils by enhancing the conversion of calcium carbonate to the more soluble calcium sulfate include various acids and acid-formers such as sulfuric acid, sulfur, lime sulfur, and iron and aluminum sulfates. Because of their relatively low cost, gypsum, sulfur, and sulfuric acid are the most common amendments for reclamation. Numerous other amendments and by-products of manufacturing and mining may also be effective but they are not used extensively because of their cost. Comparative data on the effectiveness of the more common amendments are given in Table 9.5-1.

The choice of an amendment may be influenced by the time its reaction in the soil takes and hence the time needed for reclamation. Owing to its high solubility in water, calcium chloride is probably the most readily available source of soluble calcium, but it is seldom used because of its high cost. Sulfuric acid and iron and aluminum sulfates that hydrolyze readily in the soil to form sulfuric acid are also quick-acting amendments.

The amount of exchangeable sodium to be replaced during reclamation depends on the initial sodium concentration ($Na_i$, mmol/kg of soil), the soil's cation-exchange capacity ($X_c$, mmol/kg of soil), soil bulk density ($\sigma_b$, Mg/m³), the depth of soil to be reclaimed ($d_s$, m), and the desired final sodium concentration ($Na_f$). Sodic soils are often reclaimed to a shallower depth than saline soils because of the cost of the amendment. With good salinity management after reclamation, the nonsodic zone can often be increased slowly. A completely reclaimed sodic soil will have an ESP of about 10 or less.

Once the above parameters have been determined, the amount of gypsum in Mg/ha required to obtain a final ESP of 10 can be estimated by

$$1.3 \, d_s \, \sigma_b \, [Na_i - 0.1 \, X_c], \quad \dots\dots\dots\dots\dots\dots\dots\dots\dots\dots\dots\dots \quad [9.5\text{-}7]$$

This gypsum requirement is based upon the soil chemistry model of Oster and Frenkel (1980). The amount of any of the other amendments can be estimated by multiplying the gypsum requirement by its equivalence to gypsum given in Table 9.5-1.

Generally, soil permeability is too low to allow reclamation of sodic soils in a single leaching. Field experience has shown that no more than about 10 Mg/ha of gypsum can be used effectively the first year with about 1.5-m depth of water for leaching. In this manner the soil is reclaimed to a shallow depth the first year so that a shallow-rooted crop may be grown to help offset the reclamation costs. Subsequently, annual amendment applications and leaching volumes can be applied to reclaim the entire profile over a period of a few years.

Tillage to create a rough, yet thoroughly disturbed, soil surface is a common practice for improving water infiltration. This practice often is beneficial when reclaiming sodic soils, particularly when soil surface crusting is a problem. Typically, a sodic soil is tilled prior to each intermittent water application during reclamation.

In locations where sodium-affected soils are underlain by soil containing significant quantities of gypsum or lime, deep plowing has been effective in breaking up and mixing the layers while supplying soluble calcium to aid

TABLE 9.5-1. THE RELATIVE EFFECTIVENESS OF VARIOUS CHEMICAL AMENDMENTS IN SUPPLYING CALCIUM

| Amendment | Chemical composition | Physical description | Solubility in water g/L | Amount equivalent to 100% gypsum |
|---|---|---|---|---|
| Gypsum | $CaSO_4 \cdot 2H_2O$ | white mineral | 2 | 1.00 |
| Sulfur | S | yellow powder | 0 | 0.19 |
| Sulfuric acid | $H_2SO_4$ | corrosive liquid | very high | 0.61 |
| Lime sulfur | 9% Ca + 24% S | yellow-brown alkaline liquid | very high | 0.78 |
| Calcium carbonate | $CaCO_3$ | white mineral | 0.02 to 1.0* | 0.58 |
| Calcium chloride | $CaCl_2 \cdot 2H_2O$ | white salt | 120 | 0.86 |
| Calcium nitrate | $Ca(NO_3)_2 \cdot 2H_2O$ | white fertilizer | 60 | 1.06 |
| Iron sulfate | $FeSO_4 \cdot 7H_2O$ | corrosive granular material | 30 | 1.62 |
| Ferric sulfate | $Fe_2(SO_4)_3 \cdot 9H_2O$ | corrosive granular material | | 0.61 |
| Aluminum sulfate | $Al_2(SO_4)_3 \cdot 18H_2O$ | corrosive granular material | | |

*Solubility of $CaCO_3$ is pH-dependent. Solubility varies from 0.02 to 1.0 g/L as the pH decreases from 10 to 6.

reclamation. This is often the situation along the Snake River and its tributaries in the Pacific Northwest, in much of the Salt River Valley in Arizona and the Rio Grande Valley in New Mexico and Texas, and in the trans-Volga regions of the USSR. The depth of plowing required may vary from 0.5 to more than 1.0 m, depending on the concentration and depth of both the sodic and calcium-rich layers. A procedure is available to predict the optimum depth of plowing to maintain adequate permeability during the reclamation process (Rasmussen and McNeal, 1973).

Subsoiling or chiseling, that is pulling vertical strips of steel or iron enlarged on the bottom through the soil at 0.5- to 1.0-m spacings to open channels to improve soil permeability, does not normally mix the layers sufficiently to warrant the expense. Subsoiling is beneficial for several years if indurated layers are broken.

In the Pacific Northwest, for example, more than 100,000 ha of irrigated and potentially irrigable lands are affected by inclusions of saline-sodic (slick spot) soils. These soils are typified by $\sigma_e$ values of about 2 dS/m in the surface layers and 15 dS/m in the subsoil, ESP values of 10 to 30 percent and 30 to 50 percent, and gypsum concentrations of 0 and 100 mmol/kg of soil. After deep plowing to a depth of 0.9 m, reclamation was completed within 3 to 4 yr without any amendment (Rasmussen et al., 1972). Subsoiling was not beneficial and a gypsum application of 36 Mg/ha without tillage gave only moderate improvement.

## 9.6 SUMMARY

Salinity is a major threat to agriculture whenever precipitation does not significantly exceed evapotranspiration because without sufficient leaching soils accumulate dissolved salts. Salts originate from geological weathering; they are transported and accumulated in the rootzone as a result of water movement. Some of these accumulated soluble salts must be removed by leaching to maintain a viable agriculture. The actual amount of water that must leach below the rootzone depends upon the salt concentration of the irrigation water, the soil, and the groundwater; the salt tolerance of the crop; climatic conditions; and soil and water management. Crop production is limited on about one-fourth of the irrigated cropland in the western United States because of excess salinity.

The predominate influence of salinity on plants is growth suppression. Typically, growth decreases linearly as salinity increases beyond a threshold salinity level. The maximum allowable salinity that does not measurably reduce yield (threshold level) and the rate of relative yield reduction per unit increase in salinity are presented as relative salt tolerance parameters for more than 60 crops. These parameters must be taken as tentative, however, because salt tolerance can be influenced by irrigation management, climatic conditions, stage of plant growth, crop variety and rootstock, and plant nutrition. Individual solutes, such as chloride and boron, may be toxic to certain crops at concentrations too low to cause total osmotic effects. Lists are given of crops particularly sensitive to these solutes. Also presented are techniques for evaluating sodium levels that may be detrimental to soil structure.

Guidelines for assessing soil salinity are presented to aid in diagnosing potential salinity problems. In addition to describing the laboratory analyses required for diagnosis, devices are described that monitor soil salinity in the field. For saline soils of marginal quality, various management options are discussed to improve crop production. These include changes in the frequency, amount, or system of irrigation, more suitable seed placement, proper crop selection, and soil profile modification. Likewise, both chemical and physical methods are given for improving water penetration in sodic soils.

Reclamation is required to restore soil productivity where salinity is excessive. If soil permeability and drainage are adequate, about 70 percent of the soluble salts can be removed by applying a depth of water equivalent to the soil depth to be reclaimed. For reclamation of sodic soils, the exchangeable sodium must be replaced, normally by calcium added as a amendment, and the displaced sodium must be leached from the rootzone.

## References

1 Ayers, R. S., and D. W. Westcot. 1976. Water quality for agriculture. FAO Irrigation and Drainage Paper 29, 97 p.

2 Benz, L. C., F. M. Sandoval, E. J. Doering, and W. O. Willis. 1976. Managing saline soils in the Red River Valley of the North. USDA-ARS-NC-42. 54 p.

3 Bernstein, L., M. Fireman, and R. C. Reeve. 1955. Control of salinity in the Imperial Valley, California. U.S. Dept. Agr. ARS-41-4, 16 p.

4 Bernstein, L., and G. A. Pearson. 1956. Influence of exchangeable sodium on the yield and chemical composition of plants. I. Green beans, garden beets, clover, and alfalfa. Soil Sci. 82:247-258.

5 Bernstein, L., and H. A. Hayward. 1958. Physiology of salt tolerance. Annu. Rev. Plant Physiol. 9:25-46.

6 Bernstein, L. 1964. Effects of salinity on mineral composition and growth of plants. Proc. 4th Internatl. Colloquium Plant Analysis and Fert. Problems (Brussels) 4:25-45.

7 Bernstein, L. 1965. Salt tolerance of fruit crops. USDA Agric. Information Bull. 292, 8 p.

8 Bernstein, L., and L. E. Francois. 1973a. Comparisons of drip, furrow, and sprinkler irrigation. Soil Sci. 115:73-86.

9 Bernstein, L., and L. E. Francois. 1973b. Leaching requirement studies: Sensitivity of alfalfa to salinity of irrigation and drainage waters. Soil Sci. Soc. Am. Proc. 37:931-943.

10 Bernstein, L., L. E. Francois, and R. A. Clark. 1974. Interactive effects of salinity and fertility on yields of grains and vegetables. Agron. J. 66:412-421.

11 Bingham, F. T., A. W. Marsh, R. Branson, R. Mahler, and G. Ferry. 1972. Reclamation of salt-affected high boron soils in western Kern County. Hilgardia 41(8):195-211.

12 Bower, C. A., G. Ogata, and J. M. Tucker. 1969. Rootzone salt profiles and alfalfa growth as influenced by irrigation water salinity and leaching fraction. Agron. J. 61:783-785.

13 Bower, C. A., G. Ogata, and J. M. Tucker. 1970. Growth of sudan and tall fescue grasses as influenced by irrigation water salinity and leaching fraction. Agron. J. 62:793-794.

14 Bresler, E. 1975. Two-dimensional transport of solutes during nonsteady infiltration from a trickle source. Soil Sci. Soc. Am. Proc. 39:604-613.

15 Cooper, W. C. 1951. Salt tolerance of avocados on various rootstocks. Texas Avocado Soc. Yearbook 1951: 24-28.

16 Cooper, W. C. 1961. Toxicity and accumulation of salts in citrus trees on various rootstocks in Texas. Florida State Hort. Soc. Proc. 74:95-104.

17 Doering, E. J., and F. M. Sandoval. 1976. Hydrology of saline seeps in the Northern Great Plains. TRANSACTIONS of the ASAE 19(5):856-861, 865.

18 Eavis, B. W., and E. R. St. J. Cumberbatch. 1977. Sugar cane growth in response to mulch and fertilizer on saline-alkali subsoils. Agron. J. 69:839-842.

19 Harding, R. B., M. P. Miller, and M. Fireman. 1958. Absorption of salts by citrus leaves during sprinkling with water suitable for surface irrigation. Proc. Am. Soc. Hort. Sci. 71:248-256.

20 Hayward, H. E., and E. M. Long. 1941. Anatomical and physiological responses of the tomato to varying concentration of sodium chloride, sodium sulphate, and nutrient solutions. Bot. Gaz. 102:437-462.

21 Heilman, M. D., C. L. Wiegand, and C. L. Gonzalez. 1968. Sand and cotton bur mulches, Bermudagrass sod, and bare soil effects on: II. Salt leaching. Soil Sci. Soc. Am. Proc. 32:280-283.

22 Hoffman, G. J., and S. L. Rawlins. 1971. Growth and water potential of root crops as

influenced by salinity and relative humidity. Agron. J. 68:877-880.

23  Hoffman, G. J., E. V. Maas, and S. L. Rawlins. 1975. Salinity-ozone interactive effects of alfalfa yield and water relations. J. Environ. Qual. 4:326-331.

24  Hoffman, G. J., and J. A. Jobes. 1978. Growth and water relations of cereal crops as influenced by salinity and relative humidity. Agron. J. 70:765-769.

25  Hoffman, G. J., S. L. Rawlins, J. D. Oster, J. A. Jobes, and S. D. Merrill. 1979. Leaching requirement for salinity control. I. Wheat, sorghum, and lettuce. Agr. Water Mgmt. 2:177-192.

26  Hoffman, G. J. 1980. Guidelines for reclamation of salt-affected soils. Proc. Second Inter-American Conference on Salinity and Water Management Technology. Juarez, Mexico.

27  Maas, E. V., and G. J. Hoffman. 1977. Crop salt tolerance—Current assessment. J. Irrig. & Drainage Div., Am. Soc. Civil Eng. 103(IR2):115-134.

28  Maas, E. V., and R. H. Nieman. 1978. Physiology of plant tolerance to salinity. In: G. A. Jung (ed.), Crop tolerance to suboptimal land conditions, Chap. 13, ASA Spec. Publ.:277-299.

29  Magistad, O. C., A. D. Ayers, D. H. Wadleigh, and H. G. Gauch. 1943. Effect of salt concentration, kind of salt, and climate on plant growth in sand cultures. Plant Physiol. 18:151-166.

30  Meyer, J. L., and R. S. Ayers. 1968. Boron leaching. Calif. Agric. 22(9):6.

31  Miller, R. J., D. R. Nielsen, and J. W. Biggar. 1965. Chloride displacement in Panoche clay loam in relation to water movement and distribution. Water Resources Res. 1:63-73.

32  Nielson, R. F., and O. S. Cannon. 1975. Sprinkling with salt well water can cause problems. Utah Science, Agr. Expt. Sta. 36(2):61-63.

33  Nieman, R. H., and R. A. Clark. 1976. Interactive effects of salinity and phosphorus nutrition on the concentrations of phosphate and phosphate esters in mature photosynthesizing corn leaves. Plant Physiol. 57:157-161.

34  Nieman, R. H. 1980. Bioenergetics of salt-stressed plants. J. Theoretical Biol. (in preparation)

35  Oster, J. D., and H. Frenkel. 1980. The chemistry of the reclamation of sodic soils with gypsum and lime. Soil Sci. Soc. Am. J. 44:41-45.

36  Oster, J. D., and L. S. Willardson. 1971. Reliability of salinity sensors for the management of soil salinity. Agron. J. 63:695-698.

37  Oster, J. D., L. S. Willardson, and G. J. Hoffman. 1972. Sprinkling and ponding techniques for reclaiming saline soils. TRANSACTIONS of the ASAE 15(6):115-117.

38  Pearson, G. A., and L. Bernstein. 1958. Influence of exchangeable sodium on yield and chemical composition of plants: II. Wheat barley, oats, rice, tall fescue, and tall wheatgrass. Soil Sci. 86:254-261.

39  Penman, F. 1966. Slow reclamation by tile drainage of sodic soils containing boron. Internatl. Comm. Irrig. & Drainage, 6th Congress (New Delhi, India), Question 19, p. 113-221.

40  Rasmussen, W. W., D. P. Moore, and L. A. Alban. 1972. Improvement of a Solonetzic (slick spot) soil by deep plowing, subsoiling, and amendments. Soil Sci. Soc. Am. Proc. 36:137-142.

41  Rasmussen, W. W., and B. L. McNeal. 1973. Predicting optimum depth of profile modification by deep plowing for improving saline-sodic soils. Soil Sci. Soc. Am. Proc. 37:432-437.

42  Rawlins, S. L. 1977. Uniform irrigation with a low-head bubbler system. Agric. Water Mgmt. 1:167-168.

43  Reeve, R. C., A. F. Pillsbury, and L. V. Wilcox. 1955. Reclamation of a saline and high boron soil in the Coachella Valley of California. Hilgardia 24(4):69-91.

44  Reeve, R. C., and E. J. Doering. 1966. The high-salt-water dilution method for reclaiming sodic soils. Soil Sci. Soc. Am. Proc. 30:498-504.

45  Reisenauer, H. M. 1976. Soil and plant-tissue testing in California. Div. of Agric. Sci., Univ. of California, Bull. 1879, April, 54 p.

46  Rhoades, J. D., R. D. Ingvalson, and J. T. Hatcher. 1970. Laboratory determination of leachable boron. Soil Sci. Soc. Am. Proc. 34:871-875.

47  Rhoades, J. D., and R. D. Ingvalson. 1971. Determining salinity in field soils with soil resistance measurements. Soil Sci. Soc. Am. Proc. 35:54-60.

48  Rhoades, J. D. 1972. Quality of water for irrigation. Soil Sci. 118:277-284.

49  Rhoades, J. D., J. D. Oster, R. D. Ingvalson, J. M. Tucker, and M. Clark. 1974. Minimizing the salt burdens of irrigation drainage waters. J. Env. Quality 3:311-316.

50  Rhoades, J. D. and J. van Schilfgaarde. 1976. An electrical conductivity probe for determining soil salinity. Soil Sci. Soc. Amer. J. 40:647-651.

51   Rhoades, J. D. 1979. Inexpensive four-electrode probe for monitoring soil salinity. Soil Sci. Soc. Am. J. 43:817-818.

52   Richards, L. A. 1966. A soil salinity sensor of improved design. Soil Sci. Soc. Am. Proc. 30:333-337.

53   Sandoval, F. M., and L. C. Benz. 1966. Effect of bare fallow, barley, and grass on salinity of a soil over a saline water table. Soil Sci. Soc. Am. Proc. 30:392-396.

54   Shalhevet, J., and J. Kamburov. 1976. Irrigation and salinity—A world-wide survey. Internatl. Comm. on Irrig. & Drainage, New Delhi, India. 106 p.

55   U.S. Salinity Laboratory Staff. 1954. Diagnosis and improvement of saline and alkali soils. U.S. Dept. Agr. Handbook 60, 160 p.

57   van Schilfgaarde, J. (ed.) 1974. Drainage for Agriculture. Agronomy 17. Am. Soc. Agron., Madison, WI, 700 p.

58   van Schilfgaarde, J., L. Bernstein, J. D. Rhoades, and S. L. Rawlins. 1974. Irrigation management for salt control. J. Irrig. & Drainage Div., Am. Soc. Civil Eng. 100(IR3), Proc. Paper 10322, pp. 321-338.

59   Wadleigh, C. H., and A. D. Ayers. 1945. Growth and biochemical composition of bean plants as conditioned by soil moisture tension and salt concentration. Plant Physiol. 20:106-132.

60   Wadleigh, C. H. 1968. Wastes in relation to agriculture and forestry. USDA Misc. Pub. 1065, 112 p.

61   Wilcox, L. V. 1960. Boron injury to plants. U.S. Dept. Agr. Bull. 211, 7 p.

62   Willardson, L. S. 1974. Envelope materials. In: Jan van Schilfgaarde (ed.), Drainage for Agriculture. Agronomy 17, Am. Soc. Agron., Madison, WI, pp. 179-196.

63   Willardson, L. S., and W. W. Donnan. 1978. Drain depth and spacing criteria. Internatl. Comm. Irrig. & Drainage, 10th Congress, Athens, Greece. Question 34, p. 241-246.

# chapter 10 ▉▉▉▉▉▉

## WATER USE MODELS FOR ASSESSING ROOT ZONE MODIFICATIONS

# 10

# Water Use Models for Assessing Root Zone Modification

by  R. A. Feddes, Institute for Land and Water Management Research, Wageningen, The Netherlands

## 10.1 INTRODUCTION

A crop that suffers from water shortage and that under the prevailing environmental conditions cannot transpire at the potential rate, will be reduced in its growth. We can, therefore, expect a general relationship between yield and water use defined as the transpiration of the crop. Provided this relationship is established, it must be possible to compute the variation of crop growth with time from the variation of transpiration with time. Effects of various root zone modifications and other measures upon crop production could then be evaluated according to the influence they have upon the transpiration.

In the past these effects were often measured by means of expensive and time consuming field and laboratory experiments, collecting as many data as possible, making various changes in the prevailing circumstances, and analyzing the results. With the introduction of the computer, it became possible to simulate the mentioned effects with the aid of numerical models, that in an idealized way should react in the same manner as the actual system to any changes made.

A 'model' is another term for a set of mathematical equations that describe the physical system. Thus, transpiration and production models can be either very simple or very complicated. They are a representation of the present day knowledge of the soil-plant-atmosphere system. As this knowledge is often incomplete, one has to introduce simplifying assumptions in order to obtain a practical model that is easy to use. In such models one must often use empirical relationships that have to be derived from field or laboratory experiments.

This chapter will try to answer the following questions:

1  Are there models presently available to the engineer or agronomist that can be useful in evaluating management practices?

2  How have models dealing with the plant root environment been used?

---

**Acknowledgement:** The author expresses his gratitude to Professor C. H. M. van Bavel, Department of Soil and Crop Sciences, Texas A&M University, for critically reviewing this contribution and for suggesting technical and editorial changes and additions.

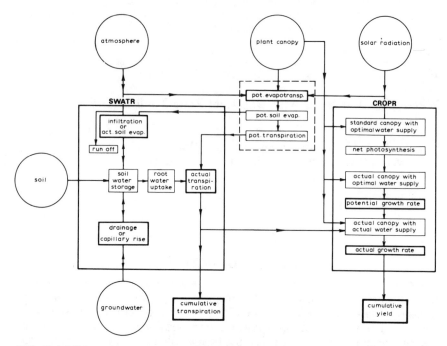

FIG. 10.1-1 Schematic representation of the processes occurring in the soil-plant-atmosphere system and the way to assess influences of root zone modification upon crop water use and crop production (after Feddes et al., 1978).

3   Can such models indicate when appropriate root zone modifications should be made, and what modifications they should be?

To answer these questions in principle we will start with the water balance of the soil-root system, as is schematically shown in Fig. 10.1-1. Irrigation or rainwater that is not intercepted by the crop will reach the soil. Part of it will become soil moisture only to be lost by soil evaporation or transpiration. The part of rainfall that does not infiltrate will be lost as surface runoff. The excess of soil moisture will percolate downward to the groundwater table and recharge the groundwater storage.

Influences of waterlogging on plants can usually be separated in long and short term effects. Long term effects are a permanent limit of metabolic activity and development of the root system, while short term effects such as oxygen deficiency and excess of carbon dioxide are the main injurious factors. Generally both effects are more serious when temperatures are high. At low temperatures biological activity is low and consequently damage to the plant small.

In w e t soils gas exchange between soil and atmosphere is often low, causing oxygen deficiency: root respiration and root volume are reduced, the resistance to water and nutrient transport increases and toxic compounds can be formed in soil as well as in plants (Wesseling, 1974; Chapters 4 and 5). These effects are often small as compared to indirect effects such as reduction in nitrification, low soil temperatures, deterioration of soil structure, and less workable soil. Depressions in yield caused by wet conditions can partly be compensated for by an additional application of nitrogen (Van Hoorn, 1958; Sieben, 1974). For wet soils, a large amount of heat is required to raise the soil temperature. For spring conditions, Feddes (1971) found a difference between well-drained and poorly drained soil of 1 to 2 °C. Because of this difference early growth will be delayed (Chapter 7). Because of the low soil water tensions in the surface layer of poorly drained soils, tillage, sowing and planting will be delayed, shortening the growing season and reducing crop yields. Wind (1960) found from literature data that under the existing climatic conditions of Western Europe for summer cereals a yield depression will occur when sowing after 1 February. For sugar beets the critical sowing date appears to be 25 March. The relationship between yield depression and delayed sowing for sugar beets and summer grains is illustrated in Fig. 10.1-2. Tillage under wet conditions may result in seed beds with an un-favorable soil structure (Chapter 2). Moreover, poorly drained, badly structured soils need deeper and more intensive tillage. Farming operations in spring such as fertilization, tillage, sowing and planting must be performed within a shorter period of time, affecting both the labor and the machinery requirements. This also holds for farming operations in the autumn. On well-drained soils the harvest losses will be smaller and the lifting date of root crops can be shifted to a later time, lengthening the growing season of crops like sugar beet and potatoes and thus increasing production. The decrease in yield and sugar of sugar beet under W-European conditions, when one harvests too early i.e. before 30 October, is illustrated in Fig. 10.1-3.

On permanent grassland the bearing strength of the soil determines the time of application of nitrogen fertilizer and manure in spring. Late application on poorly drained soils causes retardation in growth and delays the

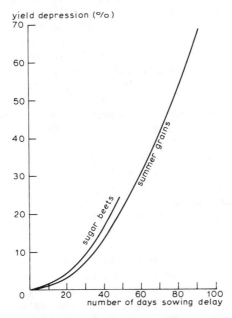

FIG. 10.1-2 Yield depression for summer grains and sugar beets in W. Europe when sowing is delayed as a result of too wet conditions in the topsoil (derived from Wind, 1960; after Feddes and Van Wijk, 1976).

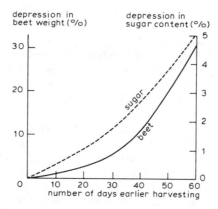

FIG. 10.1-3 Yield depression of fresh sugar beet weight and sugar content when harvesting is done too early; the yield for harvesting for W. European conditions on 30 October is set at 100 percent i.e. no yield depression (derived from Van den Hil, 1975; after Feddes and Van Wijk, 1976).

beginning of the grazing season. During summer a low bearing strength affects the losses in yield due to grazing (trampling of cattle) and to foraging (poor trafficability). During autumn the bearing strength determines the length of the grazing season.

In d r y soils, low soil moisture contents in the top layer in spring can cause retardation and even failure in germination and seedling emergence. If the soil water supply during the growing season declines, the soil water potential declines even more sharply causing stomatal closure, a reduction of transpiration and declining productivity. When large amounts of water are extracted, the soil solution becomes more and more concentrated, possibly resulting in a salinization problem. The primary effects of salts (except for toxic influences) are that the water uptake by the roots is reduced, in the same manner as a water shortage would do (Stewart et al., 1977; Chapter 9).

For soils with shallow water tables one can state that two limiting water table depths need to be considered: an upper limit to create an aerated root zone and a surface soil which is dry enough for proper cultivation practices and a lower limit set by the water requirements of the crop, with an optimum between the two.

In the following we will restrict our discussion mainly to developments occurring during the growing period of the crop, i.e. we will analyze effects of root zone modification as effects on water uptake by root systems and hence as effects on crop production as suggested in Fig. 10.1-1.

The transformation of solar radiation into actual crop yield is schematically shown in the right hand side of Fig. 10.1-1. Gross potential photosynthesis of a 'standard crop canopy' can be calculated according to a model of De Wit (1965) taking into account the height of the sun, the condition of the sky, the canopy architecture and the photosynthesis function of the individual leaves. A 'standard canopy' is defined as a canopy with a leaf area index 5 (5 ha of leaves over 1 ha of soil surface) that is fully supplied with nutrients and water. Under actual field conditions these theoretical photosynthesis rates will never be reached and corrections have to be made for such factors as respiration losses, for actual conditions of light energy flux, for air temperature, for fraction of soil cover and for amounts of roots. This procedure yields the potential growth rate of an actual canopy with optimal water supply. Finally, the actual dry matter yield can be calculated by introducing the actual water uptake of the root system. For more details, see De Wit et al. (1978) and Feddes et al. (1978).

In section 10.2 of this chapter root water uptake patterns as measured in the field and laboratory are analyzed. The importance of the upward flow from the layers below the root zone will be discussed.

In section 10.3 liquid water flow through the soil-root-plant system according to an analogue of Ohm's law is analyzed. The relative importance of the soil and plant resistances under both wet and dry conditions is considered.

Section 10.4 deals with a mathematical description of water extraction by root systems. One approach relies on the properties of a single root which can be extended to an entire uniform root system. Another approach takes into account the integrated properties of the root system by adding a sink term to the continuity equation for soil water flow. A comparison between the various water uptake models now in use will be made. The capabilities and shortcomings of the models will be discussed.

Section 10.5 is devoted to an analysis of root zone modifications by management. Examples of measured and computed effects of changes in soil moisture content, in soil profile and in groundwater table depth upon crop water use and crop production will be given. The importance of using models to quantify such effects will be stressed.

Finally, in section 10.6 the main conclusions drawn from the previous sections are presented and recommendations for future research are made.

## 10.2 ROOT WATER UPTAKE PATTERNS

When evaluating root zone modification effects upon water use, one has first to analyze the actual root water extraction patterns under various environmental conditions. Such experiments can be done either in the laboratory or in the field. In the laboratory the experiment can be performed under carefully controlled conditions that enables one especially to study details of specific interest. In the field one deals with a complex system that varies both in space and time, but that yields the actual system behavior. Both types of experiments are useful in validating root water uptake models.

Root water uptake patterns can generally be derived by applying the water balance or conservation equation to a given volume of soil. In general it is assumed that water flow in unsaturated soil takes place only in one, vertical dimension z. (Investigations of Arya et al. (1975) under a soybean row crop show that the assumption of only vertical flow of water in the unsaturated zone is not always true and that considerable deviations from this vertical flow direction may occur.) Let us consider a volume of soil, bearing vegetation with a lower boundary at $z = 0$ (for example, a groundwater table or a level with constant pressure head) and an upper boundary at the soil surface at $z = z_s$ (thus, z positive in upward direction), of unit cross sectional area in the horizontal plane. This soil volume is partly explored by roots and partly without roots. Flow through the roots can then be calculated as the measured total flow both through roots and soil, diminished with the calculated flow through the soil. The water conservation equation can then be written for every height z for a short time interval $t_2$-$t_1$, as

$$\int_0^z \bar{S}_z \, dz = \bar{q}_0 - \int_0^z \frac{\partial \theta}{\partial t} \, dz \quad - \bar{q}_z, \quad \ldots\ldots\ldots\ldots\ldots\ldots [10.2\text{-}1]$$

(roots)    (total=roots+soil)    (soil)

where $\bar{S}$ represents the time averaged volume of water taken up by the roots per unit bulk volume of the soil in unit time ($cm^3.cm^{-3}.day^{-1}$) considered positive from the soil into the roots, $\bar{q}_0$ the time averaged volumetric flux through the lower boundary ($cm.day^{-1}$), $\theta$ the volumetric soil water content ($cm^3.cm^{-3}$), $\bar{q}_z$ the time averaged volumetric soil flux ($cm.day^{-1}$). The latter can be calculated from Darcy's law as

$$\bar{q}_z = -K_z(\psi)(\frac{\partial \psi}{\partial z} + 1), \quad \ldots\ldots\ldots\ldots\ldots\ldots\ldots\ldots\ldots [10.2\text{-}2]$$

where $K(\psi)$ is the hydraulic conductivity of the soil ($cm.day^{-1}$) and $\psi$ is the soil moisture pressure head (cm). For the calculation of root water uptake patterns, one thus needs to know the $K(\psi)$ curve and both the $\theta$- and $\psi$-profiles. The latter two can be measured by well-known methods as, for example, with neutron/gamma and tensiometry methods, respectively. When the moisture retention curve is used, it suffices to measure either $\theta$ or $\psi$ and infer the unknown one via the retention curve. The flux at the bottom $\bar{q}(0,t)$ should be inferred from either measurements (lysimeter) or by means of equation [10.2-2]. Application of equation [10.2-1] gives the integrated root water uptake over a given depth interval in the root zone. Graphical or numerical differentiation of equation [10.2-1] gives the water uptake by the roots at depth z. For numerical examples see e.g. Rose and Stern (1967), Feddes (1971, Table 10) and Ehlers (1976, Table 1).

An alternative way of approaching the problem is to apply the differential form of the continuity equation, rather than the integral form

$$S_z = - \frac{\partial \theta}{\partial t} - \frac{\partial q_z}{\partial z}, \quad \ldots\ldots\ldots\ldots\ldots\ldots\ldots\ldots\ldots\ldots [10.2\text{-}3]$$

in units of $day^{-1}$.

For details about this approach see e.g. Van Bavel et al., (1968a). These authors state that the differential approach seems to work better than the integral approach.

The integral of the sink term over the rooting depth $z_r$ ($=z_s-z_b$) gives the actual transpiration of the crop, T: ($cm.day^{-1}$)

$$T = \int_{z=z_b}^{z=z_s} S_z \, dz, \quad \ldots\ldots\ldots\ldots\ldots\ldots\ldots\ldots\ldots\ldots [10.2\text{-}4]$$

where $z_b$ denotes the height of the bottom of the root zone.

In the last 10 to 15 yr a vast amount of literature has appeared that describes root water uptake investigations for various crops growing under different conditions. For f i e l d measurements of root water uptake patterns for sugar beet (*Beta vulgaris* L.) and winter wheat (*Triticum aestivum* L.), see Strebel et al. (1975), Ehlers (1976); for grasses, Rijtema (1965), Rice (1975) and Fluhler et al. (1975); for cotton (*Gossypium hirsutum* L.), Rose and Stern (1967); for alfalfa (*Medicago sativa* L.), Nimah and Hanks

(1973b); for soybeans (*Glycine max* L. Merr.), Arya (1973), Allmaras et al. (1975), Arya et al. (1975), Stone et al. (1976); for sorghum (*Sorghum bicolor* (L.) Moench.), Van Bavel et al. (1968a,b), Stone et al. (1973), Reicosky and Ritchie (1976); for a Douglas-Fir (*Pseudotsuga menziesii* (Mirs.) Franco) forest, Nnyamah and Black (1977); for maize (*Zea mays* L.) Allmaras et al. (1975); for red cabbage (*Brassica oleracea* L.), Feddes (1971). Laboratory experiments have been reported for wheat by Hansen (1974a) and Herkelrath et al. (1977); for ryegrass (*Lolium perenne* L.) by Hansen (1974b) and Belmans et al. (1977); for corn by Taylor and Klepper (1973), by Reicosky and Ritchie (1976); for cotton by Taylor and Klepper (1975); for soybeans by Hwang (1964), (c.f. Raats and Gardner, 1974) and by Reicosky et al. (1972); for pepper (*Capsicum Annuum* L.) by Gardner and Cullen (1966).

It is not feasible to discuss here all the findings of the cited experiments, but some general conclusions can be drawn. These conclusions can be elucidated with Fig. 10.2-1 (Feddes, 1971), which shows the root water uptake pattern of a red cabbage crop growing on sticky clay in the presence of a 90 cm deep groundwater table at various times. The magnitude of the root extraction rate is generally small at the top of the profile (unless the soil is wet just after rainfall or irrigation). It increases to a certain maximum zone and decreases to zero at the bottom of the root zone. As the soil dries, the zone of maximum root water uptake moves from shallower to deeper depths in dynamic correspondence with the downward progression of roots into deeper moist soil. The maximum appears to depend on the demand the atmosphere sets on the plant system, on the depths to which the roots penetrate, and on the soil moisture pressure head distribution. Later water uptake from the upper layers becomes relatively less important. Most of the water is absorbed from the zone of low tensions near the water table. Thus, a relatively small part of the root sytem can be responsible for most of the plant water uptake. It was observed (Feddes, 1971) that the roots of red cabbage close to the groundwater table grew well at gas-filled porosities of $0.01 \ cm^3.cm^{-3}$. There is some experimental evidence (Feddes et al., 1978) that, with enough air being present in the upper part of the root zone, water can be extracted by the

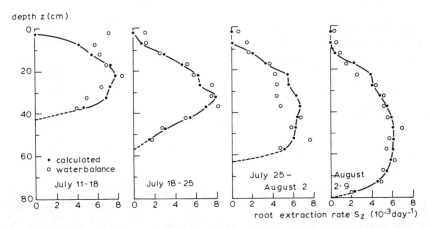

FIG. 10.2-1 Example of measured variations of root water uptake pattern with depth and time of red cabbage grown on a clay soil in the presence of a 90-110 cm groundwater table obtained from water balance studies over 4 consecutive weeks.

roots in the lower part of the root zone under nearly water saturated conditions. Reicosky et al. (1972) also show that, in the presence of a water table, about 20 to 25 percent of the root system can be responsible for about 80 to 90 percent of the uptake of water. The same experiment showed that most of the ion uptake occurred within the maximum zone of water uptake (c. f. Raats and Gardner, 1974). The last mentioned authors cite experiments of Hwang (1964) which show that root water uptake was largest at a fixed distance (about 15 cm) above the water table, independent of the depth of the water table. Feddes et al. (1978) showed from model calculations that the upward flow from a shallow water table can amount up to some 40 to 70 percent of total transpiration. Van Bavel and Ahmed (1976) have also shown from model calculations that even without the presence of a water table about 30 percent of total transpiration can originate from the layers below the root zone. Nnyamah and Black (1977) indicate that in a drying period under Douglas fir forest the water flux into the bottom of the root zone can amount to 8 to 15 percent of the evapotranspiration. An example of the results of their water uptake studies of an unthinned (1840 stems.ha$^{-1}$) and thinned (840 stems.ha$^{-1}$) site is shown in Fig. 10.2-2. The same tendency shown in Fig. 10.2-1 applies more or less to Fig. 10.2-2. In the beginning of the drying

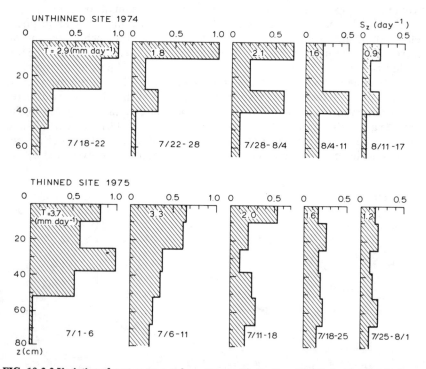

FIG. 10.2-2 Variation of root water uptake patterns with depth and time of a Douglas fir forest [Pseudotsuga menziesii (Mirb) Franco] in British Columbia at an unthinned (1840 stems.ha$^{-1}$) site and at a thinned site (840 stems.ha$^{-1}$) over two 5 wk consecutive periods during 1974 and 1975 respectively. Transpiration values T were obtained by integrating the sink term values S over the rooting depths: 0 to 65 cm at the unthinned site and 0 to 80 cm at the thinned site. Note that there always exists a continuous upward flux at the bottom of the root zones. After Nnyamah and Black (1977).

period about 80 percent of the water uptake occurs in the upper half of the root zone, while at the end of the drying period about 47 percent occurred in the lower half of the root zone. The amount of water extracted from any layer appeared to depend on both rooting density and soil water content.

In general, one can conclude that root water uptake depends on a number of factors like soil hydraulic conductivity, rooting depth, rooting density, root distribution, soil moisture pressure head, demand set by the atmosphere ('potential' transpiration) on the plant system, the presence of a water table, etc. This enumeration indicates that it is not simple to model water uptake by roots, nor to generalize on the effect of a single modification of the root zone.

## 10.3 SOIL AND PLANT RESISTANCES

### 10.3.1 General

A complete and physically-mathematically sound description of water uptake and transport by living root systems seems hardly possible. Hsiao et al. (1976) state that: 'In some species, the smaller and branch roots appear to remain alive for only a few weeks, being constantly replaced by new root networks growing into the same old soil space. Old roots, apparently suberized and low in permeability may suddenly send out numerous young laterals when the soil environment is made favorable . . . '. Despite the fact that root systems are much more dynamic than is generally realized, numerous attempts have been made to model water uptake. Water flow in the liquid phase through the entire soil-plant system can for example be described by an analogue of Ohm's law (Van den Honert, 1948). Consider a one-dimensional, pseudo-steady state flow expressed in terms of water potential and hydraulic resistance of plant and soil as:

$$T = \frac{\psi_s - \psi_r}{R_s} = \frac{\psi_r - \psi_l}{R_p} , \dots\dots\dots\dots\dots\dots\dots\dots\dots \quad [10.3.1\text{-}1]$$

where $\psi_s$, $\psi_r$ and $\psi_l$ are the pressure heads (cm) in the soil, at the root surface and in the leaves respectively; $R_s$ and $R_p$ are the flow resistances (day) in soil and plant respectively, considered as liquid-phase resistances. Therefore $R_p$ does not include stomatal resistance. When the transpiration demand of the atmosphere on the plant system is too high or when the soil is rather dry, $R_s$ and $R_p$ influence $\psi_l$ in such a way that transpiration is reduced by closure of the stomata. An analysis of the effects of $R_s$ and $R_p$ on T is important as, through their effect on $\psi_l$, stomatal closure affects the growth rate. As it is still not possible to measure $\psi_r$, one generally prefers to use an equivalent form of equation [10.3.1-1]:

$$T = \frac{\psi_s - \psi_l}{R_s + R_p} . \quad \dots\dots\dots\dots\dots\dots\dots\dots\dots\dots \quad [10.3.1\text{-}2]$$

Equation [10.3.1-2] can be applied to either the root system as a whole or to discrete horizontal layers. If one, for example, measures T (with a lysimeter), $\psi_s$ (with the neutron scattering probe and moisture characteristic) and $\psi_l$ (with a pressure chamber), and uses water uptake data of two periods one obtains two equations with two unknowns, and $R_s$ and $R_p$ can be calculated.

The relative importance of $R_s$ and $R_p$ was, and still is, an important object of study. Gardner and Ehlig (1962) and Cowan (1965) state that under wet conditions $R_s$ is close to zero. With the assumption that $R_p$ is constant, they contribute the increase in total resistance at soil drying mainly to an increase in $R_s$. Newman (1969) claims that their conclusions are mainly based on the use of rather low root density values. He showed that, when root density is not unusually low, $R_p > R_s$, except for very dry soil (close to wilting point). These findings were confirmed by experiments of Feddes, 1971; Yang and De Jong, 1971; Feddes and Rijtema, 1972; Lawlor, 1972; Hansen, 1974a,b; Taylor and Klepper, 1975; Reicosky and Ritchie, 1976; Greacen et al., 1976; Nnyamah, 1977; So, 1977; Belmans et al., 1977. Results of Molz (1975) indicate that gradients in soil water potential are small in the soil as compared with those in the roots. Usually it is assumed that the largest part of $R_p$ is located in the roots (Jarvis, 1975). Root resistance is sensitive to environmental factors such as temperature and aeration. Feddes and Rijtema (1972, Table 5) have concluded that $R_p$ increases with progressive drying of the soil and decreases when the transpiration rate is higher. This finding has been confirmed by e.g. Hansen (1974a,b) and Weatherley (1978). It must be emphasized that this is a controversial issue—many researchers believing that $R_p$ is not dependent upon soil water content nor transpiration rate. Taylor and Klepper (1975) found that the major resistance to flow is either in the root or the root-soil contact areas. Herkelrath et al. (1977), Nnyamah (1977) and Weatherley (1978) conclude from their experiments that in drying soils the decreasing area of contact between the root and soil water causes the main resistance in water uptake.

### 10.3.2 Soil Resistance Based on Single Root Model

An expression for the soil resistance can be found by considering the flow to a single root (Gardner, 1960). In the single root model the root is viewed as a hollow cylinder of uniform radius and infinite length having uniform water absorbing properties. The governing Darcy flow equation can be written in radial coordinates as

$$\frac{\partial \theta}{\partial t} = \frac{1}{r} \frac{\partial}{\partial r} \left[ r \, K(\psi) \frac{\partial \psi}{\partial r} \right] \qquad \dots\dots\dots\dots\dots\dots\dots\dots \quad [10.3.2\text{-}1]$$

where r is radial distance from the center of the root (cm). Analytical solutions of equation [10.3.2-1] have been obtained for both steady and non-steady states with the assumption of a constant flux at the root-soil interface and a known $K(\psi)$ function (Philip, 1957; Gardner, 1960; Whisler et al., 1970; for a review see Klute and Peters, 1969). For steady state conditions ($\partial \theta / \partial t = 0$) with water flowing from a distance, $r_2$, to a root with radius, $r_1$, the solution under the assumption of constant K is

$$q_r' = \frac{2\pi K(\psi_s - \psi_r)}{\ln(r_2/r_1)} \qquad \dots\dots\dots\dots\dots\dots\dots\dots\dots\dots\dots \quad [10.3.2\text{-}2]$$

where $q_r'$ is the rate of water uptake per unit length of root ($cm^3.cm^{-1}.day^{-1}$). In the field of hydrology, equation [10.3.2-2] is known as the steady-state flux towards a well per unit length of well. For a discrete soil layer of thickness $\Delta z$ and a rooting length per unit volume of soil of L ($cm.cm^{-3}$), the water uptake rate $\Delta q_r$ ($cm.day^{-1}$), can be written as

$$\Delta q_r = L \, \Delta z \, q_r' \qquad \dots\dots\dots\dots\dots\dots\dots\dots\dots\dots\dots\dots \quad [10.3.2\text{-}3]$$

With this soil layer being located at depth Z, equation [10.3.2-3] can be written as

$$\Delta q_r = \frac{2\pi}{\ln(r_2/r_1)} \; L \; K \; (\psi_s - \psi_r - Z)\Delta z, \quad \dotsc\dotsc\dotsc\dotsc \quad [10.3.2\text{-}4]$$

or, in the notation of Gardner (1960), Gardner and Ehlig (1962) and Gardner (1964):

$$\Delta q_r = BLK(\psi_s - \psi_r - Z)\Delta z, \quad \dotsc\dotsc\dotsc\dotsc\dotsc \quad [10.3.2\text{-}5]$$

where $B = 2\pi/\ln(r_2/r_1)$ represents a dimensionless geometric and root distribution factor. The soil resistance $R_s$ can then be written as

$$R_s = \frac{1}{BLK} \quad \dotsc\dotsc\dotsc\dotsc\dotsc\dotsc\dotsc \quad [10.3.2\text{-}6]$$

In a similar way an expression analogous to equation [10.3.2-5] can be derived for an entire uniform root system with rooting depth $z_r$ and root density $L_r$ (Rijtema, 1965; Feddes, 1971; Feddes and Rijtema, 1972) according to

$$T = \frac{\psi_s - \psi_r}{b/K}, \quad \dotsc\dotsc\dotsc\dotsc\dotsc\dotsc\dotsc \quad [10.3.2\text{-}7]$$

where T is in units of cm.day$^{-1}$, $R_s = b/K$ and $b = [\ln(r_2/r_1)]/2\pi \; L \; z_r$ is in units of cm. It thus appears that the root geometry and activity factor b can be written as

$$b = c \, z_r^{-1}, \quad \dotsc\dotsc\dotsc\dotsc\dotsc\dotsc\dotsc\dotsc \quad [10.3.2\text{-}8]$$

where c is a constant.

### 10.3.3 Root Resistances

It was already mentioned that most of the plant resistance, $R_p$, appears to be associated with the roots. When considering water uptake by a single root, one usually thinks of two flow components: a radial flow across the epidermis and cortex into the xylem, and an axial upward flow in the root xylem to the upper parts of the plant. Recently there has been much effort to separate the radial resistance of the root cortex from the axial resistance of the xylem vessels. This has become possible by the introduction of advanced techniques of thermocouple psychrometry (Chow and De Vries, 1973) and dew point hygrometry (e.g. Neumann and Thurtell, 1972). Analogous to equation [10.3.1-1], one can split up the various hydraulic resistances according to

$$T = \frac{\psi_s - \psi_r}{R_s} = \frac{\psi_r - \psi_{rx}}{R_r} = \frac{\psi_{rx} - \psi_l}{R_x}, \quad \dotsc\dotsc\dotsc\dotsc \quad [10.3.3\text{-}1]$$

where $\psi_{rx}$ is pressure head in the root xylem (cm) and $R_r$, $R_x$ are the root (root surface to xylem) and xylem (room xylem to leaf mesophyll) resistance (day), respectively. As it is still impossible to measure $\psi_r$, one usually rewrites equation [10.3.3-1] as

$$T = \frac{\psi_s - \psi_{rx}}{R_{sr}} = \frac{\psi_{rx} - \psi_l}{R_x}, \quad \dotsc\dotsc\dotsc\dotsc\dotsc \quad [10.3.3\text{-}2]$$

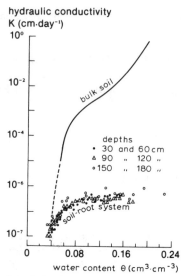

hydraulic conductivity
K (cm·day⁻¹)

**FIG. 10.3-1** A comparison of the hydraulic conductivity of a soil-cotton root system obtained at various depths with the conductivity of the soil for various soil moisture contents (after Taylor and Klepper, 1975).

where $R_{sr}$ is the combined soil and root resistance (bulk soil to root xylem). The root xylem pressure head can be measured by inserting a hygrometer (Nnyamah, 1977) or psychrometer (So, 1978) axially into the roots. By also measuring T, $\psi_s$ and $\psi_l$, the resistances $R_{sr}$ and $R_x$ can be computed. After determining the soil resistance from equation [10.3.2-6], the root resistance, $R_r (= R_{sr} - R_s)$, can be found. Nnyamah (1977) found by continuously monitoring of $\psi_s$, $\psi_{rx}$, $\psi_l$ and estimating T from the Bowen ratio/energy balance method, that $R_r \ggg R_s$, even at $\psi_s$-values of about $-10^4$ cm. Root resistance proved to be relatively constant with decreasing $\psi_s$. During the whole measurement period, the xylem resistance, $R_x$, was consistently higher than $R_r$. So (1978) also found for soybeans that both $R_r$ and $R_x$ are controlling water uptake and that the latter cannot be ignored. Taylor and Klepper (1978) came to the same conclusion for soybeans, but showed that $R_x$ for cotton is relatively small. Belmans et al. (1977) compared model calculations with laboratory measured data on ryegrass and found good results only when taking a high value for the root resistance. Expressing this resistance per length of roots they quote a value of $10^{12}$ s.m⁻¹, that is of an order of magnitude that agrees with findings of Hansen (1974b).

The root conductance for radial flow (inversely proportional to root resistance) seems to vary with pressure head difference, the rate of flow, and with the position along the root (Brouwer, 1965). For more information about root conductances, see Greacen (1977) and De Wit et al. (1978). Under conditions of poor aeration (Chapter 5), low temperature (Chapter 7) and/or water stress one may in general expect a decrease in root permeability due to increase in cytoplasmic viscosity. Because of an interacting influence of transport of water and salts, root permeability may be affected by saline conditions (Chapter 9). Young roots generally will show a higher root permeability than old roots (Taylor and Klepper, 1973). From field experiments Allmaras et al. (1975) report root conductivities for soybean decreasing linearly from $10^{-4}$ to about $0.5 \times 10^{-7}$ cm³ $H_2O$.cm⁻¹ root.day⁻¹.cm⁻¹

pressure head at soil water pressure heads varying from about $-10^2$ to $-10^4$cm. For maize they report similar results for $\psi_s$-values drier than $-700$ cm. Ritchie and Meyer (1978) mention that for sorghum grown in either soil or solution root conductivities vary from $10^{-3}$ to $10^{-4}$ cm$^3$ cm$^{-1}$.h$^{-1}$ bar$^{-1}$ and similarly to e.g. Weatherley (1978) showed that permeability increases with increasing transpiration rates.

The nature of contact between root and water is also of importance. Due to root shrinkage in a drying soil, the surface area of roots in contact with soil decreases and the root membrane resistance increases and thus a kind of 'contact' resistance may develop. Herkelrath et al. (1977) takes this resistance into account by assuming that root water uptake is proportional to the relative saturation, $\theta/\theta_{sat}$, of the soil. Hence the middle part of equation [10.3.3-2] can be rewritten as

$$T = \frac{\theta}{\theta_{sat}} \frac{\psi_s - \psi_{rx}}{R'_{sr}}. \quad\dots\dots\dots\dots\dots\dots\dots\dots\dots \quad [10.3.3-3]$$

Nnyamah (1977), using equation [10.3.3-3], found that for Douglas fir forest this contact resistance could account for up to one half of the total soil to root xylem resistance. The wheat column experiment of Herkelrath et al. (1977) was carried out with a sandy soil. The question may be raised if the concept of contact resistance also holds for heavier textured soils.

Thus, when using a root water uptake model of the kind of equation [10.3.2-5] one should, in principle, not use just the hydraulic conductivity of the soil, but an 'effective conductivity', which represents the conductivity of the soil-root interface and the root tissue (Molz, 1975; Taylor and Klepper, 1975; Molz and Peterson, 1976). In Fig. 10.3-1 results of Taylor and Klepper (1975) are shown where equation [10.3.2-2] type computed hydraulic conductivities of a soil-cotton root system, $K_{sys}$, are compared with the conductivities of the soil alone (K). At low soil moisture contents $K_{sys}$ is comparable to K, but at higher moisture contents, $K_{sys}$ approaches $10^{-6}$K (or $10^{-6}$ cm.day$^{-1}$). These data confirm the notion that the major resistance for root water uptake lies within the root or the root-soil contact areas.

### 10.3.4 Some Empirical Data and Relationships

From the foregoing discussions (section 10.3.2: Soil Resistance based on Single Root Model and 10.3.3: Root Resistances) it has become clear that at the present-day state of knowledge complete specification of all the resistances encountered in the soil-root-plant system still is difficult or even impossible for many crops. In the meantime we can use simplified models that do not require that much detail. One of the simplest models treating the soil-plant-atmosphere system as an entity is the Van den Honert approach, resulting in the equations [10.3.1-1] and [10.3.1-2]. We have seen that plant resistance, $R_p$, is dominant over the soil resistance, $R_s$. A clear example of that has been given by Hansen (1974a) who calculated $R_s$ and $R_p$ for wheat from column experiments as a function of the absolute soil moisture pressure head, $|\psi_s|$ (Fig. 10.3-2). Over the applied interval of $\psi_s$, $R_s$ increases several orders of magnitude and does not limit flow until $|\psi_s|$ is about 12 000 to 15 000 cm. Total resistance, $R_s + R_p$, increases about a factor 10 over the first 1000 cm and then remains about constant. As remarked earlier, it is known that $R_p$ decreases with increasing transpiration rate (Feddes and Rijtema, 1972). This is illustrated with Fig. 10.3-3, where $R_p$ is shown versus

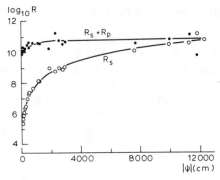

**FIG. 10.3-2** Soil resistance $R_s$ and the sum of soil resistance and plant resistance $(R_s + R_p)$, as functions of the absolute value of the soil moisture pressure head $|\psi|$ as computed from a wheat column experiment by Hansen (1974a).

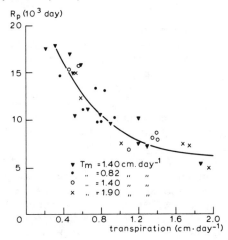

**FIG. 10.3-3** Plant resistance $R_p$ versus transpiration rate T as computed from an Italian ryegrass column experiment by Hansen (1974b) showing the inverse proportionality of $R_p$ with T.

**TABLE 10.3-1. PLANT RESISTANCE $R_p$ AND ROOT GEOMETRY FACTOR b DATA AS EXTRACTED FROM LITERATURE**

| Source | Crop | Type of experiment | Rooting depth, cm | b, cm | $R_p$, ($10^3$ days) |
|---|---|---|---|---|---|
| Gardner and Ehlig (1962) | birds food trefoil | laboratory | 60 | 0.04 | 11.2 |
| Rijtema (1965) | grass | field | 26 | 0.47 | 10.4 |
| Rijtema and Ryhiner (1966) | summer wheat | field | 60 | 0.22 | 30 |
| Endrödi and Rijtema (1969) | potatoes | field | 40 | 0.31 | 10.3 |
| Feddes (1971) | red cabbage | field | 42-82 | 0.54- 0.12 | 31.4- 36.3 |
| Yang and De Jong (1971) | wheat | laboratory | 45 | — | 9-21 |
| Yang and De Jong (1972) | wheat | laboratory | 18 | — | 31-58 |
| Hansen (1974b) | Italian ryegrass | laboratory | 22 | 0.001 | 5-17.5 |
| Reicosky and Ritchie (1976) | maize | greenhouse | — | 0.002- 0.02 | 13 |
| Reicosky and Ritchie (1976) | grain sorghum | field | 100 | 0.001- 0.02 | 11 |
| Nnyamah (1977) | Douglas fir | field | 80 | 0.03 | 25 |

transpiration rate for ryegrass grown in columns (Hansen, 1974b). This implies a so-called 'non-Darcian behavior' of water flow through plants, as the ratio of transpiration rate over head gradient is not a constant.

In Table 10.3-1 a number of data on plant resistances, $R_p$, and on the root geometry factors, b, for various crops as derived from literature data, are summarized (note that $R_s = b/K$, equation [10.3.2-7]. It is to be noted that these data are empirical and that they apply for the conditions from which they were derived. Based on data of Table [10.3-1], Feddes and Rijtema (1972) applying equation [10.3.1-1] showed that, independent of the type of crop, according to equation [10.3.2-8] b can be approximated by the relationship

$$b = 13 \, z_r^{-1}, \dots\dots\dots\dots\dots\dots\dots\dots\dots\dots\dots\dots\dots\dots \quad [10.3.4-1]$$

where b and the ('effective') rooting depth, $z_r$, are both in cm. The factor b seems to be rather constant whether it is wet or dry.

Information on values of $R_p$ in relation to crops and soil moisture conditions is still limited. It is assumed that based on data of Feddes and Rijtema (1972), a very crude approximation of the relationship between $R_p$ and soil moisture pressure head, $\psi_s$, can be used (Rijtema and Aboukhaled, 1975):

$$R_p = 0.763 \ln(-\psi_s) + 1.493, \dots\dots\dots\dots\dots\dots\dots\dots\dots \quad [10.3.4-2]$$

where $R_p$ is in bar·day·mm$^{-1}$ and $\psi$ in bar. Rose et al. (1976) do not consider the soil and plant resistances separately as in equation [10.3.1-1] but take into account the sum of the resistances $(R_s + R_p)$ as in equation [10.3.1-2]. To solve this equation they relate both T and $(R_s + R_p)$ to the plant water potential, $\psi_l$, (or the stem xylem potential) according to

$$T = T_m \, f(\psi_l) \dots\dots\dots\dots\dots\dots\dots\dots\dots\dots\dots\dots\dots \quad [10.3.4-3]$$

and

$$R_s + R_p = g(\psi_l) \dots\dots\dots\dots\dots\dots\dots\dots\dots\dots\dots\dots \quad [10.3.4-4]$$

These functions must first be experimentally determined. Assuming further that the maximum transpiration rate, $T_m$, and $\psi_s$ are known or separately obtained, equation [10.3.1-2] can be expressed as

$$\psi_l = \psi_s - T_m \, f(\psi_l) \, g(\psi_l) \dots\dots\dots\dots\dots\dots\dots\dots\dots\dots \quad [10.3.4-5]$$

## 10.4 MODELING WATER EXTRACTION BY ROOTS

### 10.4.1 Sink Term Models

The water uptake model of a single root (equation [10.3.2-2]) is often termed to be a microscopic approach and the sink term model (equation [10.2-3]) a macroscopic approach. The use of these terms is somewhat overdone. We have seen that the single root model can be used to describe water uptake for individual soil layers (equation [10.3.2-5]). When summing up the water uptake for all the layers over the entire soil profile, one obtains the uptake rate of the whole root system. Thus this model describes the extraction rate of the root system also as a macroscopic process. Moreover, many expressions of the sink term S are based on equations similar to equation [10.3.2-5]. In the following discussion we will maintain a consistent use of notation and units as much as possible, thus stressing similarities and differences between the various models. The Gardner (1964) expression (equation [10.3.2-5]) can be written as

$$S = \frac{\Delta q_r}{\Delta z} = K(\psi_s - \psi_r - Z)\, BL(z), \quad \dots\dots\dots\dots\dots\dots\dots \quad [10.4.1\text{-}1]$$

where S, $q_r$, $\psi_s$ and $\psi_r$ are all functions of z and K is a function of $\psi_s$. The coefficient of proportionality, BL, is interpreted differently by various authors and must be considered merely as an empirical entity. Whisler et al. (1968) express the sink term as

$$S = K(\psi_s - \psi_r)\, A(z), \quad \dots\dots\dots\dots\dots\dots\dots\dots\dots\dots \quad [10.4.1\text{-}2]$$

with $A(z) = k[a(z)/L_e]$, where k is a proportionality factor, $a(z)$ is the area of absorbing surface per unit volume of soil ($cm^2 \cdot cm^{-3}$) and $L_e$ some effective distance over which the water moves to the root. Thus Gardner (1964) and Whisler et al. (1968) interpret the coefficient of proportionality as the surface area of the roots per unit volume of soil and some effective distance over which the water moves, i.e. as a root effectiveness function. In a similar way, Feddes (1971) and Feddes et al. (1974) using the expression

$$S = K(\psi_s - \psi_r)\, b'(z), \quad \dots\dots\dots\dots\dots\dots\dots\dots\dots\dots \quad [10.4.1\text{-}3]$$

consider $b'(z)$ as a parameter which accounts for the length and the geometry of the roots. Feddes et al. (1974) showed that the root effectiveness function, $b'$, is proportional to root mass, w, and that both varied nearly exponentially with depth. They suggest that this function can be found by sampling the root mass with depth and assume that $b' = cw$, where the constant, c, is to be determined from model calibration.

Nimah and Hanks (1973a,b) and Childs and Hanks (1975) consider the root distribution function RDF(z) in a depth interval as the weight fraction of the roots relative to the total weight of roots. They determine this function by sampling the soil profile in the field. They also take into account the presence of salts by introducing the osmotic pressure head, $\psi_{salt}$. This osmotic component reduces the water uptake by the crop as any other water shortage would do. Specific ion or other effects are not included. Their root extraction term is thus defined as

$$S = K(\psi_s - \psi_r + \psi_{salt} - 1.05Z)\, \frac{RDF(z)}{\Delta x\, \Delta z}, \quad \dots\dots\dots\dots\dots\dots \quad [10.4.1\text{-}4]$$

where $\psi_r$ is the effective root water pressure head at the root soil surface, the term 1.05Z is a correction to $\psi_r$ at other soil depths, $\Delta x$ is the distance from the root surface to the point in the soil where $\psi_s$ and $\psi_{salt}$ are measured (arbitrarily assumed 1.0 cm) and $\Delta z$ is the depth increment.

Hansen (1974a, 1975) has developed an uptake model that uses the cylindrical root model as described earlier, but modified to steady rate. The basic, implicit expression for the flow equation is:

$$\psi_r = \left\{ \psi_s^{(1-n)} - q_r' \frac{\left[\ln(r_2/r_1) - 0.5\right] (1-n)}{2\pi a} \right\}^{1/(1-n)}, \quad\dots\dots \quad [10.4.1\text{-}5]$$

where $r_2$ is half the distance between neighboring roots, calculated from the expression $(\pi L)^{-0.5}$, and $r_1$ is root radius. The parameters a and n originate from the relationship between hydraulic conductivity, K, and $\psi_s$, according to: $K = a(|\psi_s|)^{-n}$. In combination with the Van den Honert equations and taking $R_p \approx T^{-1}$, transpiration rates are computed.

Saugier (1974) describes root water uptake from a soil layer as

$$\Delta q_r \approx K_r(z) \ (\psi_s - \psi_l) \quad\dots\dots\dots\dots\dots\dots\dots\dots\dots \quad [10.4.1\text{-}6]$$

where $K_r(z)$ is the root conductivity, taken to be proportional to the relative root biomass in that layer. Equation [10.4.1-6] is thus neglecting the soil resistance, indicators such as root length or root area are not used and root activity is taken simply to be proportional to biomass.

Goldstein and Mankin (1972) use a model based upon an electrical network analogy of a layered soil in which the rooting density diminishes with depth. The resistance between soil and roots, $R_s$, and the root resistance, $R_r$, is expressed for a soil layer of thickness, $\Delta z$, as

$$R_s = \frac{\Delta z \ \delta}{f \ I \ K}, \quad\dots\dots\dots\dots\dots\dots\dots\dots\dots\dots\dots\dots\dots \quad [10.4.1\text{-}7]$$

and

$$R_r = \frac{\Delta z/2}{\delta \ K_r}, \quad\dots\dots\dots\dots\dots\dots\dots\dots\dots\dots\dots\dots \quad [10.4.1\text{-}8]$$

where $\delta$ is cross sectional area of roots per unit area of soil, f is fraction of root surface area through which water is absorbed and I is leaf area index.

Van Bavel and Ahmed (1976) describe water uptake from a separate soil compartment of thickness $\Delta z$ as:

$$\Delta q_r = (\psi_s - Z - \psi_l) \ L' \ \Delta z \ R_p^{-1}, \quad\dots\dots\dots\dots\dots\dots \quad [10.4.1\text{-}9]$$

where $\psi_l$ is the 'effective' pressure head in the leaves, $L'$ root density expressed in cm$^{-1}$ and $R_p$ is the plant resistance (days). Equation [10.4.1-9] thus implies that the main resistance for water uptake lies not in the soil but inside the plant.

It has been shown that in the case of row crops the assumption that there exists only vertical flow in the unsaturated zone is not realistic. Neuman et al. (1975) have developed a Galerkin-type finite element model that solves two-dimensional transient flow in saturated-unsaturated soils considering soil evaporation and water uptake by roots. The latter, analogous to equation [10.4.1-3], is expressed as

$$S = K^r K_{11}^s (\psi_s - \psi_r) \ b', \quad\dots\dots\dots\dots\dots\dots\dots\dots\dots\dots \quad [10.4.1\text{-}10]$$

where $K^r$ is relative hydraulic conductivity ($0 \leqslant K^r \leqslant 1$) and $K_{11}^s$ is the principal conductivity parallel to the horizontal axis.

So far we have discussed root extraction models in terms of Ohm's law: water uptake is considered to be proportional to the difference in pressure head between the soil and root/plant interior, to the hydraulic conductivity of the soil/root system and mostly to some empirical root effectiveness function. Molz and Remson (1970), Molz (1971), Molz and Remson (1971) follow a different approach. These authors consider the sink term S at each depth to be proportional to some effective root distribution factor, $R(z)$, being either the weight or the length of roots per unit volume of soil, and the soil moisture diffusivity, $D(\theta)$:

$$S = T \left[ \frac{R(z) \, D(\theta)}{\int_0^{z_r} R(z) \, D(\theta) \, dz} \right], \quad \dots\dots\dots\dots\dots\dots\dots\dots\dots \quad [10.4.1\text{-}11]$$

where $R(z)$ is in units of g.cm$^{-3}$ or cm.cm$^{-3}$ and $D(\theta)$ in cm$^2$.day$^{-1}$. Afshar and Marino (1978), applying equation [10.4.1-11], report reasonable agreement of simulation results with field measured data of alfalfa.

Van Keulen (1975) expresses water uptake per soil layer as

$$\Delta q_r = T'_m \cdot L' \cdot \alpha \cdot \beta, \quad \dots\dots\dots\dots\dots\dots\dots\dots\dots \quad [10.4.1\text{-}12]$$

where $T'_m$ is the maximum transpiration rate per unit of root length in wet soil (cm.day$^{-1}$.cm$^{-1}$), $L'$ is root length (cm), $\alpha$ is reduction factor for soil temperature below or above the optimum root conductance temperature and $\beta$ is reduction factor for water shortages, being a function of the fraction of available water in a soil compartment, as originally defined by Viehmeyer and Hendrickson (1955).

In the field the root system varies with the type of soil and usually changes with depth and time. Thus root effectivity functions (root density, root distribution, root length, etc.) will also change with depth and time. Experimental and accurate evaluation of such functions is both time consuming and costly. Investigators dealing with the water uptake by root systems of trees will not likely be able to determine root effectiveness functions at all. Greacen (1977) remarks that '. . . direct use of measured values of root density, as an effective root distribution term in extraction models, is not satisfactory. It is generally accepted that uptake is most active in young root material but the length of young roots is not directly related to total root length'.

For the previously mentioned reasons Feddes et al. (1978) propose to use a root extraction term, S, that simply depends on the soil moisture pressure head $\psi$ alone, in the way shown in Fig. 10.4-1. It is assumed that under conditions wetter than a certain 'anaerobiosis point' $(\psi_1)$ water uptake by roots is zero (assumption 1) or quickly reaches zero (assumption 2). Under conditions drier than 'wilting point' $(\psi_2)$ water uptake by roots is also zero. When $\psi$ is below $\psi_2$ but larger than $\psi_3$, it is taken that the water uptake decreases linearly with $\psi$ to zero (assumption 3) or to a certain small value of residual water uptake (assumption 4). The existence of an anaerobiosis behavior according to assumption 2 has clearly been shown in wheat experiments of Yang and De Jong (1971) and Ehlers (1976). Although it is recognized that $\psi_2$ depends on the transpiration demand of the atmosphere (reduction in water uptake occurs at higher (wetter) $\psi_2$-values under conditions of higher demand), the limiting point is taken to be a constant. For more details, see Feddes et al. (1978). In this approach a dimensionless sink term variable

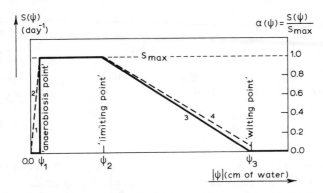

**FIG. 10.4-1 General shape of the sink term S as a function of the absolute value of the soil moisture pressure head $|\psi|$ as used in the model of Feddes et al. (1978).**

$\alpha(\psi)$ is introduced (Fig. 10.4-1).

$$\alpha(\psi) = \frac{S(\psi)}{S_{max}} \, , \, 0 < \alpha(\psi) < 1, \quad\dots\dots\dots\dots\dots\dots\dots \text{[10.4.1-13]}$$

where the maximum of the sink term, $S_{max}$, is defined as

$$S_{max} = \frac{T_m}{z_r} \quad\dots\dots\dots\dots\dots\dots\dots\dots\dots\dots\dots\dots\dots \text{[10.4.1-14]}$$

Thus the expression for the sink term becomes

$$S(\psi) = \alpha(\psi) \, \frac{T_m}{z_r} , \quad\dots\dots\dots\dots\dots\dots\dots\dots\dots\dots\dots \text{[10.4.1-15]}$$

which simply means that the maximum transpiration rate, $T_m$, is distributed equally over the easily measurable rooting depth, $z_r$, and reduced for prevailing water shortages by the factor $\alpha(\psi)$. Kowalik (1979) remarks that under conditions of a uniform root distribution over the soil profile and optimum soil moisture content the Molz and Remson model (equation [10.4.1-11]) yields the same result as equation [10.4.1-15], namely $S = T/z_r$. We must emphasize that equation (10.4.1-15] is a drastic simplification made in the interest of practicality.

The fundamental differential equation that describes the soil water balance (equation [10.2-3]) has two obstacles to its solution, namely that both the variables, $\theta$, and the hydraulic conductivity, K, are highly nonlinear functions of $\psi$. The majority of flow problems is therefore solved by numerical techniques. Analytical solutions can only be obtained for specific cases. Gardner (1958) has eliminated the nonlinear $K(\psi)$ dependence by defining a new dependent variable $\phi$ (cm².day⁻¹), as

$$\phi = \int_{-\infty}^{\psi} K(\psi) \, d\psi \quad\dots\dots\dots\dots\dots\dots\dots\dots\dots\dots \text{[10.4.1-16]}$$

Assuming further that K can be expressed as an exponential function of $\psi$, then when

$$K = K_0 \exp(\alpha\psi), \quad\dots\dots\dots\dots\dots\dots\dots\dots\dots\dots\dots \text{[10.4.1-17]}$$

where $K_0$ is saturated conductivity (cm.day⁻¹) and $\alpha$ is a constant, equation

[10.4.1-16] reduces to

$$\phi = K/\alpha, \dotfill [10.4.1-18]$$

where $\phi$ is to be considered a 'diffusivity potential' or 'matric flux potential'. Because $\partial\phi/\partial z = K(\psi)[\partial\psi/dz]$ and assuming that $\theta = \theta(\phi)$ one can write equation [10.2-3] as:

$$\frac{d\theta}{d\phi} \cdot \frac{\partial\phi}{\partial t} = \frac{\partial^2\phi}{\partial z^2} - \alpha\frac{\partial\phi}{\partial z} - S \dotfill [10.4.1-19]$$

If one may consider $d\theta/d\phi = D$ in equation [10.4.1-19] to be a constant, one has linearized the soil moisture flow equation, which may result in analytical solutions. For nonsteady state the linearized flow equation thus reads as

$$\frac{1}{D}\frac{\partial\phi}{\partial t} = \frac{\partial^2\phi}{\partial z^2} - \alpha\frac{\partial\phi}{\partial z} - S(z,t), \dotfill [10.4.1-20]$$

and the steady state equation as

$$\frac{d^2\phi}{dz^2} - \alpha\frac{d\phi}{dz} = S(z,\phi) \dotfill [10.4.1-21]$$

Solutions for steady state flow situations in a semi-infinite medium and a finite-depth medium overlying a water table have, considering $S(z,\phi)$ as an explicit function of depth z only, have been given by Warrick (1974) and Raats (1974), who also take into account the distribution of salts. Lomen and Warrick (1976) have presented analytical solutions of equation [10.4.1-21] for $S(z,\phi)$, either defined explicitly in terms of depth z or implicitly through several functions of $\phi$. Warrick et al. (1979) present solutions for two-dimensional trickle irrigation systems with S depending on z only.

Solutions for the time dependent flow in a semi-infinite medium have been given by Lomen and Warrick (1978) who describe S by a sequence of depth-dependent functions that change at specified times.

As it is impossible to list here all the sink functions used in equations [10.4.1-20] and [10.4.1-21], the reader is referred to the cited references.

### 10.4.2 Model Capabilities and Shortcomings

From the previous sections we have seen that models describing the water flow through the soil-plant-atmosphere system can be simple, as the Van den Honert model, or complex, as the various expressions for the sink term, discussed under section 10.3.1: General.

All models discussed need some input concerning the soil and plant parameters, initial conditions and boundary conditions that determine the behavior of the plant system. Usually, one is mainly interested in predicting the water use of the crop and relate it to crop production, rather than in details such as accurate prediction of soil moisture content, extraction at different depths, etc.

These details, however, are important when calibrating a model or when comparing simulated results with field measured data. In general, as far as the author is aware, there exists no single model that can predict reliable results without calibration and/or updates from measurements. Mathematical description of root water uptake is still hampered by the complexity of the uptake mechanism, the variability of transpiration demand with time, the heterogeneity of the root system varying with depth and time, effects of temperature, aeration and age on root conductivity, the presence of high resistances for water flow in the plant as compared to that in the soil, the possible presence of 'contact' resistances, the dependence of plant resistance on transpiration rate, effects of soil and plant resistance on closure of the stomata, etc. For these and other reasons one has to consider present root water uptake models as primitive tools in predicting water use of a crop. Strict meteorological methods to estimate (evapo)transpiration such as the energy balance approach, the aerodynamic or profile method, or the so-called combination method, however, seem no more accurate or useful.

Experimental verification of models thus is still a necessity. Once having determined the soil and plant parameters, one has to calibrate the model parameters to the actual field situation. From there on one can try to evaluate the effects of root zone modification of changing soil moisture, soil profile, groundwater table depth, etc. upon the water use and productivity of a crop. This is the greatest advantage of a simulation model: to estimate effects of root zone modification, rather than to determine absolute values of water use.

Having applied a model to certain conditions, the question arises: can the model be applied to different crops, soils, meteorological and agricultural conditions and provide reliable results. To answer this question one has also to apply a sensitivity analysis, i.e. to carry out numerical experiments, besides checking with field experiments. In this way one will learn how the system behaves.

Sensitivity analysis is applied to evaluate the effect of structural changes in a model and to determine the relative importance of parameters and boundary conditions. Sensitivity analysis connected with the variation of input values includes all the initial and boundary conditions and all parameters of the model. Changes in output values are a measure of the change in overall system behavior as compared with the reference case. It is useful to evaluate how an error in each parameter affects the overall system performance. Examples of sensitivity analysis for ecological system models are given by Miller (1974), Maas and Arkin (1978).

When the partial differential equation [10.2-3] is solved by a numerical procedure one can apply either a finite difference or a finite element approach. In the finite difference approach, one replaces all derivatives by finite differences to reduce the original continuous initial-boundary-value problem to a discrete set of simultaneous algebraic equations. In dealing with nonuniform compressible soils of complex geometry and arbitrary anisotrophy, and irregular atmospheric boundaries such as evaporation, infiltration and transpiration surfaces, the finite difference approach is often difficult to apply. The finite element method (Neuman et al., 1975) can overcome most of these difficulties. It is based on an integral representation of the problem, rather than on the differential form. Whatever numerical method is used, one can apply either explicit or implicit methods. Although an implicit approach is more complicated, it is preferable because of its better stability and convergence. Moreover, it permits relatively large time steps thus keeping computer costs low.

Analytical solutions are usually restricted to specific, mostly simplified flow cases. For example, when the hydraulic conductivity is an exponential function of the pressure head, the soil moisture diffusivity is constant with depth, and surface flux and sink term are specified. However, they provide exact answers for the situation investigated and can thus be used to check complex numerical schemes, for which it is difficult to detect errors.

In section 10.4.1: Sink Term Models, the reader was exposed to many different models and rather superficially to their meaning and derivation. The practicing agronomist or engineer now has to decide which approach might be best used in given circumstances for particular objectives. The practical applicability of any model depends to a large extent on the input requirements of the model. Once the problem in the field has been defined, it is known which input data for any arbitrary model are available. The availability, or more likely nonavailability, of input data already implies a strong selection in the choice among the various existing models. In general, lack of information about the properties of the plant root system will impose the strongest limitations in the use of a model. Thus, models which need information about root length, root area, root density, root distribution or root conductivity can hardly be used for practical applications. Those models containing expressions of only simply determinable quantities as relative or absolute root weight, rooting depth, etc. are much more attractive to use. Thus in practice the procedure of selecting the appropriate model for the specific problem will be rather simple. In the sections 10.5.2, 10.5.3 and 10.5.4 a selection of models has been made which, in the author's opinion, can be applied to a number of practical field problems.

## 10.5 ASSESSING ROOT ZONE MODIFICATION BY MANAGEMENT: MODEL STUDIES

### 10.5.1 Water Use and Crop Production Models

From the previous sections we have seen that there are a relatively large number of root water uptake models available to the engineer or agronomist for potential use in evaluating management practices. Root zone modifications influence the water management of the soil and, therefore, the water use and the productivity of a crop. In this context we should pay attention to relationships between water use and actual crop production.

Water use can be interpreted either as transpiration, T, the integral of the sink term, or as evapotranspiration, ET. In Fig. 10.5-1 total dry matter production of corn is plotted against evapotranspiration for the 1974 growing season in Davis, California, under different irrigation regimes and water quality treatments (taken from Stewart et al., 1977). Also inserted are corn yield data of Hillel and Guron (1973) obtained during a 5-yr irrigation experiment at the Gilat Experimental Farm in Israel. It appears that for corn there is a linear relationship between yield and ET. The intersection of the line with the horizontal axis represents the amount of evaporation from the soil. Nonlinear relationships are also reported in literature. For example, experiments of Stewart and Hagan (1969) with winter wheat show a clear convex relationship when plotting either absolute or relative yield versus absolute or relative evapotranspiration, respectively. These results can be explained by the different way water deficits influence straw and grain production. Similar curvilinear relationships obtained from 12 yr irrigation experiments are reported by Musick et al. (1976) for winter wheat, sorghum and soybeans when plotting relative yield versus soil water depletion. On the other hand, Stewart and Hagan (1969) show concave relationships for alfalfa, indicating an increasing water use efficiency until a maximum is reached. Moreover,

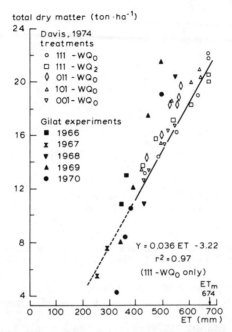

total dry matter (ton·ha⁻¹)

FIG. 10.5-1 Dry matter yield of corn in Davis, CA, in 1974 as a function of total evapotranspiration during the growing season for different irrigation regimes and water quality. Irrigation regime 111 implies that sufficient water was applied during all growth stages. The irrigation variant 101 implies that an insufficient amount of water was applied during the vegetative growth stage 1 and mature growth stage 3 but not during the pollination stage 2. The irrigation variant 001 implies that irrigation was applied only in the mature stage 3. Water of good quality is indicated with $WQ_0$, of poor quality with $WQ_2$ i.e. an electroconductivity of 2 mmhos.cm⁻¹ (after Stewart et al., 1977; also included are data of Hillel and Guron (1973) obtained during the 1966-1970 irrigation seasons in Gilat, Israel.)

water use efficiency is high in spring, medium in summer and low in autumn. These examples show how careful one needs to be in generalizing production functions without specific field studies with different species and varieties of crops.

Some of the production functions proposed by various authors are the following. De Wit (1958) relates total dry matter, Y (kg.ha⁻¹), linearly to total transpiration, $T^{tot}$, by:

$$Y = A_1 T^{tot}, \quad \dots\dots\dots\dots\dots\dots\dots\dots\dots\dots\dots\dots\dots\dots\dots\dots [10.5.1\text{-}1]$$

where $A_1$ is in units of kg.ha⁻¹.mm⁻¹. The slope, $A_1$, seems to depend on climate and latitude. For arid areas De Wit (1958) found that yield and transpiration could be related by:

$$Y = A_2 \frac{T^{tot}}{E_0}, \quad \dots\dots\dots\dots\dots\dots\dots\dots\dots\dots\dots\dots\dots\dots\dots [10.5.1\text{-}2]$$

where $E_0$ is the evaporation rate (mm.day⁻¹) of a free water surface, being a measure of the evaporative demand. Here, $A_2$ has the dimension of kg.ha⁻¹.day⁻¹. Rijtema (1969) has plotted alfalfa data, collected by De Wit (1958) according to equation [10.5.1-2]. The results are shown in Fig. 10.5-2. While water supply is limiting, there is a linear increase of Y with $T^{tot}/E_0$. If

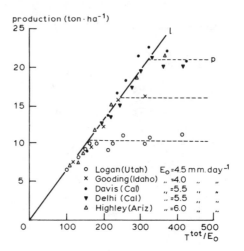

FIG. 10.5-2 Hay production of alfalfa as a function of total transpiration over open water evaporation rate $T^{tot}/E_o$ from data of De Wit (1965) as derived by Rijtema (1969).

another growth factor becomes limiting, in this case the availability of mineral nutrients, a production ceiling is observed, which is different for different locations.

Bierhuizen and Slatyer (1965) proposed a relationship based on the diffusion processes of $H_2O$ and $CO_2$ in the crop canopy. Their equation reads:

$$Y = A \frac{T^{tot}}{\Delta e}, \quad \cdots\cdots\cdots\cdots\cdots\cdots\cdots\cdots\cdots\cdots\cdots\cdots\cdots\cdots\cdots \quad [10.5.1\text{-}3]$$

where $\Delta e$ is the vapor pressure deficit of the air in mbar. Thus, A has dimensions of kg.ha$^{-1}$.mbar.mm$^{-1}$. Climatic areas with different values for $\Delta e$ will result in a different value of Y for the same $T^{tot}$.

Feddes et al. (1978) assume that actual production rate ($\dot{Y}$) is delimited by two asymptotes: one (line l in Fig. 10.5-2) that shows a proportional increase of the growth rate with increasing supply of water (Aw), and the second one (line p in Fig. 10.5-2) imposed by all growth factors together that limit the growth rate to a certain maximum or ceiling ($\dot{Y}_{pot}$). The production rate can then implicitly be written as

$$(1 - \frac{\dot{Y}}{Aw})(1 - \frac{\dot{Y}}{\dot{Y}_m}) = \xi, \quad \cdots\cdots\cdots\cdots\cdots\cdots\cdots\cdots\cdots\cdots\cdots \quad [10.5.1\text{-}4]$$

where w is taken according to equation [10.5.1-3] (w = $T/\Delta e$), the potential production rate, $\dot{Y}_m$, is determined from De Wit (1965), Feddes et al. (1978) and/or De Wit et al. (1978), and $\xi$ is a numerical constant close to zero.

Analogous to equation [10.5.1-1] one can write

$$Y_m = A_1 T_m^{tot}, \quad \cdots\cdots\cdots\cdots\cdots\cdots\cdots\cdots\cdots\cdots\cdots\cdots\cdots\cdots \quad [10.5.1\text{-}5]$$

where $Y_m$ is maximum (potential) yield and $T_m^{tot}$ is maximum (potential) transpiration. Dividing equation [10.5.1-1] through equation [10.5.1-5] and rearranging gives

$$Y = \frac{T^{tot}}{T_m^{tot}} \cdot Y_m \quad \cdots\cdots\cdots\cdots\cdots\cdots\cdots\cdots\cdots\cdots\cdots\cdots\cdots\cdots \quad [10.5.1\text{-}6]$$

Stewart et al. (1977) start with the assumption of a maximum yield level at each evapotranspiration level, varying from $Y = Y_m$ at $ET = ET_m$ (the origin of the function) and linearly decreasing for $ET < ET_m$. The 'Stewart' model thus reads as

$$(1 - \frac{Y}{Y_m}) = \beta_0 (1 - \frac{ET}{ET_m}), \quad\dots\dots\dots\dots\dots\dots\dots\dots\dots\dots\dots\dots \quad [10.5.1\text{-}7]$$

where $\beta_o$ is a dimensionless slope.

All these production models are relatively simple and generally apply to crops without water sensitive growth stages, i.e. the effects of water stress on yield during all growth stages are similar. For crops showing different effects of water stress during various physiological stages of growth, rather complicated expressions have been developed which often do not show better results than the simple models (Stewart et al., 1977).

So far models have been discussed that deal mainly with total dry matter production. Recently crop growth simulation models have been developed that are based on phenological, morphological and physiological hypotheses and data. These models simulate plant development through a complete growth cycle in great detail. The equations describing the various processes are often empirically derived from field measurements. Generally the water use of the crop canopy in these models is calculated in a submodel in a rather simplified way. For a literature list of existing simulation models of various crops, the reader is referred to Arkin et al. (1976) and Arkin et al. (1977). The ELM-model, an example of one which simulates biomass dynamics in a variety of grassland types and considers the response of the system to irrigation, fertilization, and cattle grazing, is reported by Innis (1978). In addition to yield determination these models are also intended for decision making.

Having established the relationships between water use and crop production, one then can consider the effects of root zone modification on production through water use. The following section gives a description how models dealing with the plant root environment have been applied. Examples of simulated and measured effects of root zone modifications, such as changes in soil moisture, soil profile and groundwater table depth upon water use and crop production, will be given. Most results are taken from field experiments having a practical orientation.

### 10.5.2 Models for Evaluating Change of Soil Moisture and Salinity

One of the most important means to influence the root zone directly is changing its soil moisture content by applying irrigation. In this respect the surface conditions of the soil are important. Modification of the surface zone by mechanical or chemical means is generally aimed at promoting infiltration, suppressing evaporation, and optimizing conditions for germination and early plant growth. Hillel (1973), and Hillel and Berliner (1974) have addressed this problem. From their investigations it appears that the main effect of aggregating the soil is an increase of the soil hydraulic conductivity in the wet range and a decrease in the dry range. Hydrophobic treatment of the soil aggregates can enhance the mentioned effects, and also prevent slaking and crust formation (Van Bavel, 1950). In order to quantify effects of aggregate size and depth of aggregated layer upon infiltration and evapotranspiration, one should apply those soil physically oriented flow models that allow for a detailed description of soil moisture flow in heterogenous soils (either in or not in conjunction with water uptake by roots). In this context a complete review of all existing soil moisture flow models cannot be given. For more information see Hillel et al. (1975), Hillel (1977), and Wind (1979).

The soil moisture regime in the topsoil determines to a large extent the field trafficability and timeliness of farming operations. In spring workability for seedbed preparation influences the amount of labor and machinery required, the date of sowing, hence the length of the growing season and accordingly the yield of the crop (Fig. 10.1-1). In autumn workability determines the harvest date and therefore also crop yield (Fig. 10.1-2). For the application of a numerical model in predicting machinery operation time, see Bordovski (1978).

Soil moisture flow models provide an excellent tool to simulate the soil moisture content of the top layer of the soil under varying environmental conditions and thus to evaluate the influence of soil technological measures such as drainage and profile modification on workability. Because simulation of long time series with numerical models can become quite costly, Wind (1972) developed a hydraulic analog model and Wind and Mazee (1979) an electronic analog model, which have almost negligible operation cost. Because these models use an exponential hydraulic conductivity-pressure head relationship, practical applicability of the models is confined to fairly moist conditions. Wind (1976) investigated the effect of drainage on workability in spring with a hydraulic analog for a sandy clay loam soil. He calculated moisture conditions in a topsoil from natural rain and evaporation data over 22 yr for different drain depths and drain intensities. A soil moisture pressure head of −300 cm in the top 5 cm was taken to be the limit for workability. The calculations were verified with field observations. Fig. 10.5-3 shows both the number and the distribution of workable days for 22 years, the latter being important for labor studies. The first date of soil workability and number of workable days show large variations over the years. Deeper drainage results in earlier workability. Drain intensity had hardly any effect. Feddes and Van Wijk (1980) used this analog approach for spring and autumn conditions in combination with a numerical evapotranspiration model for the growing season. With this integrated model approach they were able to estimate the optimum drain depth. Van Wijk (1979) applied the hydraulic analog of Wind to quantify playing conditions of grass sportsfields for various combinations of top subsoils, drain depths and drain intensities.

When evaluating influences of changes in soil moisture upon the root zone one needs to correctly determine the appropriate water management scheme. This includes such factors as irrigation time, amount of water to apply, the irrigation efficiency, water losses to the subsoil below the root zone, water uptake from the layers below the root zone, irrigation water quality, effects of irrigation on salinity and uniformity of irrigation. Root water uptake models can be used to investigate these factors both quantitatively and qualitatively.

One of the models already discussed in equation [10.4.1-4] is that of Nimah and Hanks (1973a,b) which has been tested in a field experiment with an alfalfa crop growing on a sandy loam soil. The input data needed for this model are: the soil moisture properties $\psi(\theta)$ and $K(\theta)$, air dry and saturated condition $\theta_d$ and $\theta_s$ respectively, root distribution function as a function of

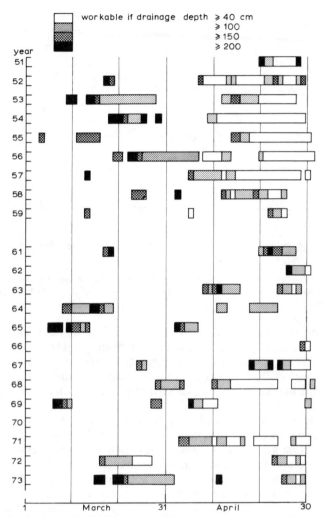

**FIG. 10.5-3 Number and distribution of workable days on a sandy clay loam soil in The Netherlands in 22 yr for four drainage depths as determined with an analog model (after Wind, 1976).**

depth and time RDF(z,t), maximum and minimum pressure head (wilting) at the root surface, $\psi_{wilt} \leqslant \psi_r \leqslant 0$, initial moisture condition, $\theta(z,t = 0)$, initial salinity profile, $\psi_{salt}(z,t = 0)$, top-boundary conditions such as potential transpiration, $T_m$, potential soil evaporation, $E_m$, (or rainfall/irrigation) and osmotic pressure head of the irrigation water, $\psi_{salt}$, all as functions of time, bottom-boundary condition such as the absence or presence of a water table, e.g. $\psi(z,t) = 0$. The output data of the model include: cumulative values of $T(t)$, $E(t)$ and $ET(t)$, profiles of $\theta(z,t)$ and $\psi(z,t)$, cumulative upward/downward fluxes through the lower boundary $q(z,t)$, values of $\psi_r(t)$.

For a validation of the model output one can compare numerical data with measured values of, for example, $\theta(z,t)$, $ET(t)$, $q(z,t)$. Here we take one example from the results of Nimah and Hanks (1973b): the variation of the soil moisture contents at the 30, 70 and 100 cm depths throughout the 116

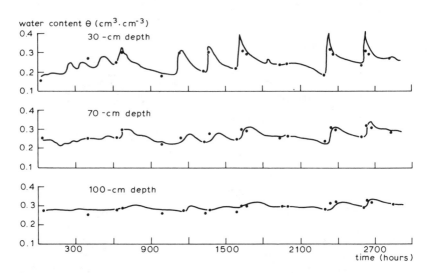

**FIG. 10.5-4 Moisture contents (solid lines) as simulated with the model of Nimah and Hanks (1973b) and neutron probe measured data (points) at depths of 30, 70 and 100 cm under an alfalfa crop.**

day growing season of alfalfa in 1971 (Fig. 10.5-4). It appears that at least for the 70 and 100 cm depths there is excellent agreement between the computed and the measured moisture data. The greatest discrepancies occur at the 30 cm depth, especially in the redistribution stage just after irrigation, which might be due to the neglect of hysteresis, of a uniform soil and to the approximate nature of the sink term.

The model of Nimah and Hanks has been checked for salinity effects by Childs and Hanks (1975) with data taken from the literature, and for these data reasonable agreement between simulation and measurement was also found. Childs and Hanks also made some simulation runs for crops with different rooting depths for different water applications and initial soil salinity. These runs were not validated with field experiments. The main results are shown in the Figs. 10.5.5 and 10.5.6. In Fig. 10.5.5 relative yield is shown as a function of irrigation plus rain for various initial salinities and three different rooting depths. Relative yield was determined by means of equation [10.5.1-6], showing a 1:1 relationship of $Y/Y_m$ with relative transpiration $T/T_m$. It is seen that, at increasing water application, yield increases and that the shallow-rooted crop is most affected. The effect of initial salinity is highest for the deeply rooted annual crop.

Since in the simulation runs a 2 m groundwater table depth was assumed, there exists considerably larger, by about a factor 4, upward flow from the water table for the deep than for the shallow-rooted crop (Fig. 10.5-6). This causes a larger effect of water application on yield for the shallow-rooted crop (Fig. 10.5-5). Consequently the effect of initial salinity must be highest for the deep-rooted crop (Fig. 10.5-6). Childs and Hanks (1975) have also investigated the influence of low and high water application rates on crop yield and salinization over several growing seasons having uniform climatic conditions. They found that low irrigation rates give a practical linear decrease in yield and corresponding increase in salinization through 4 yr with a leveling off thereafter.

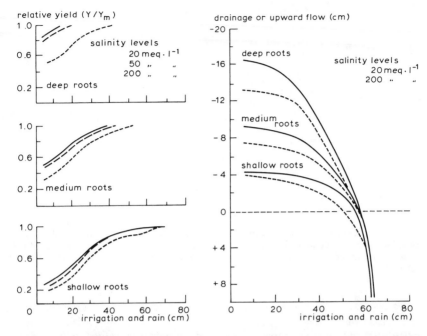

FIG. 10.5-5 Relative dry matter yield $(Y/Y_m)$ for crops with three different rooting depths and three different initial salinity levels as a function of the amount of water (irrigation + rain) applied, as computed by the model of Childs and Hanks (1975).

FIG. 10.5-6 Drainage (+) or upward flow (−) under crops with three different rooting depths and two different initial salinity levels as a function of the amount of water (irrigation + rain) applied, as computed by the model of Childs and Hanks (1975).

The previously cited results show the advantage of using simulation models that have first been tested and calibrated with experimental data. Expensive field experiments that are difficult or impossible to carry out can be replaced by models to simulate detailed effects over long time periods. In this respect the accuracy of absolute results are generally of less importance than the relative results comparing one type of root zone modification with another.

Feddes et al. (1974) used a modification of the numerical model of Nimah and Hanks. In their model the air dry moisture content at the soil surface is taken to vary with time depending on meteorological conditions. The maximum possible rates of transpiration, $T_m$, and soil evaporation, $E_m$, are calculated by considering both meteorological conditions and crop properties, unlike Nimah and Hanks, who base them merely on calculations of open water evaporation. The water uptake function used is the one described in equation [10.4.1-3], with root effectiveness function values of $b'(z)$ as determined experimentally by Feddes (1971) in the field, varying exponentially with depth, according to the relationship: $b'(z) = b'(0)\exp(-\alpha z)$. Theoretical results as predicted by the model were compared with a field experiment in which red cabbage was grown on a heavy clay soil in the presence of a water table. Water balance studies were performed with a specially designed nonweighing lysimeter in which the water table, varying between 94 and 107 cm below the soil surface, could be maintained continuously at the

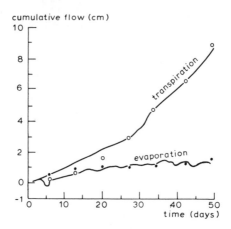

FIG. 10.5-7 Comparison of cumulative transpiration and soil evaporation (solid lines) as simulated with the model of Feddes et al. (1974) with measured data (open and closed points respectively) of Feddes (1971) for a red cabbage crop growing on clay in the presence of a water table.

same depth as it is in the surrounding field (Feddes, 1971). One example of the results of Feddes et al. (1974) is shown in Fig. 10.5-7, which compares calculated cumulative transpiration and evaporation values with measured data over a 49 days period. It is seen that, except during the first three weekly periods, there is excellent agreement between calculated and measured values of transpiration. The values of soil evaporation also show good agreement for the entire period considered. It was further shown that the proposed modifications for calculating $T_m$ and $E_m$ from meteorological and crop properties were an improvement of the model.

Feddes et al. (1978) have used the same data as mentioned above to test a model called 'SWATR' with a completely different root water uptake model, namely the one described according to equation [10.4.1-15]. For a schematic representation, Fig. 10.1-1, left hand side. Input data in the model are: $\psi(\theta)$ and $K(\psi)$ relationships for upper and lower soil layer, depth of the root zone $z_r$, critical values of the sink term as denoted in Fig. 10.4-1, initial condition $\theta(z,t=0)$, boundary conditions at the soil surface of $T_m(t)$ and $E_m(t)$, boundary condition at the bottom of a water table with $\psi(z,t) = 0$. Output data of the model include cumulative values of $T(t)$, of integrated water content over the soil profile, of upward/downward flows, of runoff, of $\theta(z,t)$, and of $S(z,t)$. One example of the results is shown in Fig. 10.5-8, where curves of cumulative flow are given: first the measured cumulative evapotranspiration ($ET_{water\ balance}$) as obtained from the lysimeter; second the cumulative transpiration, $T_{comp}$, as computed with the model by integration of the sink term over depth; third the cumulative soil evaporation, $E_{soil}^{comp}$, derived from the computed terms of the water balance.

From Fig. 10.5-8 it is seen that there is good agreement between computed and measured evapotranspiration, especially at the beginning and end of the period considered. When comparing the results of the first 49 days with Fig. 10.5-7 it appears that similar results are obtained using either the relatively complicated expression of equation [10.4.1-3] for the sink term or the simple one of equation [10.4.1-15]. Using the transpiration data of Fig.

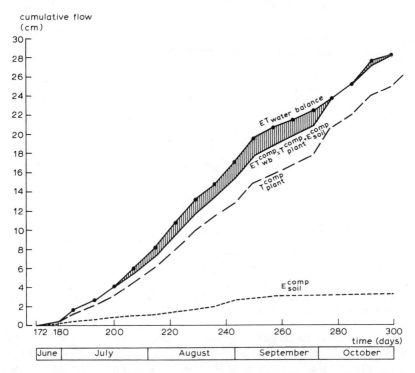

**FIG. 10.5-8 Comparison of cumulative evapotranspiration $ET_{wb}^{comp}$ as simulated with the model of Feddes et al. (1978) and lysimetrically measured data $ET_{water\ balance}$ for a red cabbage crop growing on clay in the presence of a water table.**

10.5-8 as an input for a crop production model, called 'CROPR' (the scheme is depicted at the right hand side of Fig. 10.1-1) and the production function used is that of equation [10.5.1-4], computed actual dry matter yields could be compared with measured data of Feddes (1971). Fig. 10.5-9 shows that calculations compare well with measurements. The measured data points represent weekly harvests of one plant. With a heterogeneous crop like cabbage, a large variation in dry matter production is then to be expected. The points show a random scatter around the calculated actual curve, but final yield was predicted quite well. The difference between actual yield and computed potential (maximum) yields appeared to be 12 percent. Both models SWATR and CROPR were also tested against field experiments of potatoes growing on a loamy sand. In this case reasonably good agreement was also obtained. For a complete description and list of instructions to use the programs of the two models, Feddes et al. (1978). To allow for efficient computation for areas with deep water tables, the SWATR program has recently been extended with a supplement (Wesseling and Feddes, 1979).

Van Bavel and Ahmed (1976) developed a model in which the water uptake by the plants is expressed according to equation [10.4.1-9]. This expression is applied to different layers in the root zone and is joined with a layered crop canopy model. Flow of water from the subsoil into the root zone is taken into account. The general solution method has been explained by Van Bavel (1974). The inputs to the model are hourly average values for global radiation, air temperature, dewpoint and windspeed. The description of the crop

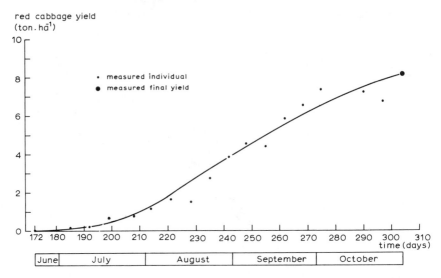

**FIG. 10.5-9 Comparison of actual cumulative dry matter yield of a red cabbage crop as computed with the production model CROPR of Feddes et al. (1978) using as an input the simulated transpiration data of Fig. 10.5-8, with measured actual yield data.**

canopy consists of its height, roughness parameter, leaf area index, and the distribution of leaf angle in each layer. The root system is described by its total depth and the distribution of the root mass within it. The soil is defined by the functions $K(\theta)$ and $\psi(\theta)$. Also required is a function that relates leaf resistance to leaf water flow in the plant.

Outputs are the hourly transpiration rate, the hourly rate of soil evaporation, hourly profiles of water content in the soil, and of leaf temperatures and water potentials. The latter values are calculated separately for sunlit and shaded leaves, in each of the canopy layers. Daily totals of evapotranspiration and of drainage or upward flow of water from or into the root zone are also obtained.

Simulation with the model was carried out for a 20 day-drying period with a constant diurnal weather pattern for a fully developed sorghum crop. One example of the results is shown in Fig. 10.5-10. This figure shows that, after a few days of constant evapotranspiration, a monotonic decline occurs, which is principally attributed to a decrease in transpiration, as soil evaporation remained almost constant during the whole period. The upward flow into the root zone amounts to about 30 percent of total evapotranspiration, a figure that agrees with values found in field experiments by Van Bavel et al. (1968a). From the simulation it was found that during the night redistribution of moisture occurs inside the root zone: the upper part of the root zone becomes wetter, the lower part drier. The authors interpreted this phenomenon mainly as a result of water transfer and distribution by the roots themselves. If this is true it should be found from measurements. Molz and Peterson (1976) have shown experimentally that flow from roots to soil is rather small, thus casting doubt on the theoretical inference.

We will now consider an example obtained from analytical solution of the soil moisture flow equation. Lomen and Warrick (1978) solved the linearized moisture flow equation (equation [10.4.1-20]) for the case of a specified, time dependent surface flux at the upper boundary, i.e. the soil

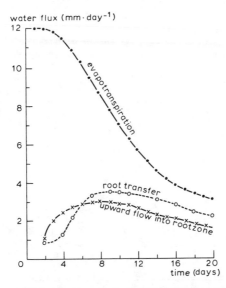

**FIG. 10.5-10 Comparison of the evapotranspiration, upward flow from below into the bottom of the root zone and water transfer through the roots from the lower to the upper part of the root zone, for a fully developed sorghum crop as simulated by the model of Van Bavel and Ahmed (1976).**

surface, a deep water table as the lower boundary and the sink term as a specified, time dependent function of depth. In order to be able to solve equation [10.4.1-20] the factor $\alpha$ needs to be removed from the term $\alpha(\partial \phi / \partial z)$. Therefore a dimensionless depth coordinate $\zeta$ and time coordinate $\Gamma$ can be introduced, which are defined as:

$$\zeta = \frac{\alpha z}{2}, \qquad\qquad\qquad [10.5.2\text{-}1]$$

and

$$\Gamma = \frac{\alpha^2 \, Dt}{4} \qquad\qquad\qquad [10.5.2\text{-}2]$$

Substitution in equation [10.4.1-20] yields

$$\frac{\partial \phi}{\partial \Gamma} = \frac{\partial^2 \phi}{\partial \zeta^2} - 2 \frac{\partial \phi}{\partial \zeta} - \frac{4}{\alpha^2} \, S(\zeta, \Gamma) \qquad\qquad [10.5.2\text{-}3]$$

Applying the appropriate boundary and initial conditions, equation [10.5.2-3] can be solved analytically (for details, see Lomen and Warrick, 1978). Input data required for this model are: empirical constant $\alpha$ in the exponential $K(\psi)$ relationship, constant diffusivity value D, soil surface flux $q(0,\Gamma)$, sink term $S(\zeta,\Gamma)$, deep water table $\phi_\infty$. For the output one obtains the distribution of the matrix flux potential $\phi(\zeta,\Gamma)$. One example of the results of Lomen and Warrick (1978) is shown in Fig. 10.5-11. In this flow situation $\alpha$ = 0.015 cm$^{-1}$, D = 4000 cm$^2$.day$^{-1}$ and $\phi_\infty$ = 33.3 cm$^2$.day$^{-1}$. The surface irrigation flux is periodic, being 3 cm.day$^{-1}$ for the first 24 h and 0 for the next 24 h, continuing in 2-day cycles. The sink term is expressed as S = a

| surface flux | $q_1 = 3$ | | $q_2 = 0$ | | $q_3 = 3$ | | $q_4 = 0$ | | $q_5 = 3$ | | $q_6 = 0$ cm·day$^{-1}$ | |
|---|---|---|---|---|---|---|---|---|---|---|---|---|
| plant uptake | $a_1$ | $a_2$ | $a_3$ | $a_4$ | $a_5$ | $a_6$ | $a_7$ | $a_8$ | $a_9$ | $a_{10}$ | $a_{11}$ | $a_{12}$ day$^{-1}$ |
| | 0.04 | 0 | 0.04 | 0 | 0.04 | 0 | 0.04 | 0 | 0.04 | 0 | 0.04 | 0 |

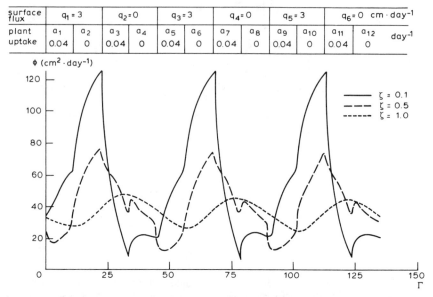

Fig. 10.5-11 Variation of matric flux potential $\phi$ at three dimensionless depths $\zeta$ of 0.1, 0.5 and 1.0 (i.e. actual depths of 13.3, 66.6 and 133 cm respectively) with dimensionless time $\Gamma$ for a cyclic irrigation pattern and a cyclic root water uptake pattern as computed by the analytical model of Lomen and Warrick (1978).

exp ($-0.02$ z) and root water uptake is made cyclic by setting a $= 0.04$ day$^{-1}$ for the first 12 h period in each day, and a $= 0$ for the next 12 h in a 1-day cycle. The $\phi$-curves in Fig. 10.5-11 are depicted for dimensionless depths $\zeta =$ 0.1, 0.5 and 1.0, corresponding to actual depths z $= 13.3$, 66.6 and 133 cm, respectively. The matrix flux potential $\phi$ reaches its highest values under zero root water uptake, just before irrigation stops. It reaches its lowest values after irrigation ceases and root uptake proceeds. The decrease in amplitude and shift in phase with depth is clearly shown. Analytical models of the type discussed are advantageous, because of the few input data they need and the negligible computation costs. The disadvantages are in the restrictive assumptions that are needed. For example, the use of a constant soil moisture diffusivity restricts practical application to the case of Fig. 10.5-11, i.e. high-frequency irrigations, so that the soil moisture content does not vary appreciably.

### 10.5.3 Models for Evaluating Change of Soil Profile

In many soils water shortages of crops are caused by a too shallow rooting depth. The reasons for a restricted rooting depth can be many, such as poor aeration, mechanical impedance, chemical impediments, etc. These aspects have been discussed in detail in the previous chapters. To the author's knowledge, there are very few results of model applications for evaluating effects of soil profile changes on crop water use. Ehlers (1976) has investigated water use of crops in zero-tilled soils. He found that root water uptake in soil layers with low porosity is smaller than in adjacent layers with higher porosities. However, it appeared that total water use is only slightly affected, as roots extending to deep layers compensate for the decline of activity of roots in top layers. In the following we will therefore restrict ourselves to results from field experiments and one result of a numerical experiment taken from investigations in The Netherlands. In these experiments relatively shallow water tables are present.

In cases where it is impossible to bring water to the roots, e.g. by raising the groundwater table artificially, one can do the reverse, namely increase the rooting depth by changing the soil profile. This offers the advantage that for good workability conditions the groundwater table in spring can remain deep, >1 m minus soil surface, and the moisture supply can still be maintained. The latter occurs in two ways: the availability of moisture in the rooted zone is increased because water is extracted by the roots from a larger soil volume, and because the distance between root zone and groundwater table is decreased, a greater upward flow towards the root zone takes place.

Results of investigations of De Vries (1974) on sandy soils show that one cause of restricted rooting depth is a high density of the mineral sublayer. Roots cannot penetrate and are restricted to the organic top layer. Increasing the rooting depth by loosening the subsoil yielded an available moisture increase of 5 to 8 mm per 10 cm of soil layer. More important, however, is the decrease of the distance from the root zone to the water table. When this distance is 60 cm, the upward flow amounts to about 1 mm.day$^{-1}$. If the distance increases, e.g. to 80 cm, then the capillary rise decreases to some 0.2 mm.day$^{-1}$. This behavior indicates that when the groundwater table is lowered, by pumping for domestic use for example, the benefit of the increase in rooting depth by the soil improvement measure is easily lost.

There are soil profiles where a layer of clay overlies a subsoil of sand. Rooting depth is there restricted to the clay layer because of the high density of the sublayer. Mixing of the sand subsoil can increase the rooting depth (Wind, 1969). In this case groundwater tables should be maintained relatively high.

Wind (1969) and Wind and Pot (1976) describe profile modifications in a peat area by soil mixing which results in a deeper rooting depth and in an increased soil moisture availability and supply. Moreover, the organic matter content of the top layer is reduced, thus increasing the bearing strength of the soil and reducing the danger for night frosts.

Feddes et al. (1978) carried out numerical experiments with respect to changes in rooting depth and soil profile. In Figs. 10.5-8 and 10.5-9 results were described of a simulation of transpiration and crop production of a red cabbage crop grown on clay soil in the presence of a water table. The rooting depth for this crop varies with time. It was 22.5 cm after planting and gradually increased during 42 days to 82.5 cm, and remained at this depth for the rest of the growing season. In the first numerical experiment the rooting depth was kept constant at a depth of 42.5 cm during the whole growing season. Transpiration was computed with SWATR and used as an input to CROPR. The yield computed with the latter is shown in Fig. 10.5-12, as the curve, clay over clay. When comparing this yield curve with the observed one in Fig. 10.5-9, it appears that, due to the reduced rooting depth, a relatively strong reduction in yield occurs. This is caused by the limited transport of water in this soil from the groundwater table up to the shallow root zone. A second numerical experiment was carried out. A sandy loam soil was substituted for the clay subsoil, keeping the rooting depth again continuously at 42.5 cm. The result is shown as the clay over sandy loam curve in Fig. 10.5-12. A subsoil of sandy loam has better water transmitting properties. Comparing the computed yield of the clay covered sandy loam with 42.5

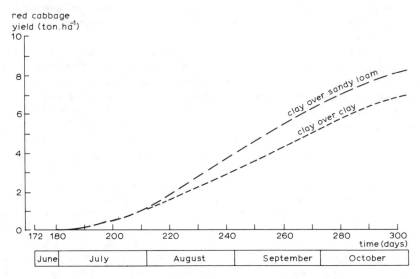

**FIG. 10.5-12 Dry matter yield of a red cabbage crop having a restricted rooting depth of 42.5 cm for two different soil profiles as simulated by the root water uptake and crop production models of Feddes et al. (1978).**

cm rooting depth with the yield curve obtained on the clay soil with 82.5 cm rooting depth (Fig. 10.5-9), we see that there is practically no difference. This finding is supported by experimental results of Feddes (1971). This example again shows the advantage of using models. Soil profile influences that are difficult and very costly to evaluate experimentally can be studied relatively simply with a simulation model. Alternatives can be computed and compared rapidly.

### 10.5.4 Models for Evaluating Change of Groundwater Level

The groundwater supply is subject to demand from various interest groups, such as agriculture, recreation, industry and municipalities. Excessive withdrawal may result in a lowering of the groundwater table, that can damage the supply of water to these users. To compensate for such a drawdown, measures that will recharge the groundwater must be taken. In groundwater recharge problems, the unsaturated-saturated zone should be considered as one.

A model that can deal with these problems was developed by Neuman et al. (1975) which takes the root water uptake term into account according to equation [10.4.1-10]. This model has been applied to field situations by Feddes et al. (1975). Included here is one example to demonstrate the capabilities of the finite element approach in treating complex two-dimensional field flow problems. It is taken from the subirrigation experimental field 'De Groeve' in The Netherlands (Feddes and Van Steenbergen, 1973). The purpose of this example is to demonstrate the capabilities of the numerical approach, and not to verify the model. The field was 16 ha of peaty soil, having an average thickness of 1.4 m. The peat is underlain by sandy soils to a depth of about 10 m. The bottom of the sand is separated from an underlying aquifer by 2 m of sediments having a relatively low hydraulic conductivity. Outside the experimental field, the aquifer is

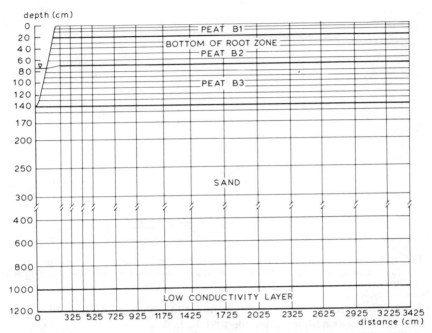

**FIG. 10.5-13 Cross section of a two-dimensional field system with superimposed finite element network. This system consists of five anisotropic soil layers and is influenced by evapotranspiration from a crop canopy, subirrigation from unlined ditches and extraction from a pumped aquifer (after Feddes et al., 1975).**

penetrated by wells that intermittently extract water for domestic supply. The field is traversed by several unlined ditches in which the water level is controlled. In this case, flow takes place under a cropped field through a five-layered anisotropic soil, the boundaries of which include two ditches on the side and a pumped aquifer at the bottom. In addition to the effect of these boundaries, fluctuations of the water table occur because of water uptake by the crop, the roots of which grow with time, and evaporation at the soil surface.

Only one-half of a vertical cross section between two ditches is considered due to the symmetry of the field. This cross section, together with the superimposed finite element network, is shown in Fig. 10.5-13. According to laboratory water retention data, the peat soil has been divided into three layers having distinct material properties. The peat and sand layers are assumed to be anisotropic, having horizontal conductivities 10 times as large as the vertical ones.

The crop grown in the field is potatoes. Initially the depth of the root zone is taken to be 40 cm. The root effectiveness function, $b'$, is assumed to vary with depth and time in the same manner as it did in the case of red cabbage (Feddes et al., 1974).

The water level in the ditches is maintained at a constant depth of 74 cm below the soil surface. This is also the initial depth of the water table. Initially the water is under a static condition (uniform head) whereas later, no flow is allowed to take place across the vertical boundaries of the system.

The hydraulic head in the pumped aquifer at the bottom of the system, as well as the maximum allowable rates of plant transpiration and soil

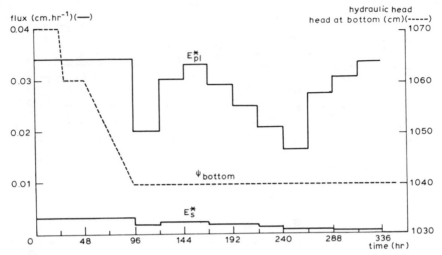

**FIG. 10.5-14 Variation with time of hydraulic head in underlying aquifer and potential rates of soil evaporation and transpiration in the two-dimensional system of Fig. 10.5-13 (after Feddes et al., 1975).**

evaporation, vary with time in the manner illustrated in Fig. 10.5-14. A time step of 24 h was adopted after the first 12 h.

The cumulative volume of water leaving the soil surface via evapotranspiration, the volume of water leaking out of the system into the underlying aquifer, and the volume of water infiltrating into the system from the ditch are shown in Fig. 10.5-15. It is noted that the loss of water due to evapotranspiration and leakage exceeds the inflow from the ditch, a fact which is reflected in the lowering of the water table by as much as 27 cm.

This example demonstrates the flexibility of the finite element approach and its capability in treating complex situations encountered in the field.

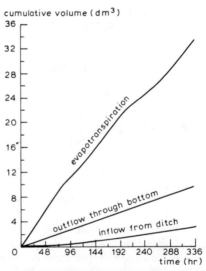

**FIG. 10.5-15 Cumulative evapotranspiration, infiltration from the ditch and leakage into the underlying aquifer of the two-dimensional system of Fig. 10.5-13 with the boundary conditions of Fig. 10.5-14 as computed by Feddes et al. (1975).**

## 10.6 STATE OF THE ART AND RECOMMENDATIONS FOR FUTURE RESEARCH

The effects of root zone modifications upon water use and crop production can be evaluated by numerical and/or analytical water uptake models. These models have to be tested and calibrated with laboratory or field experiments. In general, it is not as important to determine absolute effects of root zone modification as it is to evaluate the differences in water use and crop production that are caused by differences in root zone treatments. Water uptake models can be valuable in evaluating existing as well as proposed soil, water and crop management practices.

Most of the models discussed need detailed information about root distribution, root densities, conductivities of the soil-root system, soil and plant resistances. Often these parameters vary with soil type and, also, with depth and time. Their experimental evaluation is sometimes impossible, and always time consuming and costly. Moreover, investigations have shown that the parameters mentioned do not always adequately describe the complex root water uptake processes. Therefore, less detailed models have been proposed that are easier to handle and, despite the rough approximation of the problem, might serve the practical needs of the agronomist and engineer. The decision of which model to use will largely depend on the amount and extent of input data available and on the specific goal. In this respect it is emphasized that research is needed for verification of the models, for example by using nondestructive moisture measurement techniques to evaluate the root system behavior under various treatments.

Effects of soil aeration, soil temperature and soil fertility upon root growth and water uptake by roots have been dealt with in many separate laboratory/field experiments. Although some relationships have been found, the understanding of the entire complex system is still rather poor. Therefore in root growth and water uptake models corrections for temperature and aeration are made according to very simple relationships. Generally one speaks in terms of 'minimum', 'optimum' and 'maximum' conditions for growth and uptake, with linear effects assumed between these points.

Models can be separated into numerical and analytical ones. The first group needs more input data and are generally more costly in use than the second group. Analytical models provide exact answers, but are more restricted to specific, simplified cases, e.g. an exponential relationship of the hydraulic conductivity curve. They can be used to check complicated numerical schemes.

Effects of salinity on crop production have been quite well investigated. Detailed knowledge of effects of salinity on root water uptake, however, is still lacking and needs more experimental investigation. For example, it is not known if different water uptake patterns exist under saline and non-saline conditions.

The models can only be used if we are well informed about the physical properties of the soil. We must emphasize that the soil moisture retention and hydraulic conductivity curve should be determined from undisturbed soil samples. This is usually done for measurements in the relatively wet range. For the dry range however, one often uses disturbed samples. In this request the application of the so-called hot-air method that determines the hydraulic conductivity curve from undisturbed samples is recommended. This method was developed by Arya (1973) in the USA and is now used in Europe. The method is simple, fast and covers a large soil moisture range. The use of undisturbed samples is important, because small differences in the soil profile may have large influences on both water flow in the soil and water uptake by the roots. Even when taking undisturbed samples, the variation of soil properties within a small 'homogeneous' region may be such that interpretation of the data becomes difficult. This problem of spatial variability has been addressed by Warrick et al. (1977) and Nielsen et al. (1978).

Most of the models discussed have been verified with laboratory or field experiments for a local situation. The question then arises, can the model be extrapolated to different crops, soils, meteorological and agricultural conditions? How sensitive are the results of a model for changes in physical parameters, in initial and boundary conditions? These questions emphasize the need for a sensitivity analysis, to evaluate the effect of structural changes in a model and to determine the relative importance of the various input data upon the output of the model. Numerical experiments therefore are a necessity.

In investigations dealing with groundwater recharge problems it is not sufficient to study only the saturated zone. The unsaturated zone has to be taken into account, because processes occurring in one zone directly influence those in the other zone. When controlling groundwater recharge one is required to use an integrated unsaturated-saturated zone approach.

## References

1  Afshar, A. and M. A. Marino. 1978. Model for simulating soil water content considering evapotranspiration. J. Hydrol. 37:309-322.

2  Allmaras, R. R., W. W. Nelson, and W. B. Voorhees. 1975. Soybean and corn rooting in Southwestern Minnesota: I. Water-uptake sink. II. Root distributions and related water inflow. Soil. Sci. Soc. Am. Proc. 39:764-777.

3  Arkin, G. F., R. L. Vanderlip, and J. T. Ritchie. 1976. A dynamic grain sorghum growth model. TRANSACTIONS of the ASAE 19(4):622-626, 630.

4  Arkin, G. F., C. L. Wiegand, and H. Huddleston. 1979. The future role of a crop model in large area yield estimating. Proc. Crop Model. Worksh., Columbia, MO, Oct. 3-5, 1977, United States Department of Commerce, NOAA, Environmental Data and Information Service.

5  Arya, L. M. 1973. Water flow in soil in presence of soybean root sinks. Thesis University of Minnesota, Minneapolis. Bull. 60. 163 p.

6  Arya, L. M., G. R. Blake, and D. A. Farrell. 1975. A field study of soil water depletion patterns in presence of growing soybean roots: III. Rooting characteristics and root extraction of soil water. Soil Sci. Soc. Am. Proc. 39:437-444.

7  Bakker, J. W., S. Dasberg, and W. B. Verhaegh. 1978. Effect of soil structure on diffusion coefficient and air permeability of soils (in preparation).

8  Belmans, C., J. Feijen, and D. Hillel. 1977. Comparison between simulated and measured moisture profiles in a sandy soil during extraction by a growing root system. In: Modeling, identification and control in environmental systems, Van Steenkiste, ed. IFIP, North Holland Publishing Company:317-320.

9  Bierhuizen, J. F., and R. O. Slatyer. 1965. Effect of atmospheric concentration of water vapor and $CO_2$ in determining transpiration-photosynthesis relationships of cotton leaves. Agric. Meteor. 2:259-270.

10  Bordovski, J. P. 1978. Predicting farm machinery operation time with a soil moisture model. MS thesis. Agric. Eng. Dep., Texas A&M Univ., College Station, TX.

11  Brouwer, R. 1965. Water movement across the root. Symp. Soc. Exp. Biol. 19:131-149.

12  Childs, S. W., and R. J. Hanks. 1975. Model of salinity effects on crop growth. Soil Sci. Soc. Am. Proc. 39:617-622.

13  Chow, T. L., and J. de Vries. 1973. Dynamic measurement of soil and leaf water potential with a double loop peltier type thermocouple psychrometer. Soil Sci. Soc. Am. Proc. 37:181-188.

14  Cowan, I. R. 1965. Transport of water in the soil-plant-atmosphere system. J. Appl. Ecol. 2:221-239.

15  De Vries, Th. 1974. Grondverbetering: waarom, hoe en met welke werktuigen. I: Landb. Mech. 25,02: 155-163, II: Landb. Mech. 25,03: 221-225.

16  De Wit, C. T. 1958. Transpiration and crop yields. Agric. Res. Rep. 64.6 Pudoc, Wageningen. 88 p.

17  De Wit, C. T. 1965. Photosynthesis of leaf canopies. Agric. Res. Rep. 66.3. Pudoc, Wageningen. 57 p.

18  De Wit, C. T. et al. 1978. Simulation of assimilation, respiration and transpiration of crops. Simulation Monographs, Pudoc, Wageningen. 141 p.

19  Ehlers, W. 1976. Evapotranspiration and drainage in tilled and untilled loess soil with winter wheat and sugarbeet. Z. Acker- and Pflanzenbau 142:285-303.

20  Endrodi, G., and P. E. Rijtema. 1969. Calculation of evapotranspiration from potatoes. Neth. J. Agric. Sci. 17,4:283-299.

21  Feddes, R. A. 1971. Water, heat and crop growth. Thesis Comm. Agric. Univ. Wageningen 71-12. 184 p.

22  Feddes, R. A., and P. E. Rijtema. 1972. Water withdrawal by plant roots. J. Hydrol. 17:33-59.

23  Feddes, R. A., and M. G. van Steenbergen. 1973. (in Dutch) 'De Groeve'. Nota 735 Inst. Land Water Man. Res., Wageningen. 60 p.

24  Feddes, R. A., E. Bresler, and S. P. Neuman. 1974. Field test of a modified numerical model for water uptake by root systems. Water Resour. Res. 10(6):1199-1206.

25  Feddes, R. A., S. P. Neuman, and E. Bresler. 1975. Finite element analysis of two-dimensional flow in soils considering water uptake by roots: II. Field applications. Soil Sci. Soc. Am. Proc. 39:231-237.

26  Feddes, R. A., and A. L. M. van Wijk. 1976. An integrated model-approach to the effect of water management on crop yield. Agric. Water Man. 1:3-20.

27  Feddes, R. A., P. J. Kowalik, and H. Zaradny. 1978. Simulation of field water use and crop yield. Simulation Monographs. Pudoc, Wageningen. 189 p.

28  Fiscus, E. L. 1972. In situ measurement of root water potentials. Plant Physiol. 50:191-193.

29  Flühler, H., F. Richard, K. Thalmann, and F. Borer. 1975. Einfluss der Saugspannung auf den Wasserentzug durch die Wurzeln einer Grasvegetation. Z. Pflanzenern. Bodenk. 6:583-593.

30  Gardner, W. R. 1958. Some steady-state solutions of the unsaturated moisture flow equation with application to evaporation from a water table. Soil Sci. 85:228-232.

31  Gardner, W. R. 1960. Dynamic aspects of water availability to plants. Soil Sci. 89:63-73.

32  Gardner, W. R. 1964. Relation of root distribution to water uptake and availability. Agron. J. 56:35-41.

33  Gardner, W. R., and C. F. Ehlig. 1962. Some observations on the movement of water to plant roots. Agron. J. 54:453-456.

34  Gardner, W. R., and E. M. Cullen. 1966. Water movement in the soil profile and uptake by plant roots, p. 2.1-2.12. In: Technical report ECOM 2-66 R-A, Fort Huachuca, AZ.

35  Goldstein, R. A., and J. B. Mankin. 1972. PROSPER: A model of atmosphere-soil-plant water flow. p. 176-181. Proc. Summer Computer Simulation Conf., San Diego, CA.

36  Greacen, E. L. 1977. Mechanisms and models of water transfer. p. 163-196. In: J. S. Russel and E. L. Greacen (eds.) Soil factors in crop production in a semi-arid environment. Univ. of Queensland Press.

37  Greacen, E. L., P. Ponsana, and K. Barley. 1976. Resistance to water flow in the roots of cereals. p 86-100. In: Water and Plant Life. Ecological Studies 19. Springer Verlag, Berlin.

38  Hansen, G. K. 1974a. Resistance to water transport in soil and young wheat plants. Acta. Agric. Scand. 24:37-48.

39  Hansen, G. K. 1974b. Resistance to water flow in soil and plants, plant water status, stomatal resistance and transpiration of Italian ryegrass, as influenced by transpiration demand and soil water depletion. Acta. Agric. Scand. 24:83-92.

40  Hansen, G. K. 1975. A dynamic continuous simulation model of water state and transportation in the soil-plant-atmosphere system. 1. The model and its sensitivity, Acta. Agric. Scand. 25:129-149.

41   Herkelrath, W. N., E. E. Miller, and W. R. Gardner. 1977. Water uptake by plants: I. Divided root experiments, II. The root contact model. Soil Sci. Soc. Am. J. 41:1033-1043.
42   Hillel, D. 1973. Some aspects of the field water cycle affected by soil surface conditions. In: Med. Fac. Landbw. R. U. Gent 37(3):1114-1131.
43   Hillel, D. 1977. Computer simulation of soil-water dynamics; a compendium of recent work. Ottawa, IDRC. 214 p.
44   Hillel, D., and Y. Guron. 1973. Relation between evapotranspiration rate and maize yield. Water Resour. Res. 9(3):743-748.
45   Hillel, D., and P. Berliner. 1974. Water proofing surface-zone soil aggregates for water conservation. Soil Sci. 118(2):131-135.
46   Hillel, D., C. H. M. Van Bavel, and H. Talpaz. 1975. Dynamic simulation of water storage in fallow soil as affected by mulch of hydrophobic aggregates. Soil Sci. Soc. Am. Proc. 39(5):826-833.
47   Hsiao, T. C., E. Fereres, E. Ecevedo, and D. W. Henderson. 1976. Water stress and dynamics of growth and yield of crop plants. p. 281-303. In: Ecological studies, water and plant life. Vol. 19. Springer Verlag, Berlin.
48   Innis, G. S. 1978. Grassland simulation model. Ecol. Studies, vol. 26. 298 p. Springer Verlag, New York, Heidelberg, Berlin.
49   Jarvis, P. 1975. Water transfer in plants. p. 369-394. In: D. A. de Vries and N. H. Afgan (eds.) Heat and mass transfer in the biosphere: I. Transfer processes in plant environment. Scripta Book Co., Wash. DC.
50   Klute, A., and D. B. Peters. 1969. Water uptake and root growth. p. 105-134. In: W. J. Whittington (ed.) Root growth. Butterworth, London.
51   Kowalik, P. 1979. Model for simulating soil water content considering evapotranspiration—Further comments. J. Hydrol. 40:389-391.
52   Lawlor, D. W. 1972. Growth and water use of lolium perenne: I. Water transport. J. Appl. Ecol. 9:79-98.
53   Lomen, D. O., and A. W. Warrick. 1976. Solution of the one-dimensional linear moisture flow equation with implicit water extraction functions. Soil Sci. Soc. Am. J. 40:342-344.
54   Lomen, D. O., and A. W. Warrick. 1978. Time dependent solutions to the one-dimensional linearized moisture flow equation with water extraction. J. Hydrol. 39,1/2:59-67.
55   Maas, S. J., and G. F. Arkin. 1978. Sensitivity analysis of a grain sorghum model. ASAE Paper No. 78-4035, ASAE, St. Joseph, MI 49085.
56   Miller, D. R. 1974. Sensitivity analysis and validation of simulation models. J. Theor. Biol. 48:345-360.
57   Molz, F. J. 1971. Interaction of water uptake and root distribution. Agron. J. 63:608-610.
58   Molz, F. J. 1975. Potential distributions in the soil-root system. Agron. J. 67:726-729.
59   Molz, F. J., and I. Remson. 1970. Extraction term models of soil moisture use by transpiring plants. Water Resour. Res. 6(5):1346-1356.
60   Molz, F. J., and I. Remson. 1971. Application of an extraction term model to the study of moisture flow to plant roots. Agron. J. 63:72-77.
61   Molz, F. J., and C. M. Peterson. 1976. Water transport from roots to soil. Agron. J. 68:901-904.
62   Musick, J. T., L. L. New, and D. A. Dusek. 1976. Soil water depletion—yield relationships of irrigated sorghum, wheat, and soybeans. TRANSACTIONS of the ASAE 19(3):489-493.
63   Neuman, S. P., R. A. Feddes, and E. Bresler. 1975. Finite element analysis of two-dimensional flow in soils considering water uptake by roots. I. Theory. Soil Sci. Soc. Am. Proc. 39:224-230.
64   Neumann, H. H., and G. W. Thurtell. 1972. A peltier cooled thermocouple dew point hygrometer for in situ measurement of water potentials. p. 103-112. In: R. W. Brown and B. P. van Haveren (eds.) Psychrometry in water relations research. Utah Agr. Exp. Sta., Logan.
65   Newman, E. I. 1969. Resistance to water flow in soil and plant. I. Soil resistance in relation to amounts of root: theoretical estimates. J. Appl. Ecol. 16:1-12.
66   Nielsen, D. R., J. W. Biggar, and C. S. Simmons. 1978. Analysis of water and solute movement in field soils. EGS-ESC 5th meeting Strasbourg, 29 Aug.-1 Sept.
67   Nimah, M. N., and R. J. Hanks. 1973a. Model for estimating soil water, plant and atmospheric interrelations. I. Description and sensitivity. Soil Sci. Soc. Am. Proc. 37:522-527.
68   Nimah, M. N., and R. J. Hanks. 1973b. Model for estimating soil water, plant and atmospheric interrelations. II. Field test of model. Soil Sci. Soc. Am. Proc. 37:528-532.

69  Nnyamah, J. U. 1977. Root water uptake in a Douglas fir forest. Thesis University of British Columbia, Vancouver. 154 p.

70  Nnyamah, J. U., and T. A. Black. 1977. Rates and patterns of water uptake in a Douglas fir forest. Soil Sci. Soc. Am. J. 41:972-979.

71  Philip, J. R. 1957. The physical principles of soil water movement during the irrigation cycle. Proc. Int. Congr. Irrig. Drain. 8:124-154.

72  Raats, P. A. C. 1974. Steady flows of water and salt in uniform soil profiles with plant roots. Soil Sci. Soc. Am. Proc. 38:717-722.

73  Raats, P. A. C., and W. R. Gardner. 1974. Movement of water in the unsaturated zone near a water table. p. 311-357. In: Jan van Schilfgaarde (ed.) Drainage for Agriculture. Agronomy 17. Am. Soc. of Agron., Madison, WI.

74  Reicosky, D. C., and J. T. Ritchie. 1976. Relative importance of soil resistance and plant resistance in root water absorption. Soil Sci. Soc. Am. J. 40:293-297.

75  Reicosky, D. C., R. J. Millington, A. Klute, and D. B. Peters. 1972. Patterns of water uptake and root distribution of soybeans in the presence of a water table. Agron. J. 64:292-297.

76  Rice, R. C. 1975. Diurnal and seasonal soil water uptake and flux within a Bermuda grass root zone. Soil Sci. Soc. Am. Proc. 39:394-398.

77  Rijtema, P. E. 1965. An analysis of actual evapotranspiration. Agric. Res. Rep. 659. Pudoc, Wageningen. 107 p.

78  Rijtema, P. E. 1969. On the relation between transpiration, soil physical properties and crop production as a basis for water supply plans. Versl. Meded. Cie Hydr. Onderz. TNO, 's-Gravenhage: 28-58.

79  Rijtema, P. E., and G. Endrödi. 1970. Calculation of production of potatoes. Neth. J. Agric. Sci. 18:26-36.

80  Rijtema, P. E., and A. Aboukhaled. 1975. Crop water use, 5-57. In: Research on crop water use, salt affected soils and drainage in the Arab Rep. of Egypt. FAO, Rome.

81  Ritchie, J. T., and W. S. Meyer. 1978. Dynamics of water conductance in sorghum roots. The soil-root interface symposium, Oxford, England.

82  Rose, C. W., and W. R. Stern. 1967. Determination of withdrawal of water from soil by crop roots as a function of depth and time. Austr. J. Soil Res. 5:11-19.

83  Rose, C. W., G. F. Byrne, and G. K. Hansen. 1976. Water transport from soil through plant to atmosphere: a lumped-parameter model. Agric. Meteorology 16:171-184.

84  Saugier, B. 1974. Transports de CO₂ et de vapeur d'eau à l'interface végétation-atmosphère. These d'Etat. USTL, Montpellier. 156 p.

85  Sieben, W. H. 1974. On the influence of drainage on nitrogen delivery and yields on reclaimed loamy soils in the IJssel Lake Polders. Van Zee tot Land 51. 180 p. (in Dutch).

86  So, H. B. 1977. In situ measurement of root water potentials. Aust. Soil Sci. Soc. Inc. ACT Branch Conference, Canberra, May 19-21.

87  So, H. B. 1978. Water potential gradients and resistances of a soil-root system measured with the root and soil psychrometer. The soil-root interface symposium, Oxford, England.

88  Stewart, J. I., and R. M. Hagan. 1969. Predicting effects of water shortage on crop yield. J. Irr. and Drain. Div., Vol. 95(1):91-104.

89  Stewart, J. I. et al. 1977. Optimizing crop production through control of water and salinity levels in the soil. Utah Water Res. Lab. Report PRWG 151-1, Utah State Univ., Logan. 191 p.

90  Stone, L. R., M. L. Horton, and T. C. Olson. 1973. Water loss from an irrigated sorghum field. I. Water flux within and below the root zone. II. Evapotranspiration and root extraction. Agron. J. 65:492-497.

91  Stone, L. R., I. D. Teare, C. D. Nickell, and W. C. Mayaki. 1976. Soybean root development and soil water depletion. Agron. J. 68:677-680.

92  Strebel, O., M. Renger, und W. Giesel. 1975. Bestimmung des Wasserentzuges aus dem Boden durch die Pflanzenwurzeln im Gelande als Funktion der Tiefe und der Zeit. Z. Pflanzenern. Bodenk. 1:61-72.

93  Taylor, H. M., and B. Klepper. 1973. Rooting density and water extraction patterns for corn (Zea Mays L.). Agron. J. 65:965-968.

94  Taylor, H. M., and B. Klepper. 1975. Water uptake by cotton root systems: an examination of assumptions in the single root model. Soil Sci. 120:57-67.

95  Taylor, H. M., and B. Klepper. 1978. The supply of water to plant roots. Adv. Agron. 30:99-128.

96  Van Bavel, C. H. M. 1950. The use of volatile silicones to increase water-stability of soils. Soil Sci. 70:291-297.

97    Van Bavel, C. H. M. 1974. Exploratory simulation of the depletion of water reserves in the plant root zone and its consequences. p. 279-282. Proc. Summer computer simulation conf., Houston, TX.

98    Van Bavel, C. H. M., G. B. Stirk, and K. J. Brust. 1968a. Hydraulic properties of a clay loam soil and the field measurement of water uptake by roots. I. Interpretation of water content and pressure profiles. Soil Sci. Soc. Am. Proc. 32:310-317.

99    Van Bavel, C. H. M., K. J. Brust, and G. B. Stirk. 1968b. Hydraulic properties of a clay loam soil and the field measurement of water uptake by roots. II. The water balance of the root zone. Soil Sci. Soc. Am. Proc. 32:317-321.

100    Van Bavel, C. H. M., and J. Ahméd. 1976. Dynamic simulation of water depletion in the root zone. Ecol. Modelling 2:189-212.

101    Van den Hil, J. 1975. Premium for early and late delivery. Maandbl. Suikerunie 9(7):8-10 (in Dutch).

102    Van den Honert, T. H. 1948. Water transport in plants as a catenary process. Discuss. Faraday Soc. 3:146-153.

103    Van Hoorn, J. W. 1958. Results of a groundwater level experimental field with arable crops on clay soil. Neth. J. Agric. Sci. 6:1-10.

104    Van Keulen, H. 1975. Simulation of water use and herbage growth in arid regions. Simulation Monographs. Pudoc, Wageningen. 176 p.

105    Van Wijk, A. L. M. 1980. A soil technological study on effectuating and maintaining adequate playing conditions of grass sportsfields. Agric. Res. Rep. 903, Pudoc, Wageningen.

106    Viehmeyer, F. J., and A. H. Hendrickson. 1955. Does transpiration decrease as the soil moisture decreases? Trans. Am. Geoph. Union 36:425-428.

107    Warrick, A. W. 1974. Solution to the one-dimensional linear moisture flow equation with water extraction. Soil Sci. Soc. Am. Proc. 38:573-576.

108    Warrick, A. W., G. J. Mullen, and D. R. Nielsen. 1977. Scaling field-measured soil hydraulic properties using a similar media concept. Water Resour. Res. 13(2):355-362.

109    Warrick, A. W., A. Amoozegar-Fard, and D. O. Lomen. 1979. Linearized moisture flow from line sources with water extraction. TRANSACTIONS of the ASAE 22(3):549-553, 559.

110    Weatherley, P. E. 1978. The hydraulic resistance of the soil-root interface: a cause of water stress in plants. The soil-root interface symposium, Oxford, England.

111    Wesseling, J. 1974. Crop growth and wet soils. p. 7-32. In: J. van Schilfgaarde (ed.) Drainage for Agriculture. Mon. 17, Am. Soc. Agron., Madison, WI.

112    Wesseling, J. G., and R. A. Feddes. 1979. Introduction of suction as a lower boundary condition in program SWATR (for deep water tables). Nota 1127, Inst. Land Water Man. Res., Wageningen. 13 p.

113    Whisler, F. D., A. Klute, and R. J. Millington. 1968. Analysis of steady state evapotranspiration from a soil column. Soil Sci. Soc. Am. Proc. 32:167-174.

114    Whisler, F. D., A. Klute, and R. J. Millington. 1970. Analysis of radial, steady state solution, and solute flow. Soil Sci. Soc. Am. Proc. 34:382-387.

115    Wind, G. P. 1960. Landbouwk. Tijdschr. 72:111-118.

116    Wind, G. P. 1969. Grondverbetering. Cult. techn. Verh., Staatsuitg., 's-Gravenhage: 189-222.

117    Wind, G. P. 1972. A hydraulic model for the simulation of non-hysteretic vertical unsaturated flow of moisture in soils. J. Hydrol. 15:227-246.

118    Wind, G. P. 1976. Application of analog and numerical models to investigate the influence of drainage on workability in spring. Neth. J. Agric. Sci. 24:155-172.

119    Wind, G. P. 1979. Analog modeling of transient moisture flow in unsaturated soil. Agric. Res. Rep. 894, Pudoc, Wageningen. 123 p.

120    Wind, G. P., und R. A. Pot. 1976. Bodenverbesserung in den hollandischen Veenkolonien. Z. f. Kulturt. und Flurber. 17:193-206.

121    Wind, G. P., and A. N. Mazee. 1979. An electronic analog for unsaturated flow and accumulation of moisture in soils. J. Hydrol. 41:69-83.

122    Yang, S. J., and E. de Jong. 1971. Effect of soil water potential and bulk density on water uptake patterns and resistance to flow of water in wheat plants. Can. J. Soil Sci. 51:211-220.

123    Yang, S. J., and E. de Jong. 1972. Effect of aerial environment and soil water potential on the transpiration and energy status of water in wheat plants. Agron. J. 61:571-578.

# chapter 11

## SYSTEMS CONSIDERATIONS AND CONSTRAINTS

# 11

# ROOT ZONE MODIFICATION: SYSTEMS CONSIDERATIONS AND CONSTRAINTS

by    Gerald F. Arkin, Blackland Research Center, Texas Agricultural Experiment Station, Temple, Texas; Howard M. Taylor, Agronomy Department, Iowa State University, Ames, Iowa

## 11.1 INTRODUCTION

Uncertain benefits of modifying the root environment impede coherent management decisions. Documented testimonials of the benefits accrued as a result of a management practice that modified the root environment are inconsistent. Practices appearing beneficial in some years and/or places may not be in others. Hence, extrapolating management recommendations over time and space are often tenuous at best. Because neither the laws of physics nor the principles of physiology are altered over time and space, these differences in observations must result from variations in the aerial and subterranean environments. In addition, different species and cultivars respond differently to environmental stimuli. Changes in crop sensitivity with stage of development or age may elicit differing crop responses to management practices. A modeling approach that accounts for these changes in crop response is presented. The approach can be used for extending management recommendations over time and space.

Classical treatment and response studies are generally conducted for serveral years and in multiple locations. Treatment recommendations are then derived from average response observations. It is well established that maximizing resources based on averages are only marginally successful. Management strategies that reflect the expected variability associated with observed responses will likely be more successful than those based on averages. Strategies of this type developed by using a modeling approach are discussed in this final monograph chapter.

Management decisions and systems for modifying the subterranean environment cannot be divorced from other decisions and components of a total management system. Many processes of the production system are inherently co-mingled and directly or indirectly influence one another. Modifying part of a system without considering the impact on the whole system is a mistake. Using the crop model based approach, systems interactions can be evaluated.

Ultimately, economic consideration and risk acceptance dictate the management practices that are to be implemented. It is exceedingly difficult to compute cost/benefit ratios for evaluating production practices that are so strongly influenced by the vagaries of weather and the heterogeneity of soils.

The systems approach presented in the ensuing chapter can be a useful tool in many economic analyses.

In this chapter, the authors discuss general systems interactions, the role of the simulation model, and economic and conservation considerations. Considerations and constraints discussed in this chapter are implicit in the other chapters of the monograph and, therefore, only topically addressed.

### 11.1.1 Environmental Variability

Weather and climate often dictate whether or not root zone modification is needed. Weather and climate are involved directly or indirectly with each of the crop stresses previously discussed in this monograph. For example, in regions of high rainfall drainage is frequently a management consideration and conversely in low rainfall climates, salinity is often a management consideration. In some regions, stresses are persistent because of the constant nature of the climate. In those regions, for example, extremes in annual temperature and precipitation adversely affect production and modification of the root zone environment to alleviate stress is less risky and easier to evaluate than in more unpredictable environments. In the regions of highly variable climate, the risk associated with root zone modification decisions increases. In these regions, decisions should be based on prevailing weather and probabilities of expected weather events.

Strategies (long term or between season) can be planned to alleviate climate induced stress, but tactics (short term or within season) for alleviating seasonal stresses allow little time for planning and implementation and are generally very risky. Infrequently occurring crop stresses that are closely associated with fluctuations in seasonal weather often go unrecognized because their effects on growth and yield are rarely observed. Although some weather induced stresses such as inundation or drought are obvious, weather conditions that cause pathogen populations to become critical or nutrient uptake to decrease are often less obvious. Complete weather and climate records, interpreted correctly, are useful tools for diagnosing and combatting these crop stress conditions. Soil temperature and moisture measurements are directly related to the weather and as previously pointed out, can also be important indices of less obvious subterranean stresses.

Crop stresses cannot always be alleviated because of weather conditions. Conditions may be too wet or too dry for tillage operations and chemical incorporation (e.g. liming, fertilizer, fungicides, and herbicides). Fluctuations in weather and climate in a region should be considered when devising tactics and strategies for root zone modification. Needed weather and climatic information is generally available from the state climatologist in the United States or from archived weather records.

Crop stress and response to profile modification within and between fields is influenced by soil heterogeneity. Symptoms that characterize crop stress and are used as indicators of root environment problems are easier to recognize in fields having uniform soils than in fields having non-uniform soils. Timely treatment to alleviate stress in fields of non-uniform soils is most difficult. Costly spot treatments are needed, but often are applied too late to be effective because of delay in recognizing spotty crop stress conditions. Recognition of the field soil variability and a better understanding of the influence that soil variability can have on crop stress can enable more timely problem recognition. The combination of the vagaries of weather and the heterogeneity of soils in some cases makes profile modification impractical.

### 11.1.2 Crop Growth and Sensitivity.

Species and cultivars exhibit remarkable differences in their growth habits. Their aerial and subterranean organs differ in structure (size, shape, and number), function, rate and duration of growth, and time of growth. These characteristics play a significant part in determining whether or not a crop will experience stress related to its root environment. Generally, adapted crops are grown that were selected with genetically determined growth and development features enabling them to avoid or resist stress in a particular region.

In addition to the exhibited variability in growth and development, species and cultivars respond to environmental stresses in markedly different degrees. Sensitivity differences between species and cultivars have been demonstrated in the literature and discussed in previous chapters. A level, time or duration of environmental stress that adversely affects one species or cultivar is not likely to influence another in an identical manner. Therefore, conditions stressful in one situation may not be in another, especially when different cultivars or species are involved. Indicators of crop stress and corrective soil modifications are crop and cultivar dependent. This dependency complicates the recognition and treatment of root zone problems. Therefore, awareness and consideration of the variability in crop development and sensitivity is helpful in determining needed root zone modification and assessing the benefits of the modification.

### 11.1.3 Tactical Timeliness.

Clearly the weather, soil environment and crop, influence whether or not a stress condition will exist and possibilities for alleviating that condition. The dynamic weather-soil-crop system is continually changing. Short-term (tactical) root zone modification decisions are therefore vulnerable to changes in any part or parts of the system. Tactical decisions for alleviating a developing problem or one that has just arisen are therefore risky. Because of the changeable nature of the system, it is wise to consider probable scenarios, especially those dictated by weather events, before implementing any actions. If the weather permits corrective action to be taken by enabling field work, other crop-environment conditions may be such that the actions would be ineffective or not practical to implement. In addition, the benefits of many corrective practices are highly dependent on the weather events that occur thereafter. Ensuing weather events can mask the expected benefits of root zone modifications.

Not only are tactics dependent on the soil-plant atmosphere environment, but their implementation is also dependent upon the ready availability of resources (capital, labor, supplies and equipment). Most often without pre-season preparation, reallocation of resources is not likely to be an easy task nor is it likely that new resources will be available. In order to implement tactical decisions, it is generally necessary that some contingency planning be done.

Implementing tactics and strategies to alleviate stress are almost always finally decided upon by evaluating the cost of the corrective or alleviating action and comparing the cost with the benefit derived. This assessment is especially suspect when made for infrequently occurring stress conditions. Factual information on likely benefits, in terms of crop response, is generally unavailable. Quantitative relationships describing the stress alleviating practice-soil-plant-weather interactions are needed to enable more informed decisions to be made.

## 11.2 SYSTEMS DESCRIPTION

### 11.2.1 Role of Simulation Models

Descriptions of the plant response to stress and root zone treatments for alleviation are generally qualitative. Quantitative relationships were included in the previous chapters when available. In Chapter 11, Feddes demonstrated the utility of water use models (an assemblage of quantitative relationships for describing a system) for evaluating profile modification practices. He restricted his presentation to water use models, a subject area where quantitative relationships abound. Feddes discussed and presented models addressing only a part of the more diverse and complex production system. Researchers are developing models describing other components of the systems dealt with in the previous chapters. In addition to the existing water balance models, models are available for simulating crop growth and development (Baker, 1980), soil and water loss due to runoff and erosion (e.g. Williams, 1978) and tillage effects on soil characteristics (Linden, 1979). Many applicable techniques for building models have been described in several texts (Jeffers, 1972; Thornley, 1976; Hall and Day, 1977). By combining the various component models, it is becoming possible to adequately describe the total system and simulate crop response to changes in its environment, whether man-made or naturally occurring.

### 11.2.2 Integrated Crop Production Models.

A crop growth simulation model was only recently combined with a soil water balance model (Berndt and White, 1976) to evaluate the impact of land use practices upon water yield and quality. Several accounts have been reported since on the ability to combine crop models with other models and components of models in order to assess management options. DeCoursey (1978) showed that a cotton crop simulation model (GOSSYM) (Baker et al., 1976) could be combined with an independently developed hydrologic model to enable assessing management alternatives that include the dynamic interactions of crop growth, tillage and hydrology. Likewise, Stinson et al. (1981) combined a sorghum crop simulation model (Arkin et al., 1976) with the hydrologic and sediment models developed by Williams (1978) to quantitatively describe the effects of ratoon cropping grain sorghum on water quantity and quality and crop yield. Stinson et al. (1981) concluded that field plot studies for assessing sorghum ratoon cropping would not necessarily have provided the same results shown in their study because the year or years chosen for plot studies would reflect only short-term weather variability effects.

Dynamic models describing physical and physiological processes can be combined as demonstrated. The combined models have the ability of describing total system response to many input parameters, thereby providing information for both tactical and strategic decisions. The examples given demonstrate that models can be combined to expand systems simulation. Consideration should be given to the complexity, hardware, memory and execution time requirements for the models before they are chosen. Models requiring exhaustive calibration or initialization are generally prohibitive for decision making, although they may be useful for research needs. Data requirements for these models often cannot be obtained economically for field problem applications. Therefore, simpler or less complex models are needed. The degree of model simplicity or complexity needed is dependent on the users needs and resources. The accuracy required of the simulated information can dictate the degree of complexity and process detail of the simulation model. Accuracy requirements for field problem applications may be considerably different from those for research needs. Generally, more complex and detailed models are more difficult and costly to use. They can exceed the capabilities of microprocessors owned and operated by farmers and ranchers and thereby preclude their use.

### 11.2.3 Model Applications.

Unlimited possibilities exist for application of combined or integrated models. Baker et al. (1979) applied models of this type to several crop production practices including irrigation and tillage. They concluded that models of this type will "form the basis of risk analysis and system management in the near future." Linden (1979) developed a model (SIMTIL) to predict soil water storage as affected by tillage practices. A combined nonhomogeneous water infiltration model and a soil surface configuration model were used. His objective was to predict soil water content as a function of depth and time for various residue and tillage induced soil conditions. Soil surface sealing, residue decomposition, and changes in the soil and residue reflectivity are accounted for as dynamic properties in this model. Although still under development, SIMTIL appears to be a useable model with realistic output. A nitrogen-tillage-residue-management model (NTRM) that extends SIMTIL is also under development (Shaffer et al., 1980). Nutrient uptake, especially nitrate and ammonium can be simulated using NTRM. Crop response to nutrient stress are accounted for in the model. Other stresses that can be accounted for with this model are water, soil salinity, soil strength, soil aeration, solar radiation and air temperature.

Systems simulation models that include stresses dealt with in this monograph are becoming available. Sensitivity analyses, validation and documentation are needed before they can find general acceptance. Implementation will depend on their ability to assist in analyzing complex management decisions.

## 11.3 ECONOMIC AND RESOURCE CONSTRAINTS

Systems models can be used to determine the expected cost of either implementing or not implementing management practices to modify the root environment. Likewise, they can be used to determine optimal management practices. They offer a dynamic approach to economically evaluating management options.

### 11.3.1 Feasibility and Risk.

Systems simulations can provide probable crop response to management treatments. Zavaleta et al. (1979) employed a simulation modeling approach to evaluate economical irrigation patterns for grain sorghum production. Their approach exemplifies a procedure for combining agronomic technology through models with optimization and economic theories. This approach should prove useful for determining feasibility of and risk associated with management tactics and strategies discussed in the previous chapters of the monograph. Previous economic studies were based on the over-simplified assumption that the manager knew in advance the state of the future climatic, economic and institutional conditions for the growing season. Stochastic (real world) situations in either weather, economic conditions and/or institutional factors were considered in their study. The crop simulation model used for their study was structured with a feedback loop enabling the model to be executed with weather and field observation information up to the time of execution. As the growing season progresses, available weather and crop status information can be input to update the model. Using normal conditions or simulated probable weather conditions, outlooks can be generated. Several methods are available for generating probable weather scenarios (Crank, 1977; Richardson, 1981). This methodology lends itself to optimal resources allocation evaluation for maximizing profits. By enabling assessment of management practices under dynamic growing conditions and stochastic environments, production risks, and uncertainties can be more realistically addressed. Many different game and decision theory approaches have been suggested to deal with production practices affected by stochastic elements (Wagner, 1964; Neufville and Stafford, 1971). Most of these are static in nature. The decision maker is generally committed to his initial plan throughout the production period.

Dynamic modeling as presented not only allows for pre-season planning and risk assessment, but permits plan revision and reformulation as new data become available. Arkin et al. (1980) used updating techniques and simulated weather data to forecast grain sorghum yields. As a result of their technique, the effectiveness of management options can be re-evaluated as the production period progresses or until such time as implementation occurs. This stochastic open-loop feedback method enables the decision maker to capitalize on changing conditions during the growing season. Strategies can be evaluated before the production period and tactics implemented within the production period as dictated by developing and projected conditions.

The modeling approach thus far outlined can provide probable production expectations in relation to management options. The ability to quantify relative yield effects resulting from management options is critical for risk assessment. Realistic modeled yield response data coupled with economic data describing system inputs can provide some of the economic criteria for root zone modification decisions that are influenced by weather induced risk and uncertainty.

## 11.3.2 Energy and Cost Consideration.

Factoring in costs and energy requirements for decision making is exceedingly difficult. Both costs and energy requirements change with changes in equipment, cultural practices, market dynamics, and environmental conditions. For this reason, we did not include extensive cost and energy information in this monograph. Such information is available on a limited basis, but is generally scattered throughout existing literature. Baseline information for cost estimates are available for some cases (Richey et al., 1977; Giere et al., 1980; Phillips et al., 1980; Clark and Johnson, 1974; Hillel et al., 1969; Fulgham et al., 1973; Doster, 1972). Energy requirements for cultural practices are also available for some cases (Triplett and van Doren, 1977; Phillips et al., 1980; Allen and Fryrear, 1979).

Because of the large differences in cost and energy requirements for implementing root zone modifications to alleviate stress, each production situation calls for a new cost benefit appraisal. The systems approach described in this chapter when combined with appropriate economic data may in the future help in unifying concepts and eliminating, to a great extent, the dilemma faced by farmers when attempting to perform cost benefit analyses. Input costs and energy requirements information will likely remain individualized situation factors, but modeled crop and system response outputs will go a long way in simplifying cost benefit appraisals.

### 11.3.3 Conservation.

New and improved herbicides, planters, cultivators, and resultant new cropping systems are enabling farmers to conserve resources. For example, reduced tillage systems have been introduced in which time and energy are saved and net returns generally are increased. Because less field traffic is required for tillage operations, soil compaction and reoccurring tillage needs are reduced. Crop yields vary in response to this practice and in some cases, are reduced; but even so, profits can be greater than those derived from higher yields in conventionally tilled fields.

Production systems that are designed to capitalize on many advantageous conservation practices such as reduced tillage are difficult to evaluate because of the many machine-plant-environment interactions that need to be considered. A modeling oriented systems approach as previously described can be useful here too. The attractive wide-bed narrow-row cultural system for cotton (Fulgham et al., 1973; Parish et al., 1973) and sorghum production (Arkin et al., 1978) exemplifies such a system (Figs. 1 and 2). Two, or as many as 5, crop rows are planted on a 2 m bed in the wide-bed narrow-row system rather than the 1 m row spacing in the conventional production system. In the wide-bed narrow-row system: (a) tractor wheel traffic is confined to specific furrows (fixed traffic pattern) to reduce soil

FIG. 1 A precision bed shaper for forming wide beds (Courtesy of L. H. Wilkes, Texas A&M University).

FIG. 2 Planting four rows on each of the three wide beds. Beds were formed with a bed shaper similar to that illustrated in Fig. 1, but modified to shape three beds at a time (Arkin et al., 1978).

compaction in the root zone, thereby reducing primary tillage and energy requirements, (b) the wide beds provide a method for precision cultivation in narrow rows, and (c) the narrower rows provide a more dense leaf canopy earlier in the growing season resulting in more sunlight being intercepted by the canopy, less water evaporated from the soil, less soil compaction from raindrop impact and reduced runoff and erosion. Concomitant with the combined attributes of the system are potential weed, pathogen, insect, fertilizer, machine and bed design problems that need to be solved. Problem solutions through model applications are presently limited. Needed are descriptive weed, pathogen, insect, fertilizer, and machine response models that can be combined with other models to adequately simulate the system response.

The wide-bed narrow-row system affords a means for controlling field traffic. Benefits of controlled traffic have been described (Gill and Trouse, 1972; Williford, 1980). The extent to which this practice is realized in practical systems is not clear. This depends on the extent to which crops respond to excessive traffic in differing soil and climatic conditions. Likewise, the extent to which increased yields (Arkin et al., 1978), reduced runoff and erosion (Adams et al., 1978) and reduced soil water evaporation (Adams et al., 1976) reported for narrow row practices would be realized in a practical system like the wide-bed narrow-row system remains unclear. Again, limited quantitative descriptions of management practice interactions with soil and climate variability, expressed in crop response, restrict the ability to extend research findings over time and space.

## 11.4 SUMMARY

Decisions to modify the root zone in order to reduce crop stress are dependent on the interactions of highly variable components of the production system. Crop response to root zone modification is not only dependent on temporally and spatially varying soil and weather conditions, but also on the stage of crop development. Changing crop sensitivity and response throughout the growing season along with changing weather and soil conditions further confound decisions to modify the root zone.

A mathematical modeling approach for systems evaluation is presented. Models that can be used to simulate crop response to environmental stimuli are the foundation for this approach wherein crop growth and development models are combined with machine, hydrologic, pest, and other models. A total system response can be simulated with appropriate weather, climate and soils data.

With appropriate cost and other relevant economic data and simulated systems response, i.e. yield; cost benefit analyses can be generated for the given year or a series of years depending on available climate data or generated weather data. Simulations for long periods of time can be used to develop probabilities of systems response and benefit. The feasibility and risks associated with modifications of the root zone can be evaluated using the computed probabilities. A similar detailed presentation of the proposed methodology for analyzing environmental modification was presented (Hahn, 1979) in the companion monograph, Modification of the Aerial Environment of Crops.

First approximations of the expected benefits from new or alternative systems, especially those aimed at conserving resources, can be developed through simulation. Attractive practices and systems determined this way may then be worthy of more detailed laboratory and field investigations.

### References
1  Adams, J. E., C. W. Richardson, and E. Burnett. 1978. Influence of row spacing of grain sorghum on ground cover, runoff, and erosion. Soil Sci. Soc. Am. J. 42:959-962.

2  Adams, J. E., G. F. Arkin, and J. T. Ritchie. 1976. Influence of row spacing and straw mulch on first stage drying. Soil Sci. Soc. Am. J. 40:436-442.

3  Allen, R. A. and D. W. Fryrear. 1979. Energy considerations in conservation tillage systems. In: Conservation Tillage in Texas. Texas Agricultural Extension Service Bulletin 1290. College Station, TX. pp. 31-45.

4  Arkin, G. F., S. J. Maas, and C. W. Richardson. 1980. Forecasting grain sorghum yields using simulated weather data and updating techniques. TRANSACTIONS of the ASAE 23(3):676-680.

5  Arkin, G. F., E. Burnett, and R. Monk. 1978. Wide-bed narrow-row sorghum yields in the Blackland Prairie. The Texas Agricultural Experiment Station, College Station, TX. MP-1377, 4 pp.

6  Arkin, G. F., R. L. Vanderlip, and J. T. Ritchie. 1976. A dynamic grain sorghum growth model. TRANSACTIONS of the ASAE 19(4):622-626, 630.

7  Baker, D. N. 1980. Simulation for research and crop management. In: World Soybean Research Conference II: Proceedings. Edited by Fredrick T. Corbin. Westview Press, Boulder, CO 80301. pp. 533-546.

8  Baker, D. N., J. A. Landivar, F. D. Whisler and V. R. Randy. 1979. Plant response to environmental conditions and modeling plant development. In: Weather and Agriculture Symposium Proceedings, Kansas City, MO. pp. 69-109.

9  Berndt, R. D. and B. J. White. 1976. A simulation-based evaluation of three cropping systems of cracking-clay soils in a summer-rainfall environment. Agric. Met. 16:211-229.

10  Clark, S. J. and W. J. Johnson. 1974. Energy-cost budget for grain sorghum tillage systems. ASAE Paper No. 74-1044, ASAE, St. Joseph, MI 49085.

11  Crank, K. 1977. Simulating daily weather variables. Research and Development Branch, Research Division Bulletin, Economics, Statistics Service, USDA, Washington, DC. pp. 1-17.

12  DeCoursey, D. G. 1978. Erosion simulation for land use management. ASAE Paper No. 78-2082, ASAE, St. Joseph, MI 49085.

13  Doster, D. H. 1972. Economics of no-tillage. In: No-Tillage Systems Symposium Proceedings. Ohio Agricultural Research and Development Center, Columbus, OH. pp. 41-54.

14 Fulgham, F. E., J. R. Williford, F. T. Cooke, Jr. 1973. Stoneville wide-bed cultural systems. Mississippi Agricultural and Forestry Experiment Station Bulletin Number 805. 22 pp.
15 Giere, J. P., K. M. Johnson, and J. H. Perkins. 1980. A closer look at no-till farming. Environment 22(6):15-41.
16 Gill, W. R. and A. C. Trouse. 1972. Results from controlled traffic studies and their implications in tillage systems. In: No-Tillage Systems Symposium Proceedings. Ohio Agricultural Research and Development Center, Columbus, OH. pp. 126-131.
17 Hahn, C. T. 1979. Risk analysis in environmental modification. In: ASAE Monograph, Number 2. Modification of the Aerial Environment of Crops (ed.) B. J. Barfield and J. F. Gerber. ASAE, St. Joseph, MI 49085. pp. 30-51.
18 Hall, C. A. S. and J. W. Day. 1977. Ecosystem modeling in theory and practice: an introduction with case histories. John Wiley and Sons, New York, NY. 684 pp.
19 Hillel, D., D. Ariel, S. Orlowski, E. Stibbe, D. Wolf, and A. Yavnai. 1969. Soil-crop-tillage interactions in dryland and irrigated farming. USDA Report. Project NO. A10-AE-3, Grant No. FG-IS-204. 300 pp.
20 Jeffers, J. N. R. 1972. Mathematical models in ecology. Blackwell Scientific Publications, London. 398 pp.
21 Linden, D. R. 1979. A model to predict soil water storage as affected by tillage practices. Ph.D. Dissertation, Soil Science Department, University of Minnesota.
22 Neufville, R. de and J. H. Stafford. 1971. Systems analysis for engineers and managers. McGraw-Hill Book Co., New York, NY. 353 pp.
23 Parish, R. L., S. M. Brister, and D. E. Mermoud. 1973. Preliminary results of wide-bed narrow-row cotton research. ASAE Paper No. 73-1578, ASAE, St. Joseph, MI 49085.
24 Phillips, R. E., R. L. Blevins, G. W. Thomas, W. W. Frye, and S. H. Phillips. 1980. No-tillage Agriculture. Science 208:1108-1113.
25 Richardson, C. W. 1981. Stochastic simulation of daily precipitation, temperature, and solar radiation. Water Resour. Res. 17(1):182-190.
26 Richey, C. B., D. R. Griffith, and S. D. Parsons. 1977. Yields and cultural energy requirements for corn and soybeans with various tillage-planting systems. In: Advances in Agronomy 29:141-182.
27 Shaffer, M. J., S. C. Gupta, J. A. E. Molina, D. R. Linden, C. E. Clapp, and W. E. Larson. 1980. Simulating crop response to tillage: an integrated approach. Agronomy Abstracts, American Society of Agronomy, Madison, WI 53711. p. 15.
28 Stinson, D. L., G. F. Arkin, T. A. Howell, C. W. Richardson and J. R. Williams. 1981. Modeling grain sorghum ratoon cropping and associated runoff and sediment losses. TRANSACTIONS of the ASAE, 24, No. 3, (In press).
29 Thornley, J. H. M. 1976. Mathematical models in plant physiology. Academic Press, New York. 318 pp.
30 Triplett, Jr., G. B. and D. M. van Doren, Jr. 1977. Agriculture without tillage. Scientific American 37:28-33.
31 Wagner, H. M. 1969. Principles of operations research with applications to managerial decisions. Prentice-Hall, Inc., New Jersey. 937 pp.
32 Williams, J. R. and R. W. Hahn, Jr. 1978. Optimal operation of large agricultural watersheds with water quality constraints. Texas Water Resource Institute, Texas A&M University, TR-96. 152 pp.
33 Williford, J. R. 1980. A controlled-traffic system for cotton production. TRANSACTIONS of the ASAE 23(1):65-70.
34 Williford, J. R., F. E. Fulgham, and O. B. Wooten. 1974. Wide-bed cultural system for cotton production. TRANSACTIONS of the ASAE 17(6):1136-1138.
35 Zavaleta, L. R., R. D. Lacewell, and C. R. Taylor. 1979. Economically optimum irrigation patterns for grain sorghum production: Texas High Plains. Texas Water Resources Institute, TR-100, Texas A&M University. 103 pp.

# SUBJECT INDEX

Resistance
  contact, 359
  flow, 355
  plant, 355, 359, 360, 361, 363
  root, 356, 357, 358, 363
  soil, 355, 356, 357, 358, 359, 363
Respiration, aerobic, 147
Rhizotron, 5
Rice, 170
Rippers, 50, 88, 92
Risk, 397
Root
  branching, 235
  conductance, 359
  conductivity, 363
  crops, 241
  density, 357, 363
  distribution function, 362
  effectiveness function, 362, 375, 383
  effectivity functions, 364
  effects, 104
  elongation, 235
  extraction, 364
  exudates, 104
  geometry, 357, 362
  growth, 32, 149
  mass, 362
  observation chamber, 4
  penetration, 124
  permeability, 358
  replacement, 159
  resistance, 356, 357, 358, 363
  surface, 355
  system, 233
    size and morphology, 104
  water pressure head, 362
  water uptake, 352, 354, 363, 366, 380, 382, 385
  patterns, 351, 352, 380
  zone depth, 78
  zone modification, 347
Rooting
  depth, 352, 357, 361, 375, 380, 382
  length, 356
Rootstocks, 318, 320
Rotations cropping, 125
Rototilled, 79
Rototilling, 73
Roughness
  random, 222, 250
  surface, 245, 250
Row placement, 119

SMP method, 278
Saline
  seeps, 306
  sodic soils, 340
  watertable, 306, 333, 335
Salinity, 315, 371, 374, 375, 385
  hazards, diagnosing, 322
  level
    threshold, 314, 318
  sensors, 323
  soil
    distribution, 310
    saturation extract, 315
Salt, 350
  affected soils, management of, 323, 325
  crop
    tolerance, 308, 314, 324
  dissolved, 306
  distribution, 309
  tolerance, 315, 318, 319
    qualitative rating, 314, 315
    ratings, 318
  water dilution method, high, 338
Sample preparation and analysis, 113
Sampling, 112
  tissue, 112
Saturated zone, 382, 386

Saturation
  Al, 282, 283
  base, 276, 277
  soil extract, 318, 320, 323, 324
  salinity, 315
Sclerotia, 201
Seed
  germination, 152
  placement, 333
Seedling
  early growth, 317
  emergence, 33
Seepage, 332
Seeps, saline, 306
Selecting genotypes, 178
Selection and breeding approach, 16
Sensitivity
  analysis, 367, 386
  crop, 395
Sensors, salinity, 323
Sewage sludge, 256
Shear vanes, 31
Shoot
  growth, 153
  root
    ratios, 8
    relationships, 4
Silica, 281
Simulation, 396
  crop growth, 396
  models, 396
Sink term, 352, 364, 365, 366, 379
  models, 361
$SiO_2$, 281
Slip
  plowing, 89
  plows, 77, 88
Sludge, sewage, 256
Snow cover, 257
Sodic soil, 338
  reclamation of, 339
Sodium, 331, 321, 323
  adjusted, 311
  adsorption ratio, 311, 323
  excess, 335
  exchangeable, 338
  percentage exchangeable, 311
Soil
  acid, 269, 279, 287
    infertility, 270, 279
  acidity, 273, 287
  aeration, 142, 210, 385
    poor, 141
  anistropic, 383
  bulk density, 21, 22, 27, 28, 76, 79, 338
  calibration, 108
  compaction, 21, 22, 24, 26, 27, 242, 254
    causes, 25
    consequences, 25
  crusts, 9, 24, 42
  equilibria in the, 100
  evaporation, 376, 378, 383
  fallow, 335
  fertility, 319
  flux, 352
  fragipan, 75
  heat flux, 223
    density, 223
  hetrogeneity, 145, 229, 394
  improvement, 381
  inhabitants, 201
  invaders, 201
  management, 146
  matrix (matric) potential, 324
  mixing, 381
  moisture, 209, 371, 372
    contents, 358
    diffusivity, 364, 380

  flow models, 372
  pressure head, 352, 360, 365
  pan, 38
  permeability, 340
  pH, 15, 269, 270, 272, 273, 274, 275, 276, 277, 282, 283, 285, 286
    critical, 274, 275
    effects of fertilizer salts on, 273
  plowpan, 76
  poorly drained, 77
  profile, 381, 382
    acidity, 270, 272
    modification, 334
  resistances, 355, 356, 357, 358, 359, 363
  saline-sodic, 340
  salinity, 15
  salt-affected, 78
    management of, 323, 325
  sampling, 107
  sandy, 80
  saturation extract, 318, 320, 324
    salinity, 315
  sodic, 338
    reclamation of, 339
  solution, 101
    $Al^{3+}$ activity, 281, 282
    $Mn^{2+}$, 286
  storage, 307
  strength, 70, 76, 77, 79, 170
  structure, 22
  surface crusting, 339
  temperature
    emergence response to, 233
    germination response to, 233
    modification, 243, 245
    optimum, 241, 242
    plant response to, 229
    varying, 232
  test
    calibration, 108
    interpretation and recommendation, 108
    norms, 109
    rating, 109
    testing, 106, 111
    value of, 110
  texture, 22
  thermal admittances, 248
  toxins, 170
Soil-borne micro-organisms, 196
Soil plant relationships, 101
Soil water
  availability, 62
  characteristic, 63
  penetration, 311, 323
  potential, 330
    energy of, 63
  storage, 62
  uptake by plants, 64
Solutes
  effects, specific, 319
  imbalance of specific, 323
Solution
  Al, 279
  $Al^{3+}$, 282
  analytical, 365, 367
  Mn, 285
  $Mn^{2+}$, 286
Sorghum, 162, 379
Sowing, 349
Soybean, 163
Specie adaptation, 124
Springtooth harrows, 48
Sprinkler irrigation, 330
Sprinkling, 327, 337
  overhaed, 320
  intermittent, 320
Standard value approach, 116
Starter, 121